National Museums

National Museums is the first book to explore the national museum as a cultural institution in a range of contrasting national contexts. Composed of new studies of countries that rarely make a showing in the English-language studies of museums, this book reveals how these national museums have been used to create a sense of national self, place the nation in the arts, deal with the consequences of political change, remake difficult pasts, and confront those issues of nationalism, ethnicity and multiculturalism which have come to the fore in national politics in recent decades.

National Museums combines research from both leading and new researchers in the fields of history, museum studies, cultural studies, sociology, history of art, media studies, science and technology studies, and anthropology. It is an interrogation of the origins, purpose, organisation, politics, narratives and philosophies of national museums.

Simon Knell is Professor of Museum Studies at the University of Leicester.

Peter Aronsson is Professor in Cultural Heritage and the Uses of History at a multi-disciplinary Culture Studies Department at Linköping University.

Arne Bugge Amundsen is an award-winning Professor of Cultural History and Head of the Department of Culture Studies and Oriental Languages at the University of Oslo.

Amy Jane Barnes is completing research in the School of Museum Studies at the University of Leicester.

Stuart Burch is a Senior Lecturer at Nottingham Trent University where he teaches museum studies, heritage management and public history.

Jennifer Carter is Assistant Professor of Museum Studies in the Faculty of Information at the University of Toronto.

Viviane Gosselin is a researcher affiliated with the Centre for the Study of Historical Consciousness at the University of British Columbia.

Sarah Hughes is Senior Lecturer in Publishing at the Oxford International Centre for Publishing Studies, Oxford Brookes.

Alan Kirwan is undertaking research at the University of Leicester.

National Museums

New Studies from around the World

EDITED BY

Simon J. Knell, Peter Aronsson, Arne Bugge Amundsen, Amy Jane Barnes, Stuart Burch, Jennifer Carter, Viviane Gosselin, Sarah A. Hughes and Alan Kirwan

Routledge
Taylor & Francis Group
LONDON AND NEW YORK

First published 2011
by Routledge
2 Park Square, Milton Park, Abingdon, Oxon OX14 4RN
Simultaneously published in the USA and Canada
by Routledge
711 Third Ave, New York, NY 10017

Routledge is an imprint of the Taylor & Francis Group, an informa business

Typeset in Perpetua and Bell Gothic by
Keystroke, Station Road, Codsall, Wolverhampton
Printed and bound in India by
Replika Press Pvt. Ltd.

British Library Cataloguing in Publication Data
A catalogue record for this book is available from the British Library

Library of Congress Cataloguing in Publication Data
A catalogue record for this book has been requested

ISBN 13: 978–0–415–54773–4 (hbk)
ISBN 13: 978–0–415–54774–1 (pbk)

ISBN 10: 0–415–54773–3 (hbk)
ISBN 10:0–415–54774–1 (pbk)

CONTENTS

FIGURES

TABLES

CONTRIBUTORS

Eugenia Afinoguénova is Associate Professor of Spanish at Marquette University, where she teaches contemporary Spanish literature and cultural studies. She is currently working on a monograph dedicated to the social history of the Prado Museum in Madrid. Her articles on the Prado have appeared in the *Journal of Spanish Cultural Studies* and *Nineteenth-Century Contexts*.

Arne Bugge Amundsen is an award-winning Professor of Cultural History and Head of the Department of Culture Studies and Oriental Languages at the University of Oslo. A former curator and archivist, his research interests range across museums and museum history, religious history, rituals and social structures and movements. Recent books include: *Revival and Communication: Studies in the History of Scandinavian Revivals; Rituals: Cultural Historical Studies* and *Norwegian Religious History*. He is a member of the Eunamus consortium.

Peter Aronsson is Professor in Cultural Heritage and the Uses of History at a multi-disciplinary Culture Studies Department at Linköping University. Among his more recent publications is 'Representing community: National museums negotiating differences and community in Nordic countries', in *Scandinavian Museums and Cultural Diversity* (Berghahn Books, 2008). Currently he is co-ordinating several international comparative projects exploring the uses of the past in national museums, including Eunamus, probably the largest study of national museums and the uses of history ever undertaken.

Mattias Bäckström is situated in the History of Science and Ideas at the University of Gothenburg, where he is currently researching how concepts such as 'folk memory' and 'love for the native country' were used in the practice of open-air museums in Scandinavia around 1900. Previously a curator, he is still very much engaged in the museum sector, both in projects and in debates.

Simina Bădică is a researcher and curator at the Romanian Peasant Museum in Bucharest where she is collecting, archiving and exhibiting artefacts pertaining to Romania's communist past. Her recent publications include articles on remembering communism in museums and through oral history.

Amy Jane Barnes is completing research into the collection, interpretation and display of the visual culture of the Chinese Cultural Revolution in contemporary British museums, in the School of Museum Studies at the University of Leicester.

Stuart Burch is a Senior Lecturer at Nottingham Trent University where he teaches museum studies, heritage management and public history. He is currently conducting research into national art museums in northern Europe as part of 'Nordic Spaces', a four-year multinational project funded by a consortium led by Riksbankens Jubileumsfond.

Jennifer Carter is Assistant Professor of Museum Studies in the Faculty of Information at the University of Toronto, where she teaches courses on the history and theory of museums, architecture and museums, and exhibition practice. Jennifer is currently researching and writing about the roles of architecture and narrative in exhibition design, and performance and scenography in eighteenth-century freemasonry and museology.

Sunghee Choi is at Pennsylvania State University where she has studied Art Education and Museum Education. She has recently completed a research project which looks at the intersection between museum and visitor narratives. Sunghee's research concerns visitors' identity construction and situated interpretation as well as post-colonialism in museums.

Chi-Jung Chu is at the London School of Economics and Political Science. She is currently researching the changing dynamics between the state and museums as revealed in policy development.

Viviane Gosselin is Curator at the Museum of Vancouver and affiliated with the Centre for the Study of Historical Consciousness at the University of British Columbia. Her current research explores the heuristic potential of a historical thinking framework to study the production and reception of museum exhibitions.

Lotten Gustafsson Reinius is Curator at the Museum of Ethnography in Stockholm and Associate Professor of Ethnology at Stockholm University. She teaches courses on material culture and museology and has recently undertaken research on the social biographies of Congolese artefacts in Sweden. Her work has been published in *The Senses & Society* and *Ethnologie Française*.

Gwenny van Hasselt is at the Erasmus University of Rotterdam where she is currently researching the role of digitisation processes and the concept of the nation in the development process of the Dutch National Historical Museum. Gwenny currently teaches History of Art and Culture.

Sarah A. Hughes is Senior Lecturer in Publishing at the Oxford International Centre for Publishing Studies, Oxford Brookes University, where she teaches design and the application of technology to publishing. She has recently completed research examining the uses of print by UK museums and their visitors, from which she is currently preparing a number of papers.

Karoline Kaluza has recently undertaken research as a Fellow of the DAAD (German Academic Exchange Service) in Poland as part of her research at the Rheinische Friedrich-Wilhelms-Universität, Bonn, examining the history of Polish museums and the role of modern art. She is currently working as an editor for the cultural project *The Promised City* at the Goethe-Institut in Warsaw.

Alan Kirwan is undertaking research at the University of Leicester on the role of Irish museums as sites of intercultural understanding in that country's developing multicultural contexts. He has worked in museum learning and interpretation for over ten years and contributed to the Attingham Report on learning provision within the heritage sector in the UK and Ireland.

Simon Knell is Professor of Museum Studies at the University of Leicester. His present research interests include the history and geography of national museums, particularly with regard to the production of institutions, the sociology of knowledge and meaning of material culture. He is also a cultural historian of science. Recent books include: *Museum Revolutions*; *Museums in the Material World*; and *The Making of the Geological Society of London*. He is a member of the Eunamus consortium, working particularly on issues of materiality, disciplinarity and transnationalism.

Ayşe Hazar Köksal is a researcher at Istanbul Technical University. She is currently researching the institutionalisation of art and the history of the art museum in Turkey as well as working on an exhibition project at the Museum of Painting and Sculpture. She regularly writes on art and exhibitions for several journals in Turkey.

Jung Joon Lee is currently a Fellow at the Center for Place, Culture and Politics at the City University of New York Graduate Center. Her current research assesses the visual experiences of social movements, focusing on perceptual and experiential changes in photographic practices in South Korea. She co-curated the exhibition, *Deadpan: Photography, History, Politics* at the James Gallery, New York, in 2008.

Cristina Lleras is Curator of Art and History at the National Museum of Colombia. She is currently undertaking research on the role of national museums in fostering access to cultural rights in the multicultural nation while dealing with demands for historical reparation and the inclusion of minorities.

Rhiannon Mason is a Senior Lecturer in Museum, Gallery, and Heritage Studies at the International Centre for Cultural and Heritage Studies, Newcastle University. Rhiannon's interests are in national museums and heritage, history curatorship, identity, memory, and new museology. Her most recent book is called *Museums, Nations, Identities: Wales and its National Museums* (University of Wales Press, 2007).

Susanna Pettersson is Head of Development at the Finnish National Gallery, where she is responsible for national research and development projects. Her latest books feature art collecting in nineteenth-century Finland, both private and public, and the future of the art museums. She is also editor for *Suomen Museohistoria* (Finnish Museum History), and *Encouraging Collections Mobility* (2010).

Donald Preziosi is Emeritus Professor of Art History at University of California, Los Angeles. His books include: *Brain of the Earth's Body: Art, Museums, and the Phantasms of Modernity*; *Grasping the World: The Idea of the Museum*; *Rethinking Art History*; and *The Art of Art History*. He has lectured and conducted seminars on art history and museology in the US, Europe and Australia.

Pille Pruulmann-Vengerfeldt is Senior Researcher at the University of Tartu Institute of Journalism and Communication and a part-time researcher at the Estonian National Museum. Her research interests are focused on online spaces as possible venues for participation in political and cultural life. She has recently published in *Journal of Baltic Studies, Journal of Computer-Mediated Communication* and *Journal of Children and Media*.

Pille Runnel is the Research Director of the Estonian National Museum. Her main research areas are information society studies, media sociology and media anthropology. She has published extensively on practices of internet usage, including children's internet use. Her current research focuses on museum audiences and inclusive and participatory practices in museum communication. She has recently published in *Journal of Baltic Studies* and *Journal of Computer-Mediated Communication*.

Ruth Scheidhauer is a Korea Foundation Fellow undertaking research at the Institute of Archaeology, University College London, where she is examining the possibilities of Korean rapprochement through inter-Korean cultural heritage activities. She is currently preparing papers on crisis and cultural policies in Korea as well as on the selected transmission of European cultural theories into Korea.

Radostina Sharenkova is a postdoctoral researcher at the Ethnographic Institute and Museum (Bulgarian Academy of Sciences) working on the concept for the museum's future permanent exhibition. She recently completed a historical study on policymaking at Bulgarian ethnographic museums during the communist and post-communist period, which is currently being worked into a monograph. She is also interested in totalitarian museum presentations concerning communism.

Karen D. Shelby is Assistant Professor of Art History at Baruch College, City University, New York. She teaches courses in Museum Studies and nineteenth-century, modern and contemporary art. She writes on nationalism, memory, material culture and the Great War. Karen is preparing a book on the role of the IJzertoren in the context of twenty-first-century Belgian politics.

Emily Stokes-Rees's research addresses national museums in Hong Kong, Singapore and Macau, focusing on topics such as architecture and identity, writing the 'national story' in the postcolonial world, and the representation of ethnic and indigenous minorities. She is currently living and working in Boston.

Taavi Tatsi is a researcher at the Estonian National Museum. His research focuses on the cultural practices of museum production, participatory culture and new museology.

Marzia Varutti is an Honorary Visiting Fellow at the School of Museum Studies, University of Leicester. She has conducted research on the museum representation of the nation in China, and on the politics of the display of cultural difference in Norwegian museums. She is currently conducting a research project investigating the museum representation of indigenous groups in Taiwan.

Christopher Whitehead is Senior Lecturer in Museum, Gallery and Heritage Studies at the International Centre for Cultural and Heritage Studies, Newcastle University. He is the author of *The Public Art Museum in Nineteenth Century Britain: The Development of the National Gallery* (Ashgate, 2005) and *Museums and the Construction of Disciplines: Art and Archaeology in Nineteenth Century Britain* (Duckworth, 2009). He is currently working on the development of 'Northern Spirit: 300 years of art in the north east' – a permanent display at the Laing Art Gallery in Newcastle, and a book on the theory and practice of art interpretation.

Chris Wingfield is an Associate Lecturer at the Open University, where he teaches a course on understanding global heritage. He recently completed research on the collections from England at Oxford's Pitt Rivers Museum. His current research concerns the objects collected in different parts of the world by the London Missionary Society.

PREFACE

This book has its origins in 'Making National Museums', or 'NaMu', a collaborative project funded under the European Commission's Marie Curie Actions. It was the brainchild of Peter Aronsson at the University of Linköping, Sweden, and Peter invited Arne Bugge Amundsen of the University of Oslo and me to join him in the project. Over a two-year period (2007–2008), NaMu funded the attendance of forty early career researchers from around the world, at six thematic conference/workshops held in Norrköping, Leicester and Oslo. At these meetings we discussed how national museums might be researched and understood, their histories and narratives, and the impacts of new technologies, multiculturalism and postcolonial thinking. Participants came from diverse disciplinary backgrounds and many became regulars; soon these meetings became high spots in the academic year for us all. We lectured and debated, visited museums, and dined, often in the company of distinguished keynote speakers including Tony Bennett, Stefan Berger, Eilean Hooper-Greenhill, Donald Preziosi, Peter van Mensch, Thomas Michael Walle, Chris Whitehead, David-Theiry Ruddel, Saphinaz Amal Naguib, Ross Parry, Lee Iverson, George Oates, Paul Marty, Alexandra Bounia, Kristin Kuutma, Peter Apor, Debora Meijers, Dominique Poulot and others.

During the course of our meetings, Peter, Arne and I became aware that we were being treated to perspectives on national museums that were hardly discussed in the literature and which deserved a wider audience. They asked me to take the lead on producing a book of the best papers. Arne then arranged the final conference – held high above Oslo at the stunning Voksenåsen – where he encouraged delegates to test and develop their ideas for the book. Meanwhile, Peter arranged funding from Marie Curie to permit the book to be published in colour.

As lead editor, I invited Viviane, Stuart, Alan, Sally (Sarah), Jennifer and Amy – all of whom had proven themselves enthusiastic contributors to the success of NaMu – to join the team. Production of the book was to be both supportive and rigorous. Peter and Arne took on roles of senior referees and judges in the various stages of the competition for entry, while the other editors reviewed early drafts and supported authors in the preliminary editorial stages of the book's production and worked on batches of chapters to refine language and content. My role was to act as a hub between authors and editors, commenting on chapters as they flowed backwards and forwards, and to manage the editorial stages. The chapters finally arrived with me after much work by both authors and editors, in late January 2010, and I then spent the next two months shaping them into the present book. Sally took on the not inconsiderable task of coordinating and curating images and captions, and Amy and Viviane assisted with some last-minute fixing.

National Museums is a delightful chocolate box of bite-sized case studies which offer an extraordinarily rich insight into the place of national museums in society. Perhaps surprisingly, none of these case studies come from North America (despite having two Canadians on the editorial team and a number of American authors), Australia or New Zealand – parts of the world which, with the UK, dominate the English-language museum studies literature. All but one of these papers comes from the Old World. This gives each of the case studies a relationship to the others, as nearly all concern nations whose independence has been challenged at some time in the past. Many of these nations have been the subject of much discussion amongst social scientists and historians concerned with nationalism and the formation of the nation-state. The national museum as a subtle and complex cultural entity has yet to fully enter that debate. Yet, as most of the papers here show, national museums have become key elements in the historical, mythological, aesthetic and political construction of the nation. They provide the opportunity for architectural and material symbolism and evidencing; cities and nations have even sought to brand themselves in particular ways. National museums have become vehicles for the development of national narratives and sites where the social challenges of political change, immigration and multiculturalism have been problematised and negotiated.

National Museums is broadly grouped into themes, but there are many relationships which cut across the various sections and chapters. In addition to the papers entered into the competition, I asked keynote speakers, Donald Preziosi and Chris Whitehead, to contribute papers. Chris's is entirely new while Donald's is as he presented it to NaMu delegates. I also asked Rhiannon Mason to contribute a paper from her work on national museums in Wales.

The book contains three 'introductions' with different purposes. The first, by me, attempts to give some of the themes and case studies in the book a broader theoretical context. In particular it considers how national museums contribute to imagining the nation. The second, by Peter, discusses the possibilities for comparative study that arise from the kinds of case studies which populate this book. This chapter incorporates ideas which Peter presented at one of the NaMu conferences and which fuelled his interest in Eunamus. The third introduction offers a philosophical overview of the kind we have come to expect from Donald

Preziosi. We hope this book reflects the conversations which took place at the NaMu meetings and that the developmental manner in which the book has been produced meets the intentions of the Marie Curie scheme. The editors would like to thank Marie Curie and all those who made NaMu such a success.

That success led to our winning EU Framework 7 funding for a follow-up project called Eunamus. Led by Peter Aronsson, with major contributions from Arne and myself, this collaboration of eight partner organisations is, we believe, the largest research project ever to be undertaken on European national museums. A study of the uses of history by national museums, it will: produce the first comparative history of the origins of national museums in Europe; examine the construction of historical narratives of various kinds and consider the role of these museums in national and cultural conflicts and tensions; investigate the materialisation of Europe, transnationalism and national museums as contributors to a cohesive Europe; interrogate the place of national museums in policy making and implementation; locate the European citizen as a participant in the production and consumption of histories in national museums; and, finally, consider how national museums and their uses of history might contribute to European cohesion in the future. This book, then, represents some initial thinking.

Simon Knell,
on behalf of the editors,
University of Leicester, April 2010

PART ONE INTRODUCTIONS AND REFLECTIONS

Chapter 1

NATIONAL MUSEUMS AND THE NATIONAL IMAGINATION

Simon Knell

> Yet it is the idea of the nation as an imagined political community and hence as a cultural artifact, at once sovereign, finite, and horizontally cross-class and moving along linear 'empty homogeneous time,' that has so tangibly caught the postmodernist scholarly fancy . . . In fact, we often find this idea detached from its Andersonian moorings. It has become a topos of the literary imagination, a metaphor for the constructed quality of all communities.
>
> (Anthony Smith 2000: 58)

OF ALL THE foundations upon which a discussion of national museums might be built, none has appealed more completely to the postmodern museological mind than Benedict Anderson's *Imagined Communities* (2006 [1983]). As Anthony Smith observes, disciples have been only too ready to adopt a concept which seems defined and explicated in the book's title, rarely probing the historical basis for Anderson's 'imagining'. It is perhaps assumed that Anderson was looking at the nation through a postmodern lens, but he was not: 'It is *Imagined* because the members of even the smallest nation will never know most of their fellow-members, meet them, or even hear of them, yet in the minds of each lives the image of their communion' (Anderson 2006: 6). More than a quarter century after this book was first published, few would presume absolute knowledge of anything; subjectivity, imagination and construction are everywhere but owe everything to Foucault, Bourdieu, Derrida and others and nothing to Anderson. Indeed, those who read a little further will be taken with Anderson's erudition and historical and cultural specificity. As Sharon Macdonald helpfully reminds us:

> It involved projecting sentiments of belonging and brotherhood way beyond those of direct experience, but only up to a specified 'edge' – the boundary

> of the national community. As individual identification with the nation-state and the numerous unknown 'brothers' could not rest on experienced *social* relations it had instead to be *cultural* – a matter of shared knowledge and practice, of representation, ritual and symbolism.
>
> (Macdonald 2003: 2)

In recent years, a considerable amount of work has been done on the museum's ritualised and symbolic practices, its representations of 'knowledge' and its political subjectivity. Situated within that fluid coupling of thought and language which stands between people and things, the museum has revealed itself to be a rich subject for the deconstruction of cultural authority (Preziosi 1998; this volume). But while external commentators have been quick to recognise the performative qualities of the museum – its poetic contributions to knowledge and reality as configurations of belief (Knell 2007c) – museum practitioners have remained true to the moral necessities of didacticism and the possibility of a neutral, or acceptable, authoritative truth. They continue to work within museum definitions which unreasonably constrain institutional conceptualisation and contradict the subjective realities of museum provision. Based on a moral positioning manufactured through acts of professionalisation, public and professionals alike continue to imagine the museum as neutral, authoritative and trustworthy; an accurate rendition of the world as it ought to be understood.

But if museums are poetic and political spaces, rather than purveyors of objectively conceived Enlightenment truths, then all they can perform are acts of cultural symbolism. With this admittedly cynical lens in view, we might imagine national museums as providing the scenography and stage for the performance of myths of nationhood. As in the theatre we might *imagine* and believe, but in the museum our imagining can be so much more believable because we are led to think that all around us has arrived objectively and all is as it seems to be; these things are not merely props. This chapter considers how the nation is performed within the national museum, the roles played by sets, scripts and actors, and some of the connections, boundaries and determinants of our national imagining. In doing so, it will reflect upon a diverse range of practices around the world including aspects of the papers presented by authors contributing to this volume. My aim is to set these contributions in a more holistic theoretical context but without suggesting that generalisation is particularly helpful to understanding the agency of national museums. Implicitly the contributions to this book argue that each nation's national museums are the product of national history and local circumstance and perform in quite particular ways.

The performance of reality

For the museum to be effective we must 'buy into' its offerings: art history, national narratives, the unassailable logic and authority of science, and so on. In doing so, we believe that museums contribute to our sense of a knowable and reproducible reality through which we can grow our own personal knowledge. But this museum reality does not come without performance. The two can never be disassociated. All who enter the museum are, however, deceived by the illusion that the museum's authority rests on its objective representation of the world. It does not. The museum exists in the civilised world because of its claim to moral authority derived from its fostering of education, knowledge, cultivation, professionalisation, and so on. It manifests and materialises the central ideologies of civilisation. But, as Bourdieu recognised, these ideologies are situated and particular; they are not a consensual representation of the thinking of the everyman but of the educated (middle) classes. This revelation that museums propagated cultural bias and contributed to the perpetuation of social division contributed to a revolution in museum provision which is still ongoing. It resulted in political acts of social inclusion within the museum but these have done little to overturn the social divisions of which Bourdieu complained: they rest upon the same constructed moral basis that shapes the museum form and authority, and are delivered by that same class that once gave citizens 'the arts'. My intention with these arguments is not to argue against these practices but rather to suggest that our awareness of them is quite fundamental if we are to do more than *imagine* how the museum contributes to the production of the nation.

It may be a surprise to some that museums were not always the cold and factual spaces which have in recent years been remade as a result of our new inclusive awareness. Mattias Bäckström's Hazelius and Jennifer Carter's Lenoir (this volume), for example, one and two centuries ago, respectively, were fully aware of the 'art' in museum making. This was in an era when the museum's legitimacy seemed to rest upon objective 'science' or scholarliness. However, these two men believed truth and meaning lay beyond scientific formulation and by means of drama and effect, and the careful deployment of real things, they felt it was possible for the visitor to experience greater – spiritual – truths.

Lenoir's and Hazelius's contemporaries, respectively at the Louvre and the Danish Folk Museum, followed a rather different path. Their museums conformed to what we now think of as the Enlightenment museum which in the staging of exhibitions sought to insert aesthetic and intellectual distance between people and things; they wished to make the relationship 'disinterested' (i.e. free from bias or self-interest) – a place where one could engage in detached thought. Of course, such objective rationalism was never entirely possible in museums where careers and material monuments were built, where claustrophobic power games became a cultural norm, where passions could be developed, where professionalisation and institutionalisation produced blinkered norms and traditions, where a blind eye was turned to the politics of the market place, and where resources were always inadequate. These pervasive peculiarities of the museum occurred, however, outside of public view, like the truths of a secret society. In order to claim authority the illusion of absolute professionalism was all. And being long-lived but with short memories, and self-legitimising, museums ensured themselves and their publics of their uncontested rationalism.

By these means the museum acquired its perceived modern form and maintains its old illusions (for which see Handler 1988; Macdonald 2003: 5). But it emerged from a variety of cultural practices and institutions (see Altick 1978; Sandberg 2003). Indeed, the modern museum did not arrive as the culmination of a progressive evolutionary journey (MacGregor 2007); its apparent uniformity of structure and form has always concealed the cultural diversity which has altered and adapted the museum to local needs. In each setting, the museum was always an idea and a set of social spaces, never simply a collection or building. For these reasons the emergence of the national museum in different national settings cannot be read as nations doing the same thing. The case studies in this book argue against such homogenising views. The national museum as it is locally produced reflects local conditions of nationalism and wealth, international connections, identity and competition, individual and corporate interests, political and economic relationships, the ideological possibilities of culture, networks of appropriation and emulation, diplomatic efforts, and so on. No museum is an exact copy of another. Even Sharjah's Natural History and Botanical Museum, which went to extraordinary lengths to imitate precisely exhibits at the National Museum of Wales, cannot be considered a mere copy. Indeed, its act of copying in itself makes it different. But how could a national museum in the United Arab Emirates be anything like that in Cardiff? Museums are never what they seem to be; never merely buildings and collections. They are places where professional and public performances are scripted and staged.

Hazelius's social engineering in Stockholm a hundred years after Lenoir's scenography and art installation in Paris simply produced kinds of the national museum; their admission of fictional elements does not mean they were operating outside any norm of museum definition. Their willingness to admit to fictions and illusions suggests that they understood the performative possibilities of the museum. In this museum performance they saw no great disconnection between the museum and landscape gardening or pageantry. What Lenoir and Hazelius produced was neither fact nor fiction, nor indeed that dubious middle line taken by modern populist non-fiction. They believed they were creating *more than* fact; a more holistic truth available to those who could combine art and science without feeling a sense of contradiction. They understood that what they were creating was a space for imagining and believing at a time when other museum workers, blind to the performance that was taking place in their galleries, were under the illusion that rationalism was attainable and was all.

The implications of this union of fact and performance could be seen more recently in Alan Yentob and Andrew Hutton's television docudrama, *Van Gogh: Painted with Words*. Here, viewers were assured that every word spoken was sourced from van Gogh's correspondence. We were led to believe – like we might believe in the museum – that in actor Benedict Cumberbatch's performance we had the personification of the great painter. But, of course, the performance was everything and took control of the facts; Cumberbatch gave a brilliant and sympathetic rendition of the tortured genius and Vincent's work leapt from the screen. Yentob and Hutton had made us imagine and believe, just as Hazelius and Lenoir had their publics, and just as every museum does. But, of course, had they selected different (but equally authentic) words, increased Cumberbatch's instability and inserted more violent expression, we might have imagined and believed something different. We might have seen less romance and less beauty. Using fact as an artistic medium, they, like all museums, were simply producing a *fabrique* while attempting to convince us it was real (Carter, this volume; see also Bruner 1994; Gable and Handler 1996).

A problem with the script

Of course in many modern narrative-based museums the material fact – the interpretable object – has disappeared or moved to a supporting role to be replaced by mere assertion. In Emily Stokes-Rees's Hong Kong Museum of History, set building and essentialised narrative shape the experience. Script and scenography have been carefully constructed to permit a singular public performance, a singular manifestation of the nation rather than the nation found through democratic negotiation. In this respect, the museum as theatre is quite the reverse of that developed by Lenoir who encouraged his visitors to dream.

The museum in Hong Kong exposes this difference between the theatre and the museum as theatre. One is merely a work of fiction which we can like, love or hate as we please; the other purports to be a work of fact and while we can also love or hate it, we also have the option to believe it (or not). It purports to be a representation of reality and truth, but the privileging of narrative and scenography over the interpretation of objects seems to shift the museum away from those technologies which permit the museum to claim moral authority. Non-fictional narrative has its basis in historical writing, not in museum building. It entered the museum, rather late in the day, in the possession of historians, whose field of study has a disdain for objects as historical evidence (Knell 2007c: 7–8), and designers, possessing storyboards and interested in the logic of visitor flow.[1]

One of the problems of 'fact-based' narrative-driven exhibitions is that there simply is no division between propaganda and the supposedly objective narrative. All narrative attempts to propagate views, values and beliefs; objects, on the other hand, are, of themselves, vague and ambiguous. In Radostina Sharenkova's Bulgaria, the communist powers used narrative to corral this ambiguity into new meanings, forcing the local population to imagine their genetic ancestors as unsophisticated Others. The museum could then evidence Soviet Man and demonstrate the social improvements that had arrived with communism. Later, these same materials were deployed by the same political regime to create a sense of national integrity and homogeneity.

Such ideological shaping of things through narrative is also seen in Sarah Hughes' analysis of the British Museum Press. Here the museum used the moral elevation of Enlightenment universalism to depoliticise and denationalise world culture, and legitimise the museum's continued possession of contested pieces. It presented the British as a noble people, above the petty bickering and possessiveness of nations. Only world powers and old nations position themselves in this way. Universalism also produced a bond of common interest with other nations – indeed with all other nations other than those which had been the source of the material culture in the first place. As Peter Aronsson explains, the Greeks and Italians then formed a resistant huddle in response.

In Simina Bădică's Romania, the dictatorial narrative of communist museology was countered by the 'antidote museum'. Feeling revulsion against single-master narratives of any kind, in the post-communist museum the audience was to be empowered rather than taught. In a move that ran counter to the obsession with didacticism in the West, objects were to do their own speaking; to resonate rather than be the subject of narrative or interpretation. As words in ever-increasing numbers seemed to be guiding and giving purpose in museums in the West, in Romania they were fundamentally distrusted. In other parts of these museums, communist objects themselves were displayed but in the manner of an art installation intentionally encouraging satire and ridicule.

Eugenia Afinoguénova's visitors to the Prado also entered into an unscripted interaction with art objects. Unfamiliar with the consumption of nudity in art, the performances of the visitors became the subject of much humorous social commentary. These performances revealed how the gallery space itself had to be understood and negotiated, and how gender, geographical and class relationships were deconstructed by those who looked on. Madrid's educated middle classes delighted in the phenomenon and the opportunities to engage in innuendo, and in doing so also said a good deal about themselves. But even they did not know how a narrative could be written that would justify the display of such art. Rather it was the art itself that spontaneously created narrative subjects in the visiting publics. Here too, but in a rather different way, the nation was being made.

The union of scriptwriters

For all its museological difficulties, in the national museum no subject is more central to the construction of the nation than history or, to put it more neutrally, the handling of the past. The past can provide a powerful anchor for nationhood:

> The nation, like the individual, is the culmination of a long past of endeavours, sacrifice and devotion. Of all cults, that of the ancestors is the most legitimate, for the ancestors have made us what we are. A heroic past, great men, glory (by which I understand genuine glory), this is the social capital upon which one bases a national idea. To have common glories in the past and to have a common will in the present; to have performed great deeds together, to wish to perform still more – these are the essential conditions

> for being a people . . . Where national memories are concerned, griefs are of more value than triumphs, for they impose duties, and require a common effort.
>
> (Renan 1996 [1882]: 52–53)

The past as it manifests itself in the museum need not, however, conform to the requirements of rigorous historiography. Drawing upon Pierre Nora's work on memory, Stefan Berger and Chris Lorenz (2008: 13), for example, remark on 'the replacement of "history" by "memory" and "heritage"'.

> Memory and history, far from being synonymous, appear now to be in fundamental opposition . . . Memory is a perpetually actual phenomenon, a bond tying us to the eternal present; history is a representation of the past . . . Memory is blind to all but the group it binds – which is to say . . . that there are as many memories as there are groups, that memory is by nature multiple and yet specific: collective, plural, and yet individual. History, on the other hand, belongs to everyone and to no one, whence it claims universal authority . . . at the heart of history is a critical discourse that is antithetical to spontaneous memory. History is perpetually suspicious of memory, and its true mission is to suppress and destroy it.
>
> (Nora 1989: 8–9, quoted by Berger and Lorenz 2008: 14)

Berger and Lorenz note that memory is in the possession of individuals and communities, while history is in the hands of professionals. A political divide separates these two engagements which is reminiscent of that between Bourdieu's (1984: 28–44) distanced and popular aestheticism. Bourdieu argued that the distinction rested upon the legitimising actions of a cultural aristocracy: the self-empowered professional middle classes. It is this same 'class' that argues for the moral and ethical legitimacy of dispassionate and evidence-based history. But when history is distilled and laid out in the museum, in a space where there is a freedom of performance and where the performance itself writes the history, and where so much has poetic resonance, then surely history is replaced by memory and imagination. Evidence becomes art. History becomes personal heritage.

The professional's disdain for memory as a legitimate alternative to history reflects the wider issue of control in museums and the professional's fear of full public participation (but see Watson 2007 for a counter-example). Pille Runnel, Taavi Tatsi and Pille Pruulmann-Vengerfeldt's exploration of the birth of the new Estonian National Museum reveals distinct and conflicting professional and public communities. For the professional desiring a landmark architectural statement, the new museum is a high spot on a career trajectory, a stepping stone to better things. This outlook encourages the professional to see the performance set on an international stage there to emulate other professionals. In contrast, the public view the new museum 'honestly'. They see and feel it as it appears to be and as it exists within the everyday; not as rationalised by abstract historiography but as it is remembered, imagined and believed. For them, the museum pulls towards the centre, towards core beliefs, towards home. Each group is in these ways using the museum to construct their own identities; both see public benefits for the outcomes they promote. But both have in mind quite different museums and quite different national imaginings.

There is, then, a quite fundamental divide between those who make museums and those for whom they are made. The efforts required to counter the effects of this deeply embedded structure need to be quite extraordinary. Afinoguénova shows that the Prado which had so entered the popular culture of Madrid was required to bend to public tastes in art and the art that was acquired had all those immediate populist qualities the controlling class despise.

Stuart Burch takes the issue of public participation further in his study of national galleries in Oslo, Stockholm and London. Here protests in Oslo put a conservative brake on directorial autonomy and creativity, forcing the museum to adhere to an embedded narrative. Elsewhere, the division between professional and public is reified in roped-off spaces and objects. Burch locates a model for participation in Pontus Hultén's freeform play area at Moderna Museet; the audience quite literally became performers and artists. But, Burch notes, this was a rare moment of permission. It could, of course, be argued that art of itself permits participation; that it cannot exist without it. Events at the Prado seem to suggest this. But, like the Estonian authors, Burch wants more – not just a palette of colours to admire but the brushes, the canvas and direct participation.

In the Netherlands, it appeared that professional control of the script was about to be lost as politicians began to echo public disquiet about the effects of multiculturalism. Gwenny van Hasselt explains that the suggested solution was the introduction of a Dutch national canon and a Dutch National Historical Museum. This made the Dutch intelligentsia fearful of the ideological deployment of history. However, this turn to the right failed to manifest itself in the museum, where the directors used arcane museological knowledge to advocate less contentious outcomes based on learning and education. The union of scriptwriters – the profession – had again achieved its ends and in doing so perhaps felt they could claim a victory against the doubtful morality of imposed national narratives.

A further implication of the professionalisation of museums has been the establishment of firm internalised systems of belief. These permit tasks to be done properly but also place blinkers on conceptualisation and walls around creative possibilities. Institutionalism can swamp creativity. This is an interesting aspect of Rhiannon Mason's discussion of the Museum of Welsh Life at St Fagans. Established more than half a century after Skansen, it arose in a period when British intellectuals were rebelling against aggressive industrialisation and the transformation of the rural landscape (Hawkes 1951; Hoskins 1955). However, the founding curator at St Fagans seems to have possessed none of that philosophical drive which fuelled the first wave of building folk museums. Instead he deploys a made object: a museum form that had already been distilled into a professionalised thing apparently unaware of the philosophical or nationalistic energy that had fuelled that thing's invention. Mason reveals a museum operating professionally as a visitor attraction and educational facility, considering visitor numbers and other operational matters, and assured of its implicit rational purpose. In this professional world the imagined antithesis is amateurism which is certainly a feature of museum prehistory and many independent ventures. This sense of a non-professional Other – the amateur – shapes views of public involvement. The oppositional pair both keeps professionals operationally minded, and prevents the imagining of other professional configurations. It keeps the script tightly in the writer's hands.

Anchoring the nation

The material, rather than the historiographic, possibilities of historical objects have been fundamental to anchoring nations in national museums. As Nora asserts, this is an act of memory rather than history. In Karoline Kaluza's Poland, for example, historical paintings took on the aura of religious relics in those years when the nation was denied free and independent existence. They became objectifications of national memory; reliquaries which permitted the devoted to keep the faith. These paintings embodied the past within their content and permitted that past to be mythologised, romanticised and idealised. Within the image, they believed, lay the sleeping nation ready to be awakened and rebuilt. When that day came, the new state possessed in these works a memory that transcended the awkward realities of history; these objects ensured that the new Polish nation was immediately old. This historicism not only gave the nation weight, it also introduced a sense of the nation possessing qualities and experiences out of reach of the modern man, woman or politician. It made the nation resilient to rational acts of denial. But in so many ways, in the performances and poetics of the museum, any rational sensibility could be overtaken by emotion. In Kaluza's Warsaw Uprising Museum, visitors come across a monument, cenotaph and flag at which to lay real and metaphorical wreaths. In other museums, one simply needed to look and think, and perhaps imagine and believe, and another kind of wreath would be laid, another monument erected.

A good deal of the historical anchoring of national identity that has taken place in national museums has buried somewhere within a primordial sense of origin in folk culture. Even if there is no appeal to shared or distinctive ethnicity, there is a sense of connection through the land, to physical and social structures, and 'traditional' ways of life. This is felt, even today, on visits to Skansen in Stockholm, Norsk Folkemuseum in Oslo and Eesti Vabaõhumuuseum at Rocca al Mare, Tallinn. In these places one senses a connection not simply to a place but to *a people*; not simply to history but to an ideal (Figure 1.1). In Alan Kirwan's Ireland, re-imagined with the end of British rule, a reawakened Celtic identity has come to define the nation in similar ways. Materialised in the National Museum of Ireland, it is perhaps better understood as a living culture celebrated in music and dance. It is in these lived

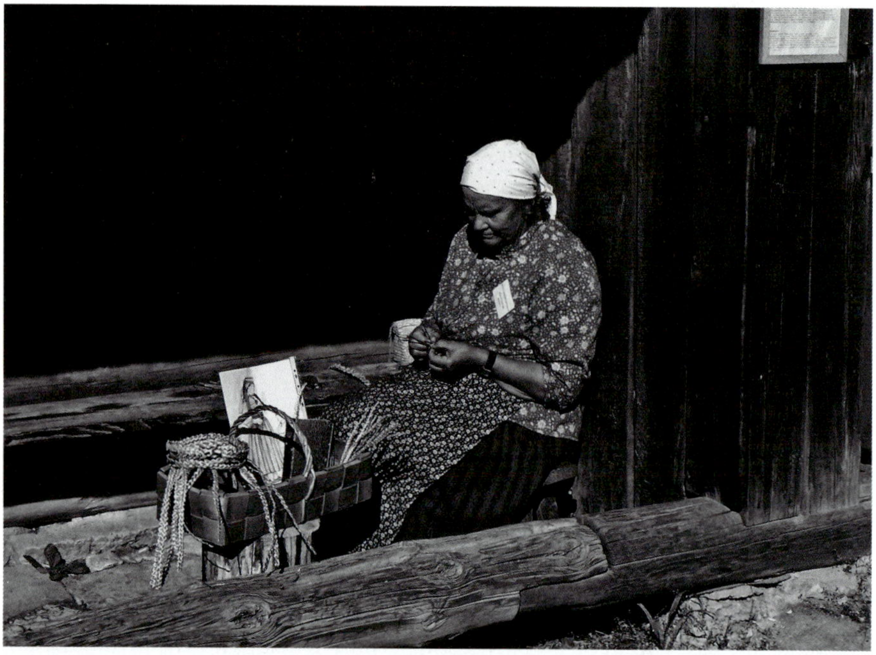

Figure 1.1 Primordial nationalism at Eesti Vabaõhumuuseum, Rocca-al-Mare, Estonia. In the warm Estonian sun, so perfect, peaceful and exotic was this manifestation of a folk past that one wanted to believe in it both as a place of origin and as a utopian ideal. Photo: author

forms no less a culture materialised: in instruments, costume, music and choreography. It is, in this respect, reminiscent of the mobile museums which took national culture into the heart of communities.

This historical anchoring is seen everywhere in national museums, but in each setting it seems to be nuanced to locate and develop the distinctive qualities of national identity: for the Poles it reawakened dynastic, heroic and religious legacies; in Germany, there remained an ethnic essentialism (Peck 1992); in Estonia, a primordial folk legacy; in Sweden, a social morality; in Bulgaria, a homogenising myth. Within each nation, the national museum provided a space for these essentialised notions to be performed and given moral legitimacy.

In Korea, there is no shortage of historical anchors but the nation remains physically and politically divided, scarred by outside impositions and unimaginable horrors, and confronted by rapid modernisation. Simultaneously, modern, old and developing, a collision of East meets West, a free nation but never completely unoccupied or sure that it is really known, South Koreans see museums as central to their yearning for a secure identity (Figure 1.2). For them history is alive. The provenance of an archaeological find, the chronology on a museum display panel – all can have political significance. Objects, their styles and forms, permit Korea to

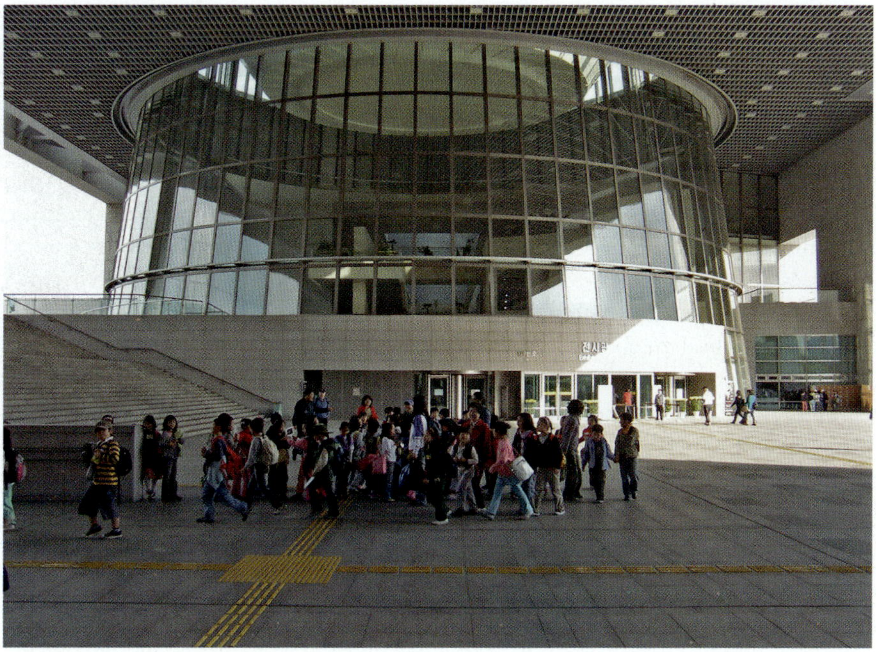

Figure 1.2 Schoolchildren outside the National Museum of Korea. A huge monolith of a building, this museum speaks to Korean citizens, to Japanese and Chinese neighbours, and to the West. In each of these conversations it is saying something different: 'this is us', 'we are not you', 'respect us and recognise us'. Photo: author

be mapped on to the historical geography of South East Asia as a continuous and influential historical entity (Kim 2005; see also chapters by Sunghee Choi, Jung Joon Lee and Ruth Sheidhauer in this volume).

Karen Shelby's Flemish nation might well empathise with South Koreans. Here Flemish identity is kept alive in the Belgium state in the IJzertoren. Amongst this museum's possessions is the Golden Painting, one of the most politically charged and contested objects in any European museum. It hangs silently as if it were simply an innocuous artefact but Belgians know it as a freedom-fighting flag. Built in imitation of the tombstones of the Flemish dead, the IJzertoren is itself a potent national symbol. The museum and monument ensure the continuing connection between Flemish identity and repressed autonomy. For the Flemish people it forms a historical anchor – a memory anchor; it denies Belgium single-nation status.

The problem of diversity

In other settings, a sense of national definition through an inherited and time-served connection to the land became challenged with the rise of multiculturalism. The Norsk Folkemuseum in Oslo, for example, added a Pakistani home in 2002.

In Cristina Lleras's Colombia ethnic diversity is even more central to the makeup of the nation, and cultural equality is a central tenet of the national constitution. Multiculturalism has replaced an earlier ideology which saw miscegenation as the means to homogenise society and eradicate the problems and tensions that often accompany diversity. In Sharenkova's Bulgaria diversity had been written into the national myth. It suggested that the nation's diverse peoples were all Bulgarians; that there were no other cultures. This changed with difficulty following the collapse of communism. In both Colombia and Bulgaria, the politics of inclusion were applied unevenly – some minorities were 'more equal' and some were burdened with a history of association which made inclusion difficult. In both countries, but for different reasons, the national museums felt unable to act decisively because the political ideology of multiculturalism had not been realised or implemented in cultural policy. It was not that the national museums wished to be politically instrumentalised but without appropriate frameworks for operation, they found themselves perpetuating old national ideals or following the superficial lead of visitor numbers.

In Colombia, the desire to remain true to diversity resulted in conversations about the allocation of space: some thought ethnic difference should be given its own artificial political geography in the museum. Sharenkova thought differently, reflecting that multiculturalism provided a means to rethink the geography of the Balkans; to reflect ethnicity in a complex, overlapping and interacting geography rather than in entrenched nationalism. She sees the potential for national museums to become bridging institutions, rather than the material basis for segregation, but also recognises the gulf that separates the ideal from an attainable reality.

Elsewhere in the world, museological techniques are used to neutralise the politics of display and iron over the creases of diversity. Most popular has been the aestheticisation of objects. Sunghee Choi's engagement with the collections in the National Museum of Korea, for example, shows how national identity can be performed through the motifs of art objects, distinguishing one nation from another and claiming that nation's right to recognition. Choi finds herself positioned as an advocate for her culture; but then understands that this national positioning is a reflection of attitudes imposed on Korea by the Japanese. She recommends a more participative approach empowering personal narratives and a capacity to look beyond national polarities and understand the self as complex and polychrome. Marzia Varutti, studying national museums in China, also considers the aesthetic turn in gallery display, suggesting that for China it permits a willing cultural and historical amnesia – a forgetting of complex, difficult and ideologically charged histories and an ethnically diverse present. The homogenising effect is to affirm the uncomplicated and reverential mythology of a single great civilisation. Like the experience in the Estonian folk museum, which might also be accused of aestheticising the past, one senses in this approach an attempt to connect past and future.

The nation on the world stage

The development of a Western canon of art history was manufactured in European national museums from the late eighteenth century. The museums Christopher Whitehead explores, positioned at the heart of the most powerful empire of the period, had no difficulty acquiring 'expensive, foreign paintings' and participating in this moment of editorship. The desire was for ready-made art esteemed by British and Continental connoisseurs; in Britain, the national gallery was to be built using foreign materials appropriated with the self-satisfaction of an old and powerful empire. But this did not mean that Britain acquired all its art through the exercising of military might. An old nation, it also possessed international ties and family connections (Figure 1.3). The British, it seems, did not worry too much about the relative lack of esteem for home-grown works as power and meaning were in possession and not in the geography of manufacture.

By their collecting efforts and approaches to display, the national museums and galleries of the older, wealthier and larger nations defined the canon of European art as an archipelago of interconnected islands (Dutch, French, Flemish, Italian, and

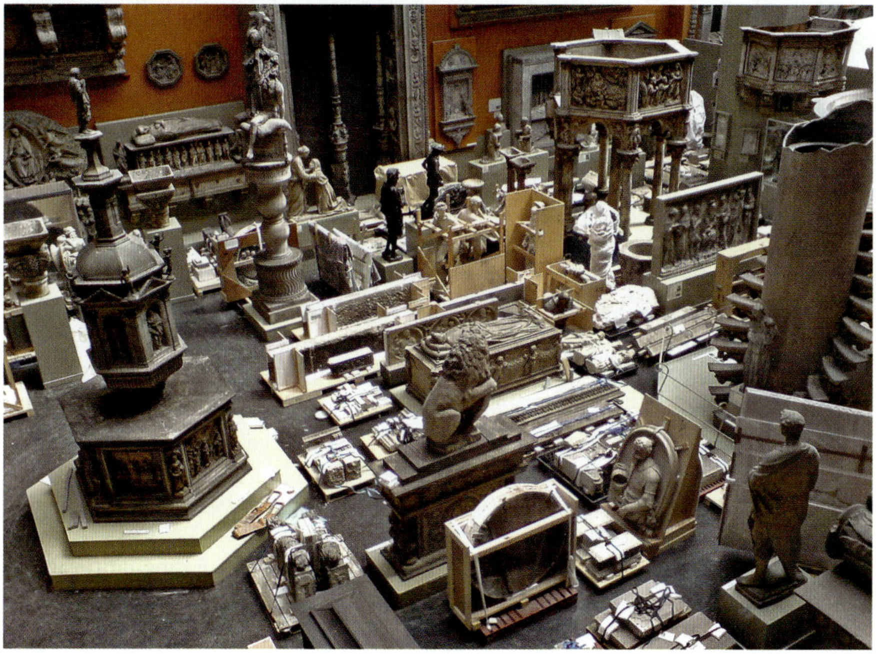

Figure 1.3 The Cast Court at the V&A, London (during renovation). This is not the example of imperial appropriation it might seem to be, but the result of commerce and cultural exchange which sought to introduce great world art to national populations of scholars and students across Europe. In 1867, fifteen European princes signed *The Convention for Promoting Universal Reproductions of Works of Art* in order to promote this practice. Photo: author

so on), set in an ocean of artistic silences. Not all nations had been empowered to contribute to this act of artistic cartography. Susanna Pettersson's Finnish Art Society, for example, recognised that Finland had no possibility of swimming in the mainstream. However, all nations engaged in acts of copying as a means to emulate and extend the canon of great art, but Finland recognised that it could not supplement this with the purchase of great works. In time, Finnish artists travelled to acquire their training, but ultimately what they produced – because the society demanded it – was something that could be called Finnish art. Inevitably, this involved the development of favoured themes, subjects, styles and artists. Finland's engagement produced both distance and implicit connection. While it contained the echoes of the Western canon, this art also, inevitably and gently, pushed the boundaries of that canon slightly towards Finland.

These artist ripples of influence have encircled the globe as a result of travel, travelling exhibitions and study abroad. It is not difficult to see, for example, in the 1980s paintings of the distinguished Taiwanese artist Ku Ping-Hsing, on display at the Taiwan Museum of Art, echoes of Cézanne, Feininger and the Futurists (Taiwan Museum of Art 1991). Yet Ku's work also reflects that island's modern art tradition, its timing, its relationship to the West, and more deep-rooted Chinese philosophies which shape technique, style and subject – as well as the relationship between master and student – rather differently. A European seeing this art for the first time cannot help but think he or she has entered an alternative configuration of reality; not a world of copies but of other imaginings.

There are, then, within national museums – and particularly national art museums – many ways in which the collections have acquired an actual or implicit international dimension. Despite its difficult-to-shake-off geographical hegemonies, the modern art market aspires to globalism and in doing so follows a counter-course to the discipline of history's growing appreciation of the indigenous and local. Art aspires to exist in an undifferentiated homogeneous world where freedom of expression is unconstrained. This internationalism turns small and provincial art museums into vehicles of nationhood. In contrast many provincial historical museums have long fought against global homogenisation, though with the rise of new media and the possibilities to celebrate local difference on a global stage, even these are coming to understand that globalisation need not produce homogenisation. They too might consider themselves as having national purpose just as English provincial museums did two hundred years ago.

Hughes' British Museum must also be delighted with the turn towards globalisation because it offers political salvation, but such international museums do not possess the world's art and antiquities simply as a result of past colonialism. Other nations, who find themselves overshadowed by powerful neighbours, such as South Korea and Taiwan, have actively promoted themselves in the West. South Korea, for example, has sponsored a number of permanent galleries in the great national museums of Europe and America (Kim 2005); now Korean celadon sits proudly in the exhibitionary equivalent of the embassy. There it stands for the antiquity and creative genius of a civilisation no less great than that of the Chinese and Japanese. Korean identity is by this means being built internationally and in doing so – by spreading the word – that nation is gaining cultural solidity and resilience: 'this is what we are, these are the things from which we are made, and we are different from these others'.

Over the last decade, governments have come to recognise that contemporary art offers a powerful medium through which to gain this kind of international foothold. South Korea, for example, has adopted the artistic calendar of biennales and other events (Lee, this volume). In Turkey, there has been a far longer attempt to connect to Europe through art. Ayşe Köksal's interest is in Turkish modern art and the role it has played in the development of the modern nation. In the post-revolutionary nation, this art permitted a break with the past and the invention of a new modern cultural tradition which nevertheless connected Turkey to artistic developments in Europe. Successive iterations of national art museums in Istanbul have attempted to mirror the West, to show Turkey's Europeanness. In Turkey this involved government intervention whereas in Poland the end of communism brought about a spontaneous eruption in contemporary art as a reflection of new-found freedoms. In both nations contemporary art is understood as politically empowering. This art found that it has a place in the nation and demanded international dialogue.

Taiwan's international engagement has been fundamentally different. With the rise of the People's Republic of China, the US saw Taiwan as a pocket-sized remnant of the old nation. As Chi-Jung Chu explains, at first benefiting from American dollars, Taiwan soon found itself abandoned and then had to begin its own diplomacy through culture. As the island embraced increasing democracy, so it grew a sense of its own identity and national museums became sites for the display of things considered Taiwanese rather than Chinese. Even the most Chinese of Taiwan's national museums – the National Palace Museum and the National History Museum – somehow managed to exhibit this new Taiwanese identity (Tseng 2003).

This process of national self-definition in art is ongoing – a process of translating the material into the nation. In 2007, this was still happening to Edward Hopper. At a major temporary exhibition of his work in the East Wing of the National Gallery of Art in Washington, Hopper's paintings, which might be thought to capture something of the American landscape, were encouraged to speak of 'America'. Perhaps inadvertently, the exhibition, its setting and everything about it seemed to be making political statements which continued to manoeuvre Hopper into the nation's heart, making his iconography the iconography of the nation. The accompanying brochure described him as 'The iconic American artist', a status that had been acquired through much post-mortem hagiography by galleries and just these kinds of statements. The video presentation of his life described him in reverential tones, and in the same words stripped him of all mortal flesh, re-inventing him as a disembodied American hero. Through this act of museumisation, Hopper was being neutralised and remade; himself an object there to sit alongside Lincoln;

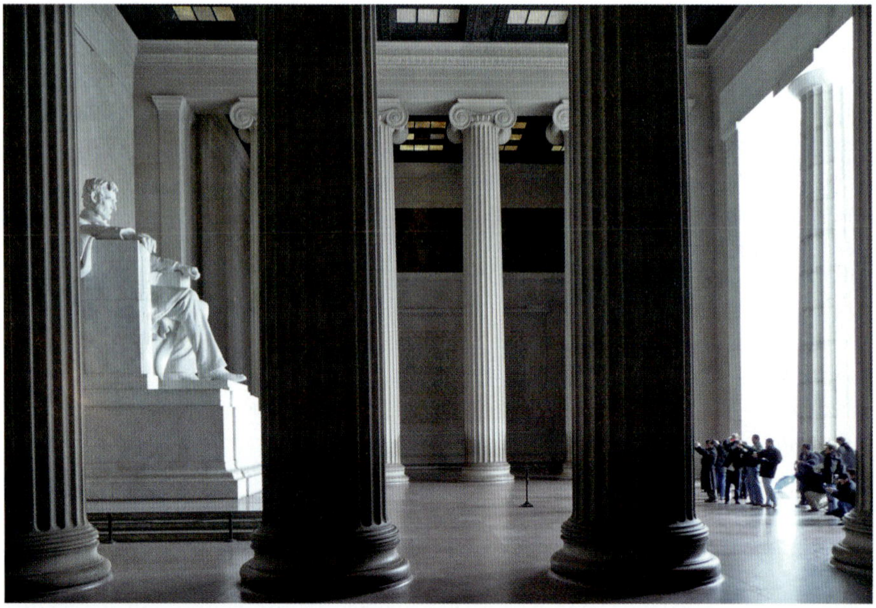

Figure 1.4 Pilgrims pay homage to Lincoln, the American colossus. The National Mall is the archetypal musealised landscape. The Lincoln Memorial is the most iconic piece in a landscape where memorials and museums are themselves objects in the materialisation of the nation. Photo: author

a naturalised citizen of the National Mall (Figure 1.4). And if Hopper's work travelled he became an American envoy, the latest in a two-century attempt to convince Europe that America possessed a culturally sophisticated civilisation.

Hopper had first come to notice in exhibitions like *Paintings by Nineteen Living Americans* at the Museum of Modern Art in New York in December 1929. In that exhibition, 'American' was more than a category. It too served to position the artist within the nation; it privileged national belonging over other categories which also might have given the exhibition logical cohesion. It was part of an ongoing art gallery process which first legitimises and later mythologises the artist. And if the artist is to remain forever in the national firmament then it is work that must be repeatedly undertaken, converting each new generation to the established canon, and ensuring collection and museum of their continued value to society.

As one of the most perfected traditional galleries in the world, the National Gallery of Art in Washington paints the nation with a wash of conservative values best felt in marble and in paintings by artists that require no introduction. Writ large in stone outside the National Gallery are the words 'United States of America' and one senses not only is the nation promoting nationalised conceptions of art, it is also proud of the American wealth that has permitted it to appropriate so many of Europe's most iconic works. The gallery is not alone in this act of carving the

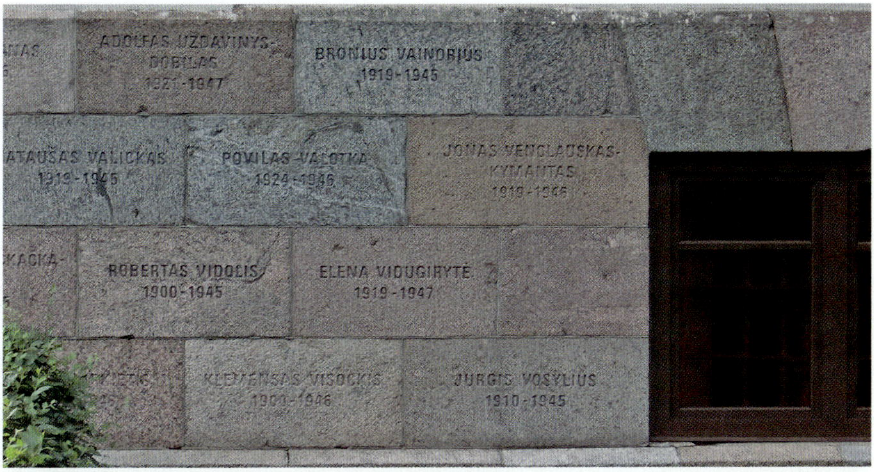

Figure 1.5 The Museum of Genocide Victims in Vilnius, Lithuania. Here the names of martyrs are written into the fabric of the building, just as they are in the IJzertoren. Like the words, 'United States of America', carved into the National Gallery in Washington, they ensure that visitors understand that a nationalistic message is being performed within the building. Photo: author

nation into the fabric of the building; others do so in the names of esteemed or martyred citizens (Figure 1.5).

The National Gallery of Art not only presents the American nation but also presents Europe's artistic nations to America. Following the European convention of providing separate galleries of Dutch, French and Italian art, it is nevertheless possible to drift through the galleries oblivious of these national distinctions. The gallery, however, is not afraid to imagine national stereotypes when it publicises this art to the nation (Figure 1.6).

With its pride in the flag, the USA covets the patriotic ideal and in this respect produces national museums and galleries which are subtly different from those in much of Europe. One senses that if Hopper is being repeatedly reaffirmed in the national museum, so too is America. Indeed, 'America' and 'American' are used with surprising frequency in these museums though we have grown so accustomed to it that it goes unnoticed. *The Art of the American Snapshot 1888–1978* on display at the National Gallery in 2007 is a case in point (Figure 1.7). We could, of course, merely attribute geographical categorisation to the use of the word 'American' here. But these snapshots were on display in the National Gallery. They did not possess any aesthetic or artistic merit; there was nothing about them that made them distinctly American – like all snapshots they were, except for those who possess them, simply banal. Here, however, 'American' is a binding force. Isolated from the world of photographic snapping, American visitors might well believe through the shared social experience of the exhibition that there is something truly American here

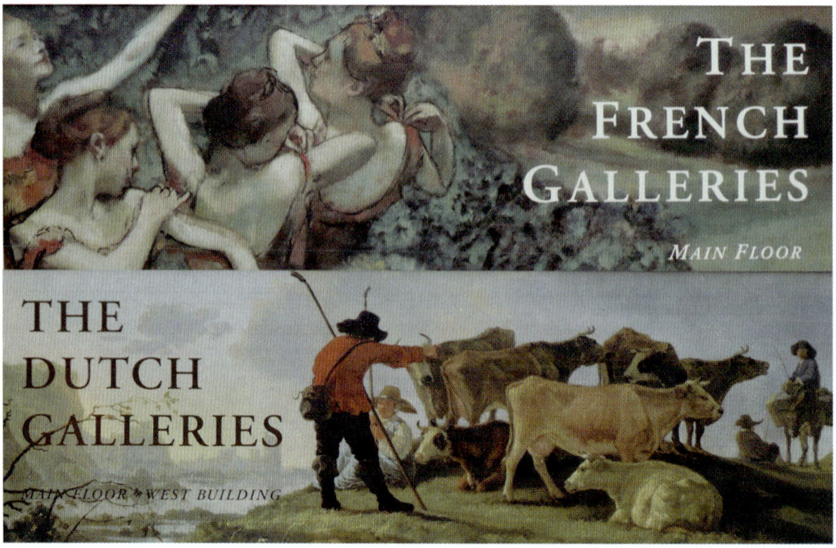

Figure 1.6 Essentialised representations of nations at the National Gallery of Art in Washington. These hoardings deal in stereotypes which are easily overturned if we remember the rural qualities of France or think of Amsterdam. Photo: author

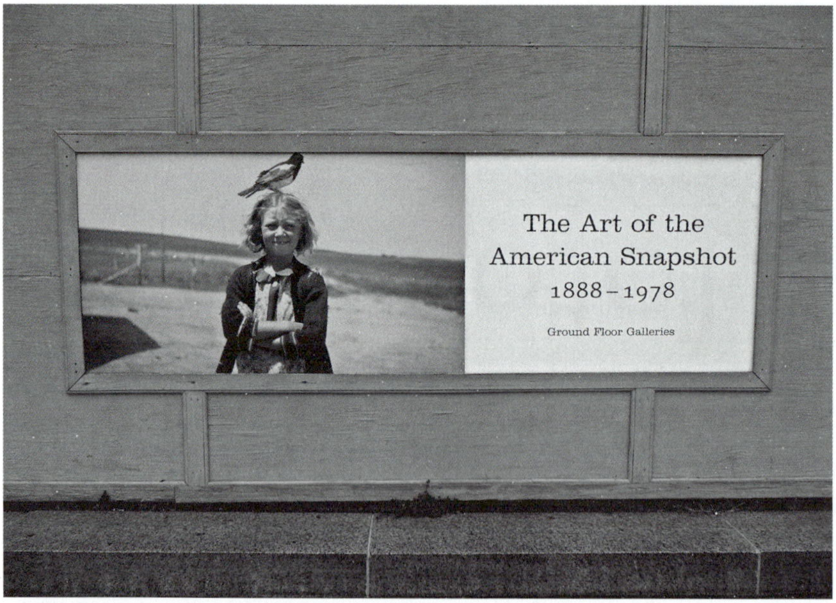

Figure 1.7 The American snapshot at the National Gallery in Washington. Is the term 'American' used here to create the illusion that in these images one might locate a national – as distinct from a universal – practice? Photo: author

which unites image and people. The term, 'American', is used to give significance to something which would struggle to achieve it by any other means. This act of aggrandisement is also seen in the new iconic buildings within which art is placed. These give the contents gravitas and permit nations to say 'we belong', 'we are like you'. One of the performances national museums give, universally, is that of grandness and weight even if to do so implies a certain superficiality of outlook.

The material boundaries and the national imagination

Much of what I have discussed in this chapter concerns the manner in which national museums contribute to imagining and defining the nation, both for citizens and wider international communities. It is possible, I suggest, to see the national museum's performance configured within a definable multidimensional space (Figure 1.8); to suggest that the museum seeks to shape a *particular* imagining of

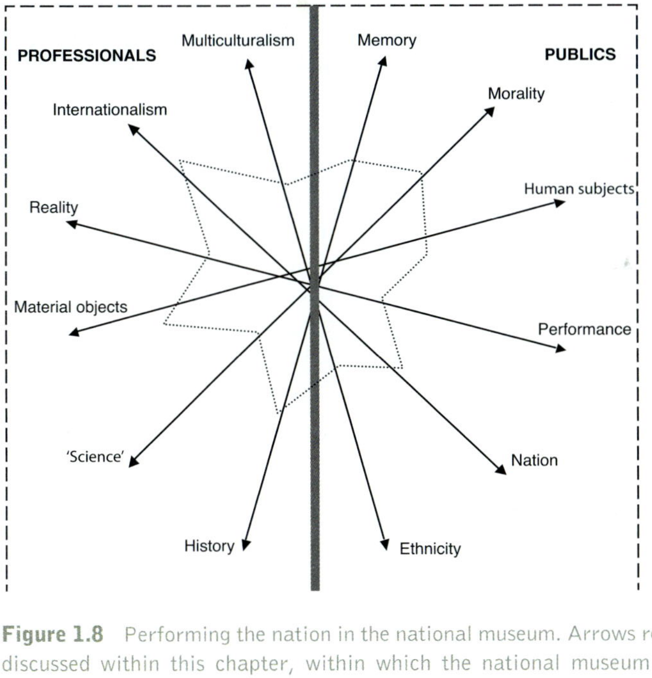

Figure 1.8 Performing the nation in the national museum. Arrows represent six dimensions discussed within this chapter, within which the national museum constructs the nation (represented by the irregular star shape in the centre). The dimensions do not form discrete categories, nor are they comprehensive. The diagram also suggests that publics and professionals (including academics) are divided in their approach to and desires for the national museum. The terms used remain ambiguous but can be given more specific meaning when associated with these two groups. Certain museum professionals may place themselves in the area of memory, morality, performance and human subjects. They might in these roles be considered 'audience advocates' as distinct from the public. The term 'science' is here used in the sense it is applied outside the English-speaking world, to mean academically rigorous.

the nation through its deployment of space and objects according to embedded values and perceptions. Figure 1.8 also suggests that these deployments reveal an unresolved tension between the public and professional.

One might argue, by extending and abusing some of Susan Pearce's (1995: 203) arguments, that museums are implicitly sites of object fetishism. Their outlook is oddly discriminating. This oddness is visible if one compares the exhibition gallery's view of the world with that of any popular snapshot. The difference, of course, is that the photograph is centred on the person. Ironically, the museum may believe that its subject is also people but, for professional and aesthetic reasons it resists the temptation to make people into objects of the disinterested gaze. The metaphor of the camera is helpful in understanding the peculiar manner in which the museum views the world. If we simply open the camera lens for long enough, the crisp image of the human subject begins to blur, then streak, ultimately to disappear, to be replaced by the more durable and immobile traces of the material world. Like this open lens, the museum compresses time and represents its human subjects through the material culture with which they have engaged. The reality of passing through a series of distinct and often unrelated 'nows', as J. B. Priestley (1964) put it, becomes altered into a single unrealistic – but somehow comprehensible – 'then' and 'there'. Thought and flesh – the ghosts of people – become implicitly and objectively embodied in object and architecture, and material things can now be made to speak for them. Identities and nation-states are by these means concretised and sentimentalised (Handler 1988; Macdonald 2003: 5). Thus when the Congo arrived in Stockholm in the early twentieth century, as discussed by Lotten Gustafsson Reinius, a small and recently disempowered nation found reassurance in the superior state of its civilisation. A distant people had arrived in the form of primitive things which appealed greatly to a Swedish sense of improving and homogenising patronage. The human subjects of the exhibition, however, remained a blur – not to be known, only to be imagined. The objects themselves were exotic – clearly they were not 'us, the Swedes' or indeed not even 'us, the Europeans'. Amy Barnes shows a similar relationship between the British and Chinese. The British frequently felt a sense of superiority but had to do rather more work to both admire Chinese achievement and yet deny that nation's modernity.

By these means we locate or imagine national disconnections through discontinuities in material things. We also, in the manner in which communities are both cohesive and Othering, locate layers of connections through these material things: local, regional, national, European. Across the national museums of Europe, there is an implicit and connective language of material things – objects and architecture – dominated by the motifs of Classical culture. Repeatedly valued and imported by nations across Europe, the ancient cultures of Greece and Rome perform as a trans-European language in defining as different those who live beyond its borders. We perform this sense of connection in the museum and in the surrounding landscape. Take a stroll through the Classical worlds of Greece and Rome at the Pergamon Museum in Berlin, for example, and it is easy to lose track of which of those worlds one is now in. But enter the market gate of Miletus and you exit through Ishtar Gate and into Babylonia, out of Europe, so one feels, and into another world (Figure 1.9).[2] A Europe that has always possessed mobile populations and multiple ethnicities, and almost constant political upheaval and reconfiguration, has been defined by the repeated appropriation of the Classical world. If there is an implicit language of things that unites or creates this European sensibility, then are national museums merely representative of that sensibility or do they give it to us (Figure 1.10)?[3] It is a sensibility that had to be made and made to persist. Chris Wingfield notes how entrenched respect for the Classical had become in the British Museum and reveals that it was only overcome when Darwinian ethnographic study legitimised the search for the origins of civilisation. The primitive British could be then encompassed within the evolution of civilisation – they too could be placed on the map and timeline. Nevertheless, the idea of a primordial civilisation played out in folk museums across Europe does not contribute to a sense of Europeanness; these museums conceptualise the indigenous and local.

One of the contributions of national museums to national imagining is to define and transcend geographical boundaries in complex and subtle ways. I am not thinking here of more obvious acts of cartography through collecting (Duclos 2004). In

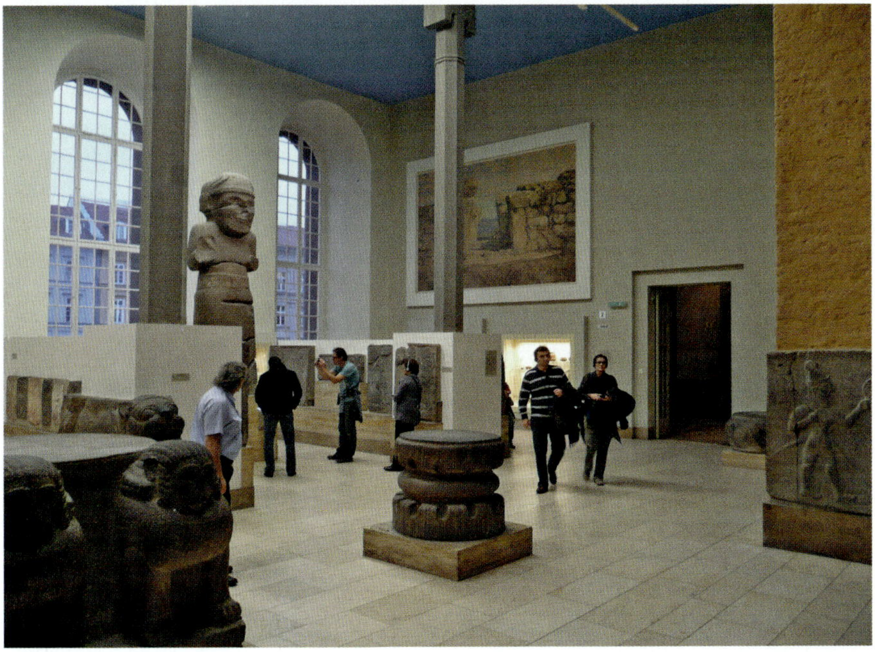

Figure 1.9 The Pergamon Museum, Berlin. Here a European can feel the connective tissue of Classicism but also experience the sense of leaving it and entering into the world of an Other. Photo: author

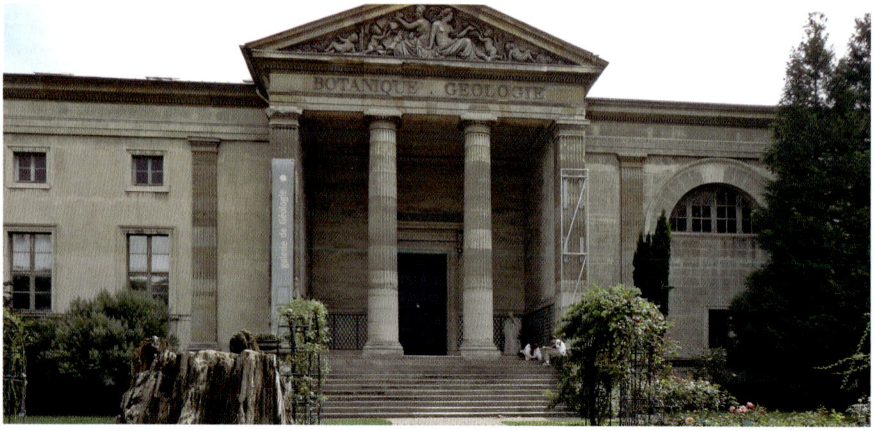

Figure 1.10 Muséum national d'Histoire naturelle, Paris, one of the most influential museums in the world. Classicism did not enter the European psyche merely via collections of antiquities. Disciplinary knowledge also has its foundations in the Classical world and has been shaped and performed in national museums. This too defines and connects Europe. Photo: author

the construction and positioning of the nation, there is both a tangible and imagined connection between the material thing within the museum and that without; between movable material culture and the immovable landscape. Again, I am not here thinking simply that these things are all heritage but rather that the manner in which they materially make the nation is the same; it is museological.

William Wordsworth wrote, with pre-nation-state sensibility, in his *Guide to the English Lakes* in 1835, 'Neither high-born nobleman, knight or esquire, was here; but many . . . humble sons of the hills had a consciousness that the land, which they had walked over and tilled, had for more than five hundred years been possessed by men of their name and blood' (quoted in Thompson 2010: 49). The barns and field systems (and for Wordsworth the deeply weathered farmhouses), now preserved across the upland landscapes of North Yorkshire and Cumbria bear material witness to this human connection; like objects in the museum, they have endured. Indeed, the musealised landscape needs to be understood as an eversion of the museum in which we can see the interior processes of the museum (taxonomy, representation, narrative construction, professional possession, and so on) spilled out and into our conceptualisation of place (Figure 1.11). In our performed imagining of the nation, a material, cultural and psychological connection links J. M. R. Turner's *Malham Cove* (1810) and James Ward's massive *Gordale Scar* (1815), both in the collections of Tate Britain, to the landscape which they represent. Tate's interpretation of Ward's painting is suitably nationalistic – 'Working in the last years of the Napoleonic wars, Ward aimed to depict a national landscape, primordial and unchanging, defended by "John Bull" in animal form'[4] – but this is only one element at play here.

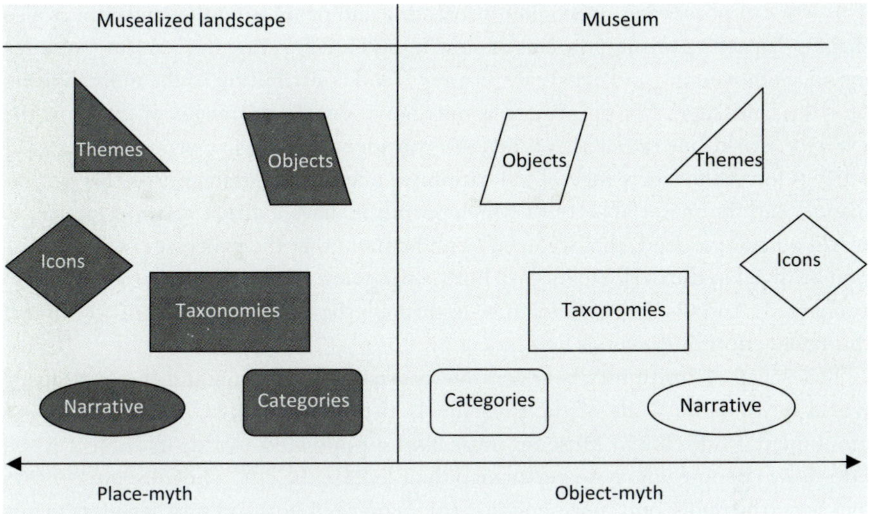

Figure 1.11 Material continuities: materialising the national imagination in museums and musealised landscapes. Reading the interpreted landscape as an eversion of the museum, it is possible to locate a continuity of constructive processes that transgress the conventions of cultural institutionalisation. Place-myth and object-myth are in constant connection, imaginings of the one affecting the other.

The key to understanding the relationship between things within the museum and those without, between the museum object and landscape, lies in a denial of what Sue Pearce referred to as the movement of objects in and out of circulation in society. Pearce argues that entry into the museum moves objects out of the circulatory system. However, the object as performed and imagined is always in circulation unless completely concealed (whether in a museum or as private property). An object (or landscape) redolent of nationhood – whether an image, imagined, held or observed – circulates freely within our national imagining regardless of the medium through which it reaches us. The object worked upon by artistic and scientific processes may gain a special place in this national making but it may also retain geographical connections beyond the museum. To use John Urry's (1995: 194–98) notion, the place-myth which produces value in the landscape has in the museum its counterpart in the object-myth. Thus Malham Cove finds its nation-imagining qualities in its association with Turner, the *English* watercolour movement and Turner's romantic rendition of Englishness but also in the significance that landscape already possessed and which drew Turner to travel there to represent it. The allegorical significance which seems to exist in the artwork may actually have preceded its manufacture, existing in subject and psyche beforehand. The artwork adds and adjusts through the performances in the museum but all the while its exterior subject may also be being worked on in other ways. Malham Cove – which has nothing to do with the sea – is possessed by the *National* Trust and is situated in a *National* Park.

This work of possessing, managing, protecting and promoting the landscape is also shaping Turner's painting; the acts of viewing and interpreting this painting, of valuing an art movement and artist, are shaping – and contributing to the musealisation of – the landscape. My point here is not, however, about images of things in the exterior world but rather that when we consider museums as providing material settings for the performance of the nation we need to understand that this performance, and the material settings which permit it, have a direct relationship to the world outside; indeed, that we need to understand that the processes of musealisation involved in the performances in both settings are essentially the same – but also connected. And we do not need to walk through the ruins of an ancient city to feel this connection; it is everywhere.

This sense of continuity between the internalised museum and the musealised world beyond the walls of the museum is also felt in Tokyo. The modern and monumental Edo-Tokyo Museum musealises a populated world; its subject is the Edo-Tokyo people who have performed their lives in the city. The museum consists of actors (the public and representations of former publics) and sets. Japanese prints have here been used as historical documents for reconstruction of the city in vast

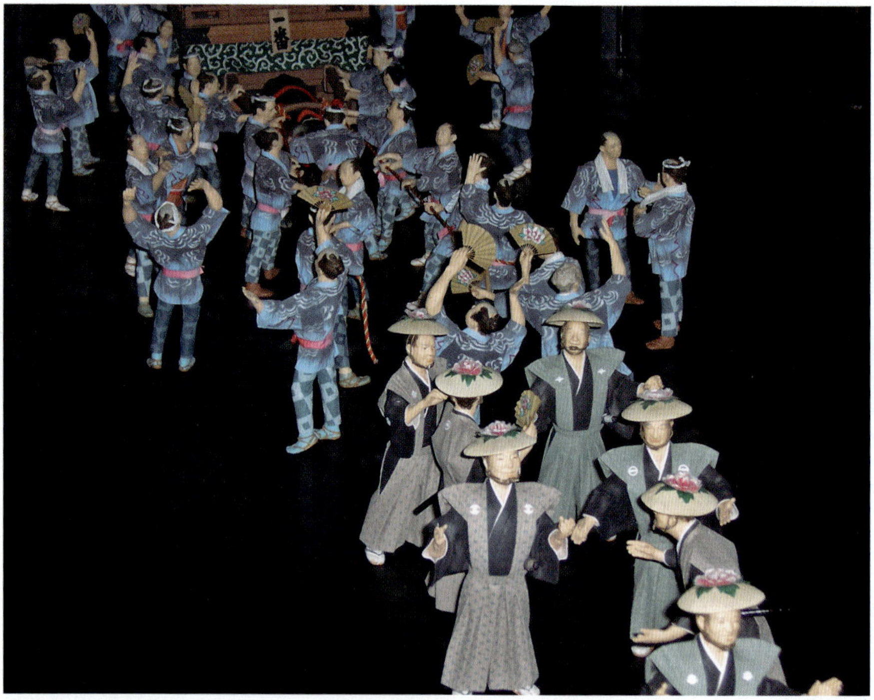

Figure 1.12 Figures dance at the Edo-Tokyo Museum. In Tokyo 'All the world's a stage' and this is simply one performance. Materialised, observed and labelled, social performances in the modern city are intimately connected to those displayed in the museum. Photo: author

dioramas: hundreds of tiny figures dance through old Edo (Figure 1.12). In the shop, the tourist finds books on Salary Men, Kanji, Cosplay, Sushi, Anime and other aspects of modern Japanese life which act as labels to the exterior world. When one exits the museum, then, one does not cross a firm boundary but enters a modern set – the musealised landscape of popular culture which seems to be simultaneously normal and extraordinary (both to tourists and the local Japanese). Like other national museums, the visitor might sense that the nation of the past is connected to the future, that the nation that is imagined and performed within the museum's walls is also performed outside, and indeed that both performances affect each other – both are connected.

Notes

1 My concerns here are about written narrative rather than the more poetic notion of spatial narrative.
2 This thought occurred simultaneously to me and to my colleague Sheila Watson as we walked through the arch.
3 This is the subject of research being carried out as part of the EU FP7 project Eunamus 2010–2013.
4 J. M. R. Turner (1775–1851). James Ward (1769–1859), *Gordale Scar (A View of Gordale, in the Manor of East Malham in Craven, Yorkshire, the Property of Lord Ribblesdale)*, *c.*1812–1814, Tate Collection, Tate Online, http://www.tate.org.uk/collection/ (accessed 12 April 2010). John Bull was a contemporary personification of Great Britain, and particularly England, regularly featuring in cartoons and caricatures.

Bibliography

Altick, A. D. (1978) *The Shows of London*, Cambridge, MA: Belknap Press.
Anderson, B. (2006 [1983]) *Imagined Communities: Reflections on the Origins and Spread of Nationalism*, London: Verso.
Berger, S. and Lorenz, C. (2008) 'Introduction: national history writing in Europe in a global age', in S. Berger and C. Lorenz (eds) *The Contested Nation: Ethnicity, Class, Religion and Gender in National Histories*, Basingstoke: Palgrave Macmillan.
Bourdieu, P. (1984) *Distinction: A Social Critique of the Judgement of Taste*, Cambridge, MA: Harvard University Press.
Bruner, E. M. (1994) 'Abraham Lincoln as authentic reproduction: a critique of post-modernism', *American Anthropologist*, 96: 397–415.
Duclos, R. (2004) 'The cartographies of collecting', in S. J. Knell (ed.) *Museums and the Future of Collecting*, Second Edition, Aldershot: Ashgate Publishing, pp. 84–101.
Gable, E. and Handler, R. (1996) 'After authenticity at an American heritage site', *American Anthropologist*, 98: 568–78.
Handler, R. (1988) *Nationalism and the Politics of Culture in Quebec*, Madison: University of Wisconsin Press.
Hawkes, J. (1951) *A Land*, London: Cresset Press.
Hoskins, W. G. (1955) *The Making of the English Landscape*, London: Penguin.

Kim, K. C. (2005) 'Korea as seen through its material culture and museums', unpublished PhD thesis, University of Leicester.

Knell, S. J. (2007a) 'Introduction', in S. J. Knell, S. MacLeod and S. Watson (eds) *Museum Revolutions: How Museums Change and are Changed*, London: Routledge, pp. xix–xxvi.

Knell, S. J. (2007b) 'Museums, fossils and the cultural revolution of science: mapping change in the politics of knowledge in early nineteenth-century Britain', in S. J. Knell, S. MacLeod and S. Watson (eds) *Museum Revolutions: How Museums Change and are Changed*, London: Routledge, pp. 28–47.

Knell, S. J. (ed.) (2007c) 'Museums, reality and the material world', in S. J. Knell (ed.) *Museums in the Material World*, London: Routledge, pp. 1–28.

Knell, S. J., MacLeod, S. and Watson, S. (eds) (2007) *Museum Revolutions: How Museums Change and are Changed*, London: Routledge.

Macdonald, S. (2003) 'Museums, national, postnational and transcultural identities', *Museum and Society*, 1: 1–16.

MacGregor, A. (2007) *Curiosity and Enlightenment: Collectors and Collections from the Sixteenth to the Nineteenth Century*, New Haven, CT: Yale University Press.

Nora, P. (1989) 'Between memory and history: les Lieux de Mémoire', *Representations*, 26: 7–25.

Pearce, S. M. (1995) *On Collecting: An Investigation into Collecting in the European Tradition*, London: Routledge.

Peck, J. M. (1992) 'Rac(e)ing the nation: is there a German "home"?', *New Formations*, 17: 75–84.

Preziosi, D. (1998) *The Art of Art History: A Critical Anthology*, Oxford: Oxford University Press.

Priestley, J. B. (1964) *Man and Time*, London: Aldus Books.

Renan, E. (1996 [1882]) 'What is a nation?', in G. Eley and R. G. Suny (eds) *Becoming National: A Reader*, New York: Oxford University Press, pp. 42–55.

Sandberg, M. B. (2003) *Living Pictures, Missing Persons: Mannequins, Museums and Modernity*, Princeton, NJ: Princeton University Press.

Smith, A. D. (2000) *The Nation in History: Historiographical Debates about Ethnicity and Nationalism*, Cambridge: Polity Press.

Taiwan Museum of Art (1991) *Modern Painting Exhibition by Ku Ping-Hsing*, Taipei: Taiwan Museum of Art.

Thompson, I. (2010) *The English Lakes: A History*, London: Bloomsbury.

Tseng, S. (2003) *The Art Market, Collectors and Art Museums in Taiwan since 1949*, Taipei: Sanyi Cultural Enterprise.

Urry, J. (1995) *Consuming Places*, London: Routledge.

Watson, S. E. R. (2007) 'History museums, community identities and a sense of place: rewriting histories', in S. J. Knell, S. MacLeod and S. Watson (eds) *Museum Revolutions: How Museums Change and are Changed*, London: Routledge, pp. 160–72.

Chapter 2

EXPLAINING NATIONAL MUSEUMS

Exploring comparative approaches to the study of national museums

Peter Aronsson

WHY ARE THERE national museums all over the world? How and why do they differ in year of establishment, division of labour and the narratives they pursue? Such questions seem breathtaking in their naïveté, but yet possible to ask of institutions which identify themselves and are identified by others as *national* museums, despite huge variations in ideology, government interference and the scope and focus of their engagement. This chapter explores how a comparative methodology might permit a general theoretical framework for understanding these institutions. Three prominent models, used to understand the birth of national museums – the spread of the Enlightenment, the nationalism born out of Napoleonic conquest and the shaping of subjects into citizens – are here used and challenged through systematic, empirical and comparative enquiry. The argument developed in this chapter is based on two complex examples taken from outside the common paradigmatic cases of Britain and France and composed of a number of interacting nations with a common past: Scandinavia, on the one hand, and Italy, Greece and Turkey, on the other. The analysis will combine a brief history of the establishment and development of these institutions with reflection on visitor or citizen experience, and will draw upon the literature and the author's own ethnographic fieldwork. Plans, rhetoric and hopes will thus be balanced with the experience of heterogeneity and polyphony caused by historical layers of cultural investment, the plurality of actors and the outcome of negotiations that can be detected but not detailed at this level of comparative research. The exploratory and explanatory power of a comparative approach will be circumscribed and assessed. The overall purpose is to guide the development of future comparative enquiries into the institutional settings, and epistemological and political negotiations, of national museums with a view to locating variation but also similarities within this institutional diversity.

Trajectories

Systematic collection started out from a more general desire to order the world, grasp the unknown and share splendour and knowledge with peers. According to MacGregor (2007), the first expansion of the concept of the museum occurred in the shift from early modern aristocratic treasury museums to cabinets of curiosities, which moved attention from the representation of splendour to the possibilities of a space of enquiry. The spread of the Enlightenment and the political transformation from the conglomerate states of *ancien régime* into nation-states concluded this development, producing paradigms of the national museum in the British Museum, as the iconic model of a universal museum of the Enlightenment, and the Muséum National d'Histoire Naturelle, Musée des Monuments français and the Louvre, as Napoleonic role models for the national museum. Both London and Paris were subsequently imitated by small and large nations, nations formerly occupied and those that wished to be counted amongst the great imperialists. The material basis for these institutions came from polite society, royal collections, systematic mapping

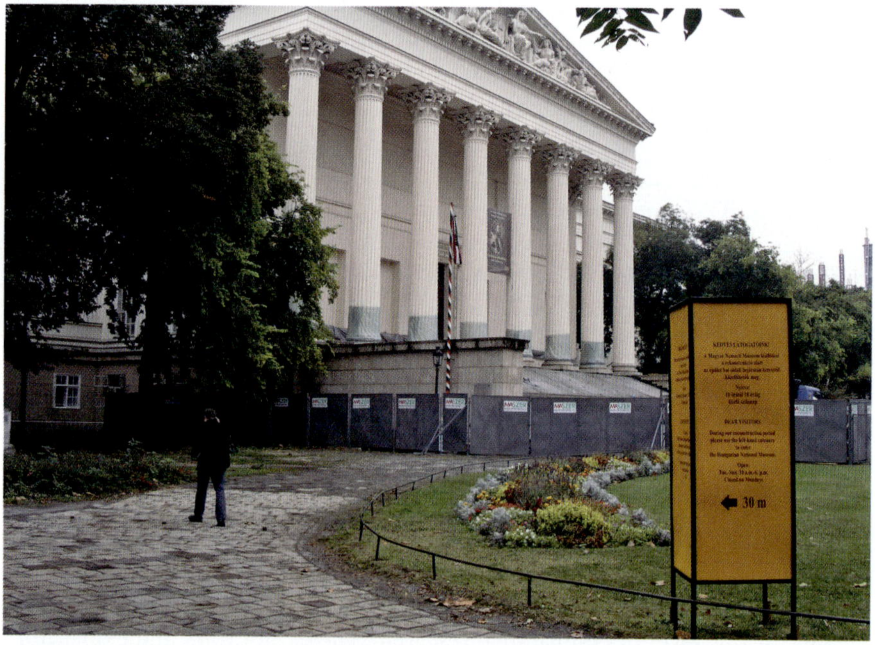

Figure 2.1 The Hungarian National Museum, founded in 1837, was created to advance a state-making process. It was a symbol and space for the formation of national freedom within the Habsburg monarchy. Photo: author

and enquiry, and looting and territorial expansion. Through their incorporation into these prestigious new institutions, they permitted the building of new identities which made reference to ancient civilisations, monarchical continuity and civilising splendour (Figure 2.1).

Although the museum was introduced around the world as a result of colonialism and international relations (trade, diplomatic visits, emulation, and so on), the first global boom in *national* museums took place in the decades after 1870, at the high-tide of imperial expansionism with a second wave taking place during a phase of post-colonial nationalism after the Second World War (Lewis 1992; Prösler 1996). The need for national display to complement the political process then seemed to overtake the desire for knowledge as a driving force for the development of these institutions. However, it is dangerous to oversimplify here – Nair (2007), for example, shows how museums entered India with a strong modernisation mission.

But also if one examines contemporary Europe, it is possible to question a simple modernisation thesis based on expanding secularisation, individualisation, peaceful bureaucratisation, multicultural integration, tourism and a growing cultural industry. War, power, religious difference and ethnicity also seem to be strong forces shaping state-making, cultural policy and museum-building. The traditional grand narratives of national museums are built out of embedded ideas about the linearity of history, the evolutionary possibilities of institutions and teleological conceptions of state-making trajectories. This is low-resolution history, satisfying the need for order and a safe direction for history both inside museums and in historiography, but does it stand up to the tests of reality; to grounded and situated analysis?

We will now continue along our road to a more definite model by outlining the establishment of national museum systems in two politically complex regions on the European periphery. They have been chosen as each of these regions possesses nations and museums acting on a common past and claiming overlapping material bases for their argument. These will help clarify the range of possibilities for dealing with the past as an active negotiation of identity in national museums. The two cases focus on (1) the Mediterranean (the nationalisation of Classical heritage in nineteenth-century nation-making) and (2) nationalised Scandinavian culture (prehistory and medieval heritage of a non-Roman territory). These two regions will be considered in turn, beginning with that corner of the Mediterranean occupied by Greece, Italy and Turkey. For the purposes of this discussion, national museums are taken as institutions of national collection and display, which claim and are recognised as being national and which articulate and negotiate national identity.[1]

Antiquity used and contested

Antiquity has been a resource for self-reflection and representation in Europe since the establishment of early historiography in the fifth century BC. In early modern Europe, the artefacts of classical antiquity were the prime objects of royal and private collections and became a core aspect of early national museums. They signalled the universal values of goodness (virtue), truth (reason) and beauty (aesthetics) as well as a historical tradition of state-making.

In Greece, culture is strongly nationalised and framed as a state responsibility – a national position on heritage which reflects the struggle for independence from the Ottoman Empire which began in the 1820s. The Acropolis was, as part of the establishment of the Greek state in the early 1830s, reinvented and successively cleaned of its historical Christian, Muslim, administrative and military functions in order to represent a glorious fifth century BC which stood for art, democracy and reason (Figure 2.2). The National Archaeological Museum which displays this ancient Greek heritage is still, as it was when it opened, divided into epochs of civilisation and material types. Work had begun on the construction of this museum in 1866 but the museum only opened in 1893. More recently, in 2009, the long inadequate Acropolis Museum was moved from the top to the foot of the hill and refocused on the missing Parthenon (Elgin) Marbles which remained on show in the British Museum. Its opening exhibition, produced jointly with Italy and inaugurated by the two presidents to mark its importance, concerned the restitution of objects and celebrated the successful return of many to their 'true' national contexts. This, then, is the narrative of continuity between the Hellenic culture and present Greek

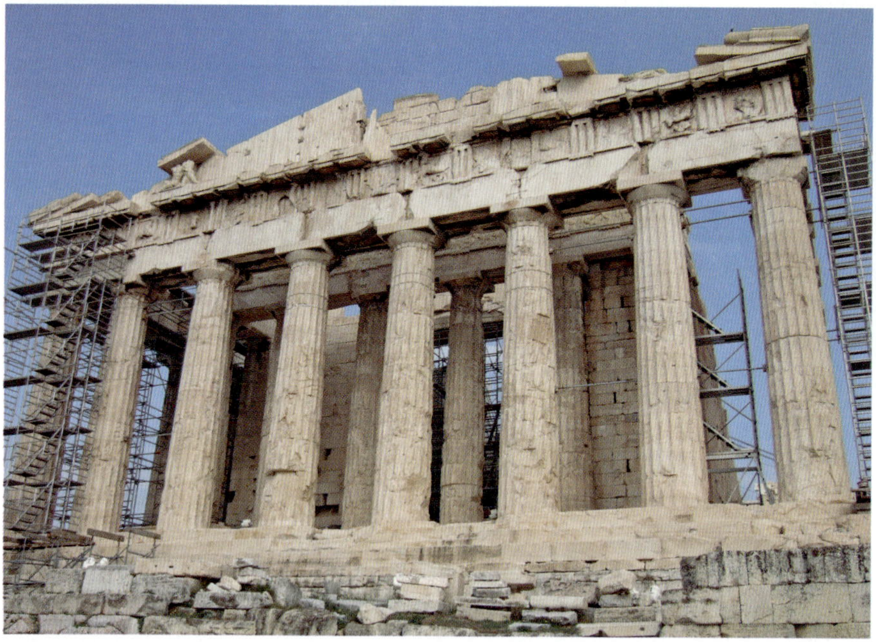

Figure 2.2 The Acropolis of Athens. This heritage site has undergone waves of reconstruction, beginning in the 1830s. A multi-layered building complex, it was cleaned of its Ottoman and Byzantine heritage and reconstituted as the materialisation of the Golden Age of Greek science and democracy in the fifth century. Photo: author

state that has long greeted tourists and speaks to the wider political community around the world (Boardman 2002; National Historical Museum 1994).

The National History Museum, opened in 1882 and now housed in the old parliamentary building, presents a narrative centred on the teleological creation of the Greek state as an inevitable outcome of the existence of a responsible centre collecting and uniting scattered but homogeneous Greek communities in the Eastern Mediterranean area. The museum expresses the nation's sad disappointment at the lack of international support at the time of the 'terrible events' in Smyrna in 1922, when Greeks were killed or expelled as a consequence of the making of the Turkish state (Hodne 2010; National Historical Museum 1994). This internal and patriotic story of the Greek nation is populated with named heroes and their artefacts, and historical paintings. It is part of a well-taught narrative found in schools, public celebrations and museums covering art, popular culture, archaeology and religion, including the National Gallery (1899), the private Benaki collection (1931), the Museum of Greek Folk Art (1918) and, through its Christian orthodox heritage, the Byzantine Museum (1914, but modernised in 2008). Together, these museums help establish the image of regional diversity united through a shared culture, identity and history which makes the modern nation-state inevitable.

The trauma of history is constituting the silent narratives in Greek public historical culture, not even highlighted by the National War Museum which opened in 1975, which is here clearly directed towards citizens rather than tourists. Similarly, the Greek Civil War (1945–1949) – an early expression of Cold War tensions – is absent, not to mention the more recent military Junta (1967–1974). These components of Greek social memory live on mainly in private stories, memories and political conflicts; they have not been incorporated into public discussion or a conscious process of coming to terms with the past, or what in Germany is called 'Vergangenheitsbewältigung'. These difficult histories constitute, by their absence from national museums, the silent narrative of public amnesia. In contrast, the conflict with Turkey over Cyprus is patriotically narrated both here and in the counter-narrative in its sister institution in Istanbul.

Ancient Greece is a peculiarity of national histories; a proud chapter for the nation but one which since the early years of the Enlightenment has been a foundational, defining and unifying element for the whole European continent. It shapes the 'Western' concept and is deployed by the EU in its political narrative of the past. This is achieved through a whole set of musealised landscapes in Greece, and particularly Athens, which communicate Greece as the birthplace of Western democracy, philosophical reflection and artistic splendour. The two main national museums in Greece representing this heritage are the National Archaeological Museum and the Acropolis Museum, but that heritage is also visible in various sites in the capital and throughout Greece and is, of course, supported by a host of national and international organisations and institutes.

In many aspects the Greek institutions have counterparts in Rome. An even later and more troubled state-making project, Italy also has a special place in the Western imagination which shaped the national museum landscape. With an internal struggle complicated by regional strengths and tensions and the presence of the Holy See, it is a nation where public monuments and museums cannot be explained away in a simple narrative of causality. Here, heroes of the Risorgimento are displayed beside the forerunners of the European Enlightenment. The statue of Giordano Bruno, when it was erected in 1889 on the Campo dei Fiori where he was burnt to death in 1600, was a 'monumental' indictment against the Vatican in the contemporary fight over power. State-making nationalism became secular after 1849, and the conflict with popular Catholicism then became institutionalised (Hodne 2001). But how could the Italian state hope to compete against the overwhelming richness of the Vatican Museum – a conglomeration of early modern collections transformed into a national museum communicating several dimensions of the power, grandeur and universal endeavours of the Holy See?

Museums are plentiful in Rome, but lack both a superficial logic and the possibility of representing the Italian state in a long and comprehensive narrative. Like Athens, Rome possesses two distinct stories, one of ancient Roman heritage and the other concerned with modern history, but the idea of cultural homogeneity and continuity is here less forcefully constructed.

Attempts have been made to create a comprehensive national narrative. An articulation of this for an international audience was the establishment of the national gallery, Galleria Nazionala d'arte moderna, in 1883 as part of preparations for the World Fair in Rome in 1887. More than half a century later, Mussolini attempted to show his well-organised fascist state, with its Roman descent, in Esposizione Universale Roma or EUR as it became known. Constructed in the suburbs of Rome for the planned exhibition in 1942, the Museo della Civiltá still displays models of Rome made originally for the grand Augustus display in 1938. On each side, a traditional popular art and culture museum (Museo Nazionale delle Arti e Tradizioni Popolari) talks to a corresponding prehistoric and ethnographic museum (Museo Preistorico ed Etnografico 'Luigi Pigorini') on the other side of the street. Here the sense of national cultural belonging interacts expressively with the production of cultural otherness. The museums in EUR were an attempt to narrate a long history of the nation but in the end produced fragmented highlights of the past and the Empire. EUR is still used to educate the citizens in Italian culture, but the museums seem to play a minor role in catching the tourists' attention.

The Museum of Risorgimento – housed in the fantastic monument to Victor Emanuel, inaugurated at Piazza Venezia in 1911, and recently refurbished and revitalised by presidents Ciampi and Berlusconi – evokes heroes of war, regional diversity, divinities of the Roman Empire and universal allegories in a magnificent and overloaded representation of the nation (Hodne 2001, 2005). It is complemented by a more modest and less political display at the National Gallery representing national aesthetic schools and epochs but in a more didactic framework. And while the Museum of Risorgimento ends with the First World War, the National Gallery reaches further and makes one of few museum attempts to deal explicitly with the darker side of twentieth-century Italian heritage in its display of fascist art.

The city's centre is itself promoted as the world's largest open-air museum by the tourist authorities. Established as the great destination in the age of the Grand Tour from the late eighteenth century onwards, it was later transformed into a major destination of mass tourism by low-price airline tickets. The equivalent of the Acropolis is in Rome, of course, the Forum Romanum or Roman Forum, heavily restored and emphasised by Mussolini who filled in didactic gaps with, for example, new statues of Caesar. Today the Forum is part of the general musealised landscape of tourist Rome, but it is also the place where, for example, neo-fascists celebrate the death of their 'meta-model' on the day of Caesar's death (Figure 2.3). These special interpretations are, however, only minor tunes in the choir of classical antiquity which overlays various historical tropes of golden days and subsequent decline, of bad old days of excess and subsequent civilisation, of the continual city and the existential forces that have operated over millennia. Feeding different audiences and different agendas, these layers resonate within each visitor. Museums provide vital evidence which demonstrates that the narratives formed by school-books, politics and even more by films and novels reflect reality and truth.

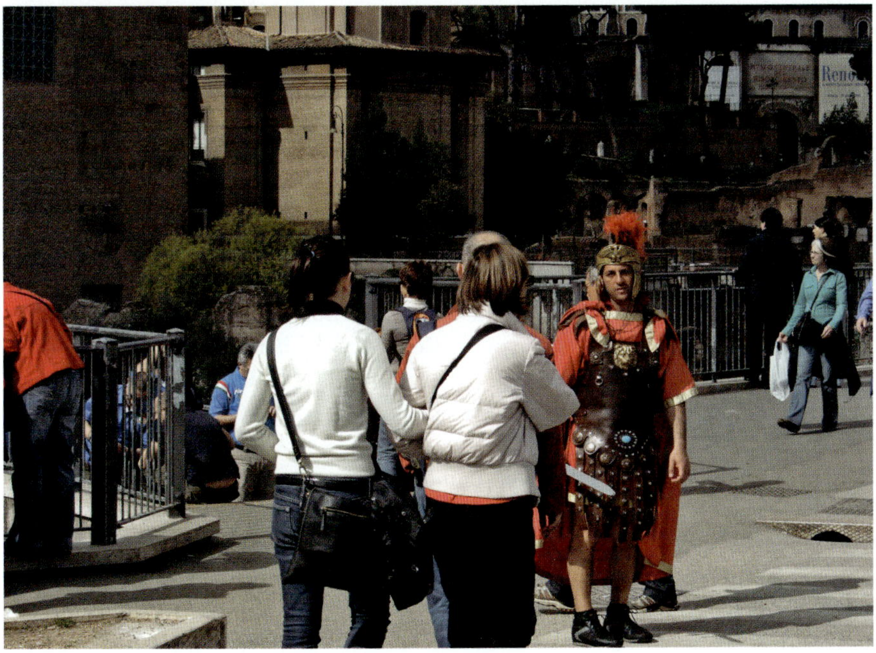

Figure 2.3 Tourism shapes the modern engagement with national museums. At the Roman Forum a multitude of re-enactments and other mediated images of a distant past drown out neo-fascist sentiments. Photo: author

One major point of difference of the museum experience in Rome, as compared to that in Athens, is that there are fewer references to Italy as a territorial whole, and no strong narrative connecting either antiquity to Risorgimento or that era to our own time. The lack of a twentieth-century discourse is in both cases a reflection of similar troubles; memories of fascism and war remain historical difficulties for Italy that live on in contemporary controversy. They have not yet been opened up for cultural negotiation in the museums. The history of the last century only has comprehensive representation in the National Gallery of Modern Art. The creation of a unified Italian language was a very late development and the unification of the strong city-states is still not fulfilled in the sense that strong regional parties, regional identity, a localised nationalism labelled *paese Italia* and a strong informal economy are so alive that any attempt to write a strong national history would contradict more than it would unite (Lappalainen 1999).

The third state, to be discussed in this comparative example, is that of Turkey. It was created in 1923, following the prolonged decline of the Ottoman Empire and in the wake of the First World War. When Turkey was close to being wiped out as a nation, Kemal Atatürk led a successful uprising against the Treaty of Sèvres in 1920 and a strong secular state dominated by military values was installed. The cult of Atatürk as an icon of the state itself was and is the prime locus of nationalism known as Kemalism. The bloody conflicts with Armenians in 1915, the ongoing conflicts with the Kurds as well as the rising forces of Islamism, all challenged and continue to challenge the national consensus.

On the official date of the defeat of the Ottoman Empire, the Topkapi Palace in Istanbul and principal religious centre in Konya, Mevlevihane, were declared museums as acts of victory and statements of a new era. The Museum of Anatolian Civilisations opened in the castle at Ankara in 1921, with the intention of presenting Hittite culture as part of an evolving Kemalist cultural policy. The city also possesses the war memorial to the dramatic turning point of Gallipoli, the War of Independence Museum in the former national assembly building and the mausoleum of Atatürk. Together they signpost the recent and heroic creation of the state, and are complemented by the Museum of the Republic which documents advances since then. These museums provide a native national public with a straightforward narrative of the nation, very similar in form and content to the second level of narrative present in Greece.

In the more cosmopolitan city of Istanbul other layers of national history are represented. The impressive Istanbul Archaeological Museum was built as the Imperial Museum back in the late Ottoman period, 1881–1910. Today, it houses Assyrian, Hittite, Greek, Roman, Byzantine and Ottoman culture; high cultures of earlier inhabitants of the Ottoman past which give splendour to the current rulers in ways similar to those deployed by Ottomans when they first founded the museum. They had 'used symbols not to identify the unique properties of a nation but to produce affiliations with other nations and, through such affiliations, to differentiate Ottoman culture from an existing world order that increasingly relied on European hegemony as much as on overt European colonial power' (Shaw 2003: 218–19). Establishing a museum in a neo-classical building was in itself a cultural encounter between the East and West, negotiated in many details, even if there was at that time no awareness of the Hellenic aspects of the Eastern heritage. The Imperial Museum had been located at the foot of that then emptied symbol of the traditional Ottoman Empire, Topkapi Palace. The royal family had by then moved to the European-style setting of Dolmabahçe. When, later, Atatürk musealised the whole of Topkapi he made a concrete separation with the past of the old regime. The Kemalist historical culture emphasised the short national history of the contemporary regime and marked a break with the weak preceding regime, similar to developments in neighbouring Iran during the Pahlavi regime (Mozaffari 2007).

The Military Museum of Istanbul began with systematic arms collections and the founding event of Istanbul, the fall of the Byzantine or Eastern Roman Empire in 1453. The iron chain that blocked the Bosporus lies majestic on the floor, and an audio programme envisions the tableau vivant of the battle. On the top floor, history is twentieth century and connects to contemporary reality. The War of Independence, Armenian traitors and the battle against terrorism (i.e. Kurds) are made meaningful through aggressive statements. These exhibitions follow the political fluctuations and were closed for entry in the spring of 2009, when negotiations in Eastern Turkey opened up political, rather than military solutions, to Turkey's relationship with the Kurds. Here, Turkish participation in peace operations in Korea, Cyprus and Bosnia showed international responsibility. The atmosphere in each room is radically different as the exhibitions subject the visitor to alternating experiences of conflict or relaxation.

As in Greece there are other displays presenting similar national narratives, but here alternative strategies for modernisation become more visible: for example, Miniatürk for the casual tourist or the cultural exhibitions of ambitious financial institutes like YapiKredi bank for the modern shopper. In the entertainment park, Miniatürk (2003), there are miniature representations of a selection of buildings representing cultures worth mentioning – Greek/Hellenistic, Byzantine, Ottoman and Turkish in three sections: Istanbul, Anatolia and Ottoman abroad – with, for example, the bridge in Mostar (Figure 2.4). The park illustrates the growing importance of tourism, Istanbul confidence, closer relations with the EU and UNESCO's World Heritage designation affecting most of central Istanbul. Miniatürk park is supplemented by a vivid Victory Museum, celebrating the republic, which shows: a wall with images and a quotation from Atatürk reassuring that state sovereignty and economic growth are the main goals of the republic; models of everyday life in the Turkish countryside; and everyone being defended by the heroic battles at Gallipoli where models of victorious Turkish soldiers (one is wounded) are massacring the invaders at the shores (vividly accompanied by audio bombardments).

Similar shifts in acknowledgement of empire, territory and diversity are applicable to other museums (Yilmaz and Uysal 2007). The opening of the cosmopolitan Santralistanbul contemporary art museum in a musealised power plant at the private

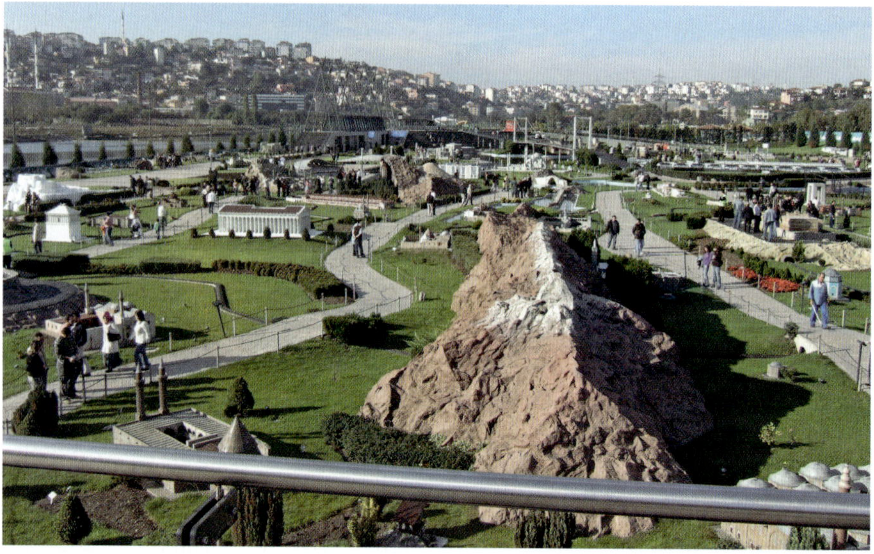

Figure 2.4 A new kind of national museum? At Miniatürk in Istanbul, family audiences negotiate the relationship between city, nation and Europe in a display of built heritage. Photo: author

Bilgi University and introduction of the Istanbul Biennial brought urban post-modern art to the city. At Istiklali, the main shopping street in Istanbul, an ambitious exhibition of the Anatolian history entitled *12000 Years Ago: Civilisation Begins to Spread into Europe from Anatolia* tells a much longer story, competing with the classical narrative of Rome and Greece as the point of origin for civilisation, cultivation and trade in a multicultural Europe (Schmidt 2006). The exhibition is based on quite recent excavations at Göbekli Tepe, in the southeast of present-day Turkey. Here the argument for progress and long-term continuity is reinforced, emphasising nation as the ancient cradle for contemporary neo-liberal values.

In Turkey, there is a heritage of short national narratives in the Atatürk mode, directed mainly at the Turkish citizen. This is updated mainly in the Military museum with the strong argument on the internationally infected issues of Armenians and Kurds. The Cyprus issue overlaps with its parallel in Athens, where the Istanbul version is more low-key than the Greek one (both seem to be tempered by the EU presence). The dramatic birth of the state in part explains the creation of a 'short national story', in spite of the possibilities for enveloping successively Roman, Byzantine and Ottoman imperialism. It also helps to keep a large number of conflicts out of sight. Instead of the presentation of pre-Turk cultures as ancestors, they are introduced as earlier inhabitants of Anatolia, representing not continuity of policy but a successive row of changing civilisations.

In a rigid institutional structure, the message reflects older layers of negotiations while the contemporary ones are more easily articulated through establishment of new arenas for national display where the visitor becomes more central in bringing the narrative together before political conclusions and institutionalisation can take place. Art museums are more flexible than military museums; private actors more open to new interpretations to negotiate new identities and strategies for modernisation than government-run ones. All use museum displays as tools for changing society while explicating more neutral values of knowledge, participation or development.

The national museum scene in Istanbul seems to be the most multi-vocal, actively negotiating contemporary issues. Private initiatives actively challenge state interpretations. Things are more directed in Athens, and hence less dynamic and more defensive. In Rome, paradoxically the longest standing place for religious and secular pilgrimage, the sheer amount of heritage and the weak nation-state makes the museums, as guardians of the past, less in contact with contemporary national issues. Together these three cities show how historical trajectories of nation-building, national museum institutions and contemporary politics blend together to explain the actual structure and function of national museums. Negotiating cultural and political integration is still a benchmark for all national museums.

The Nordic Sagas

Denmark and Sweden, as old conglomerate states, had royal, proto-national collections in existence prior to the Napoleonic wars. The bombardments of Copenhagen in 1801 and 1806 directly triggered their formal establishment. The new Napoleonic dynasty – the Bernadotte – installed in Sweden in 1809, had an urgent need to build a relationship with its subjects (Widén 2009). In Iceland, Norway and Finland, the establishment of national collections created pre-state institutions serving and presenting the developing nationalistic movement.

In Denmark, national museums evolved out of the universalist royal collection: the Museum of Nordic Antiquity opened in 1819; the Museum of Fine Arts in 1825; and the first ethnographic museum in the world in 1841, evidence of enduring imperial ambitions and complemented by the latterly established, openly patriotic, Museum of Danish National History in Frederiksborg (1878). All were clearly affected by the reaction to traumatic losses of territory to Germany in the nineteenth century, reinforced by subsequent occupation during the Second World War. These museums explicitly narrated, and continue to narrate, a much more coherent chronological and canonical story of the Danes in the world, than that represented by their Swedish counterparts.

The loss of Finland in 1809 was equally traumatic to Sweden, even if the country was compensated with a loose union with Norway (1814–1905) at the expense of Denmark. In the first half of the nineteenth century, the process of creating a comprehensive national museum like the Danish one was pursued in Sweden. The opening however was postponed until 1866, due to the reluctance of the peasant estate in parliament. Later, the unitary voice was split, by the success of Artur

Hazelius, who had founded both Skansen, the icon of open-air museums and living history, and a Nordic Museum of cultural history, which competed with the Department of Archaeology and History at the National Museum (Rentzhog 2007; Widén 2009). Increasingly professionalised, archaeology, as a discipline, directed its efforts towards scientific collection, but lost out to the more popular patriotic approach of Hazelius and has been confined thus, from 1919 until the present day, to dealing with prehistory and medieval ecclesiastical material (Aronsson 2008b; Aronsson *et al.* 2008; Hillström 2006).

Scandinavia's regional/national complications reflect a fairly ordinary European experience of conflict and war, in spite of its recent peaceful Welfare/Nordic Model reputation. Three main types of nations are represented: Sweden and Denmark have continuously existed since medieval times; Norway is an historic nation that has experienced an interrupted state history; and Finland is a new nation with no preceding independent state. Iceland was colonised in AD 874 but developed an independent society before it was incorporated with Norway/Denmark. Its independence evolved during the First and Second World Wars. The Faroes and Greenland, however, remain remnants of the old Danish conglomerate state. These histories are not only intertwined by mutual war but also with an idea of a shared Scandinavian culture, a kind of heathen parallel with the Mediterranean legacy of Hellenism. In both seventeenth-century Gothicism and twentieth-century European welfare models, there were and are national versions of the idea of shared Scandinavian culture and distinctive national histories. The conversion in this narrative, from aggressive revanchism to mutual brotherhood, was one of the decisive contributions to the end of Balkanisation and beginning of peaceful secessions in Scandinavia. Each successive success of this model has relaxed the narrative of Scandinavianism into banal Nordism, and the emphasis of European continental exchange as part of the new integrative project.

The museum structures set up in these various countries evidence a certain stability and inertia, not only in the development of their collections, but also in terms of narratives told. The contemporary Danish exhibitionary narrative, as presented at the National Museum, reproduces a rather unproblematic national discourse of an eternal Danish nation evolving to ever-higher levels of civilisation. The setting of the historical exhibition at the National Museum is holistic in its presentation of the complete timeline from the end of the ice age to the recent demise of Pusher Street, the (in)famous alternative community in Christiania. The prehistory exhibition, which opened in 2009, begins and ends with the iconic tenth-century Jelling rune stone, sacred because it is a statement of Denmark and Norway unity and Christianisation. The rune stone has been on the UNESCO World Heritage list since 1994 and is, of course, part of the official Danish canon, even represented on its passport (Wienberg 2008).

Each gallery is similarly structured: a chronological and epochal entry point, a theme dealt with on a long narrative wall corresponding to the seemingly old-style systematic exhibits of numerous objects opposite, but, in fact, giving plenty of room for aesthetic considerations. In the middle are the highlights you should try to see if you have only an hour to spend (Figure 2.5).

The problematics of identifying 'Danishness', thousands of years before the state or nation actually existed, is not touched upon:

> In a lake at Koelbjerg on Funen a woman drowned 10,500 years ago. She is the oldest known Dane.

The changing territories of the empire and its decline are referenced, and now used as a narrative device by which to redraft the bounds of Danish cultural heritage. This

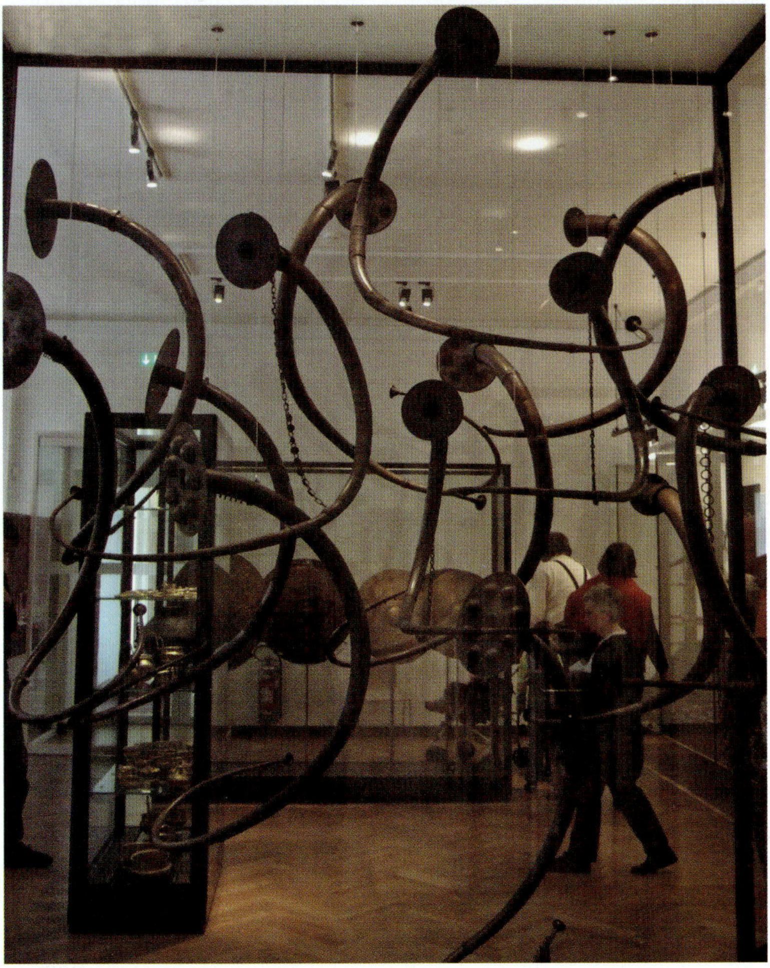

Figure 2.5 Danish history has always been here. The exhibition at the Nationalmuseum in Copenhagen emphasises national icons that conform to the official canon. Photo: author

position is more directly expressed in the Danish-language version of the interpretive material, than in the English translation:

> The history of the Danes also comprises the history of dwellers in the earlier Danish provinces of Skåne, Halland, Blekinge, South Schleswig, Holsten, Norway and Iceland. Also the Faroes, Greenland and the former colony of the Danish West Indies.

Furthermore, via ethnographic collections, the museum places Denmark in the centre of the world, Us and the Others: former colonies or targets for colonisers, adventurers and researchers. These messages are reinforced by the recent instalment of an official Danish cultural canon and the aggressive nationalistic influence of the Danish People's Party (Kikkenberg *et al.* 2008; Danmarks Kulturministerium 2006).

To present a fair comparison with Sweden we would have to encompass several museums, like the Nordic Museum and the National Army Museum in Stockholm. The latter has a strong national narrative securely locked in the distant past (seventeenth century) of Sweden's 'Age of Greatness', which the former lacks with its mostly fragmented old-style ethnological exhibit. In Sweden, the absence of a single, central narrator reinforces the contemporary multicultural discourse of equal individual and historical value, best exemplified in the National Historical Museum, but which also dominates contemporary political rhetoric. At the beginning of the Viking exhibition, it is stated that:

> A thousand years ago neither Swedes nor Sweden existed.

In Swedish, but not in English, prehistory starts with a subtle statement that:

> People have lived on the site we call Sweden for thousands of years. They have grown up, worked, formed families and died. What can we learn from their lives? What are the similarities between these people and you and me?

Here, the ethnic argument is not made openly, but the elements of long existence, shared heritage, working the land and identity are still included, but hidden, in more universal modes of expression (Figure 2.6). Most significantly, all the objects are from the territory of present-day Sweden, placing 'ethnic Swedes' within the limits of contemporary national boundaries, even though the text does not. This message is adjusted to the audience: the territorial 'us' is articulated in Swedish and more universal issues put to the fore in English.

The overarching narrative is a dialogue between the contemporary 'us' and the prehistoric 'them', but it is not framed by a reconstruction of historical development. We are presented with distant fellow human beings. We try to understand their concerns, but we are unclear as to why they have been chosen from the Swedish territory, except for the curiosity of the unknown. The exotic is reinvented and accompanied by a huge number of question marks. Historical change is not one of

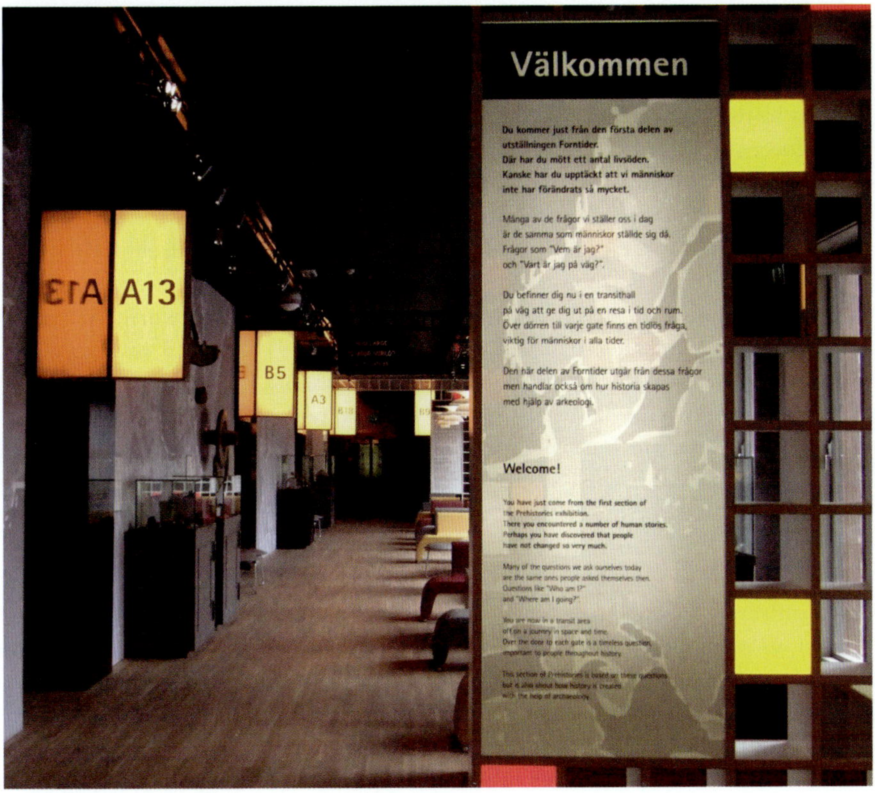

Figure 2.6 Swedish prehistory at the Historiska Museet is deconstructed using the metaphor of airport gates. Here visitors confront a past framed by universal human rights. Photo: author

the leading questions or mysteries to be explored; hence it becomes hidden *and* naturalised (Aronsson 2008b).

The declining Swedish empire transformed rapidly into a small, modernist state in the early twentieth century. Its claim to glory was no longer based on historical re-enactment, but to the working of the land and its long continuity, allowing for a uniform embracement of modern welfare-state rationality combined with radical individualism. History and ethnic community became obsolete standards revived not in national museums but in popular history, creating a tension in historical culture (Aronsson 2004).

Nation-building

Most of the variations in the creation of these national museums are, in one way or another, related to the nation-making process. This is, thus, a contextualising, comparative framework that needs to be analysed. There are two major research traditions of relevance to the creation of the modern state (Gustafsson 2006). The first deals with the establishment of the early modern state, and the second is oriented towards the modernisation period dealing with nation-building. They discern quite different dynamic forces. But, at the most simplistic level, emphasis is given to the influence of war on early state-making. The precise negotiation of social forces to achieve military prowess – and thus state survival – determines the variety of forms for early modern state-making and the creation of a modern bureaucracy (Tilly 2004). For the later phase, the ideological investments are given more space in the well-known approach of nationalism research. Often the modernist approach is connected to the development of a market economy and sometimes more critical credit is given to the evolving global world system of unequal exchange (Wallerstein 1980, 1989).

There were some major comparative projects in the late 1990s that give us reason to believe that power analyses of the early modern period underestimate the level of ideological investment in religious structures, but also in political dialogue. Some have argued that absolutist rhetoric has disclosed a lack of control and that parliamentary rhetoric has disclosed a firm grip on society, already in early modern times (Blickle 1997; Gustafsson 2006). Furthermore, it seems that propaganda for the masses was part and parcel of governmental practice in the seventeenth and eighteenth centuries. Scholars have posed the question of how negotiation of identity on various arenas was essential for understanding the formation of nations, even before the nineteenth century (Kidd 1999).

The transformation of the Louvre into a national institution by Napoleon is put forward as the archetype, but the fact is that Versailles had already been opened up to visitors, with directions given in guidebooks. In the early modern period, artefacts were gradually being perceived as pieces of art in their own right, in addition to their symbolic roles as representations of religion and power. Similar observations can be made in the European periphery, in Scandinavia. Already, in the seventeenth century, contemporaries perceived this transformation, from allegoric classicism to more secular modernism, as a crisis of representation (Burke 1998). The radicalism of the Napoleonic Louvre was its claim to liberate art from history and to free it from its historic boundedness. In this respect, and in its state-centred organisation, it was a completely different project than the British Museum (Déotte 1995; Preziosi and Farago 2004).

This line of research presents a more useful analytical toolkit than the reified positions between primordialists and constructivists that often structure debates on the nature of nationalism. In particular, it gives a background to the importance of reflecting on the negotiation of identities, also when dealing with such seemingly scientific and today, perhaps, obscure matters as museums of geology and fossils (Knell 2007). The sometimes revolutionary but, more often than not, gradual movement of royal collections into the realm of national collections becomes less of a watershed and more of a gradual transformation, lending glamour to the old mode of representation and legitimacy to both modern and late modern modes of community and uses of knowledge. When that happened, beginning around 1800, it opened a new arena where the more general process of negotiation of power and knowledge could take place. This was not new, but presented a general dynamic of political/cultural negotiation creating the context of knowledge, worldview and legitimacy urgent for societal consensus. Countries that passed through the 'Age of Revolutions' had the benefit of old royal collections, added to through the efforts of universities, royal academies and societies. Absolute states like the Vatican, Russia, France and Denmark tended to lead the transformation of representation rather than more parliamentarian countries like Sweden and Britain, where more complex negotiations and recollection of resources were needed.

For the later development, there is an interesting argument by Janet Coleman on the interdependent relationship between the simultaneous development of a strong state and the development of an increasingly viable concept of individualism (Coleman 1996). The early development of Greek individuality did not however have a modern state to support its expressions. Neither did the modern nation-state erase old feudal bonds entirely, but established a new layer of citizenry. Hence the relationship between the modern state and the development of a public sphere is not a necessary one, but the museum is one of the arenas negotiating the creation of both, as well as the nature of their relationship.

Further steps might be taken to connect these collective processes to individual appreciations of trust and well-being and to correlate this to variations in representation and the practice of museum institutions. This perspective opens the possibility that the development of modern individuality and its relationship to community is a much more open-ended process, hence the product of negotiations between formation of corporations, individual desire and contemporary challenges. They could and can be met differently. The national museum is part of the arena where these forces are negotiated, no doubt with a certain tendency to articulate the communality and the virtue of the national community, but at the same time hinting at what the desired virtues of individual citizens ought to be, giving a broad audience an ambitious opportunity to participate in the making of individual identity and community by the practice of making the museum: discussing, projecting, building, visiting, used as narrative and metaphor for self-understanding (Inglehart and Welzel 2005).

It is necessary to clarify the possible relationship between museum-making and state-making. A general line of thinking in political studies (Albert Hirschman) and historical sociology (Manuel Castells) may be utilised here. The classic option of loyalty, voice and exit (as defined by Hirschman) as strategies towards change can also be interpreted as collective efforts, or even structural features of museum-making. The most prominent option is of course 'loyalty', museums acting to promote the contemporary political context as a natural, functional response to challenges. But there are always options for various futures, hopes for restitution of grandeur that might prompt 'voice'; the proactive remapping of the world, if not outright utopianism. The third strategy of 'exit' may be demonstrated by extreme silence, or a nostalgic response that creates an alternative world of comfort and beauty without demand on realisation in the outside world.

Is the net effect of national museum representation the expression and viability of developing existing community trust? Or is it a persuasive move to create something that does not exist in reality, a futile construction, working contrary to other and stronger structures? The examples above show how these strategies are all deployed to address different aspects of the past and different audiences. In either case, they do matter and play a role, but have to be interpreted and read differently in relation to the context for a future to appear, be it a reactionary force trying to halt the dynamics of change, the footprints of an existing nation, or the state of knowledge or a project, more or less viable. A robust framework for explaining the role of national museums is to view them as arenas for complex negotiations and representations that follow changing scenarios. The creative futures role is mostly conservative or functionalist, but at certain moments progressive, anticipating new relations of community and knowledge in the establishment of new nations.

This suggestion of an interpretative framework does provide us with a complex starting point for a theoretical analysis of the external dynamic of museum foundation, but does not, however, answer more detailed questions regarding major differences in the place and structure of established national museums in specific countries. Exactly how this negotiation of knowledge, community, individualisation and globalisation is handled is not only a matter of different resources or local adjustments, but it feeds political strife with a variety of strategies chosen by the compound of elites making national museums a reality. Today, the old political strategy of highlighting the grandeur of a country has turned into an economic resource for attracting a good proportion of the rapidly growing tourism industry (Dicks 2003; Kirshenblatt-Gimblett 1998; Löfgren 1999). Using the comparative approach might also help us to refine the explanation of variation beyond the model's expectations.

The observation of simultaneous ideals of universal objectivity and claims on absolute uniqueness also shows the double need for a comparative perspective as an analytical approach, and an open mind to the fact that ideas and ideals do travel: transfer and *histoire croissant* are not opposite poles, but have to be observed as such so that the methodology of comparing national entities does not become a major tool for constructing them, but rather for historicising their existence as part of certain typical vehicles for modernisation through national integration (Berger 2006).

Comparing paths to nationhood and national museums

Structural similarities occur between the cases discussed here. The old empires of France and Britain or new post-colonial states have been the focus of many studies on national museums past and present. In this chapter, two other groups of states, smaller nations with long continuity in Scandinavia and emergent nation-states with relationship to classical continuity have been discussed. What are the possibilities and limitations of using these groups as analytical frameworks for comparative reflection?

The growing literature on individual national museums (not produced by museums themselves) provides good material on which to base a comparative investigation

(Raffler 2008). Comparative attempts to reflect on national museums are not manifold, but form an emergent field of inquiry at the pace of the reinvigoration of these institutions in the wake of 1989 and 2001 and the hopes and fears of globalisation (McIntyre and Wehner 2001). Some critical reflection on their dependence on twentieth-century traumas has been established in relation to the massive expansion of Holocaust memorialisation and studies (Rolland and Murauskaya 2009).[2] One might shape these comparisons around the characteristics of nation-states:

(A) Empires and conglomerate states: These are large entities and ambitious enough to view themselves as representing universal *Homo sapiens* and they are multicultural. This circumstance is recognisable in the capitals of Great Britain, France, Habsburg, the Ottoman Empire and Spain, the Netherlands and Belgium. Corresponding forms of national museums would originate in royal collections added to by collections from a thriving civil society, which produces a rich and diverse jungle of museums as illustrated by the museumscape of London. England is now the only nation in Great Britain without an explicit national museum structure. This is not proof of a humble approach but rather demonstrates how a universalist approach is identified with a successful national power and reinforced by the sheer magnitude of its collections. Partners with more marginal positions act otherwise. Both Scotland and Wales have revitalised and created new bodies to display the national ambitions of the 1999 secession (Aronsson and Nyblom 2008).

(B) Smaller states with a long nation-state history: Sweden, Denmark, Portugal and Switzerland. The first three had early imperial ambitions but were steadily in territorial decline during the emergence of early national museums. Less impetus was given to nation-making through national museums. Threats from abroad, societal change and a need for cohesion provided ample opportunity for building institutions gradually and defining the nation as far back in history as possible, displaying their scientific and aesthetic grandeur. This sense of pride served to balance the territorial decline and loss of imperial ambition.

(C) New emerging nation-states: Emerging states need to produce a national narrative in the face of a lack of history. This group includes all former colonies and nations emerging from the devolution of empires: the USA, Finland, Norway, Hungary, Germany, the Baltic States, Poland, Greece, Italy and Turkey. More recent examples are South Africa, the Balkans and the devolution of Czechoslovakia to the Czech Republic and Slovakia. Here we would presumably find the most conscious and explicit national narrative, precisely because of the short history of these nations. Included in this group are the nations without states: the Basques, Catalan, Scotland, the Faroe Islands, etc.

States tend to move across these categories, hence giving fuel to new uses of national museums. They are simultaneously bound to their earlier history of dependency and the necessity to interpret their revolutionary agenda. The investment in institutions such as museums becomes important either as an asset or as a problem. Austria's main assets were built during their Empire. Turkey, on the other hand, rejected the Ottoman Empire, characterised by decline and the Muslim culture. The US was a liberated colony and built the National Mall in Washington as a major and explicit investment in national pride. It could be said that the US is now part of an empire. The Soviet Union has devolved, much like the Habsburg Empire. The UK has less pronounced tendencies but is creating a diversity of national museums in Wales and Scotland, which challenge English control of the Union's heritage.

With this background and the comparative examples presented above, we can model a pattern of institutional action to be grouped roughly according to (1) the chronology and drama of the establishment of an independent state; (2) the relation between state and nation(s) – conglomerate, empire, nation-state; (3) the room for autocratic or democratic decision-making, academic, and professional influence, civil society and the public sphere; and (4) the challenges to be faced from external or internal enemies such as class, gender, region and occupation.

The narrative relates to these realities and informs what is collected and exhibited. It examines the democratisation process and more subtle definitions of citizenship, regional diversity and class divisions. Simply, we might deal with three groups of states, provisionally accepting at this point a communal road to modernisation while expecting different approaches in the construction of national museums.

In relating museums to nation-making, a major research question is delineating the importance of the *historical narrative* and its symbolic representations compared to other sources of legitimacy, such as a democratic system, effective economy and welfare provisions. A sound hypothesis might be that the power of historical repre-entation tends to increase in periods of crisis and is more relaxed in times of peaceful progress. Observers interpret conflict as a product of 'too much history' (Mason 2007: 26). Here efforts have been made to introduce comparative examples from the latter two groups in order to revise the imagery often based on France, England and the USA. How successful is this approach?

Explanatory powers

Why do we have national museums, and what underpins their differences? The common reply is that national museums are institutions where knowledge is transformed, negotiated, materialised, visualised and communicated with national identity politics. This answers what the nation was, is and ought to be. No doubt this mediation process involves a language and grammar specific to museums in general and to national museums in particular. The language is open-ended and can be used in many ways, hence the need for a more subtle analytical approach. This allows variation while acknowledging that the national museum is a cultural form of expression defined by a certain anatomy and stability. One way to achieve this is to adopt the comparative methodological framework.

The brief comparative analysis presented above reveals several methodological possibilities. First, they qualify the *generalisation* of an intuitive concept of the impor-tance of state-making trajectories in the creation of national museums. This puts in perspective the self-understanding of many museums as simply exponents of enlightenment, truth and science.

Second, comparing Denmark and Sweden with larger empires such as Britain and France indicates a need to develop systematic categories to sort larger groups of states. Categories such as 'conglomerate empires in early decline' might be necessary to acknowledge *variations* in development and scope, without specifying national historiographies referring to national uniqueness as explanations of variation.

Old and stable nations as well as ambitious new nations invest in the building of national museums and encouraging institutionalised cultural heritage. These projects require substantial resources not easily accessible through market mechanisms. For example, the weakness of the national elite in the Baltic States has resulted in the late institutionalisation of national museums compared with central Europe. A strong parliamentary representation of the third estate, such as the peasantry in Sweden, might also delay spending on national museums even in the face of military trauma. The establishment of a comprehensive national museum system has been much faster in absolutist Denmark than in parliamentarian Sweden. Supported by the French paradigm, this suggests that national museums are easier to promote at the state level in a centralised state than in a more democratic and pluralistic state. In Greece, the state has to be involved before musealisation takes place. In Rome, the Vatican acts as an insurmountable challenge. In Turkey, the revolutionary changes in the early 1920s created musealisation as a means to create distance from the past as much as homogeneity in the contemporary diversity. The process of museum-making shows patterns of difference relating to variations in these trajectories, but is also affected by the accessibility of tangible heritage – that is, the accessibility or inaccessibility of older royal or private collections.

The possibility for other cultural and religious investments to perform similar functions varies within time and between societies: education, universities, religious institutions and mediated culture are other important fields of representation that complement and compete with museums for funding and attraction. What we do know is that the national museum is a distinct arena for the negotiation and integration of knowledge, politics and pleasure. No modern state denies this; hence, the need and scope for better forms of assessments of the national museum's role.

I note that museums have a major impact on guiding the selection of what becomes heritage. We can detect changing political orientations in new heritage politics, for example, when Turkey embraces the earlier despised Ottoman past as a multicultural and more European dimension or when Denmark's and the Netherlands' strengthening of the national canon can be considered as a response to the intensification of migration. In Italy, the ancient heritage and strong regional forces put limitations on the national use of a classical capital city than is the case in Greece. In Turkey, the choice of a 'short national narrative' leads to similar results, combined with hostility towards alternative narratives with political implications related to Kurdistan and Armenia. In the North on the other hand, the lack of classical heritage and a weaker city structure resulted in an emphasis on peasant culture and the use of open-air museums.

The integrative thrust of the national museum is the common ground but political values and strategies differ on how to create communities. The heritage at hand to display this vision varies, as do the cultural processes instigated by national museums. The narratives presented by national museums individually and as an ensemble are some of the more interesting aspects of the process. They vary according to the master-narratives of each nation. Similarities in institutional roles, the use of different fields of knowledge and narrative structures in the context of multivocality suggest a frame for future observations: the story for the nation's citizens, the one for international scrutiny, with a third striking a balance between silence and explicit references to conflicts.

National museums project the need to produce knowledge and public representation of truth outside ideology and individual taste. Hence outright political national museums are scarce after the fall of the Soviet Empire. Closest to these are the military museums in nations where the military is a vital political force, as in Turkey where enemies of the state are deemed necessary to legitimatise the military presence. Here external enemies and those of the distant past are more palatable than internal and more recent ones.

Remaining research questions need to address the role of different fields of knowledge in the museum, their institutionalisation and the creation of national master-narratives in the shaping of national identity. Investigating differences in the construction of nationalism between Denmark and Sweden or between and Greece, Italy and Turkey will help identify the role of museums in establishing strong trajectories and traditions of national master-narratives through political culture and relations to immigrants and neighbours both historically and today.

Archaeological museums literally ground the narrative in the very distant past and in the soil of the territorial state. Cultural museums transform class and cultural variations into a rich heritage; they normalise the present borders and order of the community. Art museums show the quality of taste and skills, comparing them with the established canon of international museums. At the same time, art museums negotiate the distance between the creativity of the lower classes, confined to arts and crafts or ethnological museums, and the (primitive) others shown in ethnographic exhibitions established by national adventurers or missionaries.

The exact division of labour and strength of various institutions cannot be investigated here. The conclusion is rather that there is a need for further comparative endeavours to develop a definition of national museums that include a range of important museums, since the narrative is often dependent on more than one museum. This is truer in Sweden than in Denmark, in Italy and Turkey than in Greece.

The Mediterranean example demonstrates that the choice of a long or short national narrative has repercussions for the creation of national unity. The difficulty of establishing continuity is balanced against a well-established heritage. Shared heritage presents options of more or less open claims on restitution or the cultural sanction of peaceful cooperation. The centrality of museum strategy, however, seems to be more decisive in Greece than in Turkey. If this holds true in closer examination, the next question is: how can national museums be understood through differences in the scope of narratives, relation to the modern world and European heritage, and roles played by museums in the particular culture? This illustrates one way to use a comparative method, not just to reinforce a general thesis but to complicate that very generalisation. Comparing differences helps to tease out further questions.

Much work is required to refine this analytical framework. This analysis has combined a structural framing of institutions with a reconstruction of the narrative message read from a visitor's perspective rather than the reflection of the museum professional. It is possible to construct intelligible narratives out of museum representations. Questions must be raised about their relation to sound historical research, their political standpoint and their efficiency as creative, integrative or mobilising political tools. The ability to analyse silences as active repression or as a banal presence of the phenomenon requires contextualised knowledge of history and society as a starting point.

The performance of museum stakeholders is determined by the power they hold in the struggles for government support and grants to develop their strategic professional scene: archaeologists, ethnologists, art historians and other actors emerging from civil societies. This essay calls for a contextualised approach; one that analyses the relationships between the sciences, civil society, patriotism and nationalism as well as their enactment in national museums.

Comparing cultural histories

Comparative methodology has a long standing in political science, and in historical and sociological research. Scholars like Marc Bloch, Theda Scocpol, Charles Tilly, Stein Rokkan and Ronald Inglehart have inspired empirical research and methodological reflection. A rigorous comparison would demand at least as many cases as there are variables to consider, and the ability to hold other parameters fixed while they vary. This is not possible in most cases where complex cultural phenomena like national museums are studied. Too many parameters are beyond isolation or simply not sufficiently well known to be given a proper position in the theoretical framework. Methodologies vary from taking a strong deductive theoretical standing when testing models, to more empirical, heuristic and inductive ambitions. If a strong theoretical frame is at hand, the investigation might be able to handle culturally quite diverse phenomena and vice versa.

In my opinion, there is much to gain in the middle ground between the grand theory and heuristic and individualising efforts. Indeed they are more dependent on each other than is often recognised. While starting from a theoretical presumption, comparative examples give new thoughts for generalisation. In this case, for example, one can examine the role of state-making trajectories in relation to the construction and usability of relations to a distant past to frame the national narrative in national museums as parts of public culture. On the other hand, quite straightforward empirical comparative explications highlight the need for explaining similarities and diversions in other manners than when they are framed in an isolated institutional or national setting, like results of ingenious management or unparalleled uniqueness of the national past. The relative contingency of the contemporary nation-states, readily demonstrated by changing borders on a twentieth-century map, contrasts with the lack of perspective on the use of shared pasts, Scandinavian or Hellenic, in the explicit or more often banal nationalisation of culture in national museums.

Most importantly, the comparative task is determined by the questions posed. In this case, opening from a broad enquiry on the invention and function of national museums we move from the more empirical, heuristic end of the continuum and reach out for generalisations more in the form of hypotheses on several dimensions rather than for one general theory. The general quest here is as much about the usability of comparative methods as tools for explaining national museums – hence the choice of a more open mode of comparison.

Comparative lines of inquiry help to uncover both the more structural forces of national museum-building connected to state- and nation-making dynamics and the force of institutional path-dependencies, and to clarify the space for entrepreneurs, strategic mishaps and even more the political space of choice that does exist at various constellations in history. Hence both generalisation and understanding specificity is benefited by using an explicit comparative approach.

By exploring the diversities in a more comprehensive study and in greater detail, we will be better positioned to answer questions on the relative importance of national museums in the making of political culture in the past and in contemporary society.

Notes

1 A more complex and challenging model for defining the national museum is in Aronsson 2008a. Thanks to Simon Knell, Viviane Gosselin, Amy Jane Barnes, Magdalena Hillström, Bodil Axelsson, Peter Stadius and Bjarne Hodne for help and advice on earlier versions of the text.

2 By bringing disciplinary and multi-disciplinary fields together with a steady and comparative focus on national museums the NaMu program (2007–2008) formed a new departure for understanding and analysing European diversity in this central institution by connecting 200 researchers in a series of international workshops. This article builds on the work of that programme (www.namu.se and Flacke 2004).

Bibliography

Aronsson, P. (2004) *Historiebruk – att använda det förflutna*, Lund: Studentlitteratur.

Aronsson, P. (2008a) 'Comparing national museums: methodological reflections', in P. Aronsson and A. Nyblom (eds) *Comparing: National Museums, Territories, Nation-building and Change*, Linköping: Linköping University Electronic Press, pp. 5–20.

Aronsson, P. (2008b) 'Svenska historiebruk – nationellt och lokalt', in S. Beckman and S. Månsson (eds) *KulturSverige 2009: Problemanalys och statistik*, Linköping: Sörlins förlag AB, pp. 37–42.

Aronsson, P. and Nyblom, A. (eds) (2008) *Comparing: National Museums, Territories, Nation-*

building and Change: NaMu IV, Linköping University, Norrköping, Sweden 18–20 February 2008: Conference Proceedings, Linköping: Linköping University Electronic Press.

Aronsson, P., Fulsås, N., Haapala, P. and Jensen, B. E. (2008) 'Nordic National Histories', in S. Berger and C. Lorenz, (eds) *The Contested Nation: Ethnicity, Class, Religion and Gender in National Histories*, Basingstoke: Palgrave Macmillan, pp. 256–82.

Berger, S. (2006) 'National historiographies in transnational perspective: Europe in the nineteenth and twentieth centuries', *Storia della Storiografia* 50: 3–26.

Blickle, P. (ed.) (1997) *Resistance, Representation and Community*, Oxford: Clarendon Press.

Boardman, J. (2002) *The Archaeology of Nostalgia: How the Greeks Re-created their Mythical Past*, London: Thames & Hudson.

Burke, P. (1998) *The Demise of Royal Mythologies*, Oxford: Clarendon Press.

Coleman, J. (ed.) (1996) *The Individual in Political Theory and Practice*, Oxford: Clarendon Press.

Danmarks Kulturministerium (2006) *Kulturkanon*, København.

Déotte, J.-L. (1995) 'Rome, the archetypal museum, and the Louvre, the negation of division', in S. Pearce (ed.) *Art in Museums, New Research in Museum Studies*, London: Athlone Press, pp. 215–32.

Dicks, B. (2003) *Culture on Display: The Production of Contemporary Visitability*, Buckingham: Open University Press.

Flacke, M. (Hsgb.) (2004) *Mythen der Nationen: 1945 – Arena der Erinnerungen: eine Ausstellung des Deutschen Historischen Museums, Bd 2*, Berlin: Deutsches Historisches Museum.

Gustafsson, H. (2006) 'Den gamla staten på nya äventyr: Synpunkter på nyare statsbildningsforskning', *Scandia* 72(1): 93–117.

Hillström, M. (2006) *Ansvaret för kulturarvet. Studier i det kulturhistoriska museiväsendets formering med särskild inriktning på Nordiska museets etablering 1872–1919,* Linköping: Department of Culture Studies, Linköpings University.

Hodne, B. (2001) 'Roma og det offentlige rom: Et eksempel på politisk pedagogikk', *Norveg* 1–2: 73–101.

Hodne, B. (2005) '"Nasjon" – et begrep med skiftende innhold: Eksemplet Italia', *Tidsskrift for Kulturforskning* 4 (1–2): 29–54.

Hodne, B. (in press) 'Museum, gjenstand, ideologi', in S.-A. Naguib and B. Rogan (eds) *Materiell kultur og kulturens materialitet*, Oslo: Novus forlag.

Inglehart, R. and Welzel, C. (2005) *Modernization, Cultural Change, and Democracy: The Human Development Sequence*, New York: Cambridge University Press.

Kidd, C. (1999) *British Identities before Nationalism: Ethnicity and Nationhood in the Atlantic World, 1600–1800*, Cambridge: Cambridge University Press.

Kikkenberg, B., Nielsen, P. O. and Thøgersen, J.-E. (2008) 'Den nye Danmarks Oldtid: Om nyoppstillingen 2005-2008', in *Nationalmuseets Arbejdsmark 2008*, København: Nationalmuseet, pp. 219–35.

Kirshenblatt-Gimblett, B. (1998) *Destination Culture: Tourism, Museums, and Heritage,* Berkeley: University of California Press.

Knell, S. J. (2007) 'Museums, fossils and the cultural revolution of science: mapping change in the politics of knowledge in early nineteenth-century museums', in S. J. Knell, S. Macleod and S. E. R. Watson (eds) *Museum Revolutions: How Museums Change and are Changed*, London: Routledge, pp. 28–47.

Lappalainen, T. (1999) *Italien*, Stockholm: Bonnier.

Lewis, G. (1992) 'Museums and their precursors: a brief world survey', in J. M. A. E.

Thompson, *Manual of Curatorship: a Guide to Museum Practice*, Oxford: Butterworth-Heinemann, pp. 5–21.

Löfgren, O. (1999) *On Holiday: A History of Vacationing*, Berkeley: University of California Press.

MacGregor, A. (2007) *Curiosity and Enlightenment: Collectors and Collections from the Sixteenth to the Nineteenth Century*, New Haven, CT: Yale University Press.

Mason, R. (2007) *Museums, Nations, Identities: Wales and its National Museums*, Cardiff: University of Wales Press.

McIntyre, D. and Wehner, K. (eds) (2001) *Negotiating Histories: National Museums, Conference Proceedings*, Canberra: National Museum of Australia.

Mozaffari, A. (2007) 'Modernity and identity: the national museum of Iran', in S. J. Knell, S. Macleod and S. E. R. Watson (eds) *Museum Revolutions: How Museums Change and are Changed*, London: Routledge, pp. 87–104.

Nair, S. P. (2007) 'Economic logic versus Enlightenment rationality: evolution of the museum-zoo-garden complex and the modern Indian city, 1843–1900', in S. J. Knell, S. Macleod and S. E. R. Watson (eds) *Museum Revolutions: How Museums Change and are Changed*, London: Routledge, pp. 61–71.

National Historical Museum (1994) *National Historical Museum,* Athens: Historical and Ethnological Society of Greece.

Preziosi, D. and Farago, C. J. (2004) 'General introduction: What are museums for?', in D. Preziosi and C. J. Farago (eds) *Grasping the World: The Idea of the Museum*, Aldershot: Ashgate, pp. 1–9.

Prösler, M. (1996) 'Museums and globalization', in S. Macdonald and G. Fyfe (eds) *Theorizing Museums: Representing Identity and Diversity in a Changing World*, Oxford: Blackwell, pp. 21–44.

Raffler, M. (2008) *Museum – Spiegel der Nation*, Wien: Boehelau Verlag.

Rentzhog, S. (2007) *Open Air Museums: The History and Future of a Visionary Idea*, Stockholm: Carlsson.

Rolland, A.-S. and Murauskaya, H. (2009) *Les Musées de la Nation: Créations, Transpositions, Renouveaux Europe, xixe–xxie siècles*, Paris: L'Harmattan.

Schmidt, K. (2006) *Sie bauten die ersten Tempel: Das rätselhafte Heiligtum der Steinzeitjäger: Die archäologische Entdeckung am Göbekli Tepe*, München: Beck.

Shaw, W. M. K. (2003) *Possessors and Possessed: Museums, Archaeology, and the Visualization of History in the Late Ottoman Empire*, Berkeley, CA: University of California Press.

Tilly, C. (2004) *Contention and Democracy in Europe, 1650–2000*, New York: Cambridge University Press.

Wallerstein, I.M. (1980) *The Modern World-System*, vol. II: *Mercantilism and the Consolidation of the European World-Economy, 1600–1750*, New York: Academic Press.

Wallerstein, I.M. (1989) *The Modern World-System*, vol. III: *The Second Great Expansion of the Capitalist World-Economy, 1730–1840s*. San Diego: Academic Press.

Widén, P. (2009) *Från kungligt galleri till nationellt museum: Aktörer, praktik och argument i svensk konstmuseal diskurs ca 1814–1845*, Hedemora: Gidlund.

Wienberg, J. (2008) 'Kanon, mindesmærker og oldtidsfund', in L.-E. Jönsson, A. Wallette and J. Wienberg (eds) *Kanon och kulturarv: Historia och samtid i Danmark och Sverige*, Göteborg: Makadam.

Yilmaz, S. and Uysal, V. S. (2007) 'Miniaturk: culture, history and memory in Turkey in post-1980s', in P. Aronsson and M. Hillström (eds) *NaMu, Making National Museums Program: Setting the Frames, 26–28 February, Norrköping, Sweden*. Linköping: Linköping University Electronic Press, pp. 115–25.

Chapter 3

MYTHS OF NATIONALITY

Donald Preziosi

Sunt Lacrimae Rerum

(Lucretius, *De Rerum Natura*)

THE USE OF material objects to exemplify ideas or lessons about virtually anything and everything is common to most societies: every object is always potentially an object-lesson, and the artefactual environment of every culture is semiotically organized and articulated, however materially minimal or extensive. Museums are uniquely powerful semiotic and epistemological instruments for the creation, maintenance, and dissemination of meanings by fielding together and synthesizing objects, ideas, bodies and beliefs. The significance of any object can be made to appear a uniquely powerful 'witness' to past or present events, and to the character, mentality, or spirit of a person, people, place, or time.

Generally speaking, the stagecraft and dramaturgy of the modern museum rests upon two historical foundations. First, upon a 2500-year-old history in Europe of what had once been called the 'arts of memory'; instrumental technologies addressed to the production, formatting, storage, and retrieval of knowledge using material objects and their assemblages. And, secondly, upon an equally long history of philosophical and religious problems and controversies, often distilled into legally enforced doctrines, regarding the proper meanings and functions of objects in individual and social life. In short, the modern museum is an epistemological technology, producing knowledge by fielding relationships amongst objects that ostensify individual and collective human relationships. But these two foundations, the technological and the philosophical, are only nominally distinct, for they have always functioned together and neither can be adequately understood in isolation.

One of the principal products of museum stagecraft is a belief in the independent existence and agency of what its objects are taken as signifying. Indeed, it might be claimed that museums exist in the first place to *manufacture belief* in what is collected, and that assembled contents are staged as implying or exhibiting. In fact, museums manufacture a *twofold* belief: belief in what its contents or collected objects signify, and belief in the independent existence or agency of what is signified. The latter can be as various as the soul, spirit, character or mentality of a single individual, or that of an entire tribe, gender, class, race, nation, climate or species. If artefacts collected and organized together in a narrative or historical fashion suggest a certain common style or mentality in the products of a person or people, or of a time and place, then it could seem reasonable to believe that the objects are the actual product and effect *of* that (presumably pre-existing) spirit or mentality which seems to suffuse the artefacts so staged.

This is one of the central articles of faith in modern disciplines such as art history, art criticism, archaeology, aesthetic philosophy, anthropology or museology, where it is not uncommonly believed that there is or should be a common style or spirit in the products of a person or people. It is essential again to stress the word 'belief' here, for all of this rests upon foundations that are essentially theological and ethical – a point I'll discuss in more detail below.

But how reasonable are such conclusions or beliefs, and what other conclusions can be made from the ways in which museum stagecraft operates? There are in fact at least two different problems here: that of the potential *arbitrariness* of given readings of a particular exhibitionary staging, and the potential arbitrariness (and the national, cultural and historical specificities) of notions of 'reading' as such. Both problems coexist, and they reflect problems that are both deeply philosophical and religious and also very ancient and intractably controversial in a number of societies. While these controversies are quite ancient, in fact they continue to play a powerfully decisive role in the social and political struggles in the current era of transnational corporate neo-feudalism euphemistically called globalization.

Any productive consideration of the problem before us – such as the critical investigation of the phenomenon of 'national museum narratives' and how versions of them may or may not differ – must engage with its historical, philosophical, and religious foundations if it is to be more than superficial and disingenuously innocent academic taxonomy.

Because of its complex nature as both a kind of thing and a way of using things; as an artefact operating on other artefacts so as to fabricate stories which are then made legible as causal agents of artefacts themselves, the problem before us today is an especially complex and difficult one. How can we engage in truly effective critical investigations of the problems – the mythologies, in fact – of 'national museum narratives' without simultaneously engaging in a critical interrogation of the epistemological and ideological presuppositions of which modern museum and collecting practices are themselves both product and producer? In other words, any critique of what is represented is inseparable from a more fundamental critique of what is understood by *representation* itself. But then what we understand by representation is itself a culturally specific matrix of mutually reinforcing theories and practices with a long and complex and varied history in many societies. Things simply 'mean' differently in different societies, and also in the same society at different times and places and for different meaning-makers. Nonetheless, the problem of representation and representability with respect to objects is the central problem of museology, and has been so before the era of modern museums, national or otherwise.

The object in and of the museum

I'd like to begin this first part of the chapter by returning to the point where my introduction began – the problem of the nature and significance of objects, which underlies everything to do with museums and museology. I'll start by quoting a remark by the architectural critic and historian Robert Harbison, who some three decades ago observed rather laconically in the Foreword of his *Eccentric Spaces* (1977) that

> Art is troublesome not because it is not delightful, but because it is not *more* delightful: we accustom ourselves to the failure of gardens to make our lives as paradisal as their prospects.

It is the very *troublesome* nature of things – as the ancient Roman philosopher Lucretius termed it, the inherent sadness of things (*lacrimae rerum*) – that in one sense can be claimed to be the deeper underlying problem being confronted by the notion of national museum narratives. By this I mean the indeterminacy, fragility, mutability or mortality of the relationships between objects and their meanings and between what are problematically distinguished as form and content. The fact that significance is evasive, that meanings evolve, or slightly or radically drift or shift with different perceptions or viewers and users or with the same individuals over time and circumstance. The fact, in other words, that the relationship between forms and meanings is neither one of autonomy nor determination, but one of semi-autonomy and situational determinacy.

This phenomenon is at the same time at the heart of the socially sanctioned need (or even the legally enforced obsession in many societies) to discipline and 'fix' the meanings of things; to give each entity (or indeed any person) an appropriate historical, political, chronological, geographic and ethical locus or 'home'. And to do so ideally by embedding objects, persons, and peoples in a story or narrative; in an unfolding evolutionary history or teleology. Creating through a linked series of entities an overall 'shape' to the time of a time and place. This has always been a facet of a social and political *decorum* with respect to the appropriateness of claimed links or connections between what are epistemologically and ideologically distinguished as form and content or meaning. And of the racial essentialisms and social Darwinisms structurally built into (and perpetually difficult to remove from) disciplines such as art history and museology, no less powerful today in an age of globalization.

The subject of museum narrativity is of very great importance, touching as it does upon some of our most fundamental beliefs, and not only about institutions such as museums. National social and political narratives, as spectacled in and by both modern museums and in their co-dependent professions such as art history, anthropology, and archaeology, or the fashion, tourist, and heritage industries, have long been linked to ideological beliefs which amplify and perpetuate earlier religious narratologies and essentialisms, to be considered in a moment.

In the canonical dramaturgy of the modern historical museum, a rather extraordinary fusion takes place in institutional practice between historical time rendered in space as *sequence and succession* and a certain sense of time as *aspect*: time as a syntactical relation between events connected in a causal relation of *incompletion* and *fulfilment*. In this regard, the 'past' of the artefact/relic is not uncommonly staged as an incomplete manifestation or a prologue to what has now come to pass in the present place of observation. Every artefact is thus the relic of an absence: of an absent past which at the same time *pre-figures* our present, which in turn fulfils, completes, or 'proves' what we imagine the past imagines to be its future. That is: us, here and now, in and as our current social, political, economic, and religious circumstances. The present as artefact and effect – as witness – of absent causes that remain palpable to us *in and by their remaining as relics*. We imagine ourselves to be what our historical relics (both individual and communal) can be read as implying we have long been in the process of becoming. The management and control of how this is to be made legible is centrally and crucially important to any society's sense of its own present and future. This social management of memory and desire is the central business of the modern museum.[1]

The production of that from which a society wishes to be descended (say, an essential Englishness or Norwegianness or Chineseness) is in fact precisely what museum narrative stagings seek to demonstrate and naturalize as if it were truth: the facticity of its fictions. This is precisely the mythological core of all modern museum narratives, and of the uncanny space–time of museology. It is very deeply grounded in the structural logic and rhetoric of monotheistic religiosities, with their particular map of time, history and causality as anticipation, fulfilment, and the vanquishing-whilst-retaining of what has been superseded. Where, in other words, the past (of a people or nation) is articulated, framed, or co-opted as a precursor of what it will have become in the religious, national, social, or political order of things which constitutes our present time.

There is a unique facet of this which is directly related to the birth and development of the museum as an epistemological technology and instrument designed to produce the modern nation-state, and which came to specify a particular attitude towards the past and its relics. It duplicates and uncannily repeats the founding gestures of the new post-tribal monotheism that came to be called Christianity, a gesture also echoed in the self-designated successor to both Judaism and Christianity, namely what called itself 'Submission' (to the will of an immaterial being), or Islam.

Every political regime, both ancient and modern, has always been fundamentally devoted to managing and controlling collective memory. This normally entails a retrograde fabrication and projection of 'cultural memories' focused upon lost or obscured entitlements claimed as having been granted by eponymous ancestors or even imaginary beings such as one's tribal gods. Such entitlements are still commonly used to justify the 'ethnic cleansing' of an unwanted indigenous population from land wanted by present colonists, countless examples of which abound across the contemporary world on every continent.

But historically there is a specifically theological dimension to this which has had a decisive impact on the foundations, motivations, and functioning of museums as peculiarly European inventions. And it relates to the subject I began with, the problem of the nature and significance of objects, and it also concerns the phenomenon of *collecting* in both ancient and modern times. And of recollecting, and also of what might be called *de-collecting* – by which I mean the deliberate erasure of past meanings and functions through the liquidation or sequestering of what are taken as their relics or monuments. An obvious recent example is the deliberate vandalism in parts of South and Central Asia of Buddhist, Christian, Hindu, and other non-Islamic imagery or artistry, directed against what are defined by those wielding or wishing to claim power as evil fetishes or 'idols'. Which is of course an obvious yet ironic instance precisely of inverse idolatry; of investing power – taken to be somehow 'evil' – as magically *actually existing in* images or monuments, requiring their destruction, and commonly also the murder of individuals or even whole populations who 'believe in' or 'worship' such evil idols (and thus who are imagined to 'worship (the) evil' imagined as inhabiting those idols which the iconoclasts paradoxically both do and do not themselves 'believe' in.

My point is that collection or collecting always co-exists, and is co-defined and co-determined by, recollection and de-collection, and that it is with a clear appreciation of this matrix of attitudes toward the representational nature and significance of objects that may make it possible to critically investigate the phenomenon of the 'national museum narrative'.

Our task here is basically twofold: *First*, to critically investigate the inherently mythological structure of national museum narratives and, *secondly,* to articulate possible ways to cure such pathologies by articulating potentially effective alternatives. My use of the term pathology is neither accidental nor incidental, but is intended to signal that national narratives are always, everywhere, simultaneously positive and negative in their effects and implications – as with the ancient Greek term *pharmakon*, which meant both poison and cure. The fabrication of any identity or social reality is a function of its imagined relationships to alternative identities, and so may rightly be understood as a function or artefact of its imagined othernesses. Creating an identity simultaneously masks, overrides, or erases others, and consequently each co-exists as a kind of artefact or effect of what is 'othered'.

Yet this is not exactly what museums do and this is what connects this set of issues to the question of the theological substrate of modern museology. This is the *retention*, as part of the larger story of what is being historically superseded, of its antecedents or 'past'. In a crucially important sense, the *non-erasure* and the *re-framing* of the past by the *retention of what is rejected* or historically or theologically superseded as its memorial or monument is essential to making legible the legitimacy of what one wants the present to be understood as being. The museum's contents thus also serve to legitimize what is outside the museum: its contemporary social contexts.

I mentioned that in the modern museum different senses of the meaning of time and chronology are held in balance: a dominant, temporal chronology is superimposed upon an aspectual meaning of past and present, wherein the present corresponds to the fulfilment of an (incomplete) past. This tense balance is at the core of monotheistic religiosities, and most especially of Christianity, whose claims to historical sovereignty centrally entailed the retention of the continued existence of the Jews as abject biopolitical figures (whether in a national geographic homeland 'promised' by a tribal god, or as a dispersed and fragmented population) in order to demonstrate the historical truth of Christianity as not simply the successor to or replacement of Judaism but as its messianic *fulfilment*. Christianity's mission – especially so after it had been transformed by the Roman emperor Constantine into the official imperial state religion – was to make legible history and the relationships between earthly politics and an immaterial (divine) universal order. In co-opting and transforming Judaism and its population into what in effect was a *monument* of its own state of unfulfilment – its own messianic longing – that monument was to be framed as a permanent reminder and warning – a complex confluence of ideas in fact embedded in the word monument itself, with its double etymology as that which reminds and that which warns or cautions.

In a sense, this retention and continuing visibility of what is superseded as a reminder not only of one's past but of one's own past as an anticipation or antecedent of one's present is in no small measure a powerfully enabling metaphor for the very idea of the museum as a modern institution. The nation-state or republic as messianic fulfilment of something longed for evokes the relicts (left-overs) of its past (for example, the rich possessions of a dethroned aristocracy or a defrocked clergy) as relics which encapsulate signs and portents of the current glories of the new nation, seen as its fulfilment and continuation. Retaining artefacts of a monarchical past for use in legitimizing and justifying the new political order of the nation infuses those objects with a semiotic 'aura' (or 'style') *anticipating* that of the present: the watch on my wrist as a permanent reminder of the timelessness of ancient originary Swiss ingenuity. Objects, in short, are invested with a 'date-and-time-stamp', pointing towards the future; every artefact referencing its own past and potential future as a comment on its particular place and time. The fashion and heritage industries are manifestations of the same process, fashion being perhaps the most ludicrous caricature of this historical phenomenon; this phenomenon which is a philosophical statement (and epistemological technology) about alleged truths of time and history.

Whole institutions and professions also dating from the Enlightenment period in Europe were co-constructed to realize this essential museological mission: art history, anthropology, archaeology, aesthetic philosophy – each of which evolved as disciplinary practices whose aims and methodologies supported this fundamental hypothesis: *that the significance of objects would be historically legible in the style of their material formation*. The belief not only that the form of an object was a figure of its 'truth' or being, but that truth (or essence or spirit or soul) was the antecedent 'cause' of its material form and its style. Chronology as ideology: an essentialist agenda and 'object marketing device'.

History thus becomes more than one-dimensional sequence but instead becomes *teleology*: a story (a work of artifice or artistry in its own right, of course) with a direction and purpose. In the modern museum, time is arrowed. The museum is grounded in a variety of ancient practices and beliefs which in fact bridge several ancient societies and cultures and several often incompatible ideological and theological traditions. The most significant of these differences, philosophically and culturally, was that between the Greco-Roman Indo-European-speaking world and the Near Eastern Semitic-speaking world, with their contrastive world views (philosophies and theologies). It was this gap that was negotiated, and syntheses or resolutions sought, by the earliest Christian thinkers, and most especially Paul, who set himself the mission of resolving the opposition between 'Athens or Jerusalem'; between Greek philosophy and Semitic religion, whose (philosophical-religious-political) synthesis is the new religious system of Christianity, which, beginning as one of many local, ethnic, or provincial cults of Europe and the Middle East, was made into the official state religion of the Roman Empire (banishing all competition, including that of the original official religion of Rome itself) four centuries later. The invention of Islam six centuries later followed a similar trajectory, fabricating itself as the completion of a long line of prophecy spanning and connecting Judaism and Christianity: a distinctly museological narrative mythology of incompletion and fulfilment. Which raises a much larger and deeper question, the pursuit of which may be beyond the scope of this chapter, namely the notion that religion itself is a species of museology; a system of belief with regard to the decorum of relations between form and content, matter and spirit.[2]

It may be that an effective critique of national narrativities (museological or otherwise) would be tantamount to articulating alternatives to the modernist myth of the nation-state itself, or beyond this to the (largely Western) ideology of modernist individuality on which the former is founded. Yet the latter is itself rooted, at least in the West, in pre-modern religious notions of personal responsibility, discipline, and duty, and on the putative life of the individual's soul or spirit imagined and projected as having an (immaterial) existence transcending that of the (material) body it appears to inhabit (the body as the temple of its spirit, etc.).

It is essential to avoid the (structurally prefabricated) trap of 'ghost-catching' – the temptation of presupposing that 'national narratives' have an existence independent of the institutions and occasions in which we imagine them to be 'staged'. That is, as if they somehow 'pre-exist' the institutions and occasions in which they are to be 'found' by our on-site analyses – as if, say, a national British identity-narrative were the ghost (or mythological abstraction or spirit) in the museum's machinery.

The more responsible but more difficult task before us will be to articulate these as *co-constructed and co-evolving*, each facet feeding back on the other. This should be an opportunity to investigate the assumptions made about representation itself (since, after all, everything we're dealing with here fundamentally concerns how we imagine and articulate what we call 'representation' in the first place). Are museum narratives convincing because they are theatrically or dramatically effective? And if they are (or even if they're not) what does this say about their 'truth-value': is it not a product and effect of their theatricality; their artifice?

If there is anything haunting the galleries of our museums, it is a certain subterranean circular reasoning: on the one hand, museums assemble and compose artefacts so as to fabricate and construct what it is as reliquaries they are at pains to prove: the actual existence of the identity, mentality, spirit, soul or style of a nation or people. When you begin to appreciate that the perceived cogency, realism, or even truthfulness of a given historical exhibition is in no small measure an artefact of its theatricality; that its facticity is a function of its ficticity, then it becomes imperative that at the heart of our researches must be an interrogation of precisely this very real paradox.

I will devote the last part of my chapter to this central paradox – which goes beyond the institution of the museum and back to the problem at the heart of all and any museology: the problem of the nature and significance of the object (and of our relationships to it), and also the problem of the promise of art or artifice itself.

Plato's dilemma

All of these issues are related to a very specific ancient discussion and enduring debate in the history of Western philosophy, art, and religion. It may be remembered that in his arguments in the text we call *The Republic* (*Politeia*) for what would constitute an ideal city-state, Plato called for the banishment of the representational or mimetic arts from his ideal community. He was nonetheless ambivalent about this, not least because, to paraphrase his own words, we are all of us subject to the beguiling and magical powers of art. For Plato, these powers were nonetheless seen as deeply dangerous, inducing in the soul of the viewer or listener what he called 'divine terror' (*theios phobos*).

In modern times, Plato's banishment of the arts is commonly misunderstood, since art in modernity is rarely conceived (except by some of the most extreme religious fundamentalists) as being life-threatening. Thanks to modern institutions and professions such as art history, aesthetic philosophy, art criticism, or museology, which all had their origins as service industries in the development and early evolution of the modern nation-state, what is referred to as art is a largely domesticated industry; something connected to – while essentially separate from – daily life. Plato wanted to banish (what we call) the arts for a specific and deeply important reason, having to do with the powers of representation as such, which is at the heart of our concerns here. His reasoning was very straightforward: if one's city, nation, community, identity or ethnicity is understood to be a product of human artifice; a human fabrication, rather than, say, the gift of a god or a product of Nature or Geography or genetics, then this might also lead to the realization that *things could be otherwise* – potentially, a disastrously destabilizing problem from the perspective of those holding power. If, say, the 'sacredness' of a being (Zeus or Apollo or Christ or Buddha or Mohammed) is represented or portrayed in one way, the very fact of its portrayal is precisely what makes it possible to imagine it in other ways as well. Which then opens up the possibility of imagining other ways of being human or divine. Hence the strictures against artistic representation by various religious ultra-orthodoxies. The charges of blasphemy by some fundamentalist Muslims after the publication of the infamous 'Mohammed cartoons' by the Danish newspaper *Jyllands-Posten* in 2005 had less to do with the negativity of the imagery itself – although in fact most of it was quite obviously racist – than with the fact of representation as such – a logic of blasphemy no different in kind than in the prohibition of, for example, the full written representation of the complete Jewish name of its ethnic divinity, or the prohibition of representations of the figure of Christ by Byzantine Christian iconoclasts in ninth-century Constantinople.

The basic point is that representation is fundamentally a double-edged sword. Very much the same goes for any 'national narrative' as well. If that narrative history is represented by what are staged as its material relics (as in the case of a museum), then unless that narrative is convincingly portrayed as being securely fixed to something unquestionably non-artefactual, such as biology, genetic makeup, divinity, or Nature itself, then the mortality and fragility of what is portrayed by that narrative becomes visible and evident – namely its existence precisely *as* mythology and ideology – in other words, as artistry or artifice. Its 'truth-value' in other words is *contingent* – dependent upon complex sets of personal beliefs, biases, inclinations, or tastes, which are always mutable or in flux. The paradox is that in tying artifice to what is not only non-artefactual but even more fundamentally what is non-material, for example, the putative will of an immaterial being or god, what is simultaneously revealed is the artifice or fabricatedness of what is claimed as non-fabricated. In other words, the 'god' itself.

This is the ultimate irony of any text claimed by those promoting a religion as 'sacred', and thus allegedly beyond question or critique – which of course is the point. This was precisely what Plato's philosophical text about the justification for banishing art made stunningly clear 2500 years ago, to the consternation of religious believers, amongst his fellow Athenians, in the 'literal' truth of texts claimed by those in power as 'sacred' and thus putatively beyond question. Plato made absolutely clear the antithesis of democracy and theocracy.

Whatever else they set out to accomplish, great national museums exist precisely in order to foster and perpetuate the belief in the truth of abstractions such as national identity, character, mentality, or ethnicity: after all, right there in front of one's eyes, as the evidence right there in the vitrine you're standing in front of, is the actual sacred relic of some person, people, place, or time. The evidence for the abstraction (say, 'England' or 'Greece' or 'Sweden' or 'The Netherlands') is precisely in its representations, effects, and products.

Consider (as did Plato in his own way 2500 years ago) both the *simultaneous power and fragility of representation* and of what is thereby imagined to be 're-presented': the basic arbitrariness of signs. You can well understand why Plato's ideal city-state was definitely not a parliamentary republic or a democracy, but rather a monarchy ruled over by a philosopher-king who, it was claimed, was invested with the power to interpret the meaning of the universe and the Will of the gods. Plato's ideal city-state was in fact theocratic, ruled over by individuals who claimed to be the true and authentic 'voice' of all-powerful immaterial beings or gods. Ventriloquists of a divine Will.

So the task before us of investigating and articulating the mythology of a national narrative will ultimately be one of interrogating the implied abstraction of an identity as apparently (and successfully or unsuccessfully) re-presented by objects *staged to be read as if* they were relics or effects of that abstract identity. More often than not, that abstraction is claimed as being manifested by a certain consistency of 'style' in the evolving body of objects taken as national relics. But then again, what is commonly referred to as 'style' is itself precisely such an abstraction (central to professions and institutions such as art history or art criticism or museology) imagined as being demonstrated or 'found' in and amongst the massive corpus of relics of any society's past. But all such 'data' are in fact *capta*; stylistic consistency being a selection out of a mass of disparate stuff from a past that is only known by its relics, even continually viable and functioning relics. Such a selection, such a pruning and culling, is what constitutes what came historically to be called 'art history' – based on the idea (the mythological or religious belief, in fact) that artistic time or stylistic chronology itself has a shape, evolution, or trajectory. And belief in an essential bond between an individual and his or her work; where the basic analytic unit has always been the man and/or his (or her) work.

One of the things one comes to appreciate in studying museum history, for example, is how what we imagine to be the characteristic signature or style of the artefacts of a time, place, or people is the product not simply of an excavation of evidence for a consistency but equally the culling or erasure or destruction (the 'de-collection') of objects deemed as confusingly disparate. In many cases, this was specific museum policy in many countries, for example, in late nineteenth-century Egypt, where objects not quite fitting into a curator's determination of what was genuinely Pharaonic were destroyed, sold off to tourists, put in deep storage, or transferred to museums devoted to what were seen at the time as other 'ethnic' groups (distinctions which were as much if not more the product of a European colonial imagination regarding ideals of racial purity as they might have been of indigenous perceptions of sectarian or cultural characteristics). The result is a certain homogeneity or purity of a patrimony or legacy, which can be 'demonstrated' as developing progressively over time along a particular stylistic trajectory.

And if the art of a people has a time-shape to it, even a teleology; if there can be evidenced an arrowed evolution of character, mentality, cognitive sophistication, technological aptitude, etc., in a nation's products, then it can be imagined further that this national or ethnic or racial character somehow 'really exists' independent of its manifestations or 'expressions'. This is of course an essentially theological notion, dependent upon the religious belief that an individual is composed of a 'material' body and an 'immaterial' spirit, intelligence, or 'soul', the last fantasized as capable of 'outliving' what is simultaneously co-constructed as that which it 'outlives': the 'mortal part' of the self. Such a fantasy rests of course upon a semiotic and grammatical illusion – that of the existence of a meaning or a signified magically 'independent' of a signifier or expression. But as the nineteenth-century British philosopher John Stuart Mill (1843) once wryly observed,

> The tendency has always been strong to believe that whatever received a name must be an entity or being, having an independent existence of its own. And if no entity answering to the name could be found, men did not for that reason suppose that none existed, but imagined that it was something particularly abstruse and mysterious.

Any exploration aimed at 'locating a nation in museums' is of course never an innocuous, apolitical, or purely instrumentalist exercise, entailed as is the very notion of the nation with phantasmatic fabrications purportedly represented (that is, *re-presenced*) by collections of materials in fact precisely chosen for their perceived abilities to witness that which they purportedly are intended to prove. As I said at the beginning, everything we understand museums to be or want to be rests on a set of expectations or beliefs about objects and how we want them to mean and what we want them to witness.

The notion that the *character* of individuals or peoples is more than circumstantially entailed with their products and possessions is a reflex of the lingering theological desire that there *should* exist a concordance between them, as between all things in a world imagined as if it were the artefact of a divine Artificer.

What deeply problematizes all this is what brings us back to what I called 'Plato's dilemma', which signals the semi-autonomy and arbitrariness of what are syntactically distinguished as form and meaning; signifier and signified. This concerns the essential duplicity of representation: the fact that once an entity is articulated, its fabricatedness becomes apparent and its fragility must in effect be 'cured' by disciplining or fixing its meanings as if they were tied to some unquestioned or unquestionable foundation; to an imaginary realm of unaccountability beyond the play of signs. The dilemma or paradox is that what art or artifice creates is no 'second world' alongside the world in which we live our daily lives; what art creates *is* the very world in which we really do live our lives. Plato's ambivalence about this is also our own, 2500 years later, as we try to articulate the space between presentation and representation – a cognitive dissonance which renders any attempts to give voice to mythological ghosts (such as national narratives) in the museum's machinery as essentially unsatisfying – in a way similar to that expressed in my quote above from Robert Harbison about the failure of art to fully live up to expectations. Leaving us with the necessary acknowledgement of the *ethical* responsibilities of the aesthetic practices of fabrication and interpretation, so that we may learn to see ourselves as amongst others, as local examples of the richly diverse forms that human life has locally taken, as worlds among worlds.

The narratives crafted by modern museum institutions are universally indebted to a deeply disingenuous belief in the inherent power of objects to convincingly bear witness to the origins and evolution of state polities. All of which is made possible by an epistemological and semiotic hypothesis or belief inherited from pre-modern religious doctrines regarding the *proper* nature of relationships between subjects and objects – in which there was imagined to be a direct or determinate connection between the intentions of a maker and the perceived meanings of a made object or artefact. In the face, then, of the indeterminacy and mortality of objects staged as re-presentations of absent truths, investiture in the truth of a national narrative (or in its multiculturalist repetitions and alternatives) is tantamount to a pathology founded in what can perhaps best be termed a *coy idolatry* – one that lurks beneath the surface of the effect of maintaining such fictions as if they were re-presencing pre-existent (and independently existing) truths.

It is with the acknowledgement of such idolatries and of our civic responsibilities that we may then finally begin to deal cogently, effectively, and ethically with what exactly national and other museums are made of, and with why we should continue to care about them. Rather than perpetuating the management of phantasmatic 'cultural heritages', seriously caring about culture entails above all its de-sanctification.

Acknowledgement

This paper was originally written for, and presented at, the NaMu 'National Museum Narratives' meeting in Leicester on 18 June 2007. It is that version of the paper which is reproduced here with only minor editorial modification. Following the NaMu meeting, the paper was first published, in slightly revised form, as 'Narrativity and the museological: myths of nationality', *Museum History Journal*, 2(1): 37–50 (2009). We are grateful to Left Coast Press for permission to republish this paper here.

Notes

1 This contrast between different senses of time and duration recalls the structural and epistemological distinction once made by Michel de Certeau between history-writing and psychoanalysis; see Certeau (1986); also Nirenberg (2007) for a discussion of syntactic differences between Greek and Semitic notions of past and present.

2 These issues are pursued in a forthcoming volume on the relations between art and religion, tentatively titled *Enchanted Credulities: Art, Religion, and Amnesia*.

Bibliography

Certeau, M. de (1986) 'Psychoanalysis and its history', *Heterologies*, trans. Brian Massumi, Minneapolis: University of Minnesota Press, pp. 3–16.

Harbison, R. (1977) *Eccentric Spaces*, New York: MIT.

Mill, J. S. (1843) *A System of Logic,* London: Parker.

Nirenberg, D. (2007) 'The politics of love and its enemies', *Critical Inquiry*, 33: 573–605.

PART TWO ORIGINS AND IDEOLOGIES

Chapter 4

LOADING GUNS WITH PATRIOTIC LOVE

Artur Hazelius's attempts at Skansen to remake Swedish society

Mattias Bäckström

AT 10 O'CLOCK, on the morning of 1 June 1893, five shots were fired from the old guns of the Cantonment of King Karl XII at Skansen, the open-air museum. Newspapers reported the shots rumbling over Stockholm, drawing the attention of its inhabitants to the fact that the first Spring Festival of Skansen was to commence (Figure 4.1).[1] The museum's founder, Artur Hazelius, envisioned this cannonade as a vivid and lyrical calling to the people to 'wake up'. This was Hazelius's cause par excellence. As he had remarked in a letter, just three years before: 'what our nation *especially* needs is to be *roused* from its indifference to its native country' (Stridsberg 1915: 9).

At the open-air museum, the cantonment did not contain weapons of war and politics, but of love and culture. Hazelius was reactivating, in new ways, the early nineteenth-century Romantic Movement in Scandinavia. What had been formed in the minds of estranged and competitive men was here to be reinvented in the essence of womanhood and the possibilities of creating family-like bonds in society. The cannonade inaugurated a feminine social and emotional space for patriotic love, reverence and sympathy. It was to appeal to people of all classes and find its public home at this open-air museum during the Spring Festivals of the 1890s: 'At this sanctified area for patriotic memories and feelings, high and low may come together in brotherly unity!'[2]

In this chapter, the Nordic Museum, including its open-air department, Skansen, is explored as a socially reforming institution (Figure 4.3). Its conception has much to do with Hazelius's idea that Swedish society should be founded as an organic 'folk community' and that he, as one already awoken, should lead this reforming work. It is argued that Hazelius created legitimacy for his public museum through the

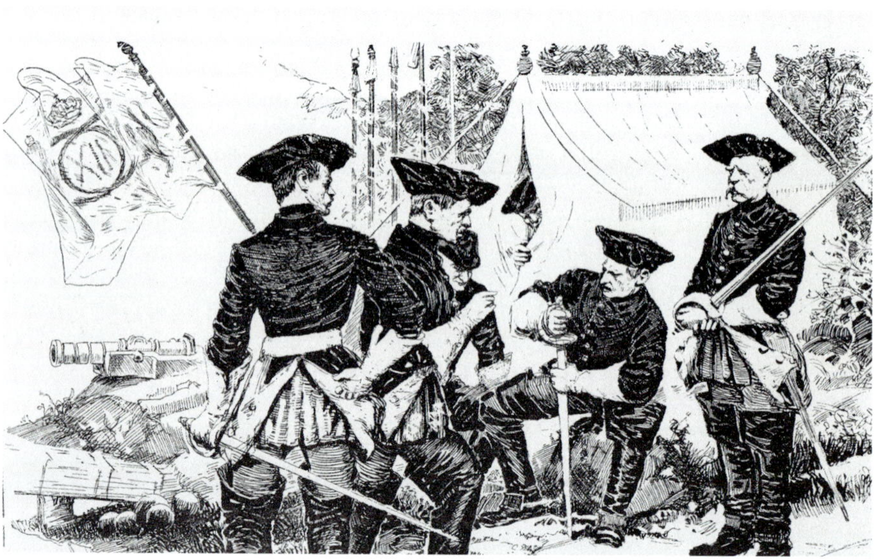

Figure 4.1 A cultural historical gun crew of patriotic love at the Cantonment of King Karl XII during the Spring Festival of 1894. Drawing: Per Hedman, 1894

Romantic idea of an organic relationship between himself, his institution and the Swedish people, and not through the museum's representation of the Swedish state.

Hazelius's museum vision and practice has been examined from a range of perspectives over the years. While earlier studies have researched the ways in which scholarly ideas and the Romantic Movement influenced Hazelius (Böök 1923; Berg 1933; Björck 1946), more recent work has considered identity, gender and ethnic issues and the museum as a specific kind of medium and a certain kind of aesthetic, social and political public space (Broberg 1982; Mårtelius 1987; Bohman 1997; Sandberg 2003; Hillström 2006). This chapter draws attention to concepts central to the idea of a 'folk community' at the intersection of notions of 'folk', 'social', 'patriotic love' and the 'essence of womanhood' (Lindberg 2009: 84–89; Lundgren 2003: 25–33). It focuses on border disturbances between the low and high in society, women and men, the intimate and the public, the social and the political, as well as on ethics, science and cultural history. It reveals a female, organic, social and emotional space; the Hazelian vision of a good society.

A loving people, folk memory and folk future

In a folk community, the social fabric had its foundation in 'folk' – a specific people filled with love for its native country. This folk concept, however, could also be described as a nationally awakened gestalt, and a crossroads between nature and cultural history. It contrasts with a late nineteenth-century view of, on the one hand,

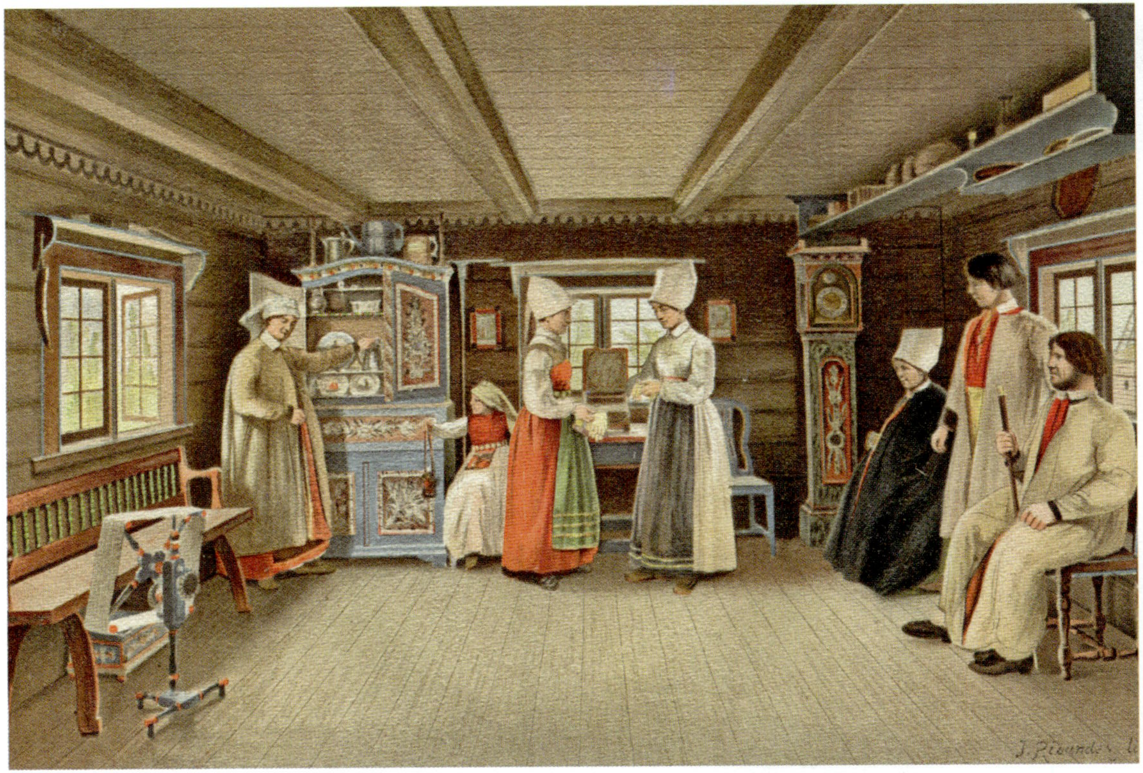

Figure 4.2 Diorama at the Nordic Museum showing the feminine emotional relationship in the family, here shown in a 'nice peasant home', the Vingåker Cottage. Womanhood was to be the foundation for the Hazelian communal society. Diorama: Artur Hazelius, 1873; lithography by J. Risander in *Memories from the Nordic Museum*, 1902.

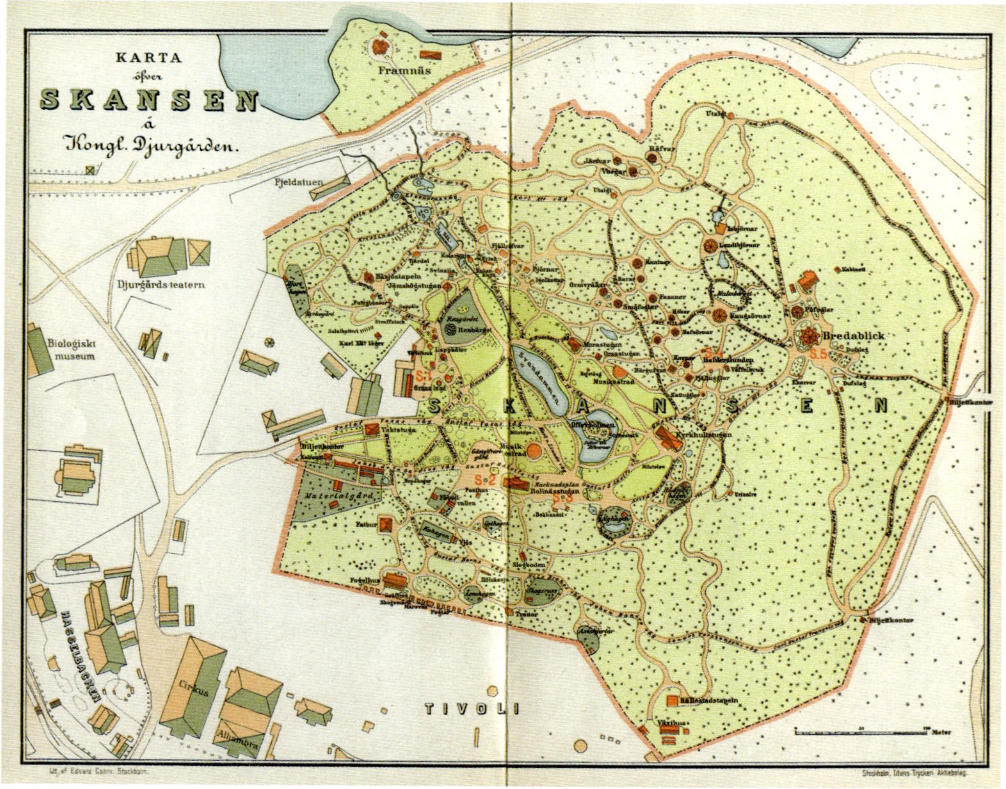

Figure 4.3 The open-air museum, Skansen, was inaugurated on Sunday 11 October 1891. Published two years later, this first public plan of Skansen shows the museum area with the first few historical buildings.

bourgeois society characterised by educated men in public service and private enterprises, and women in their homes, and, on the other, uneducated masses of working men and women, whose unguided actions were considered detrimental to society (Habermas 1962: 83–88).

In Hazelius's vision of a folk community, this division was irrelevant. Within his organic community every man and woman from every class contributed specific functions and different callings *in public*. Compressed into the old cottages and farmhouses at Skansen this community resembled the patriarchal order of peasant society, but Hazelius's vision was for a noble and reformed communal society (Figure 4.3). A new respect for the interpersonal relationships and everyday work of ancestral communities was fundamental at Skansen. It sought to locate commonality in history and womanhood and in so doing defuse the dangers of the present day, of rapacious men and fiery masses, and convert this threat into a positive force centred on a folk future. Here, the cultured woman was seen as a societal atavism in which ancestral virtue and supreme social and emotional experiences could be found and performed.

Hazelius worked in the medium of the museum and created an organic social sphere at a time when a different modern institutional vision was being pursued in the political sphere. As a token of this difference, Hazelius completed the Folk Hall of the Swedish people (*folkhallen* or *allmogehallen*), considered by him as the most important part of the Nordic Museum, at that very moment when the House of Parliament was built in Stockholm. During the first Spring Festival at Skansen, Hazelius also established 6 June as the 'Swedish National Day'.[3] The National Day, the dramatic climax of the six-day festival, commemorated the election of King Gustav Vasa – the great redeemer of the nation – in 1523. It also inaugurated a vernal and symbolic rebirth of the nation as a folk community. This double purpose of the open-air museum, of historical reflection and societal reformation, was clarified by Hazelius in a letter to his colleague Bernhard Olsen, the founder of the Danish Folk Museum in Copenhagen: 'Maybe I should direct your attention to the fact that Skansen does not only have a major cultural-historical significance, but also an important *national* and *social* one.'[4]

With rumbling cannons and ringing church bells, national speeches and national songs, those attending the festival were left in no doubt that this was rather more than a museum visit. Indeed, these events were all performances of the folk concept, and the cannonade of the old guns of Skansen was the noisiest of them all. It spoke of a patriotic love bound to cultural history and not national chauvinism (Trägårdh 1990: 31–32). The relationship between patriotic love and cultural history was also part of the object of knowledge of the museum: 'folk memories' (*folkliga minnen*). This object had one of its roots in the Romantic Movement. In 1835, for example, the Swedish Romantic historian and poet-philosopher, Erik Gustaf Geijer, had made clear that love was hidden in every profound memory, but also that it had a dark side just as love had its opposite in hate. This conflict was however raised in the essence of memory – the conscience as core in all human beings (Geijer 1835: 178).

To Hazelius, 'folk memory' was always profound, always about life and love, and, as such, constitutive of the cultural history of Sweden, which could be saved and explicated by the museum curator. But it was also a reflection of the great historical losses associated with the social fragmentation resulting from contemporary society. In the conscientious work of the curator this conflict of life and loss was raised; the 'folk memory' became a memorialised promise for a future built by a firmly rooted and loving people, a 'folk'. This double feature of 'folk memory' centred on the curatorial role: the collecting and ordering of the memories of the good old days, through which the 'living pictures' of Skansen could be created, there to awaken the minds of the sleeping. The key was the Romantic ideas of the genius and of animated nature. Hazelius, who was considered a genius, could come into contact with the life-force, the essence of 'folk', which was expressed in many historical things. He could make his exhibitions come alive. His friend, the writer Zacharias Topelius, described the museum in 1885: 'the spectator leaves the collections with a total impression of the people, which now breathes and acts in the district, in the native country, in the Scandinavian North' (Topelius 1885: 47).

Here Hazelius dealt with both idealistic and naturalistic ways of explaining reality. In one sense he evoked Friedrich Schelling's notion of art as a way to conjure up the inner life of nature; to locate the spiritual content of truth in the physical form of reality (Liedman 2006: 258–60). In another sense, he deployed a Darwinian approach to connect these objects and people to a scientific reality. Thus, the museum cottages at Skansen were ideal and real at the same time, a crossroads between art and science; between Romantic idealism and Darwinian naturalism. In an article published by Hazelius in 1897, Nils Edward Hammarstedt, curatorial assistant at the Nordic Museum, elucidated the purpose of Skansen as a quest for truth and essence: 'In our opinion, the main task of the open-air museum is more to assert the right of the *essential* to exist than the *accidental*, more to strive for *truth* than for *reality*, the latter concept to which Dr Müller, in his thesis, has made himself too much of a slave' (Hammarstedt 1897: 179; Müller 1897: 683–700). Archaeologists, like Sophus Müller, Head of the National Museum in Copenhagen, sheltered typological exhibitions inspired by a positivist and evolutionistic naturalism. In contrast, cultural historians, like Hazelius and Hammarstedt, defended a holistic idealism in which naturalistic research was only one aspect of truth as the loving and living whole. Earlier, in 1862, Hazelius had remarked: 'It is with our spiritual development, as it is with everything within the whole area of nature, it is a product of the forces that work together' (Böök 1923: 121). In 1880, he wrote to Olsen placing patriotic love and its practical applications in the foreground: 'What is needed is to a lesser degree learning and thorough scientific education, but, above all, patriotic love, a warm interest in the cause, a force which knows how to defy difficulties, and a clear and practical gaze.'[5]

The Romantic ideal of the loving and living whole was essential to every aspect of Hazelius's work. In 1880, when Hazelius established the Nordic Museum, he gave his private collections as a gift to the Swedish people and wrote that he would do everything in his power to establish a secure position for the museum. He felt a sense of urgency as history was fast disappearing, yet he also worried that the museum might become mired in the politics of state, in 'proposals submitted in one session of the Parliament after another, and then, if one of them is approved and the museum becomes the property of the state, to be subjected to endless discord during the adjustment of the claims of different state institutions on its belongings' (Hazelius 1880: 1).

Here, Hazelius established civic ownership for his museum (Hillström 2006: 244–45). It represented, in his mind, a different relationship between a national museum and a people, in which ownership was legitimised by the museum's ability to represent the Swedish people organically through patriotic love and cultural history, not through the modern institutions of the Swedish state. Political and economic relations could not constitute the foundations of the Nordic Museum as they did not foster a harmonious community but a society of discord and selfishness (Hazelius 1896: 13c). The relationship between the emotional and historical was quite fundamental. In 1888, for example, Hazelius asked one of his collectors to make such an emotional appeal: 'See to it that you get old articles of clothing for free. The dwellers of Orsa should be a little bit more willing to sacrifice. Ask them to think of their native country!' (Hammarlund-Larsson 1998: 187).

It was, in this Romantic setting, through love, and its organic relation to eternal truth, on the one hand, and to practical life, on the other, that the museum curator could distinguish between rubbish and 'folk memories' worthy of preservation. This rift in existence between memory and rubbish reflected science's separation of spirit and reality and the commercial separation of spirit and goods. A museum curator, like Hazelius, who accepted the Schellingian worldview of nature as 'slumbering Spirit', thus saturated with life, could locate the spiritual life in folk artefacts and folk stories in a way that he could not in dead mass-produced things. This locating of the valuable was made possible by the fact that 'folk memories' contained the same truth as within himself; they were soul mates.

At the turn of the century, Scandinavian museums found themselves battling over their intellectual and ethical purposes. Truth or reality? Love or typology? Idealistic holism or naturalistic realism? It was during this struggle that Hazelius gave the order to load the old guns of Skansen with 'patriotic love'.

A Romantic story: love, good and womanhood

In 1835, the Swedish Schellingian and Romantic poet-philosopher Per Daniel Amadeus Atterbom discussed the relationship between love and the will to do good (Atterbom 1835: 4, 9). To Atterbom, it was only those people who have been 'awakened', defined by him as 'folk', who had the ability to love and therefore to act socially. Indeed, love was the emotional force which made people aware of external reality. Love was the beginning and end of all human experience, 'an inclination and ability to leave, to elevate oneself, to liberate oneself from one's Self, to give oneself to something Else' (Atterbom 1835: 6).

The good deed was the worldly and practical extension of the idea of eternal Love. It could only be carried out after man's transformation, when human life had been uplifted from its presence as raw state, as 'mass', and had become the organic communal gestalt, 'folk', that is the worldly expression of the primordial spirit with its eternal Love and Truth. This platonic love triangle, *love–will–good deeds*, was the idealistic foundation of a communal folk society for the future. And as a pure emotion directed from a people to its native country – as a natural location but also the country being a historical accumulation of good deeds – patriotic love was a rousing expansion of eternal Love.

Others, like Danish poet-philosopher Nikolai Frederik Severin Grundtvig, possessed similar conceptualisations of 'folk' and 'folk life' (Grundtvig 1832: 395–96; 1848: 345). He made womanliness an ideal for a folk community, and characterised the Danish people as feminine. Spirituality, he wrote, passed here through the heart and not through the head, and, furthermore, here happiness had its home. This feminine people did not pursue worldly recognition, but an insightful understanding of the multiplicity of human progress (Grundtvig 1836: 47, 52). These publications of Grundtvig and Atterbom, together with those of Geijer, Bremer and Key discussed below, all appeared within Hazelius's lifetime. They all presented the view that, in a fragmented society of men and masses, a primary and uncorrupted communal potentiality could be found. Grundtvig made the cultured woman the bearer of such a potentiality, embodying a folk morality and the ideal for a good community; in 1840, Geijer elevated her to custodian of pure humanity. As educator of a rising generation and holder of an emotional motherliness, she incorporated the primary position of the home. This was, Geijer stated, a specific moral calling more fundamental than politics. The 'true meaning of female *emancipation*', he continued, was to liberate the educated woman from politics so that she could perform her higher motherly duties to society, though only within the private home (Geijer 1840: 5).

In *Lif i Norden* (*Life in the North*) (1849), the writer and feminist, Fredrika Bremer, recognised the public motherly function of woman: 'It was the departing of the motherly from private life into the public, forming a new home' (Bremer 1849: 11). This transition was, however, limited to nursing and care roles. Similarly, the Swedish writer and pedagogue, Ellen Key, in 1896, argued that women should find their work in fields which were in harmony with their female essence and which create higher values than men's work. As 'the social mother', the educated woman should reshape the home, the school and society (Ambjörnsson 1974: 252–53).

Hazelius took these earlier ideas and mixed them with Darwinian ideas to produce a holistic community with scientific legitimacy – a communal society in harmony, between the parts and the whole, between yesterday and today, and across the classes and genders. In his vision and museum practice, Hazelius gave a solution to the strong social and regional conflicts evolving in a rapidly developing industrial society (Aronsson 2011 (in preparation)). It was a vision that saw the public female social role as shaping and consolidating customs, of integrating and bringing up people, and of reshaping home, school and society. This was not a role limited to educated women. Rather, this presumed natural ability to manage and foster personal relations and to combine charity with a moral message formed the ideal for the social fabric of an entire nation (Ambjörnsson 1974: 55; Jordansson 1998: 163).

During the Spring Festivals, the 'Women of Spring Festival (*vårfestfruarna*)', as Hazelius called them, were all of good families, and presented at the open-air museum within a suitable and respectable female space for their public activities. Here the women were given a public role in which political, scientific and scholarly ambitions were abandoned. Instead, they emphasised the moral and emotional aspect of the open-air museum representing patriotic love and public good. In this motherly setting, Hazelius and his assistants combined a number of historically based events, which together had 'social significance' for contemporary society (Hazelius 1899: 312) (Figure 4.5). Hazelius saw this invention of national tradition as the dawn of a new way of living in the capital:

> Thus, there are ennobled public entertainments, which accordingly have been evoked, and which should be beneficial and educative, especially for the youth, and might thereby count as important contributions to the sound development of the capital. It is here a matter of creating traditions, which will never be abandoned.
>
> (Hazelius 1899: 313)

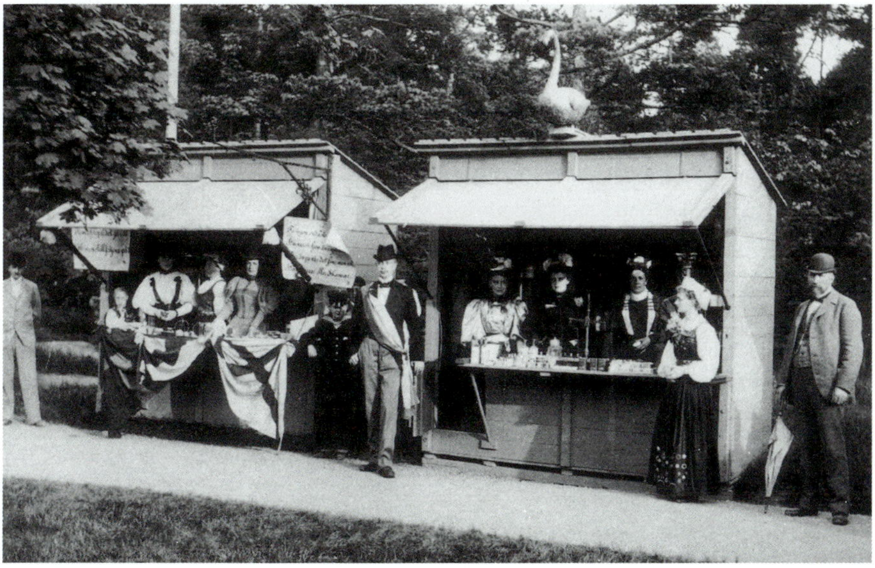

Figure 4.4 The inner, family-like, group of women and men of high society as vendors in national dresses. The young protégés are surrounded by female chaperons and a male marshal with 'blue and yellow' across his chest (1896).

The Spring Festivals, themselves, were to be one of these traditions. Sigrid Millrath, the museum assistant responsible for the festivals, described the outdoor bazaar as 'a new idea never before tested in Stockholm'.[6] She saw them as recreating the spirit of old-time fairs – with the assistance of a hundred high-born women under the leadership of Mrs Anna Wallenberg, née von Sydow. By dressing them up in authentic national dresses and parading them before the inhabitants of Stockholm, these high-born women represented the Swedish people, both peasants and peers.

The notion of the essence of womanhood contained all the ingredients of a good folk community. Through the Women of Spring Festival the idea of femininity as filled with patriotic love and a will to do good deeds for fellow countrymen was put into practice. These women incorporated both the ideal for a communal society and the female moral and caring function of it – they were to be both a social ideal and the social workers maintaining that ideal. Thus, Hazelius's practice was very similar to the public 'motherly movement' of Bremer and Key. Within this feminine society, however, the male's function was to be a scholar or politician, and in his various ways produce the cultural history of the Swedish people as well as articulate visions for the future. He did so, in part, in speeches on the 'Swedish National Day' at Skansen.[7]

Maintaining civility

These morally awakened women and men were ideal museum visitors. At Skansen, Hazelius requested:

> Thou, who enter here, leave all that is ill-bred behind you! Let no raw expressions, no curse, the appalling witness of coarseness, fall from your lips! At this sanctified area for patriotic memories and feelings, high and low may come together in brotherly unity! Here, the nobility of the soul gives the peerage.[8]

This statement was made as staff were instructed to keep the site a place of high repute. The gatekeeper, for example, was ordered to exclude unwanted persons: the brutal and the frayed, the drunken and the uneducable.[9] Individuals who gained access were by these means shaped into museum visitors whose clothing and behaviour were not disharmonious with the intentions of the museum (Bennett 1995: 51–58). The visitors were to be well dressed and well disposed and therefore receptive to the museum's message concerning folk ethics, patterns of behaviour and lines of action in a spiritual community uniting high and low in society.

> The one who has not experienced what patriotic love is will not be able to appreciate Skansen rightly; but if you love home and country, you are capable of feeling reverence and gratitude to the accomplishments of our fathers and their legacy to us, and then Skansen will stand out as among the most beautiful creations of our country.
>
> (Åkermark 1901: 489)

In the day-to-day life of the museum, this spiritual and patriotic atmosphere was sufficient to keep everything in order. The Spring Festivals, however, were a new public arena for the wives and young daughters of high society, considerably more open than earlier charity bazaars. As a new female public space, it opened up new dangers for improper meetings and other nuisances. To mitigate these moral risks, specific regulations were established for the approximately 600 participants and thousands of visitors. Young ladies – whom Hazelius thought inexperienced and vulnerable – were, for example, to be chaperoned by experienced high-born women. Male marshals then acted as outer guardians of these family-like groups (Figure 4.4). Beyond these groupings, a fine-meshed network of museum assistants, museum guards, Dalecarlian women and policemen was deployed to maintain order and civility (Linder 1918: 118).[10]

Despite the idea of public functions of women and men in a folk community, the organisation of the Spring Festivals was an image of late nineteenth-century bourgeois society with its division between the intimate life of the female and the public life of the male (Habermas 1962/1998: 83). Only in an emotional folk future would cultured women no longer have to be protected from the moral evils of public life; then they would be guardians of that homely community. During the festivals, however, the only one who was not subdued by this social division in contemporary society was Hazelius himself. As head of the museum, he examined everyone according to their specific public functions and callings, transgressing intimate and public spheres. He conducted his festival surveillance within an organic folk logic, and wanted to do so in lonely majesty, as is clear from one angry missive he sent in the middle of the festival week to Sigrid Millrath (S.M.):

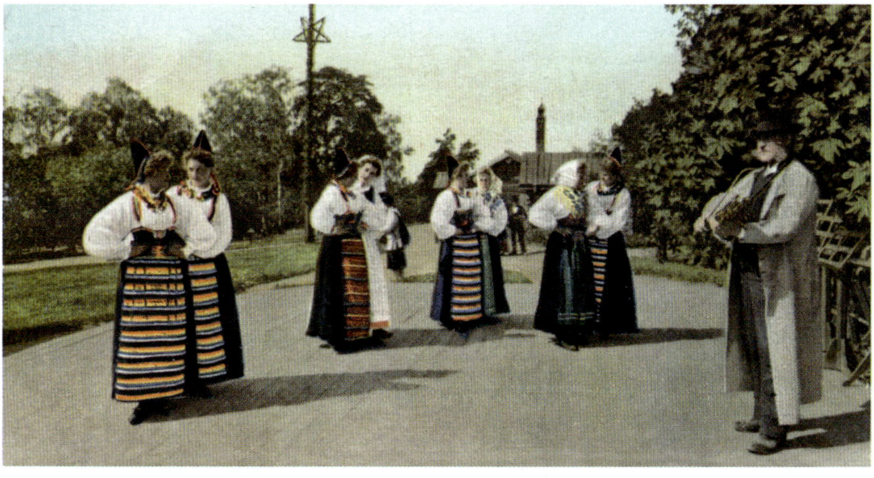

Figure 4.5 Folk-dance at Skansen, an old tradition re-actualised as a morally sound activity ennobling the city's youth. All activities for young people at the open-air museum were monitored by chaperons. Photo: Axel Lindahl, *c.*1900

> I cannot avoid giving utterance to my indignation that my specific orders, that the inner room of the Bollnäs Cottage should be at my disposal *completely,* were not in the slightest degree carried out. It is however too absurd that I during the whole festival should be surrounded by washers-up – to say nothing of other things. *And yet S. M. doubtlessly knows* that I want to do my observations and give my orders *in peace.*[11]

Hazelius had his eyes fixed on immoral elements, which might easily be missed by the surveillance network. He picked up rumours, for example, that 'Mrs Adelsköld supposedly had brought a ballet girl with her, who should not be allowed!'[12] And as all-seeing but invisible, he amazed the female participants in 1895: 'Later we learnt how impossible it was to see him during the festivals, no one knew where he was, no one had seen him, but he was everywhere and saw everything' (Gawell-Blumenthal 1931: 3).

The festival as performance of a good society

During the festivals, the dressed-up high-born ladies mingled with ordinary museum staff who had been chosen ethnographically on account of their ethnic origins. On normal days at the open-air museum, the carefully researched face and body features of the employed 'Skansen-Lapps' and Dalecarlian women were in tune with the ethnographic patterns of their dresses, the cottage buildings and camp (Figure 4.6). At the festivals, however, such scholarly principles were put aside for the sake of the 'good cause'. The legitimacy of the Lapp Camp, for example, changed from the

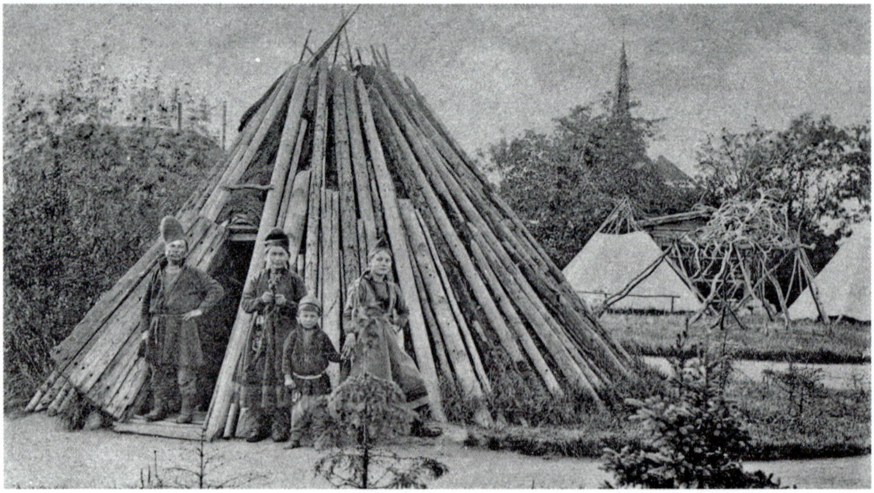

Figure 4.6 The family Macke Årén was employed as Skansen-Lapps (ethnographically representative nomads), 1892–1899. Photo: Anton Blomberg, 1890s

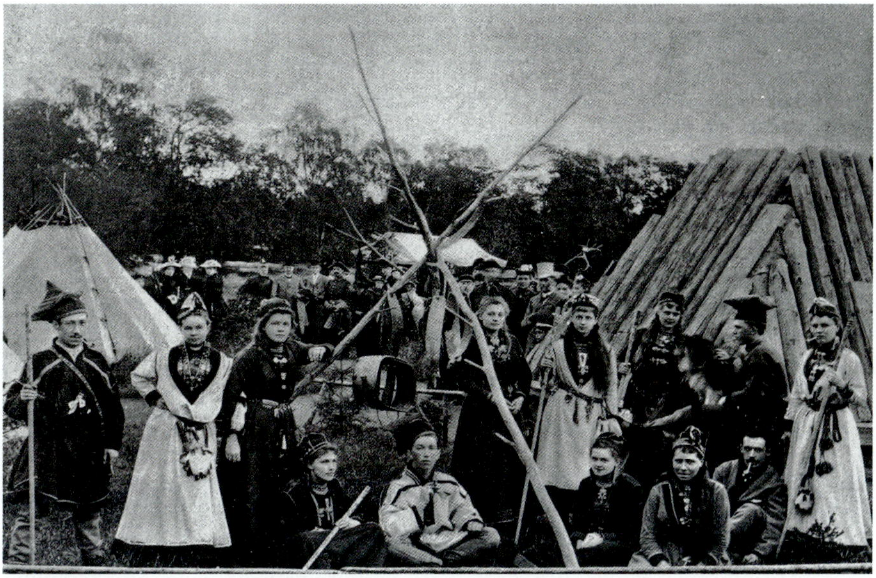

Figure 4.7 Young protégés and their chaperons dressed as Lapps at the Spring Festival of 1894. In this staging of the camp, young ladies with dark hair were picked out to read visitors' fortunes 'exactly as real Lapps'. Photo: Frans Gustaf Klemming, 1894

scientific and scholarly sphere to the ethical sphere as it was populated by Mrs Samzelius, Mrs Jacobsson, Mrs Tigerhielm and others – none of them ethnic Sami, but every one of them filled with patriotic love and good deeds (Figure 4.7).[13] Hazelius told his assistants that the appearance of Skansen and its cottages could be altered for the festivals, but must be restored to their everyday representativeness as soon as possible: 'As it has been found necessary, the cast iron stove may be placed in the brewer house, but *by threat of life penalty* it must be taken away the day after the festival.'[14]

The Spring Festivals were dramas founded in patriotic love. For this reason Countesses were persuaded to dress up in peasant garb. This unscientific state of affairs, however, confused visitors. Ida Gawell, 'Delsbostintan', a folk singer and humorist, reports the reactions of the festival audience of 1895: 'It was good fun to hear how people wondered amongst themselves if we were "dressed up or real", and actually, we had the most fun during the first days, when the audience was convinced that we were "real"' (Gawell-Blumenthal 1931: 3).

These upper-class women acted both as vendors in the old-time fair and as historical characters in the festival parade (Upmark 1916: 67–68). In a letter dated 4 March 1893, Hazelius explained his vision for the carnival:

> I have also almost decided to try to organise a grand *Swedish* carnival on the first of May. About 10 proud chariots with gentlemen and ladies in

> corresponding costumes. A festival procession, where the cultivated participate and not, as *now*, only *common people*. Each and everyone would choose to represent a historical person and acquire the costume and portrait likeness him- or her-self.[15]

Here Hazelius maintained class differences: while the members of high society impersonated significant historical figures including monarchs, geniuses and military leaders, the nameless masses performed as the lower ranks of the military in the parade (Figure 4.8). The social elite already possessed an inner likeness to the characters they portrayed: they were 'awakened' and educated, and just needed costumes, coiffures and gestures to give life to historical figures in the procession.

One of the young protégées of the first Spring Festival was 25-year-old Siri Wallenberg, daughter of Anna Wallenberg: 'What festive and sunny days they were!'[16] At the Wallenbergs, thorough preparations were made for the festival and when the big day arrived, 'we the young people were dressed in national costumes and went by landau' directly to the open-air museum. The young ladies were to be publicly visible during this trip; to be, as she says, 'a living advertisement', yet at the same time, they should also be touched by pure patriotic love. This feminine emotional understanding of Skansen's importance was also the reason why her

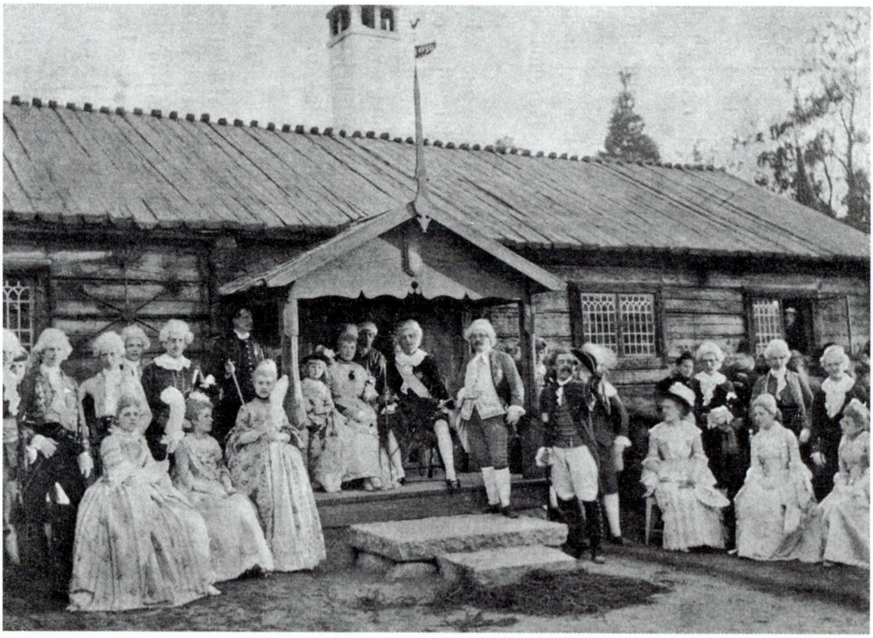

Figure 4.8 The historical procession at the Spring Festival of 1893. High-born ladies and gentlemen dressed up as King Gustaf III, Queen Sofia Magdalena and the royal entourage. Photo: Atelier Birger Jarl, 1893

mother led the Spring Festivals: 'Foreseeing and patriotic, she appreciated fully the genius of Dr Hazelius's idea.'[17]

Through their willingness to act socially for the sake of patriotism, the women of high society had proven that their family names were not empty, but had the loving and caring content of a *good* family. These married women were the mothers of a homely society, and their daughters the promise of a good future to come. For this reason their names were published in the Stockholm newspapers together with their genealogical lineage (Figure 4.9). To Hazelius these women were incorporations of the two fundamental aspects of a folk community: love and heritage. Both were

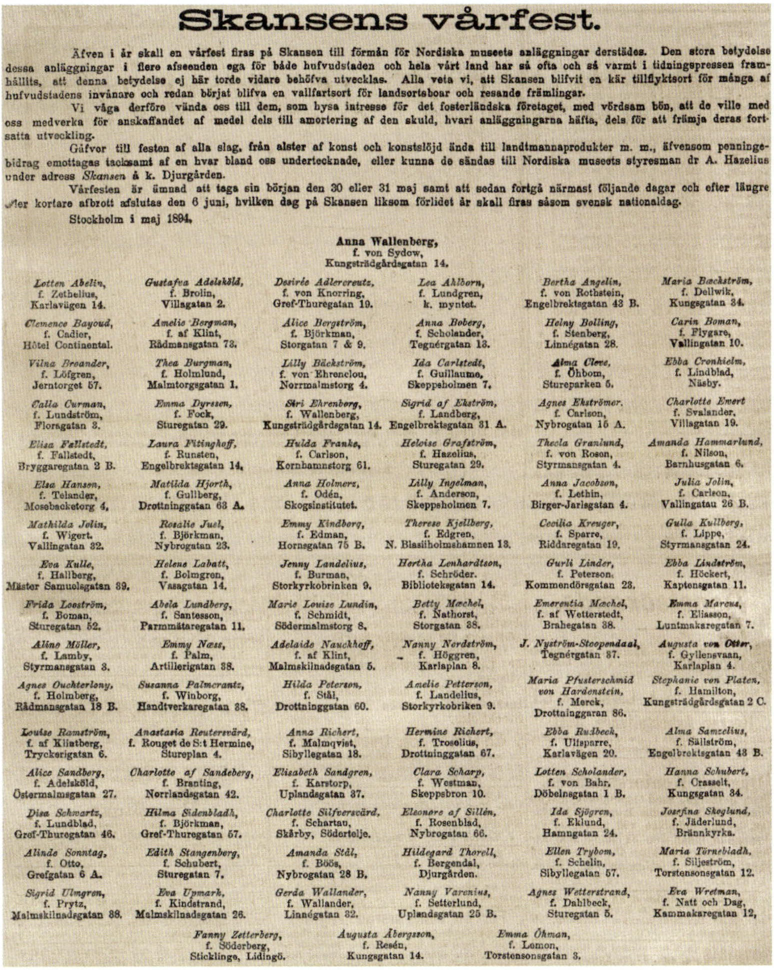

Skansens vårfest.

Äfven i år skall en vårfest firas på Skansen till förmån för Nordiska museets anläggningar derstädes. Den stora betydelse dessa anläggningar i flere afseenden ega för både hufvudstaden och hela vårt land har så ofta och så varmt i tidningspressen framhållits, att denna betydelse ej här torde vidare behöfva utvecklas. Alla veta vi, att Skansen blifvit en kär tillflyktsort för många af hufvudstadens invånare och redan börjat blifva en vallfartsort för landsortsboar och resande främlingar.

Vi våga derföre vända oss till dem, som hysa intresse för det fosterländska företaget, med vördsam bön, att de ville med oss medverka för anskaffandet af medel dels till amortering af den skuld, hvari anläggningarna häfta, dels för att främja deras fortsatta utveckling.

Gåfvor till festen af alla slag, från alster af konst och konstslöjd ända till landtmannaprodukter m. m., äfvensom penningebidrag emottagas tacksamt af en hvar bland oss undertecknade, eller kunna de sändas till Nordiska museets styresman dr A. Hazelius under adress *Skansen* å k. Djurgården.

Vårfesten är ämnad att taga sin början den 30 eller 31 maj samt att sedan fortgå närmast följande dagar och efter längre eller kortare afbrott afslutas den 6 juni, hvilken dag på Skansen liksom förlidet år skall firas såsom svensk nationaldag.

Stockholm i maj 1894.

Anna Wallenberg,
f. von Sydow,
Kungsträdgårdsgatan 14.

Lotten Abelin, f. Zethelius, Karlavägen 14.	*Gustafva Adelsköld,* f. Brolin, Villagatan 2.	*Desirée Adlercreutz,* f. von Knorring, Gref-Thuregatan 19.	*Lea Ahlborn,* f. Lundgren, k. myntet.	*Bertha Angelin,* f. von Rothstein, Engelbrektsgatan 43 B.	*Maria Bæckström,* f. Dellwik, Kungsgatan 34.
Clemence Bayoud, f. Cadier, Hôtel Continental.	*Amelie Bergman,* f. af Klint, Rådmansgatan 73.	*Alice Bergström,* f. Björkman, Storgatan 7 & 9.	*Anna Boberg,* f. Scholander, Tegnérgatan 13.	*Helny Bolling,* f. Stenberg, Linnégatan 28.	*Carin Boman,* f. Flygare, Vallingatan 10.
Vilna Broander, f. Löfgren, Jerntorget 57.	*Thea Burgman,* f. Holmlund, Malmtorgsgatan 1.	*Lilly Bäckström,* f. von Ehrenclou, Norrmalmstorg 4.	*Ida Carlstedt,* f. Guillaume, Skeppsholmen 7.	*Alma Cleve,* f. Öhbom, Stureparken 5.	*Ebba Cronhielm,* f. Lindblad, Näsby.
Calla Curman, f. Lundström, Floragatan 3.	*Emma Dyrssen,* f. Fock, Sturegatan 29.	*Siri Ehrenborg,* f. Wallenberg, Kungsträdgårdsgatan 14.	*Sigrid af Ekström,* f. Landberg, Engelbrektsgatan 31 A.	*Agnes Ekströmer.* f. Carlson, Nybrogatan 15 A.	*Charlotte Ewert* f. Svalander, Villagatan 19.
Elisa Fallstedt, f. Fallstedt, Bryggaregatan 2 B.	*Laura Fitinghoff,* f. Runsten, Engelbrektsgatan 14.	*Hulda Franke,* f. Carlson, Kornhamnstorg 61.	*Heloise Grafström,* f. Hazelius, Sturegatan 29.	*Thecla Granlund,* f. von Roson, Styrmansgatan 4.	*Amanda Hammarlund,* f. Nilson, Barnhusgatan 6.
Elsa Hansen, f. Telander, Mosebacketorg 4.	*Matilda Hjorth,* f. Gullberg, Drottninggatan 63 A.	*Anna Holmerz,* f. Odén, Skogsinstitutet.	*Lilly Ingelman,* f. Anderson, Skeppsholmen 7.	*Anna Jacobson,* f. Lethin, Birger-Jarlsgatan 4.	*Julia Jolin,* f. Carlson, Vallingatau 26 B.
Mathilda Jolin, f. Wigert. Vallingatan 32.	*Rosalie Juel,* f. Björkman, Nybrogatan 23.	*Emmy Kindborg,* f. Edman, Hornsgatan 75 B.	*Therese Kjellberg,* f. Edgren, N. Blasiiholmshamnen 13.	*Cecilia Kreuger,* f. Sparre, Riddaregatan 19.	*Gulla Kullberg,* f. Lippe, Styrmansgatan 24.
Eva Kulle, f. Hallberg, Mäster Samuelsgatan 39.	*Helene Labatt,* f. Bolmgren, Vasagatan 14.	*Jenny Landelius,* f. Burman, Storkyrkobrinken 9.	*Hertha Lenhardtson,* f. Schröder. Biblioteksgatan 14.	*Gurli Linder,* f. Peterson. Kommendörsgatan 23.	*Ebba Lindström,* f. Höckert, Kaptensgatan 11.
Frida Loeström, f. Boman, Sturegatan 52.	*Abela Lundberg,* f. Santesson, Parmmätaregatan 11.	*Marie Louise Lundin,* f. Schmidt, Södermalmstorg 8.	*Betty Mæchel,* f. Nathorst, Storgatan 38.	*Emerentia Mæchel,* f. af Wetterstedt, Brahegatan 38.	*Emma Marcus,* f. Eliasson, Luntmakaregatan 7.
Aline Möller, f. Lamby, Styrmansgatan 3.	*Emmy Næss,* f. Palm, Artillerigatan 38.	*Adelaide Nauckhoff,* f. af Klint, Malmskilnadsgatan 5.	*Nanny Nordström,* f. Höggren, Karlaplan 8.	*J. Nyström-Stoopendaal,* Tegnérgatan 37.	*Augusta von Otter,* f. Gyllensvaan, Karlaplan 4.
Agnes Ouchterlony, f. Holmberg, Rådmansgatan 18 B.	*Susanna Palmcrantz,* f. Winborg, Handtverkaregatan 38.	*Hilda Peterson,* f. Stål, Drottninggatan 60.	*Amelie Petterson,* f. Landelius, Storkyrkobriken 9.	*Maria Pfusterschmid von Hardenstein,* f. Merck, Drottninggaran 86.	*Stephanie von Platen,* f. Hamilton, Kungsträdgårdsgatan 2 C.
Louise Ramström, f. af Klintberg, Tryckerigatan 6.	*Anastasia Reutersvärd,* f. Rouget de S:t Hermine, Stureplan 4.	*Anna Richert,* f. Malmqvist, Sibyllegatan 18.	*Hermine Richert,* f. Troselius, Drottninggatan 67.	*Ebba Rudbeck,* f. Ulfsparre, Karlavägen 20.	*Alma Samzelius,* f. Sällström, Engelbrektsgatan 48 B.
Alice Sandberg, f. Adelsköld, Östermalmsgatan 27.	*Charlotte af Sandeberg,* f. Branting, Norrlandsgatan 42.	*Elisabeth Sandgren,* f. Karstorp, Uplandsgatan 37.	*Clara Scharp,* f. Westman, Skeppsbron 10.	*Lotten Scholander,* f. von Bahr, Döbelnsgatan 1 B.	*Hanna Schubert,* f. Crasselt, Kungsgatan 34.
Disa Schwartz, f. Lundblad, Gref-Thuregatan 46.	*Hilma Sidenbladh,* f. Björkman, Gref-Thuregatan 57.	*Charlotte Silfversvärd,* f. Schartau, Skårby, Södertelje.	*Eleonore af Sillén,* f. Rosenblad, Nybrogatan 66.	*Ida Sjögren,* f. Eklund, Hamngatan 24.	*Josefina Skoglund,* f. Jäderlund, Brännkyrka.
Alinde Sonntag, f. Otto, Grefgatan 6 A.	*Edith Stangenberg,* f. Schubert, Sturegatan 7.	*Amanda Stål,* f. Böös, Nybrogatan 28 B.	*Hildegard Thorell,* f. Bergendal, Djurgården.	*Ellen Trybom,* f. Schelin, Sibyllegatan 57.	*Maria Törnebladh,* f. Siljeström, Torstensonsgatan 12.
Sigrid Ulmgren, f. Prytz, Malmskilnadsgatan 38.	*Eva Upmark,* f. Kindstrand, Malmskilnadsgatan 26.	*Gerda Wallander,* f. Wallander, Linnégatan 32.	*Nanny Varenius,* f. Setterlund, Uplandsgatan 25 B.	*Agnes Wetterstrand,* f. Dahlbeck, Sturegatan 5.	*Eva Wretman,* f. Natt och Dag, Kammakaregatan 12,

Fanny Zetterberg, f. Söderberg, Sticklinge, Lidingö.	*Augusta Åbergsson,* f. Resén, Kungsgatan 14.	*Emma Öhman,* f. Lemon, Torstensonsgatan 3.

Figure 4.9 In 1894, the 100 Women of Spring Festival presented themselves as one behind the genius of Hazelius as a symbol of a harmonic society. Every woman was situated in class and history through her genealogical lineage. Photo: *Stockholms Dagblad*, 8 May 1894

congenital of motherhood as a notion shaped by the contemporary idea of essence of womanhood, and both were propagated when the cannonade inaugurated the feminine public space of work at Skansen. Participation in the Spring Festivals was almost mandatory for high-born women. As Women of Spring Festival, they received a public stamp of approval as a 'good family', and in return they provided the museum with moral reputation and legitimacy.

At Skansen, all parts of the women were to speak of patriotic love and cultural history: their bodies as vessels of good lineage, their hearts as hearths of moral virtue, their dresses as tokens of national traditions. The importance of pedigree and costumes for Hazelius is apparent in two memoranda to Millrath: 'Are we not lowering ourselves to the level of bourgeoisie with our wives. Could we not, through Mrs Abelin and Mrs Adelsköld, blend in a few Countesses and Baronesses?', and 'I am "totally furious" because national dresses were not used by the wives (especially Mrs Bergman) in the stand attended by our own assistants. *Never* more is such a thing to occur at our spring festivals.'[18]

In 1895, the Women of Spring Festival saw Skansen as an emotional refuge for the inhabitants of the capital and a solemn place where the visitors' love of the country and its people could be awakened. They wrote in the *Stockholms Dagblad* on 8 May:

> It is the silent, but mighty impression of the memories of the life of our people, which meets us all around, within a frame of the nature of one's native country, in a way as thousands have learnt to love it, that creates the, of its kind, unprecedented atmosphere, which Skansen gives away.

These women were neither considered as passive objects nor as rational subjects; they were given an in-between position as moral and emotional subjects, who founded their opinions on a feeling of certainty, that is, their love for and conviction of the value of their native country. If their inner feeling was in harmony with their outer experiences, as it was at Skansen, they could say with confidence that it had an important value. By affirming in public this feeling of affinity, they gave legitimacy to Hazelius's social and educational work of creating 'folk' for the good of Swedish society.

This was Hazelian museology: to blend the Romanticism of the early nineteenth century with contemporary Darwinism. To show the moral and emotional essence incorporated in buildings, objects and people, and, at the same time, to exhibit their evolutionary types. This essence of national character and patriotic love was also the key to a future folk community. At Skansen, the women of Stockholm's high society embodied the essence of that community and the will and practical capacity to make this vision come true. It was a vision that pointed to the future, to the construction of an inclusive welfare state – the Peoples Home (*folkhemmet*) – a concept which had both conservative and social democratic formulations before its political realisation in Sweden from the 1930s and onwards.

Notes

1 *Vårt Land*, No. 124, 1 June 1893.

2 Hazelius. F2A:1 The Private Literary Works of Artur Hazelius, 1868–1901, Early Archive of Artur Hazelius, The Nordic Museum and Skansen (EAANS), The Nordic Museum Archive in Stockholm (NMA). All translations of quotations by the author.

3 'Swedish National Day', or 'National Festival' ('svensk nationaldag' or 'nationalfest'), was the name given by Hazelius himself, see advertisements in newspapers – *Dagens Nyheter*, 6 June 1893 and 1894; *Stockholms Dagblad*, 6 June 1893; *Svenska Dagbladet*, 6 June 1894. The name of this day was changed in 1916 to the 'Swedish Flag's Day' and spread throughout the country. Not until 1983 was it given official status as 'National Day of Sweden' by the Swedish Parliament, and in 2005 it became a public holiday.

4 Hazelius to Olsen, 1897. Correspondence with the Nordic Museum, 1880–99, Archive of Bernhard Olsen, Archive of the Younger Eras of Denmark (AYED), The Archive of the National Museum in Copenhagen (ANMC).

5 Hazelius to Olsen, 1880. Correspondence with the Nordic Museum, 1880–99, Archive of Bernhard Olsen, AYED, ANMC.

6 Millrath-Behm, 1930s. Transcript of the memoirs of Sigrid Millrath-Behm, The Archive of Sigrid Millrath-Behm (ASMB), The Archive of Skansen in Stockholm (AS). The main part of this manuscript, which was written in the early 1930s, was published by the Nordic Museum in 1993. See also: 'Agenda at the meeting on 17 May 1893, at Mrs Anna Wallenberg', F7:1 Public Programmes at Skansen, 1885–1922, EAANS, NMA.

7 'Nationalfesten å Skansen: Vårfestens afslutning', *Stockholms-Tidningen*, No. 1065, 7 June 1893; 'Riksarkivarien Odhners tal', *Kalmar*, No. 87, 10 June 1893.

8 Hazelius. F2A:1 The Private Literary Works of Artur Hazelius, 1868–1901, EAANS, NMA.

9 See instructions to the gatekeepers of Skansen: F9AA Work Instructions, 1886–1901, EAANS, NMA.

10 See, concerning regulations and instructions to the museum assistants, gatekeepers, marshals, guards, Dalecarlian women, and bailiff of Skansen: F9AA Work Instructions, 1886–1901, EAANS, NMA. See also concerning chaperons: E2B:2/21 Hugo Samzelius, 1894–1901, EAANS, NMA.

11 Memoranda Notes Hazelius-Millrath, ASMB, AS.

12 Memoranda Notes Hazelius-Millrath, 1898, ASMB, AS.

13 Transcript of the memoirs of Sigrid Millrath-Behm, p. 72, ASMB, AS. See also: 'Skansens vårfest', *Stockholms Dagblad*, 8 May 1893.

14 Memoranda Notes Hazelius-Millrath, 1893, ASMB, AS.

15 Hazelius to Samzelius, 1893. Letters H:1 Artur Hazelius, 1890–1900, Archive of Hugo Samzelius, NMA.

16 E2B:7/107 Siri Oxenstierna n. Wallenberg, Incoming Documents, Anna Wallenberg, EAANS, NMA.

17 E2B:7/107, Siri Oxenstiernan. Wallenberg, EAANS, NMA.

18 Memoranda Notes Hazelius-Millrath, 1898, ASMB, AS.

Bibliography

Åkermark, K. (1901) 'Skansen', in *Boken om Stockholm i ord och bild*, Stockholm: Nilsson & Berglings Förlag.

Ambjörnsson, R. (1974) *Samhällsmodern: Ellen Keys kvinnouppfattning till och med 1896*, Gothenburg: University of Gothenburg.

Aronsson, P. (2011 (in preparation)) 'Exhibiting Scandinavian culture: The national museums of Denmark and Sweden', in S. Berger, C. Lorenz and B. Melman (eds) *Consuming the Past: Popular National Histories in Modern Europe*, London: Routledge.

Atterbom, P. D. A. (1835) *Philosophiens historia och system*, Uppsala: Palmblad.

Bennett, T. (1995) *The Birth of the Museum: History, Theory, Politics*, London and New York: Routledge.

Berg, G. (1933) *Artur Hazelius: Mannen och hans verk*, Stockholm: Bokförlaget Natur och Kultur.

Björck, S. (1946), *Heidenstam och sekelskiftets Sverige: Studier i hans nationella och sociala författarskap*, Stockholm: Bokförlaget Natur och Kultur.

Bohman, S. (1997) *Historia, museer och nationalism*, Stockholm: Carlssons.

Böök, F. (1923) *Artur Hazelius: En levnadsteckning*, Stockholm: P. A. Norstedt & Söners Förlag.

Bremer, F. (1849) *Lif i Norden, skizz*, Stockholm: C. A. Bagge.

Broberg, G. (1982) 'Lappkaravaner på villovägar: Antropologin och synen på samerna fram mot sekelskiftet 1900', in G. Eriksson (ed.) *Lychnos*, 1981–82, Uppsala: Lärdomshistoriska Samfundet.

Gawell-Blumenthal, I. (1931) 'Ett vårfestminne', in *Skansens Nyheter*, 5, Stockholm: Skansenföreningen.

Geijer, E. G. (1835) 'P. D. A. Atterboms filosofi', in J. Landquist (ed.) *Samlade skrifter*, Vol. 7, Stockholm: P. A. Norstedt & Söners Förlag (1928).

Geijer, E. G. (1840) 'Representationsfrågan', in J. Landquist (ed.) *Samlade skrifter*, Vol. 9, Stockholm: P. A. Norstedt & Söners Förlag (1929).

Grundtvig, N. F. S. (1832) 'Nordens Mythologi', in H. Begtrup (ed.) *Nik. Fred. Sev. Grundtvigs Udvalgte Skrifter*, Vol. 5, Copenhagen: Gyldendalske Boghandel Nordisk Forlag (1907).

Grundtvig, N. F. S. (1836) 'Haandbog i Verdens-Historien: Anden Deel: Middel-Alderens Historie', in H. Begtrup (ed.) *Nik. Fred. Sev. Grundtvigs Udvalgte Skrifter*, Vol.7, Copenhagen: Gyldendalske Boghandel Nordisk Forlag (1908).

Grundtvig, N. F. S. (1848) 'Menneskehed og Folkefærd betragtede med nordiske Øine', in G. Christensen and H. Koch (eds) *Værker i Udvalg*, Vol. 5, Copenhagen: Gyldendalske Boghandel Nordisk Forlag (1948).

Habermas, J. (1962) *The Structural Transformation of the Public Sphere: An Inquiry into a Category of Bourgeois Society*, Cambridge, MA: MIT Press (1998).

Hammarlund-Larsson, C. (1998) 'Samlingarna och samlandet', in H. Medelius, B. Nyström and E. Stavenow-Hidemark (eds) *Nordiska museet under 125 år*, Stockholm: Nordiska museet.

Hammarstedt, N. E. (1897) 'Dr Müllers uppsats i Tilskueren', in A. Hazelius (ed.) *Meddelanden från Nordiska museet 1897*, Stockholm: Nordiska museet.

Hazelius, A. (1880) *Nordiska museet såsom självständig anstalt*, Stockholm: Nordiska museet.

Hazelius, A. (1896) '13 c) Fil. Doktor Artur Hazelius', in *Autografier och Porträtt af framstående personer*, Häftet 13, Stockholm: Per Lindell and C. E. Fritzes K. Hofbokhandel.

Hazelius, A. (1899) 'Ur Nordiska museets historia', in *Meddelanden från Nordiska museet 1898*, Stockholm: Nordiska museet.

Hillström, M. (2006) *Ansvaret för kulturarvet: Studier i det kulturhistoriska museiväsendets formering med särskild inriktning på Nordiska museets etablering 1872–1919*, Linköping: Linköping University.

Jordansson, B. (1998) *Den goda människan från Göteborg: Genus och fattigvårdspolitik i det borgerliga samhällets framväxt*, Eslöv: Arkiv Förlag.

Liedman, S.-E. (2006) *Stenarna i själen: Form och materia från antiken till idag*, Stockholm: Albert Bonniers Förlag.

Lindberg, B. (2009) 'Folkets sammansättningar – om begreppet *folk* i historisk belysning', in M. Reinhammar (ed.) *Folkkultur i fokus*, Uppsala: Kungl. Gustav Adolfs Akademien för svensk folkkultur.

Linder, G. (1918) *Sällskapsliv i Stockholm under 1880- och 1890-talen*, Stockholm: P. A. Norstedt & Söners Förlag.

Lundgren, F. (2003) *Den isolerade medborgaren: Liberalt styre och uppkomsten av det sociala vid 1800-talets mitt*, Hedemora: Gidlunds förlag.

Mårtelius, J. (1987) *Göra arkitekturen historisk: Om 1800-talets arkitekturtänkande och I G Clasons Nordiska museum*, Stockholm: ArkitekturMuseet.

Müller, S. (1897) 'Museum og Interiør', in *Tilskueren*, September 1897, Copenhagen: Gyldendalske Boghandel.

Sandberg, M. B. (2003) *Living Pictures, Missing Persons: Mannequins, Museums, and Modernity*, Princeton, NJ and Oxford: Princeton University Press.

Stridsberg, L. (1915) 'Några bref från och till Artur Hazelius', in G. Upmark (ed.) *Fataburen: Kulturhistorisk tidskrift*, Stockholm: Nordiska museet.

Topelius, Z. (1885) 'Bomärke', in *Saga: Minnesblad för Nordiska Museet*, Stockholm: Nordiska museet.

Trägårdh, L. (1990) 'Varieties of Volkish ideologies: Sweden and Germany 1848–1933', in B. Stråth (ed.) *Language and the Construction of Class Identities*, Gothenburg: University of Gothenburg.

Upmark, G. (1916) *Skansen 25 år*, Stockholm: P. A. Norstedt & Söner.

Chapter 5

NARRATIVE AND IMAGINATION

Remaking national history at the Musée des Monuments français, Paris

Jennifer Carter

SEEKING A MODERN origin for practices that have resulted in the present-day museum trend of creating narrative templates and simulated experiences (Dernie 2006), this chapter explores the formation of the narrative history museum in the late eighteenth century. A genre of museum that departs from the scientific model being developed at that time, the narrative museum purports to represent events in a continuous and cohesive narrative environment, effected through the aesthetic, spatial, temporal and architectural *mise-en-scène* of the museological setting and the deployment of objects. By definition, the narrative museum is philosophical, not rational, and creates meaning through the combined narrative of its collections, scenography, and architecture. As a synthesized or 'total' environment, it is the museological equivalent of the *gesamkunstwerk*.

Today, this genre has undergone profound change, owing to transformations in social attitudes toward time (with emphasis on the here and now through technologies that permit instantaneous representation and the continual 'making' of historical events), representation (where a penchant for sensationalism privileges trauma, victimization and shock), and technological innovations that permit simulated rather than authentic experience. But to understand the intentions of the narrative history museum in the eighteenth century, it is necessary to restore the concept of experience and its ties to individual viewing habits developed in the eighteenth-century landscape garden.

The narrative history museum emerged in the moment that history and art – hitherto two related and mutually enforcing human endeavours – would become distanced through the creation and institutionalization of academic disciplines. Alexandre Lenoir's project of the Musée des Monuments français – which opened in Paris in 1795 and closed in 1816, and became France's first national museum of monuments – was a self-proclaimed history of nation and art.[1] This museum had the distinction of defining what was essentially a nineteenth-century phenomenon (the public museum) with an eighteenth-century sensibility. In other words, Lenoir's project was premised upon the strong belief that truth was acquired not simply from a factual or analytic engagement with material, but rather from sensory experience; the object embodied truth through the coalescence of *techne* and *poesis*. Lenoir's vision, however, was to be rapidly superseded by nineteenth-century concerns for rational and scientific displays. Displays in the arts, for example, would soon follow the model of pictorial arrangement then being pioneered by the Louvre and other institutions. In the late eighteenth century, there was no inherent inconsistency in an epistemology that understood art through history and history through art, and in this respect, it can be said that Lenoir's vision of history was that of a philosopher-historian, and his vision of art that of the poet.

The Musée des Monuments français was constructed through the coalescence of history, objects and language. History provided the overall narrative framework through which meaning was to be achieved. Objects introduced a material, temporal, and spatial dimension. And language imparted an idea and a way of being in the museum. These three elements were inextricably linked in Lenoir's work: each informed the other to create a heightened and unified experience of art. By considering these elements in Lenoir's work, this chapter seeks to unravel Lenoir's production of narrative and its meaning for the public. Amongst the objects found in this museum were Lenoir's *fabriques* which in many respects capture his artistic historical vision.

A museum of experiences

The Musée des Monuments français emerged in 1795 in the wake of the French Revolution. At that time, the institutions of science and art in Paris were completely reconstituted and, for example, the Musée du Louvre began its transformation into a public museum. The Musée des Monuments français had its origins in one of Paris's temporary storehouses for the country's newly seized 'national' collections: the confiscated objects of France's monarchy, nobility, and clergy. From mid-October 1790, and under Lenoir's guardianship, they had found temporary refuge at the dispossessed monastery of the Petits-Augustins on Paris's Left Bank. The new museum emerged not out of the interests of a learned society or institute but as the vision and creative undertaking of a single, highly motivated, and arguably idiosyncratic, individual. As founder and director, Lenoir was also its only administrator and curator. It was he who conceived of transforming this temporary storage depot into a museum, and it was he who had begun to undertake – even before the idea was officially sanctioned by the relevant governing authorities – the measures necessary to produce a more permanent, and choreographed, installation of objects. In short, Lenoir was the Musée.

Lenoir's museum aimed to recount the history of the French nation through the arrangement and aestheticized representation of sculpture, monuments, and

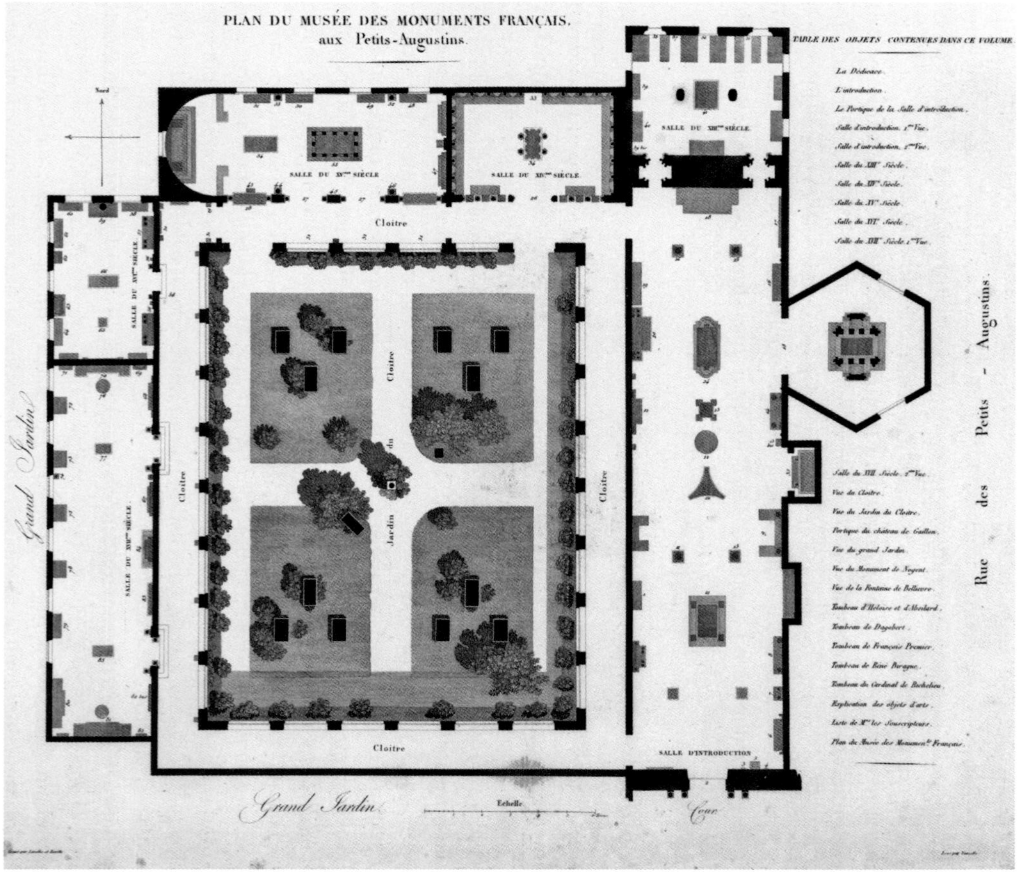

Figure 5.1 Plan of the Musée des Monuments français, in *Vues pittoresques et perspectives des Salles du Musée des monuments français*, 1816. Photo: Collection Centre Canadien d'Architecture/Canadian Centre for Architecture, Montréal

architectural fragments. To this end, Lenoir recreated a narrative *parcours* throughout the cloisters and halls, chapel and courtyards at the Petits-Augustins. Each hall was given the focus of a single century, and together they materialized French artistic history from the thirteenth century to the birth of the new republic. Here French Gothic sculpture challenged the longstanding authority of Greek and Roman. Borrowing from contemporary architectural and aesthetic practices, Lenoir gave each hall its own distinctive 'character' through the modulation of light and various scenographic techniques, completing the arrangement with an Elysium garden at the rear of the site, modelled on contemporary principles of landscape design and the picturesque (Poulot 1997b) (Figure 5.1). Unlike the halls, which used linear chronology as a structuring device, the Elysium garden suggested cyclical time. All, however, were unified by the historical and narrative possibilities of the monument.

In his design of ambient, century-specific halls, Lenoir inaugurated what is commonly considered the precursor to the period room, while his museum was one of the first in Europe and the first in France to realize a chronological sequence of galleries for the arrangement of art. Both of these display strategies would shortly afterwards be deployed more widely in 'modern' museums. Through a heightened attention to the conditions of exhibition, Lenoir disrupted the Baroque paradigm of decorative patterns of display; he radically altered the spatial structure and ideological premise of the collection of art. However, if Lenoir's scenographic interventions seemed innovative, many of them were in fact inspired by established literary and landscape contexts, most notably the *histoire monumentale* and the picturesque garden (Figure 5.2).

Despite almost single-handedly administering the Musée, Lenoir's project was nevertheless constantly and relentlessly mediated – even compromised – by several intervening forces and circumstantial factors. Both the city's and the nation's political structures underwent significant changes throughout the life of the museum as France re-invented itself following the overthrow of the *Ancien Régime*, the years of 'the Terror', the Napoleonic Empire, and the Bourbon Restoration. Various organizations, ranging from the Commission des monuments to the Commission temporaire des arts and the Commission d'instruction publique, oversaw Lenoir's work. From the outset, they were responsible for authorizing Lenoir's requests to expand the collection and give it permanence as a public museum. It was their role to the wealth of France's national heritage amongst the nation's other developing museums, including the Musée du Louvre[2] and the Musée Versailles.[3] Seen as competing national collections of fine arts, Lenoir was often instructed to transfer objects from his museum to the Louvre – much to his frustration. Visitors to Lenoir's museum were sure that he was creating something quite particular and quite different from that being produced in the Louvre. Hubert Robert, J.-E. Biet and Jean Lubin Vauzelle, for example, were among the artists who depicted views and scenes of the Musée's galleries and Elysium. Jules Michelet and Victor Hugo were among those historians and novelists who referred to formative

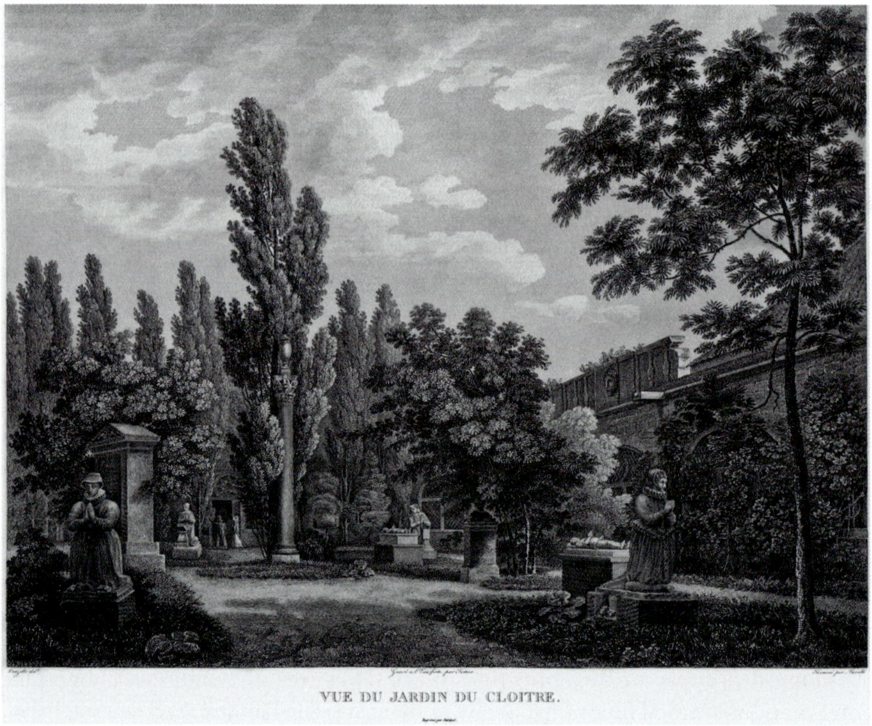

Figure 5.2 View of the Elysium garden from the cloisters, drawn by Vauzelle, in *Vues pittoresques et perspectives des Salles du Musée des Monuments français,* 1816. Photo: Collection Centre Canadien d'Architecture/Canadian Centre for Architecture, Montréal

childhood experiences at the museum. Even Napoleon was amongst those members of the visiting public who recorded impressions of their visits in letters and correspondences. Lenoir's museum provided a unique experience, even if it was built using ideas concerning history and display that were already being discussed and developed throughout Northern Europe.

History, monuments and *parcours*

The use of artefacts and monuments to narrate history was well established in such epic works as Bernard de Montfauçon's *Les monuments de la monarchie française* (1729–33) and Aubin Louis Millin's *Antiquités nationales, ou, Recueil de monuments pour servir à l'histoire générale et particulière de l'empire françois* (1790). Like Millin, Lenoir's artefact-based history focused increasingly on the achievements of accomplished individuals, especially in the garden of moral virtue that was the Elysium. Montfauçon's monarchical lineage, which belonged to an earlier era of historiography, was less prevalent but remained an important element in the interior of the Musée. Also like Millin (and unlike Montfauçon, who preferred more neutral settings), Lenoir placed his monuments within specific and defining spaces to enhance the narrative context and effect of the artefact.

Lenoir's attempt to provide a context for the object was developed on many levels; however, it was particularly enhanced by the use of the *parcours*, an idea that Lenoir borrowed from garden design. The *parcours* was as much a concept as it was a tangible construction, and it introduced the idea of the path, or narrative itinerary, to the museum's temporal logic. By its association with the planned itinerary of the garden, the *parcours* established the notion of a 'sense' or intention to the museum visit, and its use in the Musée highlighted composition, movement, and metaphors of time. Effectively, the idea of the *parcours* shifted the philosophical focus of the museum from the object to the visitor, and more specifically, to the visitor's experience of the museum.

The museum reflected those pictorial principles of the picturesque as could be observed in the paintings of such notable artists as Nicolas Poussin and Claude Lorrain and which had become established in eighteenth-century garden design. As the Englishman, Horace Walpole, wrote at the time, the landscape garden was a 'journey . . . made through a succession of pictures' (Andrews 1994: 7). The garden developed a number of new elements: the natural and untamed, and monuments, temples, and other architectural features which made mythological or historical allusions and references. By these means the garden could stimulate poetic associations, emotions and the imagination through the careful orchestration of vistas using paths, lakes and shrubbery. This transition from the regularity of seventeenth-century Classical and Baroque garden design in France signalled a major change in the era's understanding of the symbolic and formal possibilities of the garden. The pretence to authoritative narrative that had characterized André Le Nôtre's famous interventions at Versailles, for example, gave way to more personal itineraries of moral virtue as found in Lenoir's Elysium garden. This garden was to have an air of the melancholic, so as to induce the ideal state of contemplation and reflection required for its proper understanding:

> In this calm and peaceful garden, one sees more than 40 statues; tombs, placed here and there on a carpet of green, rise majestically amidst silence and tranquillity. Pines, cypresses and poplar trees surround them; propped against walls, larvae and urns compete to give this blissful place a gentle melancholy that speaks to the soul.
>
> (Lenoir 1800–1821, vol. 1: 19)

Lenoir imported similar design sensibilities inside the museum, where each hall was intended to inspire the visitor's identification with history. He sought to elicit an emotive response, using analogies of light to support narratives of progress or decline in the century's artistic practices. When describing the décor of the thirteenth-century hall (Figure 5.3), for example, Lenoir claimed that

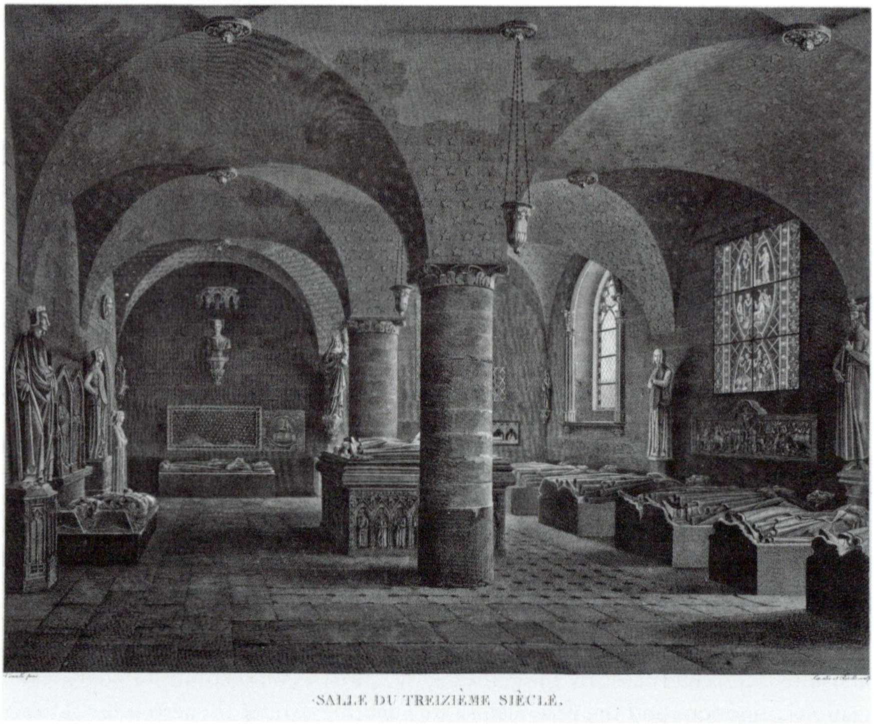

Figure 5.3 View of the thirteenth-century hall, drawn by Vauzelle, in *Vues pittoresques et perspectives des Salles du Musée des Monuments français*, 1816. Photo: Collection Centre Canadien d'Architecture/Canadian Centre for Architecture, Montréal

> The sombre light that illuminates this place is an imitation of the period; magic by which people were kept in a perpetually feeble state, people whom superstition had filled with fright. I have observed that the closer we come to our own century, the more we see light in public monuments, as if the vision of the sun could only be apparent to the educated person.
>
> (Lenoir 1800–1821, vol. 1: 181)

Conversely, for his portrait of the sixteenth century – the regenerating century of the fine arts – Lenoir adopted a much brighter and more celebratory decorative scheme, in order to accentuate the artistic accomplishments of the era.

These narrative strategies were not lost on visitors, including Napoleon and his wife Joséphine, and a future generation of Romantic writers and historians. Jules Michelet, reflecting on his childhood visits, recalled: 'It's there, and nowhere else, that I first experienced a vivid impression of history.' He continued,

> Even now I can recall the feeling, still just the same and still stirring, that made my heart beat when, as a small child, I would enter beneath those dark

> vaults and gaze at the pale faces; and would then, keen, curious and timid, walk and look, room after room, epoch after epoch. What was I looking for? I hardly know – the life of the time, no doubt, and the spirit of the ages. I was not altogether certain that they were not alive, all those marble sleepers, stretched out on their tombs. And when I moved from the sumptuous monuments of the sixteenth century, glowing with alabaster, to the low room of the Merovingians, in which was to be found the sword of Dagobert, I felt it possible that I would suddenly see Chilpéric and Frédégonde raise themselves and sit up.
>
> (Michelet, quoted in Haskell 1993: 252)

Chronology and completeness

In his own words, Lenoir described the twofold aims of the Musée as visibly demonstrating the progress of French art and history through the object. Lenoir expressed his intentions in his eight-volume compendium, *Musée des monuments français* (published from 1800–1821), and the more modest museum catalogue, *Description historique et chronologique des monuments de sculpture réunis au Musée des monuments français*, which he updated repeatedly throughout the museum's existence. In the former, Lenoir described two points of view, derived from models in Antiquity, that all museums should aspire to having: the first, political, the second, pedagogic. As a political institution, he argued, the museum should be impressive enough to attract the interest, and treasures, of other nations. Yet it was as a didactic institution that Lenoir's insight was most interesting. With a totalizing vision that aptly characterized both Enlightenment thinking and the psyche of the collector, Lenoir claimed that the museum must contain a chronological arrangement of 'all that the arts and sciences combined can offer to public instruction' (Lenoir 1800–1821, vol. 1: 51). It was chronology and completeness that made the museum a didactic institution: 'from this moment it becomes a learned school and an encyclopaedia where youth will find word for word all of the imperfections, perfections and decadences, by which the arts of drawing have successively passed' (Lenoir 1800–1821, vol. 1: 52).

Lenoir understood his art museum as the spatial equivalent of a book, and it was in the literary traditions of the encyclopaedia and the *histoire raisonnée* (literally, an analytical, or reasoned, history) that Lenoir modelled the museum. From the literary genre of monumental history and the German art historian Johann Joachim Winckelmann, Lenoir appropriated the structure and values of progress, decadence, and decline that would form the narrative basis of his museum. The encyclopaedia imparted the idea for a totalizing collection of French artistic heritage centralized in one location. Indeed, Lenoir was obsessed with locating objects that would further his historical narrative, and with transporting these from the provinces to Paris, often pursued by heavy criticism. This act of decontextualizing French heritage was one of the principal reasons for the eventual closure of the museum in 1816. Yet to Lenoir, the complementary notions of chronology and completeness justified such acts.

Lenoir's vision for the Musée as a site of comparative study was in fact intended to challenge the very hegemony of the *Académie* as the sole source of teaching for artists. A community from which he felt bitterly excluded, he never missed an opportunity to condemn it:

> Versed in the art of drawing since my childhood, I concluded that collections were far more precious in the quest for progress in the arts than schools, where students never see monuments, and in which they never hear theoretical ideas. The examples that we have before our eyes, the comparisons that we make, form good taste and are the basis of analytical study. Without this intellectual work, study is nothing more than a routine, and art becomes an occupation and devalues.
>
> (Lenoir 1800–1821, vol. 1: 46)

In a footnote, Lenoir confided that for these reasons, he intended to offer drawing classes and a course on theory within the museum, though he soon became too preoccupied with ensuring the museum's survival to pursue this ambition. Lenoir's most immediate concerns as founding director and curator were to oversee its daily operations, to secure the financial means to enlarge and complete its collection, to renovate the exhibition halls, and to restore the objects on display. He was also fighting a rearguard action against the competing interests of the Louvre and critics who saw Lenoir's curatorial blend of invention and fantasy with the object's reconstruction as unorthodox.

Revolution and *fabrique*

The French Revolution had resulted in the nationalization of royal and ecclesiastical collections, and altered concepts of conservation, preservation, pedagogy and historiography. For the first time, culture was being formally politicized, and the concept of a comprehensive and binding cultural heritage formulated (Deloche and Leniaud 1989). Vicq d'Azyr and Dom Poirier's *Quatre Instructions Initiales*, written in 1790–1791, was an initial attempt to determine an ethics and practice for governing historic monuments. It dealt with conservation, inventory, preservation and cataloguing.

It can also be argued that the democratic ambitions of the Revolution needed to be legitimized, and in a highly visible way. The emergence of the national museum at this moment fulfilled the need to divest the art object of its pre-*Ancien Régime* symbolism. At the Musée, the issue was particularly pressing; Lenoir was determining how to recycle the spaces of a former religious institution, and the objects of a regime now separated from the politics and ideologies of the deposed *Ancien Régime*. His solution was to overwrite the funerary and royal origins of much of the collection using a narrative that sought to establish a cohesive sense of nationhood in line with the Revolution's principles of liberty, fraternity, and equality over the realities of a monarchical past (Poulot 1997b). Lenoir's narrative promised a restorative poetics of time by virtue of a rewriting of national history, a rewriting that occurred in no small measure through the use of the *fabrique*.

The tradition of the *fabrique* had its origins in landscape painting in the mid-eighteenth century. The French artist and landscape theorist Claude-Henri Watelet coined the term *fabrique* in a 1756 entry in the *Encyclopédie*, though his description at that time was confined to the language of painting. It was, in essence, an invention or construction in the landscape, a cross between a confabulated ruin and an imaginary structure, often composed of disparate elements. Lenoir appropriated the concept and developed it in different ways in his sculptural creations in order to realize his own curatorial vision.

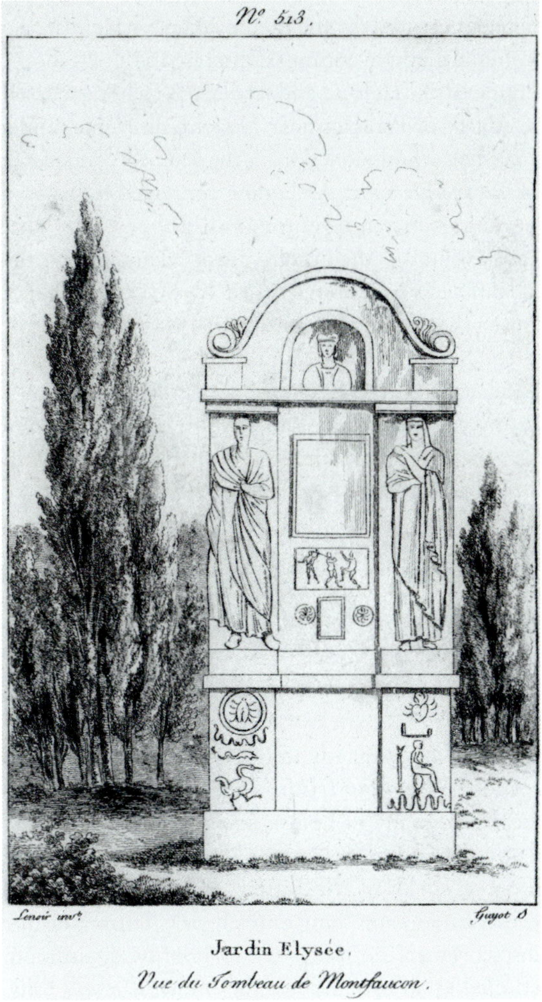

Figure 5.4 Fabrique monument to Bernard de Montfaucon, designed by Alexandre Lenoir, in *Recueil de gravures pour servir à l'histoire des arts en France, prouvée par les monumens*, 1812. Photo: Collection Centre Canadien d'Architecture/Canadian Centre for Architecture, Montréal

Although he completed reconstructions of monuments in the interior halls, it was the *fabriques* Lenoir created and placed in the Elysium that demonstrated the most outlandish and unorthodox designs. These were often odd sculptural constructions with complex and creative combinations of emblematic and symbolic iconographies. It is doubtful that any single reading was intended for these objects. Lenoir described his design for the monument dedicated to the antiquarian and historian Bernard de Montfauçon (Figure 5.4), for example, as a composite of 'hieroglyphs, Egyptian figures, Greek reliefs, figures from the late Roman Empire and remains of monuments from the first years of the French monarchy' (Lenoir 1810: 290). This was one of many monuments that Lenoir created from the remains of others and not without significant criticism in his own time as well. Although Lenoir insisted that the monuments he fabricated combined only materials from similar historical eras (thus echoing his period halls), he did not always abide by this dictum. In the very popular chapel he recreated for the medieval lovers Héloïse and Abélard (Figure 5.5), Lenoir combined a newly commissioned neo-gothic canopy, the twelfth-century funerary monuments of Héloïse and Abélard he had purchased from their original setting at the Abbey of Paraclet near Nogent-sur-Seine, and a contemporary death mask of Héloïse he commissioned from the sculptor – and later detractor of his practices – Louis-Pierre Deseine. As Lenoir confessed in his own writing, in the absence of authentic objects, an invention will do, providing that it conformed to – one might read 'evoked' – the architectural 'character' of the period. These two examples of *fabriques* given above attest to two traditions of *fabrique*, one pure fantasy, the other the likeness of an original (human being or monument).

It could be argued, as Michel Makarius (2004) has more generally, that Lenoir's *fabriques* were a form of ruin. But by virtue of their composition of fragments, of sculptural and architectural 'ruins', the *fabriques* promised the antithesis of the ruin: regeneration. They were, in effect, the anti-ruin. It is true that in the shape of the *fabrique* lay the demise of one aspect of the ruin's poetic dimension: in the object reconstituted there could be no acknowledgement of the lived past, at least not through the object itself, and therefore no contemplation of the future – not in the traditional sense. However, it is also true that Lenoir was engaged in rewriting France's past. Indeed, his entire project of the Musée was dedicated to this single goal. If 'The ruin is a dialogue between an incomplete reality and the imagination of the spectator' (Woodward 2002: 139), Lenoir's anti-ruins sought the opposite: to establish a sense of completion and wholeness on a post-Revolution, fractured French psyche. Lenoir's *fabriques* spoke of another truth.

Unlike the folly of garden traditions, many of Lenoir's *fabriques* paradoxically required something of the authentic object for their completion. Divested of the pure vestige, the anti-ruin could not speak of the future, it merely contained the past, not to erase it, but to present it as something entirely *new*. Ultimately for Lenoir, the *fabrique* was a means to reinscribe the past. To borrow again from Christopher Woodward, it functioned as an inversion of the *Ancien Régime*'s folly.

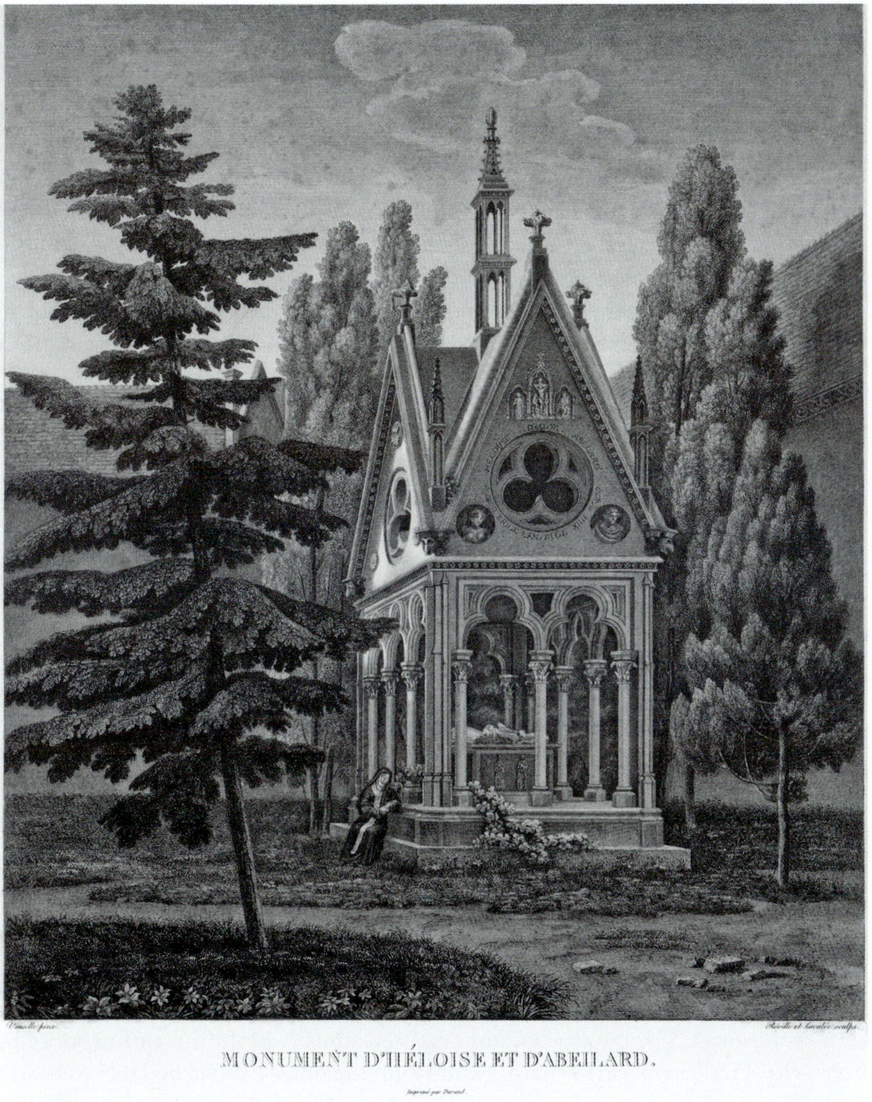

Figure 5.5 Fabrique monument to Héloïse and Abélard, designed by Alexandre Lenoir, in *Vues pittoresques et perspectives des Salles du Musée des Monuments français*, 1816. Photo: Collection Centre Canadien d'Architecture/Canadian Centre for Architecture, Montréal

If the ruin imposed a certain catastrophic image of the present, the *fabrique* inverted that image and invested it with another order, one capable of reanimating the past and the stasis of the ruin.

The *fabrique*, then, permitted Lenoir to pursue his ideal of a museum of progress by illustrating a perfected history. Contemporaneous to Constantin-François

Volney's meditative and highly influential *Ruines, ou, Méditation sur les révolutions des empires* (1791), which used the ruin as a departure point for moral reflections on the decline and decay of empires, there could be no greater antithesis than Lenoir's own project of the Musée, which similarly summed up history in empires and revolutions. Where Volney sought enlightenment in human actions through his invocation of the ruin (Zucker 1968: 198), Lenoir's was an effort to correct human (read artistic) shortcomings through his very reconstruction of the ruin. His obsession with reconfiguring the object to a pristine, though not necessarily 'authentic' condition, is also found in Jean-Jacques Rousseau's literary 'embellishments' which sought to overcome memory lapses, not to confabulate or to promote falsity but to avoid emptiness: 'and if by chance I have used some immaterial embellishment it has been only to fill a void due to a defect of memory' (Nalbantian 2003: 27). Lenoir's impetus seems to have been a post-Revolution need to fill in the void: to rewrite a cohesive history of France and to render this history visible, tangible and felt.

Emplotment

Perhaps Antonio Gramsci (1986: 3) best characterized the significance of Lenoir's work within its post-*Ancien régime* context in an unrelated passage he wrote in the *Prison Notebooks*: 'The crisis consists precisely in the fact that the old is dying and the new cannot be born; in this interregnum a great variety of morbid symptoms appears.' Not all projects undertaken in this period were of such morbid character, however; many were concerted efforts to construct a new modern self, and to this end, public spaces and objects figured prominently. This construction of the new modern self involved adopting a particular attitude to the past and to objects. Monuments in particular were poignant statements about history, but ones that could be modified. Within this context, the ruin was an ambiguous locale, where an unpleasant past could be revisited, and exorcized. The monument speaks of entitlement and disparity, hierarchy and privilege, and in Lenoir's hands it metamorphosed into a tool for democratizing history in the space of a monastery-turned-museum. The curator's ambitions were thus twofold: he endeavoured to overwrite France's decaying past, while simultaneously repairing this past and putting it to the service of the edifying ideals of the Revolution: to educate, enlighten, and lead the new, modern public toward progress.

Thus to speak of a form of emplotment in the Elysium is to recognize the role of the visitor in linking separate monuments – or episodes – into their own personal and cohesive narrative, a narrative that stood as the interior's other within the overall framework of the Musée. The garden for Lenoir presented all of the possibilities that the interior halls did not. With its predominance of *fabriques* and the variety of historical figures to which these paid tribute, through its use of the relic and the reconfigured ruin, through its characterization and alliance with contemporary landscape theory, the Elysium was neither bound by chronology nor monarchical

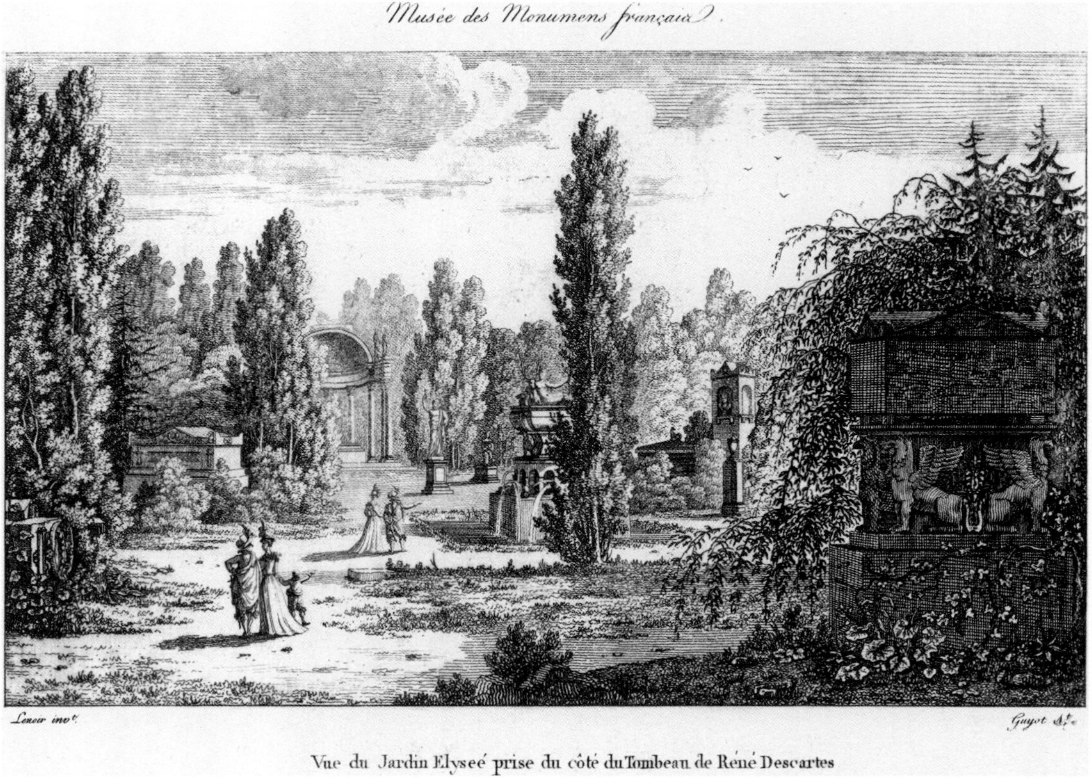

Figure 5.6 View of the Elysium garden with Descartes' tomb in right foreground, in *Recueil de gravures pour servir à l'histoire des arts en France*, 1812. Photo: Collection Centre Canadien d'Architecture/Canadian Centre for Architecture, Montréal

historiography, and in this liberated space Lenoir posited themes that lay outside of the dialectic marked by notions of progress and decline (Figure 5.6). He was free to explore history as a continuum, rather than as a linear evolution, and to this end it was in the garden that cycles of nature, time, and humanity itself were given full expression.

To consider the Elysium as in some way existing outside of time is to ignore Lenoir's most important commentary about the past. His message in the garden that history is best understood not as a chronological sequence towards progress, but as a continuum, normalized the recent Revolution as but one instance of many throughout time. By emplotting the visitor within the Elysium's *parcours*, Lenoir placed the visitor in history, creating a tangible and visceral connection to the past and through the past to the present.

Evoking narrative history

What emerged in the halls of the Musée des Monuments français was hybrid. Part monument to French accomplishment, part mythic narrative, Lenoir's project was an embodiment of Enlightenment ideals in its deliberate attempt to provide moral and didactic instruction to its visiting publics through the sequencing of objects in choreographed spaces. Lenoir's almost fanatical obsession with fragments and their reconfiguration, and his desire to evoke mythic origins and traditions, proved fertile concepts in the psychological recovery of a nation emerging from revolution and the denial of its feudal and monarchical past. Thus it was as a direct consequence of the Revolution – to which the museum owed its fortuitous origins – and in response to this Revolution, that the museum truly realized its poetic dimension as an evocative narrative of history. The Musée des Monuments français was both museum of art, and museum of history, and in the fissures of the discursive structure lay a poetic intention that served to open up other possible relationships with history through the very unity that Lenoir gave to the building and the objects this building housed.

Notes

1 The exact title of his eight-volume catalogue is *Musée des monuments français; ou, Description historique et chronologique des statues en marbre et en bronze, bas-reliefs et tombeaux des hommes et des femmes célèbres, pour servir à l'histoire de France et à celle de l'art. Ornée de gravures; et augmentée d'une dissertation sur les costumes de chaque siècle*. All quotations from Lenoir's catalogues are translations by the author.

2 The Louvre was known as the Muséum français (September 1792), the Musée de la République (July 1793), the Muséum des arts (August 1793), and the Musée Central des arts (January 1797).

3 After 1797 known as the Musée spécial de l'école française.

Bibliography

Andrews, M. (ed.) (1994) *The Picturesque: Literary Sources and Documents* (3 vols), Mountfield, East Sussex: Helm Information.

Aristotle (1997) *Poetics*, trans. S. H. Butcher, Mineola, NY: Dover.

Brès, J.-P. (1821) *Souvenirs du Musée des Monuments français: collection de 40 dessins perspectifs gravés au trait représentant les principaux sous lesquels on a pu considérer tous les monuments réunis dans ce Musée. Dessinés par M. J.-E. Biet et gravés par Mm. Normand père et fils avec un texte explicatif par M. J.-P. Brès*, Paris: Chez l'Auteur; Chez Normand fils, P. Didot aîné impr. du Roi.

Clark, H. F. (1943) 'Eighteenth-century Elysiums: The rôle of "association" in the Landscape Movement', *Journal of the Warburg and Courtauld Institutes,* 6: 165–89.

Condillac, E. B. (1754) *Traité des Sensations*, London, Paris: de Bure.

Courajod, L. (1878–1887) *Alexandre Lenoir, son journal et le Musée des monuments français*, 3 vols, Paris: Champion.

Deloche, B. and Leniaud J.-M. (eds) (1989) *La Culture des Sans-Culottes: Le premier dossier du patrimoine 1789–1798*, Paris-Montpellier: Éditions de Paris/Presses de Languedoc.

Dernie, D. (2006) *Exhibition Design*, New York: W. W. Norton.

France. Ministère de l'instruction publique (1883-1897) *Inventaire général des richesses d'art de la France: Archives du Musée des monuments français*, 3 vols, Paris: E. Plon, Nourrit.

Girardin, S. (1788) *Promenades, ou, Itinéraire des jardins d'Ermenonville, auquel on a joint vingt-cinq de leurs principales vues, dessinées et gravées par J. Mérigot fils*, Paris: Mérigot.

Gramsci, A. (1986) *The Prison Notebooks*, in G. James, *Morbid Symptoms: Arcadia and the French Revolution*, Princeton, NJ: Princeton Architectural Press.

Haskell, F. (1993) *History and Its Images: Art and the Interpretation of the Past*, New Haven, CT: Yale University Press.

Lenoir, A. (1796 or 1797) *Description historique et chronologique des monuments de sculpture, réunis au Musée des monuments français; suivie d'un traité historique de la peinture sur verre, par le même auteur*, troisième édn, Paris: Au Musée, rue des Petits-Augustins.

Lenoir, A. (1797–1798) *Description historique et chronologique des monuments de sculpture, réunis au Musée des monuments français; suivie d'un traité historique de la peinture sur verre, par le même auteur,* quatrième édn, Paris: Au Musée, rue des Petits-Augustins.

Lenoir, A. (1800–1821) *Musée des Monuments français; ou, Description historique et chronologique des statues en marbre et en bronze, bas-reliefs et tombeaux des hommes et des femmes célèbres pour servir à l'histoire de France et à celle de l'art. Ornée des gravures: et augmentée d'une dissertation sur les costumes de chaque siècle*, 8 vols, Paris: Guilleminet.

Lenoir, A. (1810) *Musée Impérial des monuments français: Histoire des arts en France, et description chronologique des statues en marbre et en bronze, bas-reliefs et tombeaux des hommes et des femmes célèbres, qui sont réunis dans ce Musée*, Paris: De l'Imprimerie d'Hacquart.

Lenoir, A. (1814) *La Franche-maçonnerie rendue à sa véritable origine, ou, L'antiquité de la franche-maçonnerie prouvée par l'explication des mystères anciens et modernes*, Paris: Fournier.

McClellan, A. (1994) *Inventing the Louvre: Art, Politics, and the Origins of the Modern Museum in Eighteenth-century Paris*, Berkeley, CA: University of California Press.

Makarius, M. (2004) *Ruins*, trans. David Radzinowicz, Paris: Flammarion.

Middleton, R. (1999) 'Soane's spaces and the matter of fragmentation', in M. Richardson and M.-A. Stevens (eds) *John Soane, Architect: Master of Space and Light*, London: Royal Academy of Arts, pp. 26–37.

Millin, A. L. (1791–1798) *Antiquités nationales, ou, Recueil de monuments pour servir à l'histoire générale et particulière de l'empire françois: tels que tombeaux, inscriptions, statues, vitraux, fresques, etc.; tirés des abbayes, monastères, châteaux et autres lieux devenus domaines nationaux*, 5 vols, Paris: Drouhin.

Montfauçon, B. (1729–1733) *Les monuments de la monarchie française: qui comprennent l'histoire de France, avec les figures de chaque règne que l'injure des tems à épargnées*, 5 vols, Paris: J. M. Gandouin.

Morel, J.-M. (1776) *Théorie des jardins*, Paris: Pissot.

Nalbantian, S. (2003) *Memory in Literature: From Rousseau to Neuroscience*, Basingstoke: Palgrave Macmillan.

Nietzsche, F. (1995) *Unfashionable Observations*, trans. Richard Gray, Stanford, CA: Stanford University Press.

O'Neal, J. (1996) *The Authority of Experience: Sensationist Theory in the French Enlightenment*, University Park, PA: Pennsylvania State University Press.

Poulot, D. (1997a) 'L'Élysée du Musée des monuments français: un jardin de la mémoire sous le Premier Empire,' *Jardins et châteaux. Dalhousie French Studies* 29 (Winter): 159–68.

Poulot, D. (1997b) *Musée, nation, patrimoine 1789–1815*, France: Éditions Gallimard.

Roquefort, B. (1816) *Vues pittoresques et perspectives des Salles du Musée des monuments français: et des principaux ouvrages d'architecture, de sculpture et de peinture sur verre qu'elles renferment; gravées au burin, en vingt estampes par MM. Réville et Lavallée, d'après les dessins de M. Vauzelle: avec un texte explicatif par B. de Roquefort*, Paris: Imprimerie de P. Didot, l'Aîné.

Stafford, B. M. (2000) '"Illiterate monuments": The ruin as dialect or broken classic', in E. Suderburg (ed.) *Space Site Intervention: Situating Installation Art*, Minneapolis, MN: University of Minnesota Press.

Stara, A. (1999) 'Lenoir, Quatremère and the hermeneutic significance of the Musée des monuments français', unpublished thesis, University of Oxford.

Thouronde, S. (1964) 'Le couvent des Petits-Augustins', *Information d'histoire de l'art* 4: 161–77.

Vicq d'Azyr, F. and Poirier, D. G. (1790–1791) 'Instruction sur la manière d'inventorier et de conserver, dans toute l'étendue de la République, tous les objets qui peuvent servir aux arts, aux sciences et à l'enseignement proposée par la commission temporaire des arts, et adaptée par le comité d'instruction publique de la Convention nationale, 25 ventôse an II', in Bernard Deloche and Jean-Michel Leniaud (eds) *La Culture des Sans-Culottes: Le premier dossier du patrimoine 1789–1798*, Paris-Montpellier: Éditions de Paris/Presses de Languedoc.

Volney, C.-F. (1835) *Volney's Ruins; or, Meditation on the Revolutions of Empires*, Boston: C. Gaylord.

Watelet, C.-H. (1774) *Essai sur les jardins*, Paris: Prault.

White, H. (1973) *Metahistory: The Historical Imagination in Nineteenth-century Europe*, Baltimore, MD: Johns Hopkins University Press.

Woodward, C. (2002) *In Ruins*, London: Vintage.

Zucker, P. (1968) *Fascination of Decay*, Ridgewood, NJ: Gregg.

Chapter 6

NATIONAL ART MUSEUM PRACTICE AS POLITICAL CARTOGRAPHY IN NINETEENTH-CENTURY BRITAIN

Christopher Whitehead

THIS CHAPTER WILL explore one of the key characteristics of nineteenth-century national museums in Britain: that of mapping the world, both geographically, epistemologically and socially. I will argue that the national museum provided an institutional technology for mapping, while in its morphology it was, literally, a multi-dimensional map which constructed knowledge spatially, connectively and divisively, to represent cultural and natural hierarchies and relations and differences between things and between people. The chapter begins with a brief exploration of the notion of the museum as map by examining the use of cartographic technology within the context of institutional collecting and display of material cultures. Then, looking at the network of national museums in mid-nineteenth-century London, the chapter will discuss the importance of geography, mobility, travel, cartography and appropriation of objects within the organizing structures of curatorial practice and knowledge construction, with predominant reference to the notion of art and the idea of the work of art. Within this, it will also look at the nature of national museum representations of the home nation. Where was Britain's place on the map? And what were the cartographical politics of national othering and selfing? In relation to the latter the chapter will also consider the national museum as a cartographic technology for social mapping through which, in a post-1832 context, the social and moral order of the new British electorate was plotted. The chapter will conclude with an account of the apparently weak expressions of nationhood in mid-century national museums, enabling a view of Britain as cultural cartographer rather than as obvious cultural territory, and opening up a way of discussing the museum as space for theorizing, whether explicitly or not, the complex political relations between places, cultures and peoples past and present.

> Mapping technologies and practices have been crucial to the emergence of modern 'views of the world', Enlightenment sensibilities and contemporary modernities. The world has literally been made, domesticated and ordered by drawing lines, distinctions, taxonomies and hierarchies: Europe and its others, West and non-West, or people with history and people without history. Through their gaze, gridding, and architectures the sciences have spatialized and produced the world we inhabit. And, indeed this is perhaps the crucial issue: maps provide the very conditions of possibility for the worlds we inhabit and the subjects we become.
>
> (Pickles 2004: 4–5)

Over recent decades the use of cartographic metaphors has become ubiquitous and potent in a variety of fields as part of what some have identified as a 'spatial turn' in the humanities and social sciences (see Warf and Arias 2009). The idea of the museum as map has been evoked in the context of museological study too, including Kirshenblatt-Gimblett's (1998: 132) view of museum collections as 'undrawn' maps 'of all the places from which the materials have come' and museum floor plans as delineating 'conceptual paths through what becomes a virtual space of travel' (Whitehead 2006: 20–22). In a similar sense, Robert Lumley notes that museums 'map out geographies of taste and values' and function as a 'means whereby societies represent their relationship to their own history and to that of other cultures' (Lumley 1988: 2). More recently, and at greater length, I have argued that the most appropriate way of 'reading' museums in the textual tradition (for which see Mason 2006) is to do so cartographically (Whitehead 2009), in that this is more sympathetic to their spatialized operation than the verbal and even grammatical readings of display attempted by authors like Ferguson (1996), Bal (1996) and Storr (2007).

This way of thinking requires an expanded definition of the practice of cartography: as *the structured, spatial representation of knowledges about the world.* Such representation is certainly not limited in the forms it might take to the conventional two-dimensional 'god's-eye view' of printed maps of land and sea (see also Pickles (2003: 15) for questions of definition). One might discuss the difference between the paper map and the museum as map, where the former may be seen to be exclusively spatial in its representative reach, while the latter is spatio-temporal. This, however, is to deny the temporal dimensions of the paper map. Akerman has noted that 'folding space into time, map into calendar' is common in pre-modern maps. Maps can also be explicit historical representations, as in the Ordnance Survey's 'period maps' of Roman Britain, Dark Ages Britain, etc. Sir Charles Close, Director General of the Ordnance Survey between 1911 and 1922 went so far as to suggest that to use a map is to travel through time (Daniels 1998: 125) while Lestringant clarifies that maps are never synchronic representations:

> Any given map was never established on entirely fresh ground, but always inherited from previous maps a not inconsiderable – even a preponderant –

> share of its information . . . The map did not reveal the state of the world at a given moment, but a mosaic of data whose chronology might extend over several centuries, the whole being assembled in a floating space. These driftings, at the same time spatial and temporal, conferred a dynamism and a prospective value on the map.
>
> (Lestringant 1994: 113)

In this context the map is the result of a kind of historical, or historiographical, bricolage (Pickles 2004: 87–89), and this is a suggestive way of thinking about museum representations such as collections and displays as a literal organization of found objects. There are inevitable differences between the respective processes of production of paper and museum maps, such as the need, in museums, to work to and with the constraints and possibilities of physical architecture and staff politics (so that the museum is always a multi-authored map), but these differences in regulation are not fundamental and mean simply that we must be attentive to diversities in cartographic agency.

This expanded definition of the map, allowing, as it does, for an understanding of the *museum* as map, brings with it the need to expand the experiential parameters of map *reading*. It goes without saying that the experience of 'reading' a paper map is very different from that of visiting a museum, for the museum is (mostly – at least prior to mass internet usage) articulated in the physical space of architecture, often in different rooms and on different floors, so that one cannot, with the same rapidity, in a single spreading gaze, survey the world shown on the map. The visit provokes different forms of physical transit, different experiences of being in space, and usually more longitudinal consumption than does the act of gazing upon a paper map. It also operates under a different scopic regime where the world is (largely) seen at the eye-height of an able-bodied adult human and not at god-height, where the particles of the world are revealed only incrementally and not all at once. Moreover, while the paper map represents the world graphically the museum does so materially, literally arranging bits of the world – be they constitutions of 'art', 'archaeology', 'nature', and so on, as if in microcosm.

But the paper map and the museum as map share communicative technologies, transcending superficial differences between graphic and institutional/architectural formats. These technologies relate to selecting, excluding, generalizing, grouping, relating, pointing, segregating, bounding, labelling and scaling (see Karrow 2007: 3–6). They work to organize literal geographies made of land masses and bodies of water, to organize peoples and cultures, to organize knowledges, cultures and material and to organize the past. And as Gieryn argues, maps do to non-geographical referents what they do to the earth:

> Boundaries differentiate this thing from that; borders create spaces with occupants homogeneous and generalized in some respect (though they may vary in other ways). Arrangements of spaces define logical relations among sets of

> things: nested, overlapping, adjacent, separated. Coordinates place things in multidimensional space, making it possible to know the direction and distance between two things. Landmarks and labels call attention to typicalities or aberrations, reduce ambiguities about the precise location of a boundary, highlight differences between spaces of things: they are reality checks, of sorts.
> (Gieryn 1999: 7)

Seen (or 'read') in this way, the museum as map may represent literal places such as continents and countries like Africa and Kenya (although these are of course human rationalizations of space). It may also represent conceptual and epistemological terrain, for example, by bounding art from archaeology, anthropology or natural history. This is done by identifying, constituting and segregating art's material constituents and stipulating – through structures like display – the ways in which 'art' should be experienced and known. Within the conceptual and epistemological terrain of art are more complex multi-dimensional geographies. These relate on one side to the locations of artistic practice, for example, the regional 'schools' of Italian Renaissance painting (Tuscan, Umbrian, etc.), and the mobility of artists. Such locations and mobilities were later refracted in the travel itineraries of art historians (many of whom were also some of the first public art museum directors), for whom, until the advent of mass photography, travel was the predominant way of building up professional competence. So in some ways it is not surprising that geography formed a central structure within the organization of art history. But the geographies of the terrain of art also relate to the identification of discrete artistic forms, media, genres, eras and movements and the intensely political search for differences (hierarchical and other) and relations between them. While maps can be seen as grossly simplified representations of the external world, generalizing the 'complexity, flexibility and mobility of human society in fixed, territorial patterns' (Cosgrove 2007: 104; see also Black 1997: 104), they can at the same time form highly complex classification systems operating spatially which are in fact constructive, as indicated in the quotation from John Pickles at the beginning of the chapter. Or, as King (1996: 16) notes, 'cultural mappings play a central role in establishing the territories we inhabit and experience as real' whether these territories are geographical, epistemological, intellectual, moral or any mixture of the above.

It is in this way that the representations of the paper map and the museum as map also share specific kinds of political operationality, in their delineation and description of terrain and in the relative attention and importance which it is accorded. The cartographic appeal to authority, its truth claim, is staked upon the indexing of an 'out-there' reality either through the graphic sign or through the organization of material objects (particles of that 'out-there' reality) in the physical space of the museum:

> [The] apparent stability [of maps] and their aesthetics of closure and finality dissolve with but a little recognition of their partiality and provisionality,

> their embodiment of intention, their imaginative and creative capacities, their mythical qualities, their appeal to reverie, their ability to record and stimulate anxiety, their silences and powers of deception.
>
> (Cosgrove 1999: 2)

Meanwhile, the ubiquitous idea of mapping as a form of territorialization ('the map precedes the territory', as Baudrillard famously put it (1988: 166)) and appropriation, and therefore of the map as an ownership claim, is rendered both more explicit and more complex in the museum. Here occupation is signalled not simply through the delineation of a graphic line and the titling of a region, be it geographical or conceptual or both, but through the physical subtraction of material from contexts. Such materials – fragments and metonyms of their contexts – are literally owned by the collecting nation, and lead to an experience of the museum as a space of suggestion in which 'other' places and 'others' pasts' can be intellectually and morally 'occupied'. Here the new host nation levels a cartographic gaze at its others, and through surveying and mapping, lays claim to the self-naturalizing right to steward, to marshal, to comprehend and to represent their cultures. Through the objective mappings of the museum, and especially the 'national' museum with its particular appeal to the authority of the state, cultural cadastres are formed. As Daniels observes, 'maps have the potential materiality to realize political claims' (1998: 129), and this is emphatic in the museum.

The relationship between mapping and colonizing has been well established. 'The production of maps of colonies is a symbolic form of conquest,' notes King (1996: 27): 'To map a territory is to stake various kinds of claim to it, to make assertions of ownership, sovereignty and legitimacy of rule.' While many of the material cultures which national museums in nineteenth-century Britain collected were not from colonized territories in the literal sense of the term (Italy and Spain, for example, were not colonized in this sense, but constituted source territories of high symbolic importance in the construction of hierarchies of art), I argue that collecting and displaying material culture from such territories represented the modulated impulse towards cultural occupation. This bore especially on the occupation of the pasts of cultures once perceived to be great, and now in political, civic, economic and (to the surveyors' eyes) artistic disarray. Museum actors like Henry Cole, the first Director of the South Kensington Museum established in 1857, recorded his views of the degeneracy of the Italian administration, religion and people in his travel journals, while simultaneously buying Italian objects (some of which were buildings) for the museum or earmarking them for acquisition and appropriation (Whitehead 2006). In 1866, Cole's employee, the curator J. C. Robinson, wrote to him to ask for money to make purchases from a Spain 'in semi-revolution, [where] money has disappeared, distress prevails and whatever there is to be sold is in the market [selling for] a fraction of what would have been formerly asked'. The clergy, Robinson noted, 'will sell their copes, cassocks and mitres even' (cited in Wainwright and Gere 2002: 56). Back in South Kensington a microcosmic account of curatorial travel was produced through the arrangement of objects appropriated from overseas and the adoption of historicist architecture, reproducing in some measure the ambience of foreign historical physical contexts to create a qualitative geohistory of art (Whitehead 2006; see also DaCosta Kauffmann 2005: 9). So, while a literal imperial intervention on Britain's part was not desirable in the context of European politics and natural resources, a cultural occupation – the staking of a claim to manage the past – was both *possible* in that economic and political conditions allowed such exploitation and subtraction of resources, and *permissible*, in having no obviously violent or illegal dimensions. This adds a new dimension to Tim Ingold's observation that 'surveying is a mode of occupation, not habitation' (2007: 87): there was no intention to people such other countries, rather, the impulse at play was to obtain symbolic control and ownership of, and authority over, foreign pasts. Macdonald takes up this theme in the context of colonialism, but again the argument has valence beyond the literal colonization of territories. For Macdonald (2003: 3), the possession of objects from other cultures on the part of a nation state, in the care of a national museum, was important in that such objects were 'signs of the capacity to gather and master beyond national boundaries'. As such, they were claims of the capacity to know and to govern; 'signs too for the visitors that theirs was a nation, or a locality, that also played on the global stage'.

Similarly, Daniels (1998: 114) has argued that state-sponsored cartographical surveys are 'key instruments for the expansion and consolidation of central government', aiding the 'internal colonization of the nation state as well as the incorporation of its territories overseas', again pointing out the importance of surveying the foreign other for the purposes of constituting the national self. Such insights can help us to understand the play of cultural power – the contest for the right to make knowledge – between countries without obvious colonial relations.

This brief account has, I hope, shown how the machinery of the museum (an 'engine of difference', as Alberti (2008: 82) has called it), can be well understood as a set of cartographic technologies, which invoke important political questions relating to ownership of cultures, cartographic agency, national identities and identity relations, and world views. To concretize these assertions we might proceed to ask: how does cartography operate in practice in national museums, whose scale and whose denomination and status as 'national' bestow particular significance on the process of mapping?

We will now examine this in relation to art in national museums in mid-nineteenth-century London. It is important to point out here that in understanding museum representations it is necessary to be attentive to practice over and between numerous museums, for their physical distinctness relates closely to the development and management of discrete knowledges and borders between them. This concerns the ways in which intellectual rationales for differential classification of material cultures were forged through interaction between museums, but it also does not ignore the importance for museum epistemology of the vagaries of state funding, interpersonal and institutional rivalries and quotidian regulations of all kinds. Indeed, such factors need to be recognized as fundamental elements of the history of knowledge. As pointed out by Pickles, any map is embedded within, and co-determined with, other texts (2003: 50), and our readings need to be sensitive to the complex aetiology of knowledge.

The British Museum, the National Gallery and the South Kensington Museum (known, from 1899, as the Victoria and Albert Museum (V&A)), exemplify well the notion of the museum as cartographic technology because of what Bennett (2006: 48) has called the 'totalizing frameworks of representation characterizing nineteenth-century museums' in their aspiration 'to render metonymically present the global histories of all things and all peoples'. The focus on art here needs to be justified, and this can be done first in relation to the brevity of this chapter and second in that the process of identifying (constructing) art and histories of art necessarily involved the act of distinguishing between art and its others – art and the rest of the world. This means that what follows is sensitive to the singularity of art as a territory, but sensitive also to the ways in which such territorialization naturally invokes and involves other territorializations. The world map to be discussed, then (one of art in the world, and in history), is thematic. Through discrete but sometimes interrelated instances of practice and discourse, we will focus briefly on various museum mappings: of knowledge, of the past, of self and other, of society.

What counts as knowledge in the museum (or, what is made to count as knowledge), and how can it be mapped? As I have argued elsewhere (Whitehead 2009: 8), knowledges (in the plural) are accounts of the world, or some aspect of the world in its material and/or spatial and/or temporal dimensions (thus including the history of x or y) as well as being *ways of accounting* for the world. The elasticity of this formulation is, admittedly, a somewhat disingenuous means of unlinking the concepts of knowledge and truth and sidestepping epistemological discussions premised on analytic philosophical traditions; but it is also an attempt to think about knowledges as both representations and representational processes. In the instance of museums, if not elsewhere, the simplest way to think of this – albeit necessarily a tautological one – is that knowledges are constituted by, and constitutive of, the objects mapped, for example, in the material-disciplinary contexts of conceptual fields (territories) such as art, ethnography, archaeology, natural history, and so on.

While some in mid-nineteenth-century London, like the art critic John Ruskin, argued that national museums could be spaces in which such knowledges would be united and their borders dissolved (see Whitehead 2009: chs 3 and 4), in reality the museums were utilized as expressions of the ultimate need to identify discrete disciplinary knowledges and to hold them separately from one another in distinct physical contexts, where such segregation was fuelled by the imperatives of conceptual differentiation. This is seen in seemingly interminable arguments in Parliament and the press about the need for discrete spaces to hold conceptually bounded collections, as well as in the splintering of what were seen by many as inappropriately heterogeneous collections, for example, in the removal of natural history collections from the British Museum to the new Natural History Museum in 1883 (an event with its roots in the 1850s), or in the gradual severing of the industrial and science collections from the art collections at the V&A which ultimately concluded with the opening of the Science Museum in 1928.

The practice of mapping in this broadest sense becomes the delineation of literal borders (institutional and architectural) between categories of material, the identification of coordinates which plot something as within one border and outside of another. What, for example, makes something 'art' as opposed to 'archaeological'? If, as might be anticipated, an object can hold the status of each, then what coordinates determine the ultimate place of the object on the map? This is not merely an incidental question about where an object happens to end up, for with the occupation of conceptual terrains such as art and archaeology come specific cultures of consumption (which themselves propagate representation) and disciplinary technologies and aesthetics of which museum visitors, through institutional and interpretive prompts, are invited to be mindful and to gaze appropriately.

This act of epistemological choice is evident in the 1855 sale of the defunct MP Ralph Bernal's collection, for bidding at which both the British Museum and the South Kensington Museum had been granted government funds. In order to work within distinct cultural territories the two institutions developed different criteria for collecting ostensibly very similar objects: the South Kensington Museum was to collect objects as works of art, while the British Museum was to collect objects which, while they might be considered works of art, had a predominantly documentary significance (Whitehead 2009: 52–56). In this act of push and pull, objects were, somewhat arbitrarily, valorized and encoded in different disciplinary settings associated respectively with art and art history on the one hand and with human history on the other.

For a starker example we may consider the cartographical location of Old Master paintings in the nineteenth-century museum. They were not displayed or interpreted in order to illuminate the social history of the time and place in which they were first produced and consumed; rather, they were more normally presented and represented as objects of consecrated cultural-aesthetic value through whose agency viewers could experience affective and moral transformation. In the discursive formation of fine art at this time, notions of consumption concerned ennoblement, quasi-spiritual appreciation, reverie, transport, and transcendence of time and place (Whitehead 2005: 6). Such notions acted as counterpoints to archaeological notions of comprehending objects in historical cultures and comprehending historical cultures through objects, and also counterpointed the predominantly intellectual understanding of the past as opposed to the moral and affective understandings associated with fine art. Such discursive formations are at work in the actions of disciplining both objects and viewers into specific museological regimes of consumption, as aesthetic or historical apprehension each become more or less possible depending on the institutional frame: the National Gallery, for example, effectively suppressed documentary interpretation of objects in favour of aesthetic apprehension.

Debates about what material objects should count as history and not as art were intense in the mid-nineteenth century, focusing on what I have called 'discrepant boundary objects' (Whitehead 2009: 97) which were used in debate to explore discrepancies in disciplinary classifications, with the ultimate purpose of establishing the disciplinary cadastres of individual museums. Where should armour (often highly decorated and perhaps little suited to the battlefield) be placed? Where should Egyptian antiquities and medieval manuscripts be located and in relation to what other material cultures, if any? When is sculpture fine art and when is it decorative, ornamental or applied art? The results of this mapping activity in the mid-century were extreme, with the brutal conceptual and physical segregation of (most) paintings in the National Gallery. This allowed for a single epistemological appeal to the imagined Gallery visitor, who, in being receptive to the ennobling effects of engaging with paintings as a highly consecrated material culture, took his place within a morally responsible new electorate. Meanwhile, the freedom of access to the National Gallery and public ownership of its collection was seen as an index of the character of the nation and a stimulus for national pride (Taylor 1999: 43–44). It was also presented as a means of democratizing the aristocratic leisure practice of the amatorial aesthetic experience, whereby inequalities likely to provoke civil unrest could be smoothed, if not eliminated. Just weeks after the Great Reform Act received the Royal Assent, the future Prime Minister Sir Robert Peel argued in liberal terms for the granting of funds for a new National Gallery building (which would ultimately open in 1838), steering a strategic path between Tory fears and Whig interests in relation to electoral reform:

> In the present times of political excitement, the exacerbation of angry and unsocial feelings might be much softened by the effects which the fine arts had ever produced upon the minds of men. Of all expenditure, that like the present, was the most adequate to confer advantage on those classes which had but little leisure to enjoy the most refined species of pleasure. The rich might have their own pictures, but those who had to obtain their bread by their labour, could not hope for such an enjoyment. With respect to the situation of the building, it was as well selected as possible, close to Charing Cross, where, as Dr. Johnson said, 'the great tide of human existence is fullest in its stream'; and, consequently, where all classes of the community would be equally accommodated. He therefore trusted that the erection of the edifice would not only contribute to the cultivation of the arts, but also to the cementing of those bonds of union between the richer and the poorer orders of the State, which no man was more anxious to see joined in mutual intercourse and good understanding than he was.[1]

The nineteenth-century discourse of the moral agency of old master paintings (whose roots, as Arthur MacGregor (1997: 32, n. 85) notes, go back at least as far as Jonathan Richardson) has been much remarked (Robertson 2004: 4), and it is a complex discourse indeed, with the power to legitimize large public expenditure and participation in international cultural competition on the art market. (There was no dissensus that expensive, foreign paintings should be bought for the nation – any disagreement regarded precisely which *sorts* of expensive, foreign paintings should be bought.) But in the context of post-1832 electoral reform the discourse of the moral agency of old master paintings takes on particular significance and the picture gallery, as the institutional arena for the ennoblement of what Robert Peel termed the 'poorer orders of state' becomes not just a map of art but a visualizing technology for the mapping of society. In this arena, anxieties about new publics could be aired, for example, in the ubiquitous complaints in the press about 'loungers' or breastfeeding in the gallery (Waagen 1853). At the same time, the vulgarities of the 'poorer orders' could be identified and corrected, and inappropriate conduct could be checked and disciplined, as in the Keeper Thomas Uwins' 1850 account of how he remonstrated with a party of country people who sat down to take a picnic – replete with servings of gin – in one of the gallery rooms; 'I represented to them,' he said, 'that those things could not be tolerated' (Parliamentary Paper 1850: 82). This patriarchal discourse of correction and instruction, of induction into appropriate social practice, is also at work in contemporary visual representations (see Figure 6.1), reflecting Peel's desire that the National Gallery might alleviate social unrest by cementing the 'bonds of union between richer and poorer orders of state'. This woodcut from an illustrated magazine in 1870 is suggestive of elite hopes and misgivings about such social dynamics: under the casual gaze of a genteel young mother, members of a Working Men's Club respond conscientiously to the teaching of their well-bred mentor in the gallery, all except one, who yawns conspicuously in the background.

The National Gallery, in this sense, was a cartographic technology intended to 'fix' the social order of the nation, in the double sense of the correction of defects and the making-permanent of an electorate both amenable and acceptable to government, ultimately working to allay the social elite's fears of sedition. This work of social fixing (with the aid of the map's 'aesthetics of closure and finality') is inextricable from the cartographic fixing-in-place of old master painting as paragon of human endeavour, within which certain specific cultures of production, such as those associated with the Italian Renaissance, were uniquely privileged for their perceived value to civilization and as exemplars for contemporary artistic production. This in turn relates to diachronic mappings of time prevalent in some art historiography of the nineteenth century, organized in Renaissances, Periods of Refinement and Periods of Decline (e.g. Wornum 1856). But where, in such qualitative cartography, did British material culture and British history figure? Where was Britain's place, in the plotting of time and space?

In fact, it has been noted that British material culture was far from predominant in museum representations in the early to mid-nineteenth century, and that in part this may have been a consequence of nervousness at the idea of evoking the 'nation' lest this should lead to unwanted assertions of rights by the as-yet largely

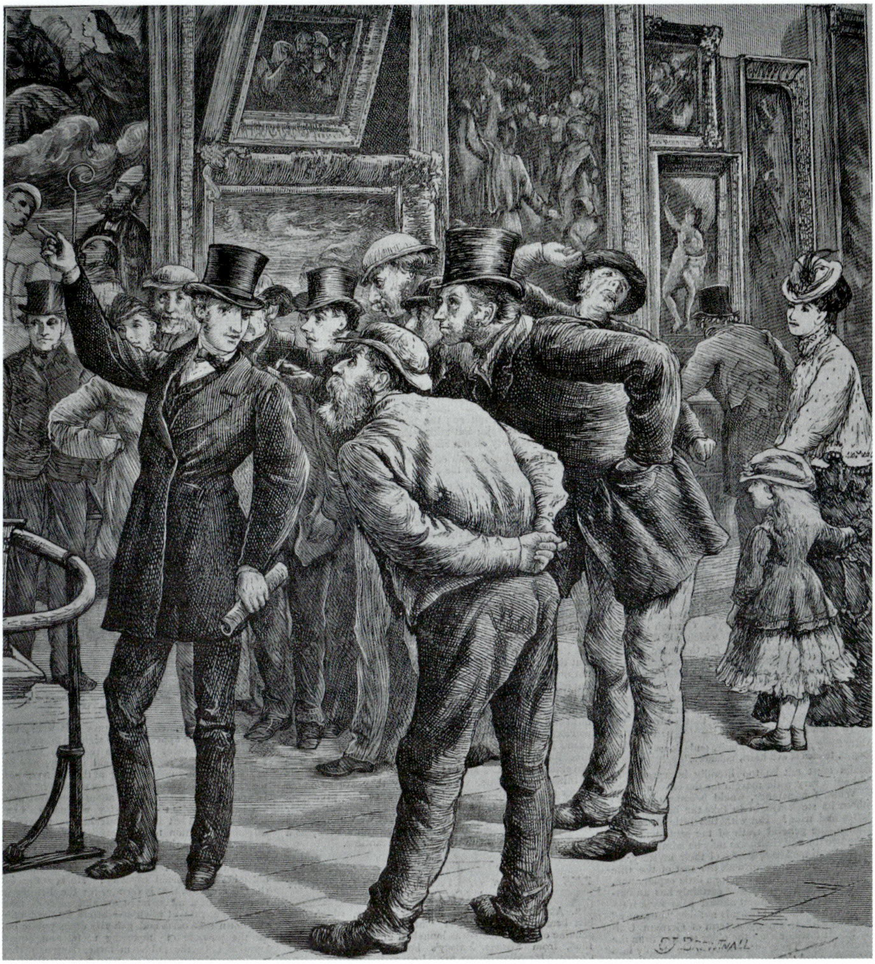

Figure 6.1 E. J. Brewtnall, woodcut illustration from *The Graphic*, 1870. Photo: author

unenfranchized populace (Duncan 1995: 42). Indeed there was a long tradition of patriotism being associated not with the state but with radical politics, whereby opposition movements to government called themselves patriotic and accused government of corruption (Cunningham 1981: 10). Mack (1997: 37) notes that there was a clear bias at the British Museum towards the antiquities of the Mediterranean and the Near East, while MacGregor points to a number of contrasting early to mid-nineteenth-century examples elsewhere in which national museums played a key role in the forging of distinct national identities emerging out of old political orders (e.g. the national museums founded in Budapest (1802), Copenhagen (1808), Prague (1818) and Nuremberg (1852)), and suggests that it was precisely Britain's insularity and long-established political stability which inhibited similar developments (1998: 125), a view to which I will offer a counter-argument, attentive to political *in*stability, at the end of this chapter. Whatever the truth, it is evident that the movement to collect British antiquities had, until the mid-century, 'met with indifference bordering on hostility' from the British Museum Trustees (MacGregor 1997: 22), who were themselves in large part made up of the political elite.

This sentiment was not shared by the antiquarian and archaeological communities in Britain, who took every opportunity to point to the lack of a collection of antiquities found on or in national soil and to argue that this was discreditable to government (MacGregor 1998: 127). This was a charge against which the Trustees – and some museum officers – defended themselves through reference to budgetary limitations and the cartographic imperative (as they saw it) to concentrate upon material culture from the rest of the world which would otherwise be lost to the nation, leaving native antiquities to find their way safely into provincial collections (MacGregor 1998: 131; Mack 1997: 38) with all that this means about their perceived relative significance. Britain's dominant role as surveyor of other cultures is also evident in relation to ethnography collections, the cultural value of which were often denigrated in museum discourse as entirely subordinate to 'artistic' and archaeological material culture (Whitehead 2009: ch. 3; Mack 1997: 40). Nevertheless, their acquisition and organization was clearly linked to the lines of exploration, trade and Empire which Britain had established, to the extent that such cartography could be seen as a national obligation. As the British Museum Keeper A. W. Franks would later put it in 1880: 'If there is a country where travellers and others have a right to expect that there should be a well organized and well arranged ethnographical collection, it is certainly in Great Britain which is in closer connexion with all parts of the globe than any other nation' (in Mack 1997: 47).

Meanwhile, Saumarez-Smith (1997: 275) notes of the South Kensington Museum that the 'original impulse behind the establishment of the . . . collections was essentially imperial', marked by a 'willingness to go on predatory expeditions, buying up medieval and Renaissance works of art in Italy, France and especially Spain, while ignoring British ecclesiastical treasures in the churches of Herefordshire and East Anglia'. This suggested in some measure the marked sense of the superiority of art from other countries in contemporary discourse, which also fuelled the museum's imperative to improve British manufactures by exposing artisans, manufacturers and consumers to foreign material culture. In this sense Britain's place was not on the map, but above it, taking the god's-eye view and *encompassing*, in the cartographic sense, 'within its reach the full spectrum of [other] world cultures' (Saumarez-Smith 1997: 275). While apparently absent from the map, Britain was in fact implied within it as the maker of knowledge about the world, as cartographer.

However, Britain was not omitted as a territory for representation for long. In 1857 the collector John Sheepshanks presented his collection, mainly of British paintings, to the nation but the terms of gift meant that it was acquired by the South Kensington Museum (Whitehead 2005: 163) and not by the National Gallery (thus irredeemably complicating the cultural cartography of the museum world in London). The collection was subsequently assembled at the museum in what was entitled the 'Gallery of British Art'. In the first guide to the South Kensington Museum, produced in 1857, the gallery is introduced as follows:

> [There] are illustrations of our national poets, episodes of our domestic life, and the scenery of our native country; and not a few of these serve to exemplify the truth that genius, despite the universality of its range, derives its happiest inspirations from the home where it has been nurtured.
>
> The peculiar interest which this collection is calculated to excite is due, therefore, not alone to its appeal to that sense of the beautiful which many possess, or to that social instinct which makes us love the delineations of social life, but also to our *home* feeling – our peculiarly national characteristic.
>
> (South Kensington Museum 1857: 7, original emphasis)

Curiously, this is the only invocation of the nation in the entire guide and also the only piece of text which assumes the characteristics of rhetorical criticism. The consecrated works of art from the Italian Renaissance and from other historical material cultures of renown were afforded no such textual frame of reference. British antiquity too was represented in the collections present at the museum in the form of fragments of and casts from Romanesque and Gothic buildings, but the accompanying text was descriptive and not rhetorical. The introduction to the Gallery of British Art (and thus also the gallery itself) formed an act of interpellation, appealing to a sense of common national identity and prompting visitors to feel content with their place in the world, identifying with and recognizing their lives in the paintings on display. Another appeal to domestic contentment of a different but perhaps related kind occurs later on in the *Guide*, when the 'Economic Museum' of household objects is introduced as being 'particularly addressed to the working classes, being designed to impart to them, in an easy and agreeable manner, the knowledge of *common things*, and to show them how much it may promote the health, comfort and happiness of themselves and their families' (South Kensington Museum 1857: 9, original emphasis). Both appeals could indeed be read as anti-seditious in the tradition of Peel's notion of the 'softening' effects of access to museums upon 'angry and unsocial feelings' (see above). Alternatively, the unusual celebration of British painting is perhaps to be accounted for in the fact that securing the Sheepshanks collection had been a significant coup for the museum, which placed itself strategically and competitively in relation to the other national museums in London, repeatedly playing for territory (Whitehead 2005: 162–72). Or perhaps it reflects something of the patriotic feeling stirred up over the troublesome, recently concluded Crimean War (Cunningham 1981: 20), or represents a new interpretive key intended to be more congenial and familiar to a working public than the structured stylistic genealogies practised at the National Gallery. Perhaps it functions as part of the didactic-ideological machinery of the South Kensington Museum to raise the quality of national manufacture by exhorting the contemporary British artisan to derive 'inspirations from the home' where he or she has been 'nurtured', and to feel him- or herself as part of a national artistic tradition. Whatever the mixture of reasoning behind this piece of rhetoric, it signals a new attentiveness to the cultures of the home nation and a willingness to look inwards as well as outwards.

This found some analogue expression in London's other national museums, albeit rather weaker. In 1866 the Trustees of the British Museum bowed to pressure from the antiquarian and archaeological community and established the Department of British and Medieval Antiquities and Ethnography, with the accomplished A. W. Franks at its head. This was, however, a heterogeneous department, comprising not just British antiquities but also Indian and Chinese collections, Mexican and other South and North American material culture, as well as the Witt collection of obscene or pornographic materials which remained reserved from public view. Franks himself noted that his department 'might with equal propriety be termed the Department of Miscellaneous Antiquities', including, amongst other things, 'prehistoric antiquities of all nations', glass from 'all ages and all countries' and 'the productions of the uncivilized races of the world'. He did not seem to bemoan this or to assume any kind of nationalistic stance with regard to the desirability of isolating the British antiquities, noting instead that 'all kinds of archaeological knowledge are useful in [the department]' (quoted in Caygill 1997: 72). As Caygill notes, the cartography of Franks's 'empire' was nearly limitless in terms of time and place, and in some respects the co-presence of prehistoric and contemporary ethnographic material within one epistemological location was indicative and constructive of a logic in which prehistory could be understood through ethnography according to the governing narrative of technological evolution; 'prehistory' was not just a 'when' for Franks, but also a 'where', for it characterized races still in existence (King 1997: 146). The political mapping of nations was subservient here to the mapping of human development, exclusive only of aspects such as western 'art', which was consecrated elsewhere.

The department, however, was unique (apart from the South Kensington Museum's 'Gallery of British Art') in explicitly *placing* Britain within the national collections. One rationale for this was the combination of the significant increase of the collection of British material culture and the 'great interest taken by the public' in that collection (Caygill 1997: 70). Quite how this interest was gauged, other than from the protests of archaeologists and the press, is not known, highlighting the difficulty of pinpointing the source of the impulse to musealize, recontextualize and display native antiquities and thus the difficulty of evaluating the reality of any public demand in the wider sense. In any case, critics have pointed out that the term 'British' in the name of the department was intended not so much as an indication of its scope than as 'a nationalistic point' to prove that the museum did now value native antiquities (Bruce-Mitford cited in Caygill 1997: 71).

At the National Gallery there were rather less pronounced evaluations of national culture. In 1867 the rebuilding of the National Gallery was being planned and an architectural competition was in progress, when the Director William Boxall and the Trustee A. H. Layard produced a report stipulating the need for an arrangement where 'the works of the old masters should be placed on one side, or in one wing of the building, and the works of the British School in the other' (cited in Whitehead 2005: 212).

Nearly ten years later the plan to develop a new building had been scrapped in favour of a modest extension which opened in 1876. It included two almost identical galleries as its largest apartments, into the lunettes of which were inscribed names of artists representing the canons of Italian and British painting respectively, as if to draw the two 'Schools' into neighbouring or contiguous territory, perhaps to suggest their qualitative equivalence. However, the cartographic impulse of the architecture was immediately overpowered, for the British room was occupied by Netherlandish and Flemish paintings, which formed a much more frequent equivalent to Italian painting in art historiography in both qualitative and structural senses. There were in fact particular difficulties in constituting a British or English 'School' of painting of historical equivalence to more celebrated foreign schools, for as C. L. Eastlake and R. N. Wornum, respectively the first Director and the Keeper of the National Gallery, made clear in their notes on the future development of the displays, 'The native school of England is of comparatively recent date', originating with Joshua Reynolds (whom Wornum would later censure in his 1856 volume *The Epochs of Painting Characterized*). They clarified, 'all important works antecedent to Charles [I] were executed by foreigners; also, during and subsequent to the reign of that monarch, foreign artists almost engrossed the higher patronage' (Eastlake and Wornum 1853: 824–25).

As a territory for representation then, Britain emerged only weakly on the museum as map in the decades around the mid-century. Until that point there was a curious expression in museums of what might be termed as two overlapping focuses – the national self-in-the-world and the national self-in-history. In this vision, Britain itself emerges as a museum nation, a kind of central hub towards which the material cultures and histories of other places are routed and organized for consumption. This predicated and organized a 'forward course of the nation' (Poulantzas 1980: 113) of improved civility, decorum, enfranchisement and technological and artistic progress.

But there is an inevitable sense of the erasure of the national past from view until the mid-century. If the modern state, as Poulantzas has argued, organizes the unity of the nation in the form of a 'historicity of a territory and territorialization of a history' (Poulantzas 1980: 114; see also Bennett 1995: 141), then what territory was historicized and what history was territorialized? Why does there appear to have been this temporal and spatial caesura in the national museums' collective cartographic accounts of the world? Here it may be possible to consider the importance in the museum of negotiating the 'disjunction between the past and the present that had been effected, or was being effected, by the triumph of capital' (Brooks 1998: 5). While certainly never expressed by policymakers or museum workers as a conscious strategy, it may nevertheless have been convenient to the running of state to banish from sight (at least in the museum) the recent memory of political, social and economic upheaval associated with radicalism and with dramatic developments in industrialization and agricultural work, from Peterloo to the Swing Riots and the Chartists. Here any reference to national history or native antiquities might have invoked uncomfortably the 'fundamental change that had occurred in the relations of production' (Brooks 1998: 5). Far safer, perhaps, to look outwards, to direct the nation's gaze on to the other. This outward gaze was no doubt abetted by the difficulty of establishing where the 'nation' began and ended, with the joint confusions of the multinationality of the British Isles and, of course, Empire. Daniels notes that official maps can be 'used and reinscribed to represent many forms of cultural identity, some of which put the nation as a social formation into doubt' (1998: 129) and this may also be said of the strong equivocation of national museum cartography in mid-nineteenth-century London. Here knowledges and cultures of the 'nation' were often fitted into the ostensibly neutral territories of disciplinary taxonomy rather than into overtly ideological expressions of identity. Meanwhile, the intrinsic value of native art to wider audiences including working people was uncertain, even where notions of the national home and individual contentment were invoked. But, as I hope this chapter has suggested, such representations are animated by the ambiguous relations between museums, places and cultures, and they both reflect and organize the varying routings between home and elsewhere, self and other, past and future, individual and nation.

Note

1 House of Commons Hansard, 1832, available at http://hansard.millbanksystems.com/commons (accessed 27 January 2010).

Bibliography

Alberti, S. (2008) 'Constructing nature behind glass', *Museum and Society,* 6 (2): 73–97.

Bal, M. (1996) *Double Exposures: The Practice of Cultural Analysis*, London: Routledge.

Baudrillard, J. (1988) 'Simulacra and simulations', in M. Poster, *Jean Baudrillard, Selected Writings*, Stanford: Stanford University Press.

Bennett, T. (1995) *The Birth of the Museum: History, Theory, Politics*, London: Routledge.

Bennett, T. (2006) 'Exhibition, difference, and the logic of culture', in I. Karp, C. A. Kratz, L. Szwaja and T. Ybarra-Frausto (eds) *Museum Frictions: Public Cultures/Global Transformations,* Durham, NC, and London: Drake University Press, pp. 46–69.

Black, J. (1997) *Maps and Politics*, Chicago: University of Chicago Press.

Brooks, C. (1998) 'Introduction: Historicism and the nineteenth century', in V. Brand (ed.) *The Study of the Past in the Victorian* Age, Oxford: Oxbow Books, pp. 1–20.

Caygill, M. (1997) 'Franks and the British Museum: The cuckoo in the nest', in M. Caygill and J. Cherry (eds) *A. W. Franks: Nineteenth-century Collecting and the British Museum*, London: British Museum Press, pp. 51–114.

Cosgrove, D. (ed.) (1999) *Mappings*, London: Reaktion Books.

Cosgrove, D. (2007) 'Mapping the world', in J. R. Ackerman and R. W. Karrow Jr. (eds) *Maps: Finding our Place in the World*, Chicago: University of Chicago Press, pp. 65–158.

Cunningham, H. (1981) 'The language of patriotism 1750–1914', *History Workshop Journal*, 12 (1) 8: 8–33.

DaCosta Kauffmann, T. (2005) 'Introduction', in T. DaCosta Kauffmann and E. Pilliod (eds) *Time and Place: The Geohistory of Art*, Aldershot: Ashgate, pp. 1–20.

Daniels, S. (1998) 'Mapping national identities: The culture of cartography, with particular reference to the Ordnance Survey', in G. Cubitt (ed.) *Imagining Nations*, Manchester: Manchester University Press, pp. 112–31.

Duncan, C. (1995) *Civilizing Rituals: Inside Public Art Museums*, London: Routledge.

Eastlake, C. L. and Wornum, R. N. (1853) *Plan for a Collection of Paintings, Illustrative of the History of the Art*, published as an appendix to the Parliamentary Paper *Report, Proceedings and Minutes of Evidence of the Select Committee Appointed to Inquire into the Management of the National Gallery; also to Consider in what Mode the Collective Monuments of Antiquity and Fine Art Possessed by the Nation may be most Securely Preserved, Judiciously Augmented, and Advantageously Exhibited to the Public*, London.

Ferguson, B. W. (1996) 'Exhibition rhetorics: Material speech and utter sense', in R. Greenberg, B. W. Ferguson and S. Nairne (eds) *Thinking about Exhibitions*, London: Routledge, pp. 175–90.

Gieryn, T. F. (1999) *Cultural Boundaries of Science: Credibility on the Line*, Chicago: University of Chicago Press.

Ingold, T. (2007) *Lines: A Brief History*, London: Routledge.

Karrow, R. W. Jr. (2007) 'Introduction', in J. R. Ackerman and R. W. Karrow Jr. (eds) *Maps*, Chicago: University of Chicago Press, pp. 1–18.

King, G. (1996) *Mapping Reality: An Exploration of Cultural Cartographies*, London: Macmillan.

King, J. C. H. (1997) 'Franks and ethnography', in M. Caygill and J. Cherry (eds) *A. W. Franks: Nineteenth-century Collecting and the British Museum*, London: British Museum Press, pp. 136–59.

Kirshenblatt-Gimblett, B. (1998) *Destination Culture: Tourism, Museums and Heritage*, Berkeley and Los Angeles: University of California Press.

Lestringant, F. (1994) *Mapping the Renaissance World: The Geographical Imagination in the Age of Discovery*, Berkeley: University of California Press.

Lumley, R. (1988)*The Museum Time Machine: Putting Cultures on Display*, London: Routledge.

Macdonald, S. (2003) 'Museums, national, postnational and transcultural identities', *Museum and Society*, 1(1): 1–16.

MacGregor, A. (1997) 'Collectors, connoisseurs and curators in the Victorian age', in M. Caygill and J. Cherry (eds) *A. W. Franks: Nineteenth-century Collecting and the British Museum*, London: British Museum Press, pp. 6–33.

MacGregor, A. (1998) 'Museums and "national antiquities" in the mid-nineteenth century', in V. Brand (ed.) *The Study of the Past in the Victorian* Age, Oxford: Oxbow Books, pp. 1–20.

Mack, J. (1997) 'Antiquities and the public: The expanding museum', in M. Caygill and J. Cherry (eds) *A. W. Franks: Nineteenth-century Collecting and the British Museum*, London: British Museum Press, pp. 34–50.

Mason, R. (2006) 'Cultural theory and museum studies', in S. Macdonald (ed.) *A Companion to Museum Studies*, London: Blackwell, pp. 17–32.

Parliamentary Paper (1850), *Report, Proceedings and Minutes of Evidence of the Select Committee Appointed to Consider the Present Accommodation Afforded by the National Gallery; and the Best Mode of Preserving and Exhibiting to the Public, the Works of Art Given to the Nation or Purchased by Parliamentary Grants*, London.

Pickles, J. (2004) *A History of Spaces: Cartographic Reason, Mapping, and the Geo-coded World*, London: Routledge.

Poulantzas, N. (1980) *State, Power, Socialism*, London: Verso.

Robertson, B. (2004) 'The South Kensington Museum in context: An alternative history', *Museum, and Society*, 2(1): 1–14.

Saumarez-Smith, C. (1997) 'National consciousness, national heritage and the idea of "Englishness"', in M. Baker and M. Richardson (eds) *A Grand Design: The Art of the Victoria and Albert Museum*, New York and Baltimore: Harry N. Abrams Incorporated and the Baltimore Museum of Art, pp. 275–83.

South Kensington Museum (1857) *Guide to the South Kensington Museum*, London.

Storr, R. (2007) 'Show and tell', in P. Marincola (ed.) *What Makes a Great Exhibition?*, London: Reaktion Books, pp. 14–31.

Taylor, B. (1999) *Art for the Nation: Exhibitions and the London Public 1747–2001*, Manchester: Manchester University Press.

Waagen, G. F. (1853) 'Thoughts on the new building to be erected for the National Gallery of England, and on the arrangement, preservation, and enlargement of the collection', *The Art Journal*, 101–03, 121–25.

Wainwright, C. and Gere, C. (2002) 'The making of the South Kensington Museum III', *Journal of the History of Collections*, 14(1): 45–61.

Warf, B. and Arias, S. (eds) (2009) *The Spatial Turn*, London: Routledge.

Whitehead, C. (2005) *The Public Art Museum in 19th Century Britain*, Aldershot and Burlington, VT: Ashgate.

Whitehead, C. (2006) 'Aesthetic otherness, authenticity and the roads to museological appropriation: Henry Cole's travel writing and the making of the Victoria and Albert Museum', *Studies in Travel Writing*, 10(1): 1–26.

Whitehead, C. (2009) *Museums and the Construction of Disciplines: Art and Archaeology in 19th Century Britain*, London: Duckworth Academic.

Wornum, R. N. (1856) *The Epochs of Painting Characterized*, London: William Clowes & Son.

Chapter 7

PLACING BRITAIN IN THE BRITISH MUSEUM

Encompassing the Other

Chris Wingfield

ESTABLISHED IN 1753, the British Museum predates the 'proliferation of museums in the nineteenth century' which Macdonald (2003: 1) has linked to the nation-state as an 'identity colossus'. Throughout its history, the British Museum has shied away from presenting a national narrative, preferring instead the identity discourse of 'civilization'. Although the concept of 'civilization' is not itself inherently national, it is nevertheless linked to particular political histories. Norbert Elias (1939: 6) once suggested that for the English, the word 'civilization' represented 'the sum of the national self-image'. It expressed the 'self-assurance' possessed by nations 'which have long expanded outside their borders and colonized beyond them' (Elias 1939: 5). However, I argue here that the notion of civilization in Victorian England excluded certain periods of the British past just as it excluded people from remote parts of Britain's Empire, and even certain residents of the British Isles. In 1866, many of these historical and geographical 'Others' were incorporated into a new Department of British and Medieval Antiquities and Ethnography at the British Museum. Far from abandoning the museum's narrative of civilization, this move sought to shift the way in which the civilized 'self' related to its 'Others'. The new department was part of a series of strategic moves to gain an institutional foothold for proponents of a new science. Ethnology and the related field of prehistoric archaeology,[1] as they developed together in the 1860s, sought to encompass the 'uncivilized' as a part, and largely an early part, of the history of civilization.

Recent statements on the 'Universal Museum' by the British Museum's current director, Neil MacGregor, might be seen as a contemporary re-visioning of the museum's ambition to tell the story of civilization. Even clearer are the words of a sign displayed at the entrance to the British Library, an offshoot from the museum, which declares that the collections include material 'from all over the world and from all periods of civilization'. The question this begs is whether periods of history and parts of the world that are not represented in the British Library's collections are by definition not civilized? The question of how to treat cultural materials that don't meet contemporary definitions of 'civilization' also raised its head in 1836 when responsibility for the British Museum's ethnographic collections and 'artificial curiosities' moved from the Natural History to the Antiquities department (Caygill 1997: 61).

'Confounding all the unities of time and place'

On 19 September 1846, a note in *The Athenaeum* announced that 'The British Museum is again open to the public; upstairs a new department has been formed called the "Ethnographical room" in which are deposited the various articles brought from the South Seas and elsewhere' (Anon. 1846). Two years later, it was described:

> Never was such a disgraceful jumble of things seen, even in a local museum supported by voluntary contributions and regulated by a batch of half-educated provincial antiquaries . . . All our early collectors, Tradescant, and Ashmole, and Thoresby aimed at something like arrangement; but here, in the nineteenth century and in a national museum, we have a collection confounding all the unities of time and place and only worthy of a retired dealer in marine stores. And yet, this is one of the first rooms a foreigner is obliged to see on entering the British Museum.
>
> (Anon. 1848; Altick 1978: 298)

Displayed together was material from many different parts of the world and periods of history, but was nevertheless united under the title 'ethnographical', a term which carries with it etymological associations of 'Otherness'.[2] This 'Ethnographical Room' seems to have functioned in the mid-nineteenth-century British Museum as a location for the anomalous. However, what stands out as particularly curious, when compared to present-day ethnographic collections, is that in the 1840s this label could be used to denote British antiquities from distant times, just as much as material from distant places. Marjorie Caygill (1998: 31) has suggested that 'By 1850 all the antiquities of ancient Britain and Gaul could be collected in four cases in one room, with a mere thirteen more cases for later British and Medieval antiquities.' It seems that this room was the 'Ethnographical Room' described above, since a letter from a donor in October 1850 complained that 'the only room in which the British antiquities are placed is filled with clubs from the South Seas, and trifling objects from China and elsewhere' (Caygill 1997: 60).

Arthur MacGregor (1998) has revealed the problematic position of 'National Antiquities' in mid-nineteenth-century Britain. He notes that in 1845, Thomas Wright (1845: 149), concerned at the destructive effect of 'Railways and other public works', argued that 'It is discreditable to the government of this country that we have no museum of national antiquities.' MacGregor also quotes a correspondent to *The Builder* who expressed regret that no place for national antiquities had been found in the new building at the British Museum, suggesting that 'Such a collection would be of higher value, and of less expense, than the blocks of granite brought from Egypt at so much trouble and cost, which, however, singular as curiosities have no claim upon the delineator of national manners.' The campaign for the establishment of a national collection was taken up in antiquarian circles at the time and forwarded by members of the British Archaeological Association and the Archaeological Institute, both founded in the 1840s (Westherall 1998). There followed questions in Parliament and a Royal Commission (1849–1850) which determined that a collection of national antiquities should be established at the British Museum. The Commissioners looked to the National Museum of Denmark, in Copenhagen, as a model (Caygill 1997: 60) and in 1851, Augustus Wollaston Franks was appointed to fulfil the task of emulating this. As he took office, complaints and frustrations continued to be voiced: 'in vain will you seek, within the walls of that now gigantic building, any collection of British remains' (MacGregor 1998: 128). In trying to explain this curious situation when compared with practices at the National Museum in Budapest (established in 1818), the National Museum in Copenhagen (founded in 1807), or even the Germanisches Nationalmuseum (established 1852), MacGregor (1998: 125) has suggested that the 'imperatives of nationalism' in Britain may have been diluted by 'its insular character and comparatively long-established political stability'. However, it seems likely that the obstructions raised by the Trustees of the British Museum that are documented in MacGregor's article, even when they were offered collections of British antiquities, cannot be adequately explained by an absence of nationalism or even by the complacency of these Trustees (MacGregor 1998: 128). Rather, it seems to provide evidence that there was a contradictory ideology at work.

The pediment sculpture above the main entrance to the new British Museum building might be read as a key for understanding the material found inside. It was made by the sculptor Sir Richard Westmacott on the theme of 'The Progress of Civilization' (Caygill 1998: 34) (Figure 7.1). The Trustees of the British Museum, as recipients of a classical education, and frequently also of noble birth, seem to have understood themselves as the inheritors of an ancient tradition of civilization.

David Cannadine (2001: 8) has suggested that the British Empire was so inflected by an ideology of status and hierarchy that it enabled the elites of colonized countries to be treated as social equals by the elites of imperial British society. He has also suggested that the manner in which status and prestige were expressed in the British Empire was through 'ornamentalism' – 'hierarchy made visible, immanent and actual' (Cannadine 2001: 122) through architecture, elaborate codes of dress and a complex series of titles distributed by a reinvented royalty. Just as this seems to have enabled the recognition of forms of status and ornament in other places, so it seems to have allowed the recognition of similar forms of ornament and display in the ancient past. The 'ornamentalism' of ancient Greece, Rome and later Babylon

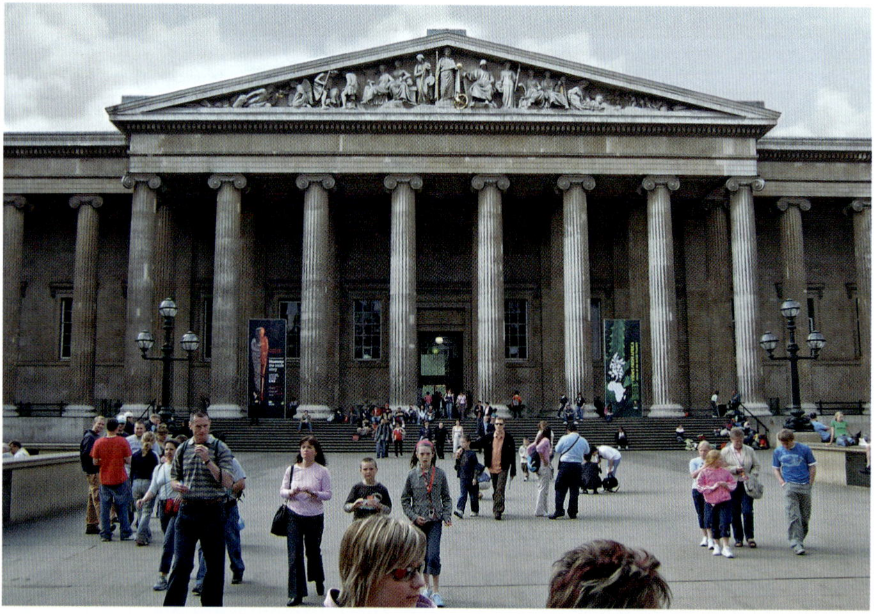

Figure 7.1 The South Front of the British Museum, showing Westmacott's pediment sculpture *The Rise of Civilization*. Photo: Simon Knell

and Egypt, qualified these ancient remains to be regarded as 'civilizations'. The British Museum's classical edifice proclaims the museum's importance in enabling the creation of 'civilization' as an 'imagined community' (Anderson 1983). South Seas objects and British antiquities were equally problematic and anomalous in these imaginings, as exemplars of the 'uncivilized'. Attempts to include them in the museum were actively resisted and frustrated by the Trustees and employees of the museum.

Barbarian antiquities

The changing departmental structures of multidisciplinary museums illustrate shifting modes of institutional classification (Wingfield 2006; Douglas 1986). The British Museum's Department of Antiquities, established in 1807, was split in 1860 into the Department of Greek and Roman Antiquities, Coins and Medals and the Department of Oriental Antiquities, following the retirement of the Keeper, Edward Hawkins (Caygill 1998: 18, 38).[3] It was the Department of Oriental Antiquities which, alongside objects from Egypt and Assyria, took possession of British and medieval antiquities, as well as Ethnography, despite the recommendations from the 1850 Royal Commission. This reorganization appears to suggest that Greece and Rome should be separated from all 'others', whether in time or space. It might have been more accurate to call the new department 'Barbarian

Antiquities' since those who controlled the museum seem to have adopted a Classical outlook which viewed the peoples from the North and East of the Mediterranean as barbarians.

Much of the responsibility for this reorganization lies with Anthony Panizzi, Principal Librarian (and effectively Director) of the British Museum between 1856 and 1866 (Caygill 1997: 66; Harris 2004). An Italian patriot in exile, he was keen to focus the efforts of the museum on the library and to limit the 'collections of antiquities to classical or pagan art' (Caygill 1997: 65). Though one might see this in terms of an Italian nationalist agenda, it seems that it also chimed well with the classically educated Trustees. Panizzi's 1857 appeal to the Trustees suggested that:

> All that space which is now occupied by medieval antiquities, by what are called British or Irish antiquities and by the ethnological collection, might thus be turned to better account. It does not seem right that such valuable space should be taken up by Esquimaux dresses, canoes and hideous feather idols, broken flints, called rude knives, and so on, to the exclusion of such objects as have been just before mentioned [classical antiquities].
>
> (Caygill 1997: 65)

Panizzi told a select committee from the House of Commons in 1860 that Byzantine, Oriental, Mexican and Peruvian antiquities were best kept in the basement (Mack 1997: 38). Solutions to the problems of overcrowding suggested to this committee included a moratorium on acquisitions of further British and Medieval material and the removal of certain collections from the British Museum altogether (Mack 1997: 38). The Ethnography and Natural History departments were the two main candidates, and in 1861 it was decided that the latter would indeed be moved out of the museum, eventually taking up residence in the new British Museum (Natural History), or Natural History Museum as it became known.

Ethnography remained, and with Panizzi's retirement in 1866, became part of the Department of British and Medieval Antiquities and Ethnography under the keepership of Franks (Caygill 1998: 66). The established narrative of civilization that had once focused exclusively on ancient Greece and Rome, had by this time begun to include the origins of cities and states in the ancient Near East. However, it still excluded large areas of the world and periods of history as 'uncivilized' and therefore 'Other'. The *Handy-Book of the British Museum* published in 1870 still devoted just six pages to the Celtic, Roman and Saxon collections from Britain, three pages to the prehistoric and ethnographic collections, eight pages each to the medieval collection with coins and medals, but devoted 80 pages to the Assyrian, 144 to the Egyptian and 129 to the Classical antiquities (Caygill 1998: 31). Nevertheless, the formation of the new department was a significant move and gave departmental and institutional recognition to a range of previously marginalized subjects and materials.

The emergence of an anthropological perspective

The dominant figure at the British Museum in the last quarter of the nineteenth century was Augustus Wollaston Franks, Keeper of the Department of British and Medieval Antiquities and Ethnography. Appointed in 1851 to establish a collection of national antiquities (Caygill and Cherry 1997), the well-connected Franks had established himself in the study of church architecture and archaeology while a student at Cambridge in the 1840s. He organized the 1850 Medieval Exhibition and was allied with the mid-nineteenth-century movement which romanticized the medieval world and Gothic architecture.[4] At the time of his appointment to the British Museum, Franks had shown little evidence of interest in ethnography. Clashing with Panizzi over his collecting activities, he had failed to be appointed Keeper in 1860 (Caygill 1997: 64–67). This setback was overcome in 1866, in part because of his role in acquiring the Henry Christy collection for the museum. It seems Franks used the significance of this collection and the financial resources which came with it as a bargaining tool in cementing his position at the British Museum.

Henry Christy (1810–1865) did not share Franks's Eton and Cambridge education. As a Quaker he was excluded from these institutions. The Christys had amassed considerable wealth as bankers, and Henry Christy had travelled widely, including to Latin America in the late 1850s with the anthropologist and fellow Quaker, E. B. Tylor. Christy was one of many Quakers actively involved in the Ethnological Society of London (Stocking 1971; Cantor 2005). A strong religious commitment to the singularity of the human species, called *monogenism,* had been emphasized during the Quaker campaigns in the anti-slavery movement. The legacy of this commitment meant that accounting for the diversity observable among humans became a central problem for many Quakers (Cantor 2005: 113). A turning point occurred in Christy's engagement with anthropology in 1858, when the discovery of chipped stone tools along with bones of extinct animals beneath a stalagmite floor in Brixham Cave suggested that humans had a history far longer than previously imagined (Harrison 2004). This sparked what Trautmann (1992) has called the 'Revolution in Ethnological Time' leading to an abandoning of a short biblical chronology and the adoption of a significantly longer timeframe in which to understand human history. This made prehistoric archaeology a field of considerable interest at the time.

By 1863, when he began a series of excavations in the French Dordogne region with Edouard Lartet, Christy already possessed significant collections. When he died two years later, his collection represented the cutting edge of archaeological science. By then archaeology had shifted some of its focus beyond the British national and medieval past, to more remote antiquity: the antiquity, so it seemed, of all humanity, and therefore also of civilization. Christy had attempted to understand the remains recovered from his excavations through a comparison with ethnographic artefacts derived from the furthest corners of the British Empire. In 1862, he had published a catalogue of his collection written by the Carl Ludvig Steinhauer, who had served as assistant to C. J. Thomsen at the National Museum of Denmark (King

1997: 136). The catalogue included 'Ancient and Modern stone implements, and other Weapons, tools, and utensils of the Aborigines of various countries' (Steinhauer 1862) (Figure 7.2). Augustus Henry Lane Fox (later Pitt Rivers) got to know Christy's collection, probably during the 1850s and early 1860s, and states that it was originally mainly prehistoric, but was extended to the 'industry of modern savages' (Lane Fox 1879: xiii). Ethnography had, by 1865, found a new role in archaeological science. In that year, John Lubbock evangelized this new science in his *Pre-historic Times as Illustrated by Ancient Remains and the Manners and Customs of Modern Savages.* A letter of thanks from Franks suggests that Lubbock sent him a copy

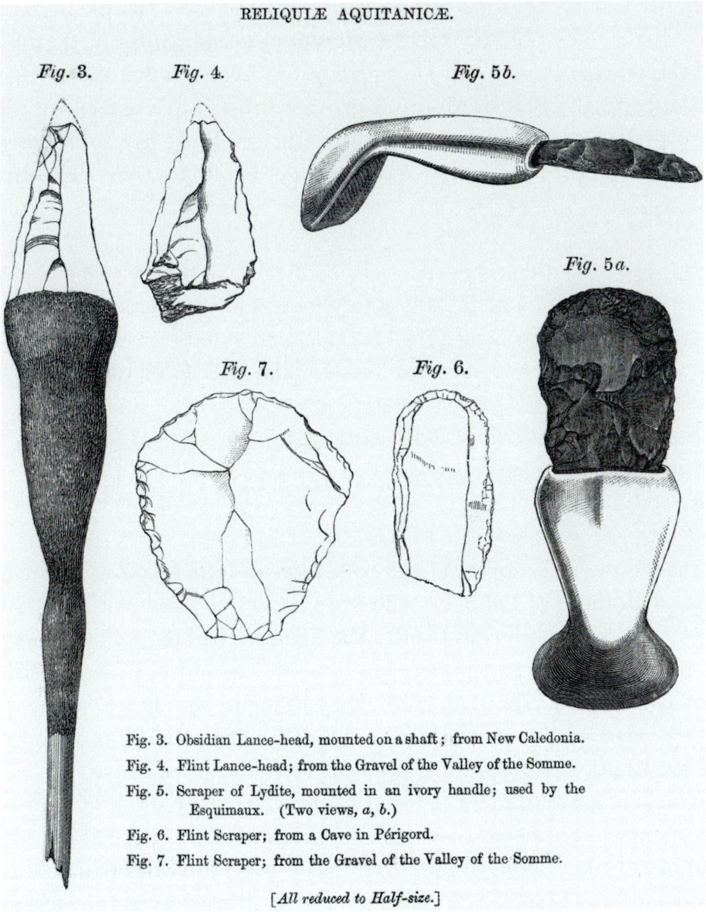

Figure 7.2 Plate prepared for Henry Christy showing the similarities between Inuit and New Caledonian stone implements and those found in the French Caves. From *Reliquiæ Aquitanicæ: Being Contributions to the Archæology and Palæontology of Périgord and the Adjoining Provinces of southern France* (Lartet and Christy 1875). Photo: Copyright The Pitt Rivers Museum

of this book.[5] This was less than three weeks after Christy had died, and Franks asked Lubbock whether he had heard any more about Christy's bequest. They had both been named as trustees along with Joseph Hooker, who became the director of Kew Gardens in that year, and Daniel Hanbury, a fellow Quaker.

In June 1865, Franks's future at the British Museum was uncertain since he had expressed a desire to resign, but he was persuaded to stay for a year without salary (Caygill 1997: 68). However, by 3 July, the date of his next letter to Lubbock,[6] Franks seems to have become actively involved with the Christy bequest, undertaking to visit the stored collections in Paris as well as making arrangements to forward Christy's Egyptian collection to Stockport. In the *Apology* of his life, Franks states that 'When Mr Christy told me that he had appointed me a trustee he said that if I wished the main collection to go to the Museum I must fight for it' (Franks 1997: 320), and this is what Franks did. He suggested to Lubbock that they 'offer it in the first instance to the British Museum provided that the Trustees would undertake to find proper space' and also that the Christy Trust pay something towards the annual £220 rent for Christy's home to 'get breathing time'. He also suggested that:

> If we play our cards right this collection may be the means of getting the Ethnological Collns put on a more satisfactory footing at the Museum or something established in this place. The multiplication of museums is not however very desirable & I hope that the Trustees may be induced to give more space.
>
> Panizzi is about to retire & many changes may take place which may lead to an improvement in those portions of the collection in which you & I take interest.[7]

It is noteworthy that Franks refers primarily to collections as 'ethnological' and also aligns his interests with those of Lubbock. Christy's collection came with £5,000 for its continued development after his death. The Trustees had to work out how best to use this money. Franks and Lubbock's 'scheme' would cost the British Museum £250 to £300 a year. The collection, they thought, should be valued at £3,000 to £4,000.[8] They operated on the principle that 'it would be unwise by making too miserly an offer to jeopardize the really most important part of our Trust viz. the permanent establishment of the collection'. Hooker and Franks proposed to Hanbury and Lubbock that they offer the museum £100 per annum for five years instead of £60 they thought they could raise from interest. Franks was anxious that 'any delay should take place in our offer to the Museum' and suggested that the offer be made to the museum with a number of conditions including the publication of a catalogue, and the hiring of Christy's home for five years, at the end of which if the collections were 'not removed to, and exhibited in, the Museum' the Christy Trustees would have the option of extending the period or resuming possession of the collection.

The British Museum Trustees agreed to accept the Christy bequest on 18 November 1865 (Cook 1997: 120), but the collection stayed at Christy's home, 103 Victoria Street, until the museum removed the natural history collections to their new home in the early 1880s. By the time it was amalgamated with the British Museum's ethnographical collections, Franks had greatly expanded the Christy collection, as well as having built up a considerable accompanying library. The result, he stated, was 'a collection of considerable size, one of the best in Europe, and such as ought to be in the National Museum of a country like England' (Franks 1997: 320).

By March 1866 (Caygill 1997: 69), Franks had applied to have his department separated from Oriental Antiquities and received Panizzi's approval. Franks had argued:

> Ethnography has assumed such a totally different aspect within the last few years; its scientific and historical value has been so fully recognised, that I feel sure that, whether the natural history collections shall be retained at the Museum or not, it will be thought desirable to keep here collections which, when properly arranged, will be highly attractive and popular.
>
> (Caygill 1997: 69)

The Department of British and Medieval Antiquities and Ethnography retained curatorial responsibility for Indian, Chinese and Arabic collections as well as American antiquities but also all non-classical European antiquities, and these included the crucial prehistoric material from the French Caves. Franks himself stated that his department might equally be called the 'Department of Miscellaneous Antiquities' (Caygill 1997: 72). What this underlines is that in many ways, despite its name, the Department became the location at the British Museum for that combination of prehistoric archaeology and ethnography that would in time become anthropology. As such it became instrumental in efforts to establish a global narrative of civilization that would tie together miscellaneous antiquities from across the world.

The rejection of Pitt Rivers

This chapter indicates the manner in which anthropology was forming at the British Museum. It was part of the same 'physical and intellectual process' as the arrival of the Pitt Rivers collection in Oxford, nearly two decades later (Larson 2008). Pitt Rivers, like Christy, displayed ethnographic objects alongside archaeological ones with a view to illustration, arguing in a catalogue of 1879:

> Acting upon the principle of reasoning from the known to the unknown I have commenced this catalogue with the specimens of the arts of existing savages, and have employed them as far as possible to illustrate the relics

> of primeval men, none of which, except those constructed of the more imperishable materials, such as flint and stone, have survived to our time. All the implements of primeval man that were of decomposable materials having disappeared, can be replaced only in imagination by studying those of his nearest congener, the modern savage.
>
> (Lane Fox 1879: xiii)

Just as the donation of a collection of objects to the University of Oxford enabled the establishment of a Lectureship in Anthropology, so the negotiations surrounding the Christy bequest enabled the creation of a museum department in which anthropology would have an institutional home at the British Museum, even if the provision of lectures was not possible under the terms of the Trust. These are not, however, mere parallels since the two events are connected in a number of ways. Pitt Rivers was heavily influenced by Henry Christy and his collection, while the lecturer appointed in Oxford was E. B. Tylor, who also had close intellectual connections with Christy.

Furthermore, in 1880, before the Pitt Rivers collection was donated to Oxford, it was offered, like Christy's collection before it, to the nation. This involved a complex negotiation with A. W. Franks, who was then preparing to move the Christy collection to Bloomsbury, and might have used the additional weight of the Pitt Rivers collection to bolster his institutional base. Pitt Rivers resisted the amalgamation of his collection with those of the British Museum and tried to argue that 'far from it being antagonistic to the B.M., it will be a most useful adjunct'. Pitt Rivers explained to Franks that if space could not be found at South Kensington for the continued display of his collection, he planned to build his own museum in or close to London and emphasized that:

> I shall have received no encouragement to leave anything to the nation. If the nation will not accept my offer now on account of a dog in the manger rivalry between the two departments I shall take good care it never gets anything from me. Science is cosmopolitan and I had rather leave everything to the United States.[9]

A committee was appointed to consider Pitt Rivers's offer. It consisted of Lubbock, Franks, George Rolleston, Linacre Professor of Anatomy and Physiology at Oxford and Thomas Huxley (Bowden 1991: 51). In July, Pitt Rivers wrote to Franks again, reassuring him that 'I am very glad you are going to be on the committee' and commenting on a suggestion that his collection might remain at South Kensington, but be attached to the British Museum, suggesting that he did not mind this as long as he would be allowed control of the collection during his life.[10] Pitt Rivers argued that otherwise government control would mean that 'fifty views and intents would have to be reconciled before any step could be taken' and suggested that Hazelius's 'Museum at Stockholm is a case in point . . . if he had been fettered by the government he could never have carried it out.' Pitt Rivers then reassured Franks that following his death the collection would be left to the nation without 'any special stipulation', suggesting it would be 'mischievous to hamper the future with the ideas of the present'. Although the committee reported favourably in 1880, Pitt Rivers's conditions for donation were ultimately found to be unacceptable.

In March 1881, after Pitt Rivers had offered his collection to Oxford, Franks was asked to write a letter that could be read at the University's Hebdomadal Council 'expressing an opinion as to the value of the collection and its usefulness at Oxford'.[11] What this underlies is the close connections of those who were working for the disciplinary establishment of anthropology in the second half of the nineteenth century, with prehistoric archaeology and ethnology as constituent parts of the new subject. Franks, Lubbock, Tylor, Pitt Rivers, Huxley and Rolleston all corresponded frequently with one another as well as serving alongside each other on the committees of the many societies to which they all belonged. Thus they formed part of a network which had its roots in the Ethnological Society of London in the 1860s (Stocking 1971; Owen 2008: 220–21) and of which Henry Christy had been a prominent member.

Cosmopolitan science and nationalist enthusiasms

In 1984, Bruce Trigger suggested that Britain in the 1850s was the location for the development of the first tradition of 'imperialist' or 'world-oriented' archaeology. He argued that:

> industrialization promoted by individual enterprise had greatly strengthened the middle class both economically and politically. By offering evidence that such progress was the continuation of what had been going on more slowly throughout human history, prehistoric archaeology bolstered the confidence of the British middle class and strengthened their pride in the leading role of Britain in this process.
>
> (Trigger 1981: 141–42)

Trigger contrasted this with 'nationalist' archaeology, which he identified as being archaeology's most common form, the primary function of which 'is to bolster the pride and morale of nations and ethnic groups' (Trigger 1984: 360). For Trigger these categories emerged from his examination of the 'similarities and differences in the questions that prehistoric archaeologists ask and the answers that they are predisposed to accept as reasonable in different parts of the world and under changing social conditions' (Trigger 1984: 355). Archaeology he suggested was 'strongly influenced by the position that countries and regions in which it is practised occupy within the modern world-system' (Wallerstein 1974: 3–11; Trigger 1984: 356). Trigger's types of archaeology were, he admitted, ideal types (Trigger 1984: 358) and as such they 'capture only certain broad features of very complex situations' and 'also fail to express the varying intensity with which the characterisations of each type are realised in specific cases'.

While it is certainly true that Britain in the last third of the nineteenth century was increasingly imperialistic, its place in the world system does not fully account for the eclipsing of a nationalist focus of archaeology in the 1840s by a world-oriented focus in the 1860s, in the case of Franks and the British Museum. Nationalist archaeology flourished in the 1840s when the new middle classes were arguably at their most confident and Britain was at its most unchallenged following the Napoleonic wars.[12] In Britain it seems that these nationalist influences were partly mitigated by the cosmopolitanism of people like Franks who travelled regularly to continental Europe to converse with colleagues. This was the cosmopolitanism of science which Pitt Rivers referred to in 1880 and it allowed many questions of local belonging to be transcended. Nationalist trends were also partly mitigated by the nonconformist or evangelical religious backgrounds of many of the new middle classes, such as Henry Christy, which committed them to a focus on the singularity of humanity over more racially determined positions. However, as we have seen in the case of Panizzi and the British Museum Trustees to whom he appealed, mid-century nationalism with a focus on northern Europe's non-classical past was also met with resistance because of the pre-existing focus for elite identity in Britain on 'civilization' in the mid-nineteenth century, which broadly built on an idealized image of Greece and Rome. The British Empire was largely modelled on classical precedents, and the classical educations of many of those who were instrumental in running Britain and its Empire enabled them to think of themselves not only as 'civilized' but as the inheritors of an ancient tradition.

The establishment of a Department of British and Medieval Antiquities and Ethnography ultimately did not depend on the debates in the 1840s about the need for a department of national antiquities, which had largely been pursued by the new middle classes. It depended instead on the development in the early 1860s of an approach that sought to explore the ultimate origins and development of civilization. Just as Darwin's *Origin of Species* depended on a principle of continuity between species in the evolutionary process, thereby challenging the distinction between humans and other animals, so the related new subject of anthropology proposed a principle of continuity by which civilization, or culture as it would be increasingly known (Tylor 1871), changed by small accretions. This assertion challenged the fundamental social and classificatory distinction between the civilized and uncivilized. However, it did not necessarily mean abandoning a narrative of civilization at the British Museum. It did mean that whereas previous attempts had been made to exclude the 'uncivilized', collections of prehistoric and medieval antiquities, as well as ethnographic collections, could become a means of illustrating the long historical process by which the current state of civilization was ultimately achieved. This change did not stop the British elite self-identifying as civilized, but it did entail a shift in what Baumann and Gingrich (2004) have called 'Grammars of identity/alterity'. These they have suggested can be used as 'guides as to how different discourses order the relationships between Self and Other' (Baumann 2004: 19).

The focus on the self as civilized in the British Museum in the first half of the nineteenth century can be understood as an 'orientalizing grammar' functioning through a reliance on the 'uncivilized' as a negative mirror image. The mid-nineteenth-century Romantic movement, by contrast, sought to emphasize the positive aspects of that mirror image.[13] The later 'ethnological' movement of the 1860s seems instead to have operated using a 'grammar of encompassment'; incorporating the uncivilized ethnographic and prehistoric past as a constitutive part of the civilized self – the earliest chapters in a new narrative of civilization. This discourse was nevertheless part of an effort to define the civilized self. Dumont (1970) has argued in relation to the Indian caste system that this mode of defining 'the self' by encompassing 'the other' is the ultimate statement of hierarchy. He illustrated his argument with the biblical account of Eve created from Adam's rib in Genesis 2, suggesting mythologically that the female 'other' is a (sub)part of the dominant male self. This suggests that uncivilized material could be encompassed rather than excluded because it was so unthreatening that it could be made a (sub)part of the civilized self – its ultimate origins. It is, as Trigger would no doubt point out, presumably no coincidence that imperialism as a political mode of encompassment, and the Department of British and Medieval Antiquities and Ethnography both flourished in the period between 1866 and 1921 when the British Empire operated at its greatest geographical extent. The 'encompassing' grammar utilized by this new narrative of civilization promoted contemporary 'civilized' Britain not simply as the inheritor of a tradition from Greece and Rome, but as the inheritor of a tradition that encompassed the whole world and all of history. This was the Whig interpretation of history (Butterfield 1931) on a global scale. British 'Civilization' was no longer defined by contrast with the uncivilized at the fringes of empire and in the remote past, but rather by positioning itself as the culmination of global history and the most developed form of 'civilization' the world had ever seen.

Notes

1 Anthropology only became more widely used as an umbrella term for these subjects after 1871. Stocking, G. W. (1971) 'What's in a name? The origins of the Royal Anthropological Institute (1837–71)', *Man, New Series,* 6, 3: 369–90.

2 The first definition of the word 'ethnic' given by the online Oxford English Dictionary is: 'Pertaining to nations not Christian or Jewish; Gentile, heathen, pagan'.

3 In 1836 a separate Department of Prints and Drawings was created out of this at around the time the new National Gallery was built.

4 This is widely associated with the names of A. W. N. Pugin and J. Ruskin.

5 A. W. Franks to J. Lubbock, 24 May 1865, British Library Manuscripts (BL): Avebury Papers MSS Add. 49641 ff. 40.

6 A. W. Franks to J. Lubbock, 3 July 1865 (BL): Avebury Papers MSS Add. 49641 ff.63. Lubbock had been absent from a meeting of the Christy Trustees, apparently due to the death of his father on 20 June 1865.

7 Ibid.

8 A. W. Franks to J. Lubbock, 5 November 1865 (BL): Avebury Papers MSS Add. 49641 ff.102.

9 Gen. Pitt Rivers to A. W. Franks, 27 June 1880, Pitt Rivers Museum Manuscripts (PRM): Foundation Volume.

10 Gen. Pitt Rivers to A. W. Franks, 1 July 1880 (PRM): Foundation Volume.

11 H. N. Moseley to A. W. Franks, 30 March 1881 (PRM): Foundation Volume.

12 Much of the inspiration for the interest in 'national antiquities' in the 1840s was inspired by developments in Denmark, but were also influenced by German romanticism.

13 This is illustrated most clearly in Pugin's *Contrasts* which cast contemporary industrial civilization in a negative light by comparison with an ideal medieval image centred around a stable social order and the church. Pugin, A. W. N. (1841) *Contrasts; Or, A Parallel Between the Noble Edifices of the Middle Ages and Corresponding Buildings of the Present Day,* London.

Bibliography

Altick, R. D. (1978) *The Shows of London,* Cambridge, MA: Belknap Press.

Anderson, B. R. O. G. (1983) *Imagined Communities: Reflections on the Origin and Spread of Nationalism,* London: Verso.

Anon. (1846) 'The British Museum', *The Athenaeum*, 986 (19 September): 957.

Anon. (1848) 'The British Museum', *The Athenaeum*, 1061 (26 February): 216–17.

Baumann, G. (2004) 'Grammars of identity/alterity: a structural approach', in G. Baumann and A. Gingrich (eds) *op. cit.*, pp. 18–50.

Baumann, G. and Gingrich, A. (eds) (2004) *Grammars of Identity/Alterity: A Structural Approach,* New York and Oxford: Berghahn.

Bowden, M. (1991) *Pitt Rivers: The Life and Archaeological Work of Lieutenant-General Augustus Henry Lane Fox Pitt Rivers, DCL, FRS, FSA,* Cambridge: Cambridge University Press.

Brand, V. (ed.) (1998) *The Study of the Past in the Victorian Age,* Oxford: Oxbow for the British Archaeological Association and the Royal Archaeological Institute.

Butterfield, H. (1931) *The Whig Interpretation of History,* London: G. Bell & Sons.

Cannadine, D. (2001) *Ornamentalism: How the British Saw their Empire,* London: Allen Lane.

Cantor, G. N. (2005) *Quakers, Jews, and Science: Religious Responses to Modernity and the Sciences in Britain, 1650–1900,* Oxford: Oxford University Press.

Caygill, M. (1997) 'Franks and the British Museum – the cuckoo in the nest', in M. Caygill and J. Cherry (eds) *op. cit.*, pp. 51–114.

Caygill, M. (1998) *The Story of the British Museum,* London: British Museum Press.

Caygill, M. and Cherry, J. (1997) *A. W. Franks: Nineteenth-century Collecting and the British Museum,* London: British Museum Press.

Cook, J. (1997) 'A curator's curator: Franks and the stone age collections', in M. Caygill and J. Cherry (eds) *op. cit.*, pp. 115–29.

Douglas, M. (1986) *How Institutions Think,* Syracuse, NY: Syracuse University Press.

Dumont, L. (1970) *Homo Hierarchicus: The Caste System and its Implications,* London: Weidenfeld & Nicolson.

Elias, N. (1939) *The Civilizing Process: The History of Manners,* republished 1978, New York: Urizen Books.

Franks, A. W. (1997) 'The apology of my life', in M. Caygill and J. Cherry (eds) *op. cit.*, pp. 318–24.

Hall, C. (2002) *Civilising Subjects: Metropole and Colony in the English Imagination 1830–1867,* Oxford: Polity.

Harris, P. R. (2004) 'Panizzi, Sir Anthony (1797–1879)' in *Oxford Dictionary of National Biography*, Oxford: Oxford University Press.

Harrison, W. J., revised by A. Bowdoin Van Riper (2004) 'Christy, Henry (1810–1865)', *Oxford Dictionary of National Biography*, Oxford: Oxford University Press.

King, J. C. H. (1997) 'Franks and ethnography', in M. Caygill and J. Cherry (eds) *op. cit.*, pp. 136–59.

Lane Fox, A. (1879) *Catalogue of Colonel Lane Fox's Anthropological Collection,* London: South Kensington Museum.

Larson, F. (2008) 'Anthropological landscaping: General Pitt Rivers, the Ashmolean, the University Museum and the shaping of an Oxford discipline', *Journal of the History of Collections,* 20(1): 85–100.

Lartet, E. A. I. H. and Christy, H. (1875) *Reliquiæ Aquitanicæ; Being Contributions to the Archæology and Palæontology of Périgord and the Adjoining Provinces of Southern France,* London: Williams & Norgate.

Macdonald, S. J. (2003) 'Museums, national, postnational and transcultural identities', *Museum and Society,* 1 (1): 1–16.

MacGregor, A. (1998) 'Antiquity inventoried: Museums and "national antiquities" in the mid-nineteenth century', in V. Brand (ed.) *op. cit.*, pp.125–37.

Mack, J. (1997) 'Antiquities and the public: The expanding museum, 1851–96', in M. Caygill and J. Cherry (eds) *op. cit.*, pp. 34–50.

Owen, J. (2008) 'A significant friendship: Evans, Lubbock and a Darwinian world order', in A. MacGregor (ed.) *Sir John Evans 1823–1908: Antiquity, Commerce and Natural Science in the Age of Darwin,* Oxford: Ashmolean Museum, University of Oxford, pp. 206–29.

Pugin, A. W. N. (1841) *Contrasts; Or, A Parallel Between the Noble Edifices of the Middle Ages and Corresponding Buildings of the Present Day,* London.

Steinhauer, M. C. L. (1862) *Catalogue of a Collection of Ancient and Modern Stone Implements, and of other Weapons, Tools, and Utensils of the Aborigines of Various Countries, in the Possession of Henry Christy, FGS, FLS, &c.*, London.

Stocking, G. W. (1971) 'What's in a name? The origins of the Royal Anthropological Institute (1837–71)', *Man, New Series,* 6(3): 369–90.

Trautmann, T. R. (1992) 'The revolution in ethnological time', *Man, New Series,* 27(2): 379–97.

Trigger, B. G. (1981) 'Anglo-American archaeology', *World Archaeology,* 13: 138–55.

Trigger. B. G. (1984) 'Alternative archaeologies: Nationalist, colonialist, imperialist ', *Man, New Series,* 19(3): 355–70.

Tylor, E. B. (1871) *Primitive Culture: Researches into the Development of Mythology, Philosophy, Religion, Art and Custom,* London: J. Murray.

Wallerstein, I. M. (1974) *The Modern World-system,* London: Academic Press.

Westherall, D. (1998) 'The growth of archaeological societies', in V. Brand (ed.) *op. cit.*, pp. 21–34.

Wingfield, C. (2006) '(Before and) after gallery 33: Fifteen years on at Birmingham Museum and Art Gallery', *Journal of Museum Ethnography,* 18: 49–62.

Wright, T. (1845) *The Archaeological Album; or, Museum of National Antiquities,* London: Chapman & Hall.

Chapter 8

PRODUCING AN ART HISTORY OF THE NATION

The origins of the Finnish National Gallery

Susanna Pettersson

IN OCTOBER 1888, a three-storey stone building called Ateneum was opened to the public in Helsinki city centre. This was to be Finland's first national gallery, and its central focus was a collection assembled by the Finnish Art Society. Built for educational purposes, the national character of the collection grew with the developments in Finnish art. As it did so, a story of Finnish art came into being, nurtured by intellectuals and academics, and influenced by the writings of German philosopher Georg Wilhelm Friedrich Hegel (1770–1831). Members of the Society read and studied many of the same books, and Hegel's books on art and aesthetics were particular favourites. Hegel believed that mankind could learn from its past by looking at art; that every work of art was the product of the thought and spirit of its age (Hegel 2004: 7–33). Hegel preached continuity – a continuous evolution of the arts without breaks – but for the Society this presented a problem. For while the Society aspired to represent this complete history, it possessed limited funds and it knew that Finland lacked its own established national history of art.

The blank canvas

The development of the fine arts in Finland was greatly affected by political upheaval in the early nineteenth century. Swedish losses in the Swedish–Russian War of 1808–1809 saw Finland become a Russian territory. This had a fundamental effect on its economic and political constitution and, more particularly, on its own self-image. The country became swept up in a national movement which was reflected in the arts. The cultural life of the nation associated with literature, theatre and music became focused on societies, the first of which was the Finnish Literature Society, founded in 1831. The Russian court approved of these nationalistic gestures, as they signalled political separation from the old mother country, Sweden.

Despite the development of a flourishing arts scene, the fine arts were slow to establish a similar base. Severed from the royal collections and art academy in Stockholm and distanced from the art treasures possessed by Tsar Alexander I (1777– 1825) in St Petersburg, Helsinki found itself in the wilderness. It possessed not a single art school or exhibition and consequently no art critique or art public. The number of private art collectors could be counted on the fingers of one hand. Put simply, the fine arts played no particular role in Finnish society and were only discussed when members of the aristocracy returned from their Grand Tours of Berlin, Paris and Rome.

This lacuna in the cultural life of Helsinki attracted the attention of a group of well-travelled academics who believed that the Finnish public needed the enlightenment of art. This could be achieved, they felt, by schooling artists, constructing a visual narrative using European models, and building structures and facilities for the display of art. This group founded the Finnish Art Society (*Finska Konstföreningen*) in 1846 using the German model of *Kunstvereins*. One of its first gestures was to establish an art school and a model collection which was to be used by the students.

The Society quickly became a powerful force which defined Finnish art along nationalistic lines. As Anthony Smith observed, culture was the proper tool for nationalists wishing to commemorate or celebrate the nation through the reconstruction of the past (Smith 1991: 92–94). In this sense the Society was composed of 'activists', which included nearly every key figure from academic, political and economical circles in Helsinki. The Chairman of the Society was particularly empowered – a kind of gatekeeper. Ultimately it was he who decided what art to support and what to include in the collection. The story of Finnish art that emerged greatly reflected the Chairman's tastes and ideals, his likes and his dislikes. Certain works would be marginalised or excluded for no other reason.

This leaves us with a number of interesting questions. For example, were specific artistic or philosophical principles followed? What were the expectations of a collection of Finnish art? However, before we can address these questions it is necessary to look at the practical purposes of collection, for the Society did not set out to produce a museum of Finnish art.

A practical collection

When first faced with questions concerning the establishment of an art collection, the board members of the Finnish Art Society had disagreed on how it should be achieved. In the 1840s, the first Chairman of the Society, Baron Carl Johan Walleen (1781–1867), an esteemed politician and experienced art collector, was firmly against starting up a collection of European art. He argued that, given the limitations of the Society's funds, dependent as they were on membership fees, it could not acquire European works of any significance that were already in the market place. He argued that market prices were high and most of the key pieces were already in public or private collections.

Despite these disagreements, the Society started purchasing art, but not for a 'real' art collection. The first works, works by the Norwegian painter Pehr Gustaf von Heideken (Figure 8.1), the French painter Joseph Auguste Desanord the elder and works by unknown artists acquired in 1847 and 1849, were meant for everyday use in the Society's drawing school (Pettersson 2008: 110).

Despite suggestions to the contrary (see Lindström 1963: 5–7), the Society was not yet producing a museum. The collection had been developed to meet the needs of the drawing school (Pettersson 2008: 109). Indeed, the Society's original intention was not to establish a museum; it wished to create an academy for fine arts instead and academies in Paris and Stockholm served as examples. This remained the plan until the 1880s when the Society and others decided to create an institution for both fine arts and applied arts. The plan included public exhibition and art education (Pettersson 2008: 154–70, 206–17).

In 1851, the Society received its first large donation; 28 works, including paintings by Lucas Cranach, Dirck van Delen and Daniel Vertangen, given by board member Baron Otto Wilhelm Klinckowström (1778–1850) (Pettersson 2008: 108–9, 310). By 1854, this 'practical' or 'model' collection had grown to 60 works (Tikkanen 1896: 67; Lindström 1963: 8). It consisted mainly of landscape artists with links to Finnish art. These included paintings by the Norwegian, Hans Gude and historical scenes by the German, Otto Mengelberg. Both had taught Finns in Düsseldorf. The

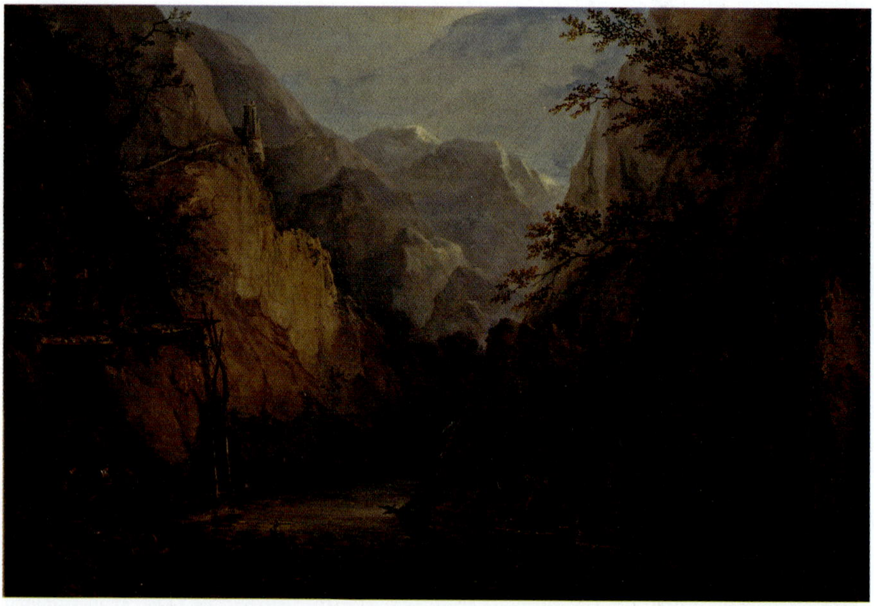

Figure 8.1 Pehr Gustaf von Heideken's *Norsk fjelltrakt* was the first acquisition to the collection of the Finnish Art Society, acquired in 1847. Photo: Hannu Pakarinen, Central Art Archives/The Finnish National Gallery

Society also commissioned works from contemporary Nordic artists such as the Swedes Joseph Magnus Stäck and Lars Theodor Billing. Although it could not afford old masters and never tried to purchase any from the auction houses in Stockholm, Paris, Berlin or London, it did commission young Finnish painters travelling abroad to make copies of key works. By 1868, the Society had acquired nearly 40 of these copies. These were mostly religious paintings and genre pictures originally by Velasquez, Delacroix, Murillo, Rembrandt, Correggio, Jusepe de Ribera, Giovanni Battista Tiepolo and Rubens (Pettersson 2008: 112–13). These copies connected Finnish students to the traditions of the central European art academies. They provided that Hegelian frame which depicted the development of Western art in different epochs, master artists and schools. Finns were taught to see themselves as part of this tradition but were required to produce a Nordic uniqueness.

When, after a long debate in 1881, the Society decided to acquire the art merchant C. Göhle's collection, it was so that the Society could present the story of the fine arts to the public better than before. The collection consisted of nearly 50 works by lesser known and partly unknown artists, small landscapes, genre pictures, portraits and religious scenes (Pettersson 2008: 128–29, 324–26). Walleen's nightmare was coming true; dabbling in the market had led the Society to acquire works of low artistic quality. There seemed to be a tension between privileging educational aspects, and the pursuit of originality and quality.

Producing a national story of art

The most influential actors within the Society were the elected Chairman, Secretary and, from 1869, Keeper of the collection. Two chairmen were particularly influential in these early years: the poet and professor, Fredrik Cygnaeus (1807–81), and his successor, Professor Carl Gustaf Estlander (1834–1910). Both represented the Society and the University. Both taught aesthetics and art history at the university, and both wrote actively about art in newspapers. Estlander even published manifestos, pamphlets and books on the fine arts (Estlander 1867, 1871, 1875). And while Cygnaeus acquired social credibility through his hosting of a salon for the artists, Estlander worked hard to establish a professional network of Nordic academics. Estlander also had a very good political eye. He may not have been an active politician himself but he knew who to turn to when he needed help and support. The two men retained good relations with the Russian court during the reigns of Tsars Nicholas I (1796–1855) and Alexander II (1818–81).

The nation story to be told in art was to utilise the imagery of paintings of historical scenes and landscapes, and portraits of heroic men and peasants going about their daily lives. They wished to construct a visual library which recorded and defined the nation, and for this reason religious and still-life paintings were of less interest to them. The myths and legends, particularly Finland's national epic, the *Kalevala*, provided an important source of inspiration which linked together fine art and Finnish literature (Figure 8.2).

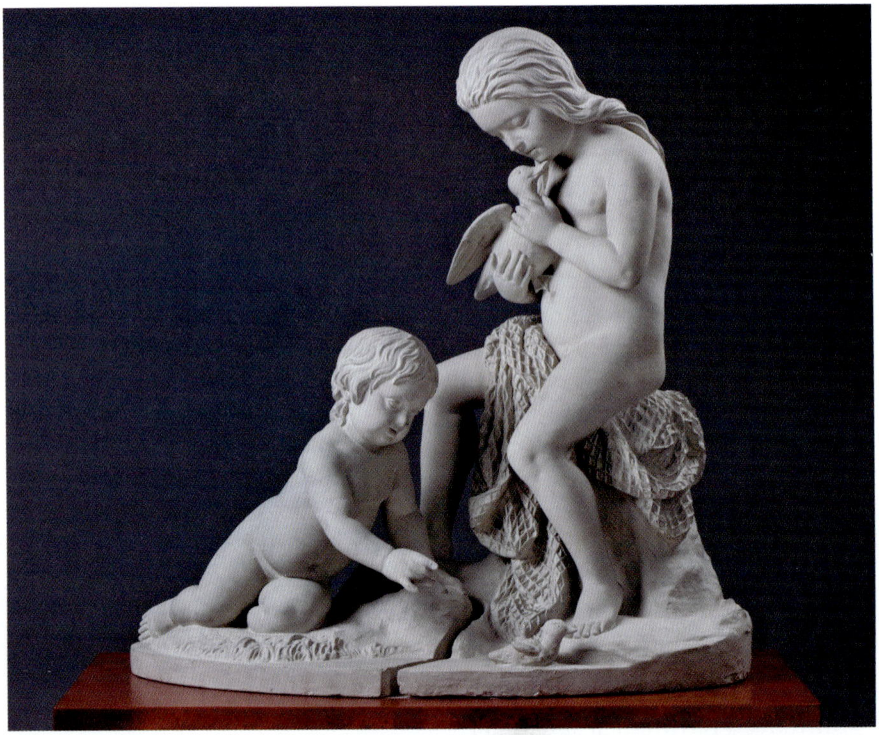

Figure 8.2 Carl Eneas Sjöstrand was amongst the first to give visual form to the national epic, the Kalevala. The work *Sotkottaret* (1869) illustrates the myth of the creation of the world; a Goldeneye lays an egg. Photo: Hannu Aaltonen, Central Art Archives/The Finnish National Gallery

There were two aspects to the collection. The first was a need to address at a basic level the history of European art so as to educate Finnish art students. The second was to establish a collection of Finnish art; it was Cygnaeus's belief that 'to flourish properly every civilised nation needs art and culture' (Cygnaeus 1887: 209). The beaux-arts were expected to combine the best of national expression with the ideals of antiquity. They were to be simultaneously strong, sincere and sublime. Pictorial content was placed above quality and value, at least at this early stage.

The outlook of the gatekeepers, Cygnaeus and Estlander, was shaped particularly by Hegel but also by German art historian and archaeologist Johann Joachim Winckelmann (1717–68). Winckelmann's evolutionary writings on art had taught these Finnish academics to value the art of the ancient Greeks without question. Classical ideals formed the basis for the evaluation of fine art and they understood that works were best displayed when arranged according to schools (Scott 2003: 170–71; Décultot 2007: 12–28). Winckelmann's ideas were so popular that they reached beyond academic circles and into the newspapers (*Morgonbladet* 1845).

Hegel's ideas had also established themselves in art circles, and had found influential support in Gustaf Waagen (1794–1868), Director of the Altes Museum in Berlin, and Friedrich Theodor Vischer (1807–87) whose book, *Aesthetik oder Wissenschaft des Schönen 1846–57*, was a mainstay of Finnish academic thinking. Cygnaeus and Estlander, and others, also travelled widely in Europe, where they studied museums, collections and displays and scrutinised the forms and methods of art education. Study trips lasted years and involved travel to several countries, but certain cities – such as Stockholm, Berlin, Dresden, Munich, Paris, Rome and London – were particularly influential. Indeed, practice in these capitals was often used to justify decisions at home.

These men also developed their tastes and connoisseurship in smaller academic circles. Estlander, for example, corresponded regularly with Nordic colleagues, including the Swedish professor, Carl Rupert Nyblom (1832–1907), and the Norwegian art historian, Lorentz Dietrichson (1834–1917). Together these men founded a Nordic journal, *Tidskrift för Bildande Konst och Konstindustri*, devoted to the fine arts and arts and crafts, and first published in 1875–1876. It expressed their ideas about the essential qualities of the arts.

The Finnish Art Society's 'official' history of art in Finland – a supplement to Wilhelm Lübke's famous *Grundriss der Kunstgeschichte* (1860) – was written by art historian and collector, Eliel Aspelin-Haapkylä (1847–1917). He excluded from this history the majority art works produced before the foundation of the Society (Aspelin 1891). This included all art produced under the Swedish regime, including portraits of the aristocracy and altar pieces in churches. The narrative being constructed in art was that of an autonomous nation but this was rather less straightforward than seems to be the case. The long connection with Sweden, in particular, created difficulties of interpretation: was, for example, a Finn who made his career in Sweden really a Finnish artist? This was a boundary to be negotiated and came particularly into focus with the acquisition of work by Alexander Lauréus (1783–1823), a painter in romantic style. A native of Turku, Lauréus had studied in Stockholm and become a court painter. When the Nationalmuseum in Stockholm came to collect its very first example of 'Swedish contemporary art' in 1845, the work chosen was a painting by Lauréus (Bjurström 1992: 110). In many respects, then, Lauréus was seen as a Swedish painter but the Finns were willing to make an exception and in doing so gain for the nation a master painter.

From the eighteenth century, the names of master artists had been used to produce the continuous history of art (Schulz 1994: 183–84). In the nineteenth century, many of these artists became the subjects of monographs and entire museums, the first examples being Antonio Canova's Gipsoteca in Possagno, Italy (1836), Bertel Thorvaldsen's museum in Copenhagen (1848) and a museum dedicated to Ingres in Montauban (1851). This reflected a more general principle that museums should only purchase works by well-established and long-dead masters. Only artistic excellence and *Galeriefähigkeit*, as Friedrich Pecht called it in 1877, would permit works to be included in a museum collection (Sheehan 2000: 95).

The Finnish Art Society worked outside these rules because its primary purpose was to support living artists. It purchased from artists that were still young, well – and alive. The Society soon found its own favourite national artists. Its first significant purchases were the idealised Finnish landscapes of Düsseldorf-trained landscape painter, Werner Holmberg (1830–60), Erik Johan Löfgren (1825–84), who had specialised in historical scenes, and the Swedish sculptor, Carl Eneas Sjöstrand (1828–1906), a board member of the Finnish Art Society, a sculptor of the *Kalevala*, and of local heroes of science and literature, and an influential teacher of the younger generation of artists. The relative lateness of these acquisitions – which began in 1858 – is to be explained by the Society's high expectations as regards the national art and artists. On the one hand, the Society was desperate to prove that Finland's artists were just as important as those of any other 'civilised' country, as Cygnaeus would have put it. On the other, the Society was painstakingly aware that the development of the fine arts required time and patience. That is not to suggest that the Society was entirely patient; in the 1840s, it sought to speed up the process by going on a talent hunt in the Finnish wilds, hoping then to hone this raw talent through intensive training. But all this did was produce a number of serious disappointments (Tikkanen 1896: 25, 37; Pettersson 2008: 83–85). The works by Holmberg and Sjöstrand formed the basis for the new Finnish collection. It is important to understand the novelty of these acquisitions; all earlier acquisitions had supported the development of the drawing school. The so-called 'national' part of the collection was only established when the new-generation Finnish painters and sculptors had something to contribute to the good cause.

The Society addressed its story to 'the nation' – a concept merely represented by the romantic ideal of enlightened peasants and happy labourers rather than something tangible. No matter how much central figures like Cygnaeus and Estlander intended to target their messages to the widest possible audience, it was the upper and middle classes who most attended. Only with Sunday opening and free admission to the members of the working class, from 1863 onwards, did this begin to change.

The collection was given official status in 1868 when the Society accepted a new set of rules (*Stadgar för Finska Konstföreningen*). A year later a Keeper of the collection was nominated: the post was given to Berndt Otto Schauman (1821–95), an art lover and connoisseur who had worked for decades at the University Library. His interest in filing and cataloguing was immediately put to good use and he was given the task of compiling the first inventory of the collection and publishing the first collection catalogue in 1873. Even then copies of European masterworks held a central place in the displays, creating a historical frame for the domestic works of art.

Private collectors became increasingly important to the Society towards the end of the nineteenth century. They donated their lifetimes' efforts. Artists and their families also joined in this act of building a national collection by donating large collections of works and sketches. The Society rarely refused such donations. This resulted in an uneven collection: instead of judging artistic qualities and using the same criteria as for the purchases, the Society considered the social and political aspects of these donations and in some cases let works of lesser significance enter the collections.

The Finnish artists supported by the Society participated in the nation-building process by painting and sculpting what was commissioned. Only towards the end of the nineteenth century did the artists achieve greater independence from the shaping opinions of the gatekeepers of the Society. This reflected a change brought about by the Artists Association, founded in 1864, which both increased the number of artists and brought in new ideas from Paris which challenged the academic view. Estlander and the board members of the Society regarded these new trends as a threat which distracted the minds of artists from the ideals of antiquity and could even endanger the idea of evolutionary art history. They were disgusted by *plein air* painting and highly suspicious of symbolism. Parisian art certainly did not serve the needs of art narrative. Now it was the artist himself that mattered and not the abstract aspirations of a nation.

The Ateneum's evolution

The building of the Ateneum in Helsinki was Estlander's initiative (Figure 8.3). It united the fine arts and applied arts under the same roof, and strengthened the need to present the story of Finnish art. Two-thirds of the building was reserved for the art schools, and one-third for the collections which still provided artistic archetypes for the students. The building housed a library and a lecture hall. At the opening, in October 1888, Estlander described the institution as 'a creation of our time': not an art museum or an art academy, even though it collected art as museums did and taught art like an academy. He saw the Ateneum as an institution that could help the nation to preserve what was ideal in the arts. He also highlighted the importance of forming a close relationship between the fine and applied arts, though practitioners were rather more doubtful about this aspect of the concern (Pettersson 2008: 260–61). However, art in Finland had changed. Artists were now Paris trained, more confident, and focused on realism. They saw the Society, with its romantic and historical ideas, as outmoded. Albert Edelfelt (1854–1905) was among the first representatives of this new generation to get public approval. His painting, *Women Outside the Church at Ruokolahti* (1887), entered the collections of the new institution, a fine example of the new way to paint the nation (Figure 8.4).

Before the opening of the Ateneum, the Society had displayed its collections at six different addresses in Helsinki, all of them apartments of various sizes. The works had been arranged in small rooms according to European models and using the best possible adaptation of the continuous narrative of the history of fine art. In principle, the Society had shown everything it had acquired. The Ateneum with its new galleries offered – even required – a different approach. A new scale and level of professionalism was required. The collection was displayed in generous third-floor spaces and in the large staircase of the building. Through careful selection and

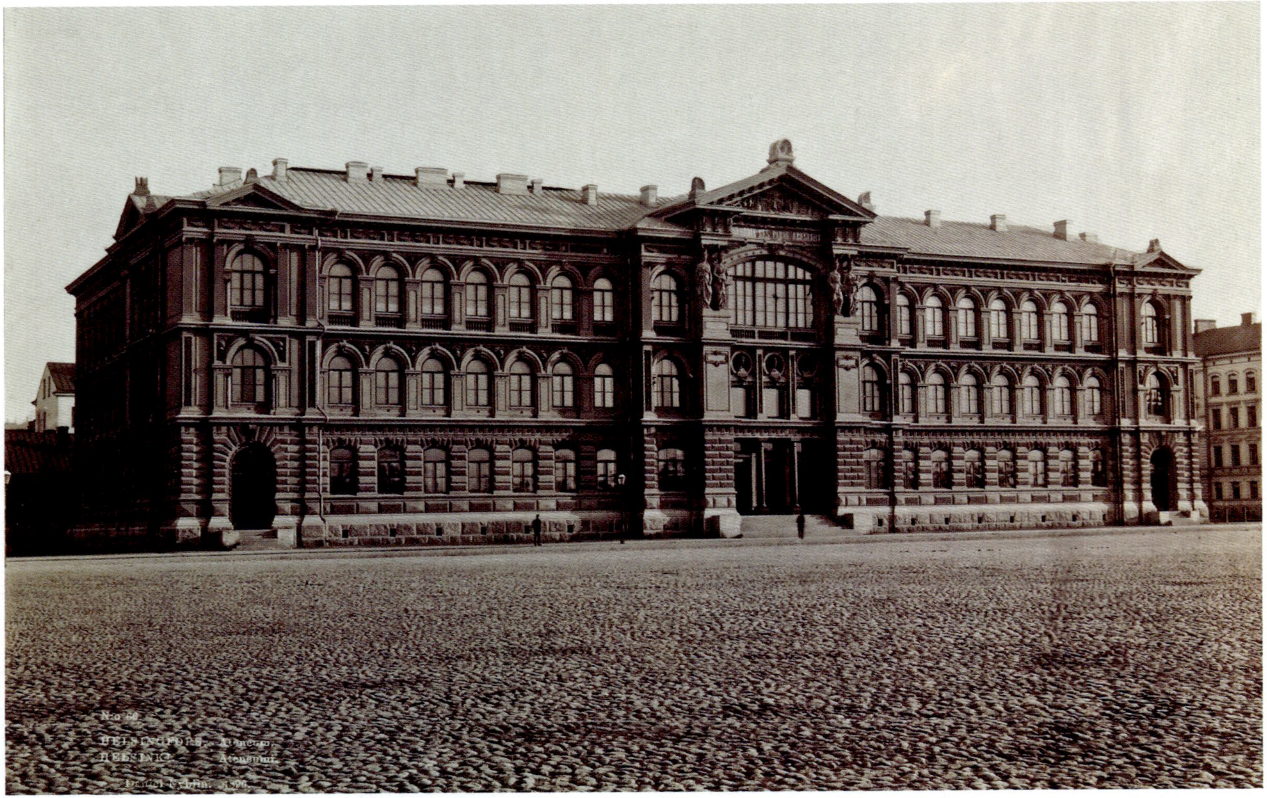

Figure 8.3 The Ateneum, 1887, the first purpose-built museum in Finland. The building was designed by the architect Theodor Höijer. Photo: Daniel Nyblin, 1890, Central Art Archives/The Finnish National Gallery

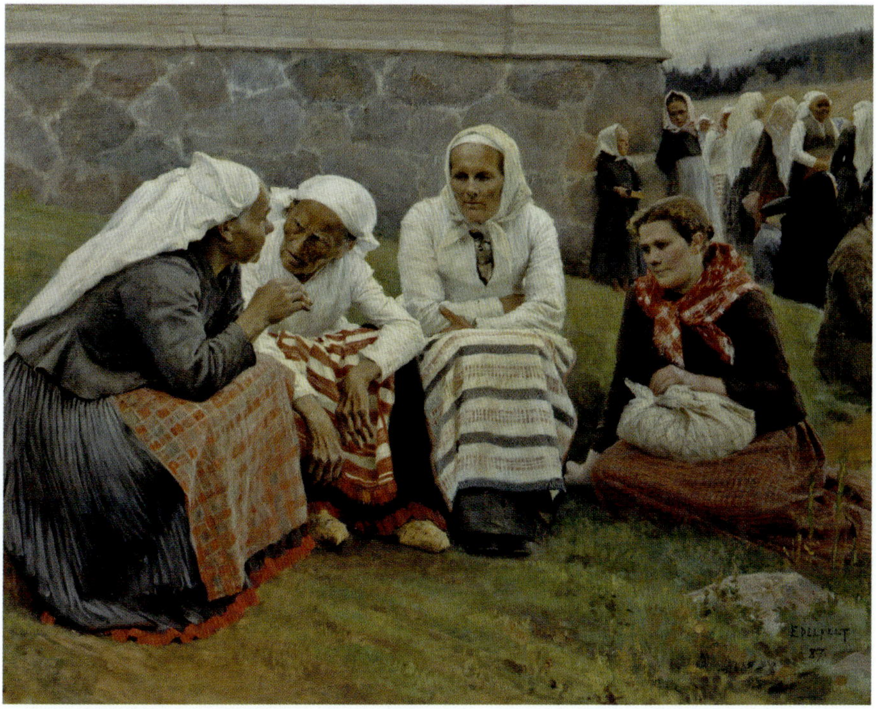

Figure 8.4 Albert Edelfelt's *Women Outside the Church at Ruokolahti*, 1887. A realistic painting, depicting church-going peasants, which in 1888 was donated by the state to the newly opened Ateneum building. Photo: Hannu Aaltonen, Central Art Archives/The Finnish National Gallery

elimination of the works, Estlander and Schauman's successor, Thorsten Waenerberg (1846–1917), organised the exhibition according to schools of art based on geographical division, a principle that had been applied in the national art museums in Berlin and Stockholm. The aim of this organisation was to ensure the best possible facilities for the members of the general public, who were expected to acquire knowledge of art and develop their taste. Now, for the first time, it was possible to exhibit and evaluate the 'story of Finnish art'.

In the new building, a whole set of new rules came into being. Works that possessed historical value were degraded. This category included artists' portraits, portraits of significant persons, and art that was older than the Society. These works were included in the inventory but displayed in the boardroom instead of the main galleries. Weak copies, poor-quality works of unknown artists and works that had been acquired only to support the artist were excluded from the collection. Estlander now shifted the focus on to artistic quality rather than quantity. Quantity had been pursued to create Hegelian continuity; a single story represented by a single collection. Now that ideal seemed beyond reach and rather less important (Pettersson 2008: 259–70).

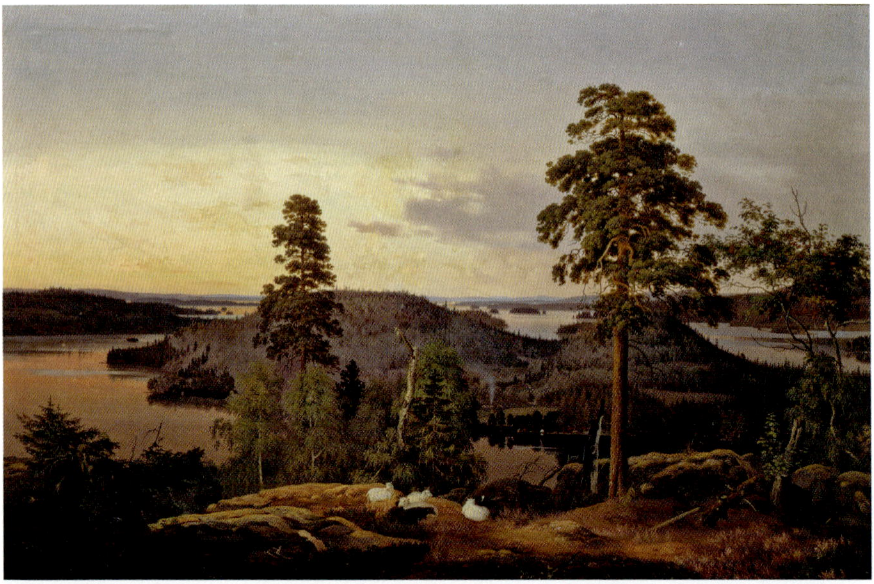

Figure 8.5 Ferdinand von Wright's landscapes were beloved interpretations of Finnish nature, painted according to the romantic ideal. *View from Haminalahti* (1853) was donated to the Society in the name of its protector, Alexander Alexandrovits. Photo: Jukka Romu, Central Art Archives/The Finnish National Gallery

European and Finnish art were to be separated, and older and contemporary art were to be displayed in different spaces. Copies were to be presented apart from originals. Sculptures were now to be displayed with paintings. National art was to be given pride of place and Estlander wanted to display its 'evolution'. At this point, the signature works of the collection began to take pride of place – genre paintings by Alexander Lauréus, landscapes by Werner Holmberg and particularly iconic works such as Ferdinand von Wright's painting *View from Haminalahti*, 1853 (Figure 8.5).

Two years after the opening of the Ateneum, Estlander had to admit that Walleen had been right to have doubts concerning the establishment of an art collection. He agreed that the true treasures of European art were either locked away in other collections or simply too expensive, and Finland possessed no royal collections which might partly fulfil this need. Telling the whole story was impossible but he recognised that the collection of copies might still have a purpose; the Ateneum's primary role was educational even if that meant sacrificing originality (Pettersson 2008: 282–83).

Despite these admissions, the Ateneum itself now began to evolve. Purchases grew in size to fulfil the museum scale of the brand new building. Estlander even spoke of 'museum-quality works' (Pettersson 2008: 281). The collection's story was no longer told to the art students and members of the Society but to the whole nation. The collection was increasingly regarded as *the* national art collection.

The need for a national museum had been a topic of intellectual debate since the 1860s, but the focus had always been on cultural history and archaeology. The role of the Society's collection had been discussed, too, but only in the context of planning an art academy. The educational role of the collection dominated the discussion but every now and then the words 'museum', 'national museum', 'national art museum' or 'gallery' were mentioned. Each time they referred to something more organised and professional than the Society's collection, which was run by academic volunteers. They also pointed to the political potential of museums as institutions for the aspiring nation.

The Finnish Art Society had had an extraordinary role in the production of a national story of art. It had nurtured the idea and masterminded and supported its production. It had seen this not simply as an act of collecting or producing a museum, but as something more dynamic and arising from educational investment. The collection acquired a national character only when the efforts of the Society had produced the first crop of new Finnish artists capable of contributing work. They had developed their skills and outlook against a backdrop of Hegelian art historical continuity and now they were to put Finland on the map of European artists. It was vital that they did so, for, as Estlander admitted and Walleen had always known, Finland was never going to be able to compete with the collecting of old European art by older and richer countries. Instead, it had all the potential of creating a collection of great national relevance.

Bibliography

Aspelin, E. (1891) *Suomalaisen taiteen historia pääpiirteissään,* Helsinki: Suomalaisen kirjallisuuden seura.

Bailey, C. B. (2002) *Patriotic Taste: Collecting Modern Art in Pre-Revolutionary Paris,* New Haven, CT, and London: Yale University Press.

Bjurström, P. (1992) *Nationalmuseum 1792–1992,* Nationalmuseum/Wiken.

Cygnaeus, Fr. (1887) *Samlade arbeten V. Literatur-historiska och blandade arbeten,* Helsingfors: G. W. Edlund's förlag.

Cygnaeus, Fr. (1889) *Samlade arbeten VI. Literatur-historiska och blandade arbeten,* Helsingfors: G. W. Edlund's förlag.

Décultot, E. (2007) 'Johann Joachim Winckelmann', in Ulrich Pfisterer (ed.) *Klassiker der Kunstgeschichte, Von Winckelmann bis Warburg*, München: Verlag C.H. Beck, pp. 12–28.

Estlander, C. G. (1867) *De bildande konsternas historia från slutet af adertonde århundradet till våra dagar,* Stockholm: L. J. Hiertas förlag.

Estlander, C. G. (1871) *Den finska konstens och industrins utveckling hittills och hädanefter,* Helsingfors: K. E. Holms förlag.

Estlander, C. G. (1875) *Vid konstflitens härdar i Tyskland, Österrike, Schweiz och Belgien. Reseanteckningar,* Helsingfors: G. W. Edlunds förlag.

Flacke, M. (ed.) (2001) *Mythen der Nationen: Ein Europäisches Panorama,* Second edition, München/Berlin: Koehler & Amelang.

Hegel, G. W. F. (2004) *Philosophie der Kunst oder Ästhetik. Nach Hegel: Im Sommer 1826*, München: Wilhelm Fink Verlag.

Henningen, B., Klein, J., Müssener, H. and Söderlind, S. (eds) (1997) *Skandinavien och Tyskland 1800–1914: Möten och vänskapsband*, Stockholm: Deutsches Historisches Museum, Nationalmuseum, Norsk Folkemuseum.

Karlholm, D. (1996) *Handböckernas konsthistoria: Om skapandet av 'allmän konsthistoria' i Tyskland under 1800-talet*, Stockholm: Brutus Östlings Bokförlag.

Kuhrau, S. (2005) *Der Kunstsammler in Keiserreich. Kunst und Repräsentation in der Berliner Privatsammelkultur*, Kiel: Verlag Ludvig.

Lenman, R. (1997) *Artists and Society in Germany 1850–1914*, Manchester and New York: Manchester University Press.

Lindström, A. (1963) *Ateneumin taidemuseo 1863–1963*, Helsinki: Ateneumin taidemuseo.

Lorente, P. J. (1998) *Cathedrals of Urban Modernity: The First Museums of Contemporary Art, 1800–1930*, Aldershot, Brookfield, Singapore and Sydney: Ashgate.

Lübke, W. (1860) *Grundriss der Kunstgeschichte*, Stuttgart: Ebner & Seubert.

Maker, W. (ed.) (2000) *Hegel and Aesthetics*, Albany: State University of New York Press.

Martinsen, R. (2006) *Helsinki Zachris Topeliuksen silmin: Kertomuksia 1830- ja 1840-lukujen Helsingin elämästä*, Helsinki: Helsinki-seura.

Pettersson, S. (2005) 'Nils Abraham Gyldén taideyhdistysliikkeen juurilla', in Tuukka Talvio (ed.) *Taidehistoriallisia tutkimuksia 31*, Helsinki: Taidehistorian seura, pp. 55–66.

Pettersson, S. (2007) 'For the people's best: Fredrik Cygnaeus and the course taken by art in Finland', in R. Suominen-Kokkonen (ed.) *The Shaping of Art History in Finland*, Helsinki: Taidehistorian seura, pp. 11–24.

Pettersson, S. (2008) *Suomen Taideyhdistyksestä Ateneumiin: Fredrik Cygnaeus, Carl Gustaf Estlander ja taidekokoelman roolit*, Helsinki: Suomalaisen kirjallisuuden seura.

Schauman, B. O. (1873) *Katalog öfver Finska Konstföreningens samling af taflor och skulpturarbeten*, Helsingfors: Finska Konstföreningen.

Schoolfield, G. C. (1996) *Helsinki of the Czars: Finland's Capital 1808–1918*, Columbia, SC: Camden House.

Schulz, E. (1994) 'Notes on the history of collecting and of museums', in Susan M. Pearce (ed.) *Interpreting Objects and Collections*, London & New York: Routledge, pp. 175–87.

Scott, J. (2003) *The Pleasures of Antiquity: British Collectors of Greece and Rome*, Paul Mellon Centre for Studies in British Art, London: Yale University Press.

Sheehan, J. J. (2000) *Museums in the German Art World: From the End of the Old Regime to the Rise of Modernism*, Oxford and New York: Oxford University Press.

Smith, A. D. (1991) *National Identity*, Reno and Las Vegas: University of Nevada Press.

Tikkanen, J. J. (1896) *Finska Konstföreningen 1846–1896*, Helsingfors.

Trodd, C. (2000) 'The paths to the National Gallery', in P. Barlow and C. Trodd (eds) *Governing Cultures: Art Institutions in Victorian London*, Aldershot and Burlington, VT: Ashgate, pp. 29–43.

Chapter 9

REIMAGINING THE NATION IN MUSEUMS

Poland's old and new national museums

Karoline Kaluza

FOLLOWING THE INTEGRATION of Poland into the European Union (EU) in 2004, a wave of new museum projects and architectural competitions swept the nation. Economic growth, especially in thriving Warsaw, coupled with the possibility of receiving funding from the EU, triggered this museum boom. It followed two centuries of political disruption and disempowerment, which had made the museum in Poland an important instrument for national preservation and reflection. This form of remembering also gave the Polish nation a particular relationship to contemporary art as a medium representing personal freedoms. This chapter will discuss the recent development of national museums in Poland and set this development in the context of two centuries of struggle. It will reveal how particular types and styles of cultural heritage have acquired national significance and how different kinds of museum are now being produced. A large number of the museums developed in and about Poland over the last two centuries have sought to preserve the memories of a nation. They have simply chosen to do so using the media considered most appropriate at the time. This chapter will begin by discussing the period when much of Western Europe gained its museum infrastructure; a period when Poland existed only as a memory.

Keeping the flame alive

In 1795, the Polish-Lithuanian Commonwealth was partitioned by Prussia, Russia and Austria. Poland disappeared from the political map of Europe for 123 years. Divided between three ruling powers, the Polish lands then began a process of cultural divergence which only ceased with the return of national independence in 1918 (Davies 2000: 105). Perhaps surprisingly, it was only in this period when state sovereignty was lost, that the Polish began to build their first 'national museums'.[1] These were founded to keep the memory of the Polish nation alive. What this memory was and precisely how it should be kept depended very much on who founded the museum, and when and where they did so.

The first such museum was founded in 1801 by Izabella Czartoryska[2] – a member of the Czartoryski family, which had given Poland its last king. With the destruction of the insignia of the Polish kings by Prussian troops in 1794 and the loss of state sovereignty, Czartoryska concentrated her collecting on rescuing Polish cultural heritage. Having previously amassed all sorts of keepsakes, artefacts and memorabilia from her sojourns in Switzerland, England and Scotland (Labuda 2006: 141–64), she now attempted to construct a genealogical memory of the high nobility connected to 'Polish kings, army commanders, military triumphs, as well as the pertinacity and glory of every dynasty of the high nobility' (Żygulski 1962). She was not alone in her attempts to preserve a record of the high nobility (see Kornacki 2004). Undoubtedly, such knowledge could be useful were the state to be re-established.

A rather different museum was founded in 1869 by Władysław Plater.[3] A Polish count in exile, he established a 'national museum' in Rapperswil, Switzerland. It contained historic relics of national heroes who had dedicated their lives to upholding the spirit and memory of the Polish nation. Arranged in chronological order, it housed weapons, paintings, uniforms, drawings, coins and maps, as well as the personal belongings of famous Poles, such as the pipe, pen and lock of hair of the romantic poet, Adam Mickiewicz.[4] The museum's board members came from the various independent political movements living in exile and in the partitioned regions (Malinowski 1979: 59), and they and others donated objects to the museum. It was a place where memories could be accumulated and stored (Schmidt 1994). In this museum, every single material object possessed great national importance and stood for the whole idea of the Polish nation. Symbolically they represented the pulling together of a scattered nation. Amongst the most powerful objects was the heart of Tadeusz Kościuszko,[5] who led the Uprising of 1794. Placed in a mausoleum within the museum it performed as a reliquary of national (rather than religious) significance.[6] Czartoryska's museum, possessing a crypt, altar and historic relics, also had a sacred character. The Polish nation truly was imagined (Anderson 1983) and both museums became sites of pilgrimage (Pomian 1992: 29).

After the Congress of Vienna in 1815, it became possible for citizens in the partitioned territories to preserve Polish national identity. But this new freedom was not an open or concrete right; everywhere it was a struggle. Unable to pursue national ideals politically, the Polish intelligentsia organized itself into societies dedicated to the arts and sciences. These new institutions became a reactive force in territories where the new rulers compelled the Polish people to take on new cultural identities and speak in the language of the ruling nation. Consequently, contemporary Polish art became highly political. Paintings showing aspects of the story of the Polish nation became particularly important and were used to awaken national feelings, to keep alive Polish history and to glorify the old republic. For those generations born into this national wilderness they were vital educational tools. Monumental and historical, Jan Matejko's[7] paintings took on particular significance, and were purchased widely across the partitioned territories and disseminated through reproductions by these societies. Even today, Matejko remains Poland's most important historical painter (see Zagrajek 1998).

The partitioned region controlled by Austria enjoyed the greatest autonomy and in 1869 Cracow was declared the only free Polish city. It was given permission to reinstate Polish as the official language, and a decade later the first public Polish national museum in the partitioned territories was established in Cracow. This museum was closely linked to the School of Fine Arts led by Matejko (Kopff 1962). The second national museum in the partitioned territories was established in Warsaw, in 1916. The museum itself – the Museum of Fine Arts – was actually much older, having been founded in 1862 (Lorentz 1985). This region, the so-called 'Kingdom of Poland', was under Russian rule but here too the Polish intelligentsia had managed to achieve some influence. In the Prussian territory, however, German culture remained strong and Polish collections and museums established in Poznań were too peripheral to possess national prestige (Ziemięcki 1879: 407; Borzykowski 1986: 176; Łepkowski 2003: 378).

After the First World War, Poland achieved its independence and the national mission shifted from fighting against complete oblivion to creating a Polish state which had to be reunited, redefined and restructured politically, economically and socially. Museums were to have a new role and the museum landscape was transformed: private and municipal museums were to become state museums; the museum system, such as it was, was to be restructured and professionalized (Treter 1922; Mansfeld 2000). This was, however, a short-lived moment as attempts to rebuild the nation were halted with the Nazi invasion in 1939. Occupation and conflict in the Second World War devastated Polish cultural heritage. After the war there was a search for stolen and sold cultural objects, and inventories were constructed to document which cultural objects had survived the war and list those that were missing or lost.[8] Negotiations for the return of this material continue to the present day.

After 1945, the now Communist Polish government brought all museums under state control. In doing so, a museum in formerly Prussian-controlled Poznań was elevated to national status. Other museums achieved national museum status in the 1970s: in Wrocław, Szczecin, Gdańsk and in Kielce (Lorentz 1971: 33–39). Each was given the name, 'National Museum', with the name of the city attached. By this means the Polish communists created the impression of a homogeneous museum landscape and symbolically extended their 'national territory'. In some senses at least, this homogeneity was, however, no illusion. Post-war Poland was a nation that had been ethnically cleansed by the Nazis. Now the Communist regime's propaganda turned this homogeneity into a national myth which it used to create a sense of togetherness. What lay beneath this myth, however, was two centuries of cultural diversity and political disruption.

New museums of national history

Poland escaped the Communist era possessing nine national museums, of which seven were city cultural history museums displaying Polish national art and historical artefacts mainly of their regions.[9] In addition to these heterogeneous collections, there were also two specialist national museums, one devoted to agriculture and the other to industry (Sołtysiak and Wierzbicka 2001). While these museums have continued to exist into the present, museum developments in the new millennium have focused on entirely new projects. Most prominent has been the proposal for a Polish History Museum capable of uniting the national narrative under one roof. Vice-Minister of the Ministry of Culture and National Heritage, Tomasz Merta, has referred to this narrative as promoting positive aspects of Polish history: the democratic tradition; its multicultural past; and its freedom fighting. The museum would also deal with the independence of 1918, times of occupation and war, the history of Solidarity (Solidarność) which led to the fall of Communism. Merta also recognized the need to confront shameful and controversial aspects of the past and to avoid an idealistic presentation of national martyrs and geniuses.[10]

Other large-scale museum projects were also launched in Warsaw, including the Museum of the History of Polish Jews,[11] the Museum of Modern Art in Warsaw[12] and the Warsaw Uprising Museum. These new museums have shifted the idea of the museum conceptually, away from treasure houses of national memory and towards places of meaning-making and knowledge networking. The focus is now the museum visitor, communication and dialogue; memory is not something kept but rather active and dynamic (Van der Meer 2001: 102). New technologies are used to challenge the way national memory is imagined and communicated.

The Warsaw Uprising Museum opened in 2004 on the sixtieth anniversary of the Uprising. The first of these new large-scale national museums, it has great symbolic significance for the Polish nation. Often confused with the 1943 Jewish Ghetto Uprising in Warsaw, the Warsaw Uprising was a bitter struggle by the Polish Home Army (Armia Krajowa) to liberate Warsaw from German occupation. It began on 1 August 1944 as part of a nationwide rebellion, codenamed 'Operation Tempest' (see Davies 2004). After the Uprising, the Germans destroyed the city. The museum attempts to reconstruct life in the occupation. At the entrance of the museum a clock is set permanently at 5 p.m., the hour when the Uprising started – the so-called 'hour W'; so the visitor enters the historical event. Although the museum is thematically divided, visitors are confronted by atmospheric depictions of Warsaw in 1944, complete with the sounds of wailing of bombers, explosions and whizzing bullets. The museum is filled with personal stories: members of the Polish Home Army, Germans, other foreigners participating in the Uprising, Jews in the Ghetto and children (Figure 9.1). To give a sense of everyday life, a real chapel has been built in the museum, named after a chaplain who perished in the Uprising. Here people can remember and pray for those who died. The visitor is both immersed and yet given the historical distance to learn. The museum may not be about materializing memories in the manner of the older museums but it still attempts to accumulate personal memories of eyewitnesses who survived the war, fought for the freedom of the city and who are today's Polish national heroes. It is a museum built upon the oral testimony of witnesses which is then conveyed through diverse multimedia presentations, films, sound clips and photographs (Figures 9.2 and 9.3). Authentic objects are used as tools to help convey the history of this event.

The museum is housed in a reconstructed tram power plant, an industrial building of the late nineteenth and early twentieth centuries. It can be seen from a distance, as it has a 35-metre-high tower looking like a ruin of the war. From the panoramic observation platform on top of the tower the visitor may view part of the city and search out surviving pre-war buildings.

The Director of the museum is appointed directly by the President of the city of Warsaw. On anniversaries of the Warsaw Uprising, the museum becomes the most important place in the city and is part of a nationwide celebration, commemorating this traumatic event and celebrating its veterans and military groups. Each year on 1 August, sirens wail across the country at 5 p.m. Even the trams and cars stop to commemorate the event. On this day Polish history in the museum becomes tangible as special events are organized to bring the youth together with the older generation which survived the Uprising. Incidents in the Uprising may also be re-enacted. For the Poles, this museum becomes a place where different generations are brought together to pay tribute to the heroes who fought for Poland's freedom.[13]

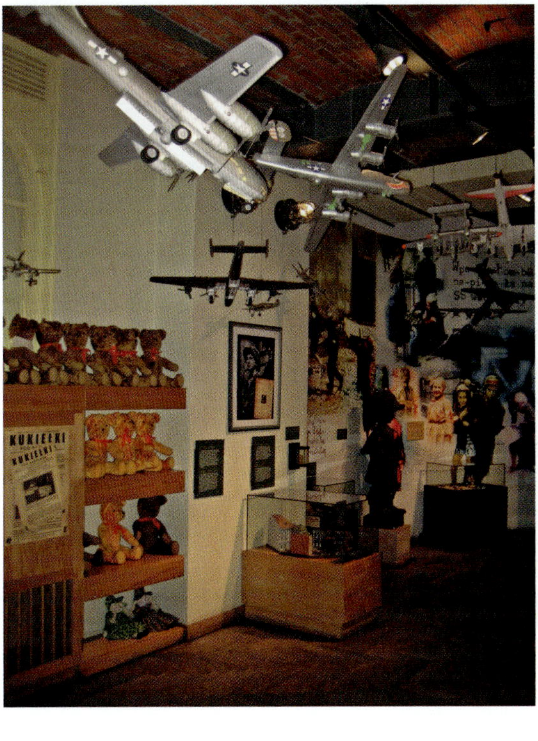

Figure 9.1 Warsaw Uprising Museum, Little Insurgents Room. A room dedicated to children with copies of pre-war toys, videos and a puppet theatre. Photo: author

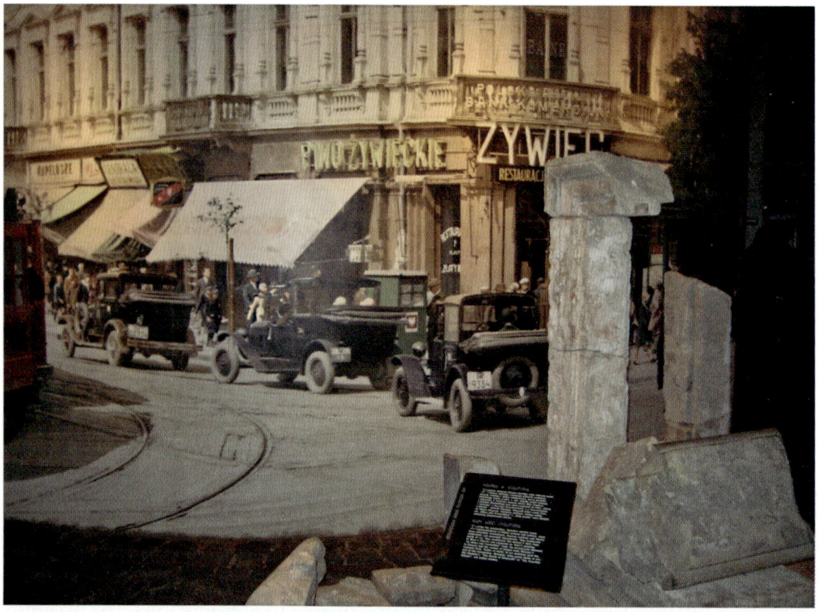

Figure 9.2 Warsaw Uprising Museum. A photograph of Warsaw taken before the war, with ruins standing in front of the picture. Photo: author

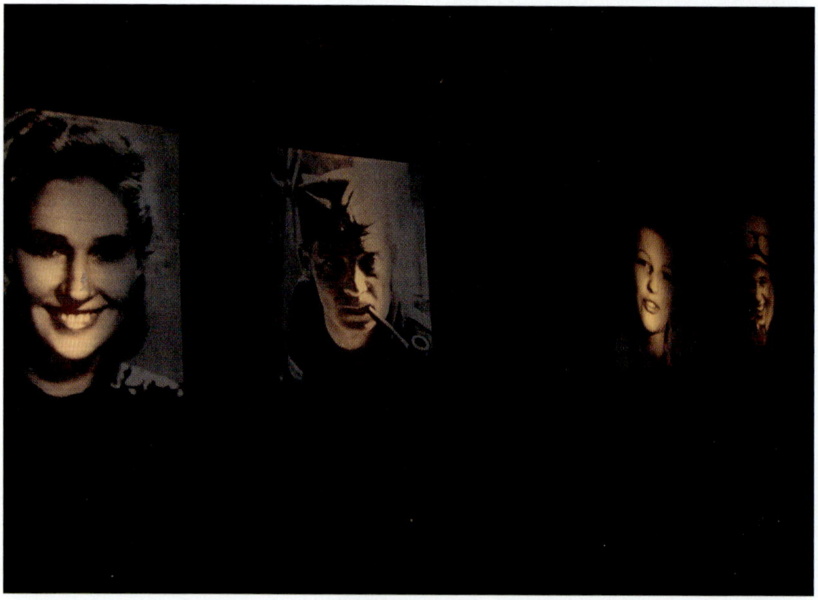

Figure 9.3 Warsaw Uprising Museum. Pictures of people who lived in Warsaw during the occupation. Photo: author

Reviving Jewish culture in Poland

The Museum of the History of Polish Jews in Warsaw is a rather different development and reflects external influence. The idea came from the Jewish Historical Institute Association in Warsaw in 1996 and was supported by Jeshajahu Weinberg, the former director of the United States Holocaust Memorial Museum in Washington. The museum will be realized through private donations and funding mainly from the USA, Germany, Great Britain and Poland. The vision for the museum places it on a par with major holocaust museums around the world and a particular goal for the Warsaw museum is to overcome the Israeli prejudice that Poland is a 'place of death'. The intention is to revive Jewish culture and memory and at the same time redefine it. This is not to be a Holocaust museum, but a museum of the history and the traditions of Jewish life in Poland (Dödtmann 2007). The museum will also foster intercultural exchange programmes between Israelis and Poles. Indeed, several exchange programmes have already started.

This museum, like other new museums in Poland, is using new technologies, both as a prelude to the physical museum but also as a place of virtual exhibitions and diverse online programmes. The website has become a central point of reference and networking regarding Jewish life in Poland. History-making in these museums is, as a result, more transparent, debated and individualized. One such project, called 'Kaddish for the departed' run by the Shalom Foundation,[14] is a database linked to the Museum of the History of Polish Jews, which allows Jewish people to search for their departed relatives or just send pictures and names of 'family members, neighbours, friends and even strangers' to the project. Another project, called 'Families separated by history', is run by the Polish History Museum. It invites everyone whose family history has been divided by war, displacement, resettlement, expulsion, political repressions or economic migration to participate. It is an attempt to grasp the history of a fragmented nation at a personal level.

Contemporary art and nation

While these historical museums are fairly obvious responses to Poland's past, there have also been considerable developments taking place in contemporary art which might at first glance appear to have nothing to do with this process of reimagining the nation in museums. It will, however, be recalled that in the nineteenth century it was contemporary historical paintings by Matejko which became iconic cultural heritage. The creation and promotion of new museums of contemporary art in the twenty-first century in these formerly Communist nations has caused Western European commentators some confusion. This is because the experience of twentieth-century art has been so different in the West. In Poland, modern and abstract art did not gain a foothold in national museums in that brief period of independence between the end of partition and the German/Soviet occupations. Only one internationally focused modern art museum had been founded, in Łódź in the 1930s: the Muzeum Sztuki or Museum of Art. The collection had been first offered to the National Museum of Warsaw but had been rejected as too modern and not sufficiently national (Ojrzyński 1998: 10). The Nazis considered modern art degenerate. The Socialist Realist doctrine of the Communist era demanded realistic paintings capable of becoming political propaganda. In Poland, then, contemporary art became something more than merely progress in the history of art; it became a political medium. This has, however, produced confusion about the meanings and value of contemporary art which is observable in the naming of the new museums of art. In Wrocław, for example, a new museum project has been launched called Contemporary Museum Wrocław – a name chosen to explicitly avoid the problems of defining a museum of modern or contemporary art or produce particular expectations amongst visitors (see Krajewski and Monkiewicz 2007).

The new Museum of Modern Art in Warsaw had begun life as the Museum of Contemporary Art but then switched its name to avoid confusion. It is nevertheless a museum of the art of now (Mytkowska 2007: 1). This project, initiated in 2004, is the first attempt to build a large-scale museum of contemporary art in Warsaw. The prime mover for this museum was the renowned Polish art historian, Anda Rottenberg, who published a preliminary draft of the museum's concept in 2006. It included a list of future partners and networks, featuring some of the most influential art institutes and museums around the world, including in the former Socialist countries. But she included 'other countries which are in transformation and looking for new values' like Kazakhstan, Turkmenistan, Uzbekistan and Kyrgyzstan, as well as countries in the Middle East and Asia, as future partners (Rottenberg 2006). It shows Rottenberg's interest in creating a global network and focusing on countries which are or might be in the future, like Poland, at a stage of transition and redefinition. A young team, including art historians and sociologists, are developing exhibitions, conferences and lectures, and discussing and testing future steps and developments.

In other cities, like Łódź and Cracow, post-industrial sites with twentieth-century architecture are to be transformed to house contemporary art. The Muzeum Sztuki in Łódź opened a new branch in November 2008, the so-called 'ms^2'. It is situated in a former factory site called Manufaktura which, following renovation, is now a shopping and entertainment complex. The site in Cracow formerly held the factory of Oskar Schindler, made famous in Thomas Keneally's Booker Prize-winning novel *Schindler's Ark* (1982) which then became Steven Spielberg's Oscar-winning film under the American book title, *Schindler's List* (1993).

This national commitment to contemporary art has also been realized in regional developments. In 2004, the government introduced its 'Signs of time – centres of contemporary art – regional collections of contemporary art'[15] programme, based on the 1983 French initiative of the Fonds Regional d'Art. This promoted the building of regional collections of Polish contemporary art in each of the sixteen Polish provinces. The programme gave state legitimacy to contemporary art.[16]

Conclusion

Poland in the twenty-first century is a country which has found multiple ways in which to deploy museums as part of the nation-making process. Nearly all reflect upon the previous century as a period of national trauma from which the new nation can be built, using what are seen as resilient national values. They are also reactions against the oppression that has been felt in daily lives, in the arts, in politics, and so on. The new museums are built using those values that were once deprived: they have been built using collective oral testimony, through processes of negotiation and participation. They, and the memories they preserve, are, however, also shaped by values of the present. Both – the museums and the memories – are what Poland needs now: 'All of this is to say no more than that the past, as it is materially embodied in museums and heritage sites, is inescapably a product of the present which organizes it' (Bennett 1995: 129).

The new Polish national museums are highly political and are tasked with raising awareness amongst the Polish people and the outside world regarding certain topics that are of national significance. As Hans Haacke pointed out, a 'museum is perforce a political institution, no matter whether it is privately run or maintained and supervised by governmental agencies' (Haacke 2004: 407). Poland is by these means to be redefined and understood anew. A similar boom in historical and art museums can also be observed in other Eastern European countries. The museum is seen as a catalyst.[17] The support from the EU has played a critical role.

Ever since Polish national museums were established, they have been connected to the mission of keeping and building the Polish nation. For much of the last two centuries this has been a task of rescuing traces. Museums in Poland were long part of a political reaction – a way for Poland to exist as a politically imagined entity. Polish national identity has been represented and preserved in various ways, from historical relics to the content of national contemporary art, but always through the materialization of memory. The present museum landscape possesses museums of many conceptions: the older regionally based national museums and the new projects. They record a changing conceptualization of how the memory of the nation should be preserved. National memory is no longer understood as something to be gathered as pieces of material culture. The act of gathering remains but today it is of intangible memories. This reflects changes in styles of history-making but also a recognition that a generation of people who lived through the political upheaval of the last century provide Poland's most important resource for recovering the past. Part of that recovery concerns the former ethnic diversity of a nation that has long lived with the myth of homogeneity.

Notes

1 In the eighteenth century there existed some plans for creating Polish museums but none were realized (See Hryniewiecki 1947: 114–18; Mniszech 1775: 66–75; Powidzki 1955: 5–9; Lorentz 1971: 17).

2 Izabela Czartoryska (1746–1835).
3 Władysław (Broel-) Plater (1808–1889).
4 Adam Mickiewicz (1798–1855).
5 Tadeusz Kościuszko (1746–1817).
6 Report of the Executive Board of the National Museum of Rapperswyl for the year 1897 (Sprawozdanie z Zarząd Muzeum Narodowego Polskiego w Rapperswylu za rok 1897).
7 Jan Matejko (1838–1893).
8 Catalogues of losses: Śrwieczyński, J. (1988) *Katalog skradzionych i zaginionych dóbr kultury*; Ośrodek Ochrony Zbiorów Publicznych (2000) *Katalog utraconych dzieł sztuki*. Cenne, bezcenne/utracone.
9 The collection from the Rapperswil museum was incorporated into the National Museum in Warsaw in 1928; Czartoryska's museum is now part of the National Museums in Cracow.
10 Merta, T. *Muzeum Historii Polski*. Online. <http://www.muzhp.pl/muzeum/o-muzeum/18/muzeum-historii-polski—poczatek-podrozy> (accessed 20 August 2009).
11 Muzeum Historii Żydów Polskich. Online. <http://www.jewishmuseum.org.pl/>.
12 Muzeum Sztuki Nowoczesnej w Warszawie. Online. <http://www.artmuseum.waw.pl/>.
13 Muzeum Powstania Warszawskiego. Online. <http://www.1944.pl/>.
14 Kadysz dla nieobecnych, Online. <http://www.kadysz.pl/>.
15 Znaki czasu – Centra Nowoczesności. Regionalne Kolekcje Sztuki Współczesnej.
16 Monika Szewczyk commenting in Kosiewski 2006.
17 Kirshenblatt-Gimblett, B. 'The museum as catalyst' in *Museums 2000: Confirmation or Challenge*, organized by ICOM Sweden, the Swedish Museum Association and the Swedish Travelling Exhibition/Riksutställningar in Vadstena, 29 September 2000.

Bibliography

Anderson, B. (1983) *Imagined Communities: Reflections on the Origin and Spread of Nationalism*, London and New York: Verso; trans. (1998) Die Erfindung der Nation: Zur Karriere eines erfolgreichen Konzeptes, Berlin: Ullstein.

Bennett, T. (1995) *The Birth of the Museum: History, Theory, Politics*, London: Routledge.

Borzykowski, J. (1986) *Inteligencja Polska w Prusach Zachodnich 1848–1920*, Gdańsk: Wydawnictwo Morskie.

Buddensieg, A. and Weibel, P. (eds) (2007) 'Editorial', *Contemporary Art and the Museum: A Global Perspective*, Ostfildern: Hatje Cantz, pp. 6–13.

Davies, N. (1984) *Heart of Europe: A Short History of Poland*, Oxford: Clarendon Press.

Davies, N. *(2000) Im Herzen Europas: Geschichte Polens*, trans. Friedrich Griese, München: Beck.

Davies, N. (2004) *Rising '44: The Battle for Warsaw*, London: Pan Books.

Dödtmann, E. (2007) 'Nicht mehr nur Ort des Todes', in *Jüdische Zeitung*. Online. <http://www.j-zeit.de/archiv/artikel.641.html> (accessed 26 February 2009).

Friedmann, B. and Krautheim, H.-J. (2001) 'Wiener Kongress – Heilige Allianz–Restauration', in Klaus Zernack (ed.) *Handbuch der Geschichte Russlands*, 2, Stuttgart: Hiersemann Verlag, pp. 994–1020.

Haacke, H. (2004) 'Museums: Managers of Consciousness', in Preziosi, D. and Farago, C. (eds) *Grasping the World: The Idea of a Museum*, Aldershot: Ashgate, pp. 400–13.

Hryniewiecki, B. (1947) 'Projekt Stefana de Rieule'a stworzenie w Warszawie Muzeum Przyrodniczego w wieku XVIII', *Wiadomości Muzeum Ziemi*: 114–18.

Kopff, A. (1962) *Muzeum Narodowe w Krakowie: Historia i zbiory*, Kraków: Muzeum Narodowe w Krakowie.

Kornacki, K. (ed.) (2004) *Muzea – Rezydencje w Polsce: Materiały sesji naukowej zorganizowanej w Muzeum Zamoyskich w Kozłówcu*, Kozłówka: Muzeum Zamoyskich.

Kosiewski, P. (2006) 'Co dalej z programem Ministerstwa Kultury "Znaki czasu"?', *Tygodnik Powszechny*, 18. Online. http://www.fundacjamuzeum.pl/media/media10.htm (accessed 26 February 2009).

Krajewski, P. and Monkiewicz, D. (2007) *Muzeum Współczesne Wrocław. Koncepcja programowa*. Online. <http://www.muzeumwspolczesne.pl.MNW_prog_pl.pdf> (accessed 26 February 2009).

Labuda, A. (2006) 'Residenzmuseen – Sammlungen und Stiftungen in Polen in der ersten Hälfte des 19. Jahrhunderts als Medium der nationalen Identitätspolitik', in Ernő Marosi and Gábor Klaniczay (eds) *The Nineteenth-century Process of 'Musealization' in Hungary and Europe*, Budapest: Collegium Budapest, pp. 141–64.

Łepkowski, T (2003) 'Naród bez Państwa', in Samsonowicz, H., Tazbir, J. and Łępkowski, T. (eds) *Polska: Losy Państwa i Narodu do 1939 roku*, Warszawa: Wydawnictwo Iskry, pp. 247–400.

Lorentz, S. (1971) *Przewodnik po Muzeach i Zbiorach w Polsce*, Warszawa: Wydawnictwo Interpress.

Lorentz, S. (1985) *Rocznik Muzeum Narodowego w Warszawie*, Warszawa: Muzeum Narodowe Warszawa.

Malinowski, K. (1979) *Prekursorzy muzeologii polskiej*, Prace Komisji historii sztuki, 9, Poznań: Państwowe Wydawnictwo Naukowe Warszawa.

Mansfeld, B. (2000) *Muzea w Drodze do Samoorganizacji: Związek Muzeów w Polsce 1914–1951*, Warszawa: Wydawnictwo DiG.

Mniszech, M. (1775) 'Myśli względem założenia "Musaeum Polonicum"', *Zabawy Przyjemne i Pożyteczne*, 6(2): 211–26.

Mytkowska, J. (2007) 'Muzeum w procesie', *Muzeum*, 1. Online. <http://www.gazeta-muzeum.pl/artykul.php?id=2> (accessed 26 February 2009).

Ojrzyński, J. (1998) 'Historia Muzeum Sztuki w Łodzi. Historia i wystawy', in Brudzyńsiki, R., Czartoryska, U., Modława, A. *et al.* (eds), *Muzeum Sztuki w Łodzi*, Łódź: Oficyna Bibliofilów, pp. 6–39.

Pomian, K. (1992) 'Museum, Nation, Nationalmuseum', in Plessen von, M. L. (ed.) *Die Nation und ihre Museen*, Frankfurt am Main and New York: Campus Verlag.

Powidzki, J. (1955) 'Muzeum Polskie Michała Mniszcha', *Muzealnictwo*, 4: 5–9.

Rottenberg, A. (2006) *Budowanie sieci partnerskiej, założenia wstępne*. Online. <http://www.artmuseum.waw.pl/dorobek/budowanie_sieci_partnerskiej_założenia_wstepne.pdf> (accessed 26 February 2009).

Schmidt, S. J. (1994) 'Gedächtnis-Erinnern-Vergessen', *Kunstforum International*, 127: 245–49.

Sołtysiak, M. and Wierzbicka, K. (2001) *Museums in Poland: Guidebook*, trans. Aleksandra Rodzińska-Chojnowska, Warszawa: Wydawnictwo DiG.

Treter, M. (1922) 'Organizacja zbiorów Państwowych Rzeczpospolitej Polskiej', *Wiadomości Archeologicznych*, 7.

Uroczystość przeniesienia serca Tadeusza Kościuszki do Mauzoleum, in Sprawozdanie z Zarządu Muzeum Narodowego Polskiego w Rapperswylu za rok 1897 (1898), Paryż: Drukarnia Polska Adolfa Reiffa.

Van der Meer, E. (2001) 'Gedächtnis', *Lexikon der Psychologie 2*. Heidelberg, pp. 99–102.

Velez, M. A. (2007) 'Stoisko uzupełniające', *Muzeum*, 1. <http://www.gazeta-muzeum.pl/ artykul.php?id=2> (accessed 26 February 2009).

Zagrajek, A. (1998) *The Jan Matejko House: A Branch of the National Museum in Cracow: a Guidebook*, Cracow: Muzeum Narodowe w Krakowie.

Ziemięcki, T. (1879) 'Zamek na Wawelu i Muzeum Narodowe', *Dwutygodnik Naukowy poświęcony Archeologii, Historii i Lingwistyce*, 2: 393–416.

Żygulski, Z. (1962) 'Dzieje zbiorów puławskich: Świątynia Sybilii i Dom Gotycki', *Rozprawy i Sprawozdania Muzeum Narodowego w Krakowie*, 7: 5–265.

Chapter 10

NATIONAL ART MUSEUMS AND THE 'MODERNIZATION' OF TURKEY

Ayşe H. Köksal

> The national mediation remains effective . . . because it's there that the relation between a structure of inclusion and what it excludes plays itself out.
>
> (Rancière, quoted in Hallward 2003: 201)

NATIONAL MUSEUMS ARE powerful 'identity-defining' machines: 'To control a museum means precisely to control the representation of a community and some of its highest most authoritative truths' (Duncan 1996: 286). Kratz and Karp (2006: 3) are not alone in seeing museums as 'one of the institutions and practices associated with modernity, part of the checklist for being a nation, a means of disparate groups to present and claim their histories and values in the public sphere and simultaneously an arena and means of constituting community identities'. Fluid and dynamic, and based upon particular myths and discourses (Foucault 1980), the identities national museums construct are, like the nation itself, reinvented and regenerated according to changes of the political position, economic change, cultural revivals and international relationships (Hobsbawn 1990). By these means each nation acquires uniqueness, which is manifested, constructed and represented in its national museums. These ideas are no longer new but their manifestation in nation states is so diverse and complex that we cannot presume to fully know them.

Turkey, the subject of this chapter, offers new insights into political change and the instrumentalization of the national museum. Possessing dynamic issues of identity and complex international relations arising out of its position between East and West, Turkey has never possessed that calmness and self-assurance seen in many Western nations. Formed in the collapse of an Ottoman Empire composed of heterogeneous components and identities, Turkey repeatedly reinvented its national identity. Its relations with the West, and its struggle to become respected as a modern state, were two constant determinants of the success identities created. Like other nations, Turkey utilized museums in order to produce the desired image of the nation.

Art objects and the aesthetic experience have been particularly important to the construction of the identity in national museums. The narrative possibilities of these objects are most obviously affected by the artist, by disciplines such as art history and by the organizational capabilities of curators but they may also be controlled by elites of various kinds. Through the invention of narratives, art objects are inscribed with legible and legitimate traces of the past as well as an intentionality to affect the future (Roberts 1997). The aesthetic experience opens up liminal possibilities that transcend mere object literacy providing 'a moment of moral and rational disengagement that leads to or produces some kind of revelation and transformation' (Duncan 1995: 14). Art is itself about politics not because of its messages and sentiments that transmit the political discourses to the world. Not even because of the way it represents the structures of the society, the conflicts or identities of the social groups. 'It is political for the distance it takes from her functions, from the type of time and space it institutes, for the way it parcels out the time and the space' (Rancière 2006: 36). Marcuse notes that art is not about reality but about its negation; it has the potential to overcome the present reality by preserving it (Kreft 2008).

Modern art was particularly important to Turkey's experience of modernization because, rather than following the periodization of the Western canon, the term *modern* referred to 'what was new and Western' (Akay 2008: 99). Thus, modern art, with its connotations of progressive, revolutionary and most of all Western, constituted a platform for negotiating the nation's (inter)national identity as modern and a way of describing its relations with the West. In Turkey, the visible or invisible history of modern art, and its representations in the museum, changed with the ideologies of each political regime. Here this change is revealed by comparing three distinct museums, each founded in a different period: the Imperial Museum in the Late-Ottoman Period (1876–1918); the Istanbul Museum of Painting and Sculpture, founded after the Turkish Republic (1923); and Istanbul Modern, established in the new millennium.

The Imperial Museum and its legacy

The emergence of nationalism, which accompanied the Ottoman Empire's attempts at modernization, dates back to the Tanzimat Period (1839–1875) of reforms and continued to affect the politics of the empire until its collapse in 1918. Nationalism provided a powerful new ideological framework for resisting the nationalist movements of minorities in the territory, which, influenced by the ideology of the French Revolution, began to claim their independency. Additionally, the rise of the European nation states constituted a major threat for the Ottoman Empire. In this new conjuncture, the government had to re-identify itself as not only modern but also inclusive for all minorities living in the territories. A nationalist discourse, which defined all subjects as equal and the Ottoman government as modern, was a solution for the Ottoman Empire to adapt and re-integrate into the new – modern – world order. The dilemma was shared by all contemporary empires but was particularly important in the Ottoman context because of weak central control, significant socioeconomic problems and structural complexity, and the dominance of Muslims encircled by Christian powers (Hanioğlu 2009).

Before the introduction of this nationalist rhetoric, the Ottoman Empire saw itself as cosmopolitan, an empire in which several ethnic groups shared the same territory, with only minor differences in terms of taxation and legal procedures. Within the empire, no single group – not even the Turkish population – was sufficiently powerful to be politicized (Karpat 2005). Nevertheless, the ideas of the French Revolution had a profound effect on the Ottoman elite, and new intellectual bureaucrats educated in Europe – the so-called 'Young Turks' – became a driving force in the modernization project. There was a renewed effort to establish a prestigious international image for the empire in order to rescue it from a growing sense of inferiority in its relations with the West (Deringil 1993).

Consequently identities within the empire changed over time. During the Tanzimat era national identity was based on Ottomanism, which emphasized a supranational politics based on the equality of all ethnic groups in order to pacify revolts within the territory. Abdulhamid II, who reigned from 1876 to 1909, promoted Pan-Islamism, which advocated the unity of Muslims under the Caliphate in order to strengthen his absolute power and fend off constitutionalists. It was only during the last years of the Ottoman Empire that Turkish identity gained political importance. During the Turkish Revolution, Islamic and Ottoman identities dissolved and were even rejected, as bourgeois bureaucrats turned to a Turkish nationalism based on European ideas of race (Karpat 2005). The adoption of these successive identities can be understood as a response to modernization and reflected the struggle to reach a level of civil sophistication that might permit the survival of the Ottoman Empire. This modernization project did not simply involve the importation of Western ideas. 'Rather, it was a complex process of acculturation, in which Western ideas, manners and institutions were selectively adopted, and evolved into different forms set in a different context' (Hanioğlu 2009: 4). The project was a reaction to the West, and thus had a contentious rather than consolidative character.

The Imperial Museum, founded in 1869, was one of the showcases of the modernization project (Figure 10.1). It was deployed to represent Ottoman national identity and was continuously refined to reflect the shifting ideologies of the Ottoman government. The seeds of the Imperial Museum were sown in 1847 when Ahmet Fethi Pasha, the Minister of War, began collecting antiquities and storing them in the Byzantine church of St Irene, then in use as an armoury (Ogan 1947). The visit of Sultan Abdülaziz to Europe, in 1867, accelerated his interest in founding a museum and particularly in displaying antiquities. Shaw (2003: 84) remarks that it is not clear whether this was a definitive moment in the Sultan's decision to support the institution of the Ottoman Imperial Museum and to enhance its

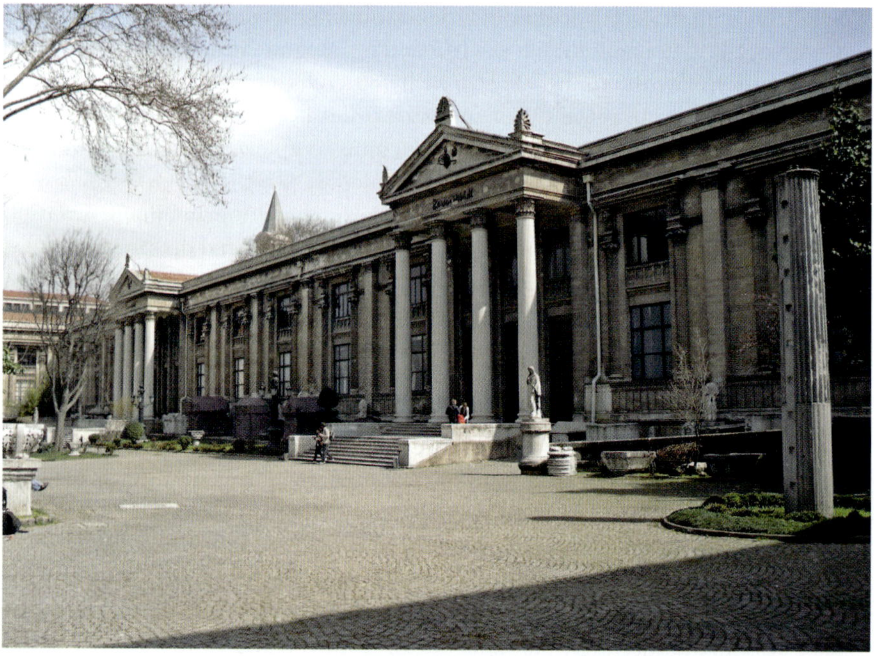

Figure 10.1 The Imperial Museum originated in the church of Hagia Sofia in 1869 but was moved into this new museum complex in 1891. This building currently houses the Museum of Archaeology of Istanbul. Photo: author

collections. But the lithographs relating to Sultan Abdülaziz's gallery visits suggested that Greco-Roman artefacts were one of the main signs of European civilization to which he was exposed in Europe – and the Imperial Museum was erected just two years after this visit. Consequently, the Imperial Museum became the new shelter for the antiquities excavated in Ottoman soil and ended the era of plunder by the Western archaeologists.

Sultan Abdülaziz not only hired European experts, but also sent his officials to Europe to undergo training in Western museum practice. The first Directors of the museum were European academics (Şapolyo 1936). However, the Sultan was not pleased with the work of these foreign Directors and turned to Turkish officials, which also served the purpose of reinforcing Ottoman hegemony over antiquities (Şinasi 1934).

The museum flourished between 1881 and 1910, under the directorship of Osman Hamdi (1842–1910), whose administrative, archaeological and legislative efforts established the norms for Ottoman museum practices (Cezar 1971). In addition to institutionalizing and expanding the museum, he also enacted laws to protect against Europeans plundering Ottoman antiquities.

The organization of display of the Imperial Museum was constantly modified in line with shifts in ideology and national identity. Before Hamdi's directorship,

Assyrian, Egyptian, Kufic and Hittite remains were in the vestibule, Helleno-Byzantine antiquities were distributed throughout the central hall, wall niches were reserved for Assyrian and Babylonian remains, and one room was devoted to antiquities from Cyprus and Rhodes (Şapolyo 1936). The objects were arranged geographically rather than chronologically and the Imperial Museum stood as a universal museum representing Ottomanism. Lacking any sense of cultural evolution or hierarchy, the museum exhibited the notion of equality for all its subjects. Additionally, the rich and multifaceted collection of objects excavated from different locations recorded the Ottoman presence in territories that had by then been lost in military defeat (Shaw 2003).

Under Osman Hamdi's leadership, the museum began to display Islamic art as well, and devoted a room to its display in 1891 (Shaw 2008). Now the museum was divided into six sections: one for Greek, Roman and Byzantine antiquities; a second room for Assyrian, Egyptian, Hittite and Phoenician antiquities, including Asian and African works; a third for Islamic fine arts; a fourth for ancient coins, and a fifth for natural history. The sixth room was a library and photography studio (Şapolyo 1936). Once more, the central focus was on the empire's Helleno-Byzantine heritage, but now Islamic art was given visibility and a sense of superiority over European heritage, in keeping with the national ideology of the time.

Although the modern art world was flourishing in Istanbul at this time, the Imperial Museum never displayed Western-style Turkish art. Since the Tulip Era of the Ottoman Empire (1718–30), Western painters had been invited to the Palace and their works had been bought for decorative purposes and collected for the Treasury (Germaner and İnankur 1989). Art classes were part of the curricula of both the Military School (Mekteb-i Harbiye) and the Civil Engineering School (Mühendishane-i Berri-i Hümayun). Even though these art classes were for the most part technical rather than creative, focusing on modern techniques of fortification, artillery and navigation, they created the basis for an interest in Western-style art (Cezar 1971). The Academy of Art (Sanayi-i Nefise), founded in 1883 as an initiative of Osman Hamdi, had a modern curriculum based on that of the École des Beaux Arts of Paris, and its students were sent to Paris for education in contemporary trends. The district of Pera (now Beyoğlu) was a centre for art exhibitions and galleries as well as a humble art market, which was the favourite district of collectors keen to buy the work of local artists (Germaner and İnankur 1989). A newspaper published by the Guild of Ottoman Painters (1908–19) confirms that Ottoman artists were actively involved in intellectual discussions on such topics as the excessive impact of the West on Ottoman art, the institutionalization of national art and Islam's philosophical approaches to art (Güler 1994).

Osman Hamdi was aware of the interest in modern art and of the need for an art museum to complement the Academy. Indeed, when the Academy of Art was founded, he emphasized the urgent need for an art museum for paintings and lithographs to institutionalize an integrated education system (Edhem 1970). A Ministry of Culture document, produced in 1909, reveals that the government did not establish this department due to a lack of funds, although the same document also records that the government sponsored the foundation of a museum in Jerusalem and the renovation of another in Bursa in the same year (Cezar 1971: 559). After the death of Hamdi, his brother Halil Edhem (1861–1938) succeeded him as Director of the museum and managed to secure a small budget from the government to establish an art collection that included copies of original paintings from museums in Paris, Berlin and Vienna, some original works of Western artists and the works of modern Turkish artists. His initial purpose was to collect art works by Western artists as well as from contemporary Turkish artists. However, the budget for the collection was small and European artworks were expensive, so in the end he decided to have copies made (Edhem 1970). This collection, Elvah-ı Nakşiye (meaning Tableaux of Painting), which was attached to the collections of the Imperial Museum, was never displayed in its galleries (Mansel 1947).[1]

Art in the Western style was excluded from Ottoman ideology; the Ottomans' need was to exhibit superior and distinct components of its society in order to re-establish their dominance in the new world order. By displaying a variety of Helleno-Byzantine antiquities – the cultures on which European nations had invented their traditions – the Imperial Museum not only claimed hegemony, but also appropriated these antiquities as part of its own national heritage. Thus the progressive narrative of display was adapted to Ottoman ideology, and the museum served to emphasize the inherent modernity of the Ottoman Empire, in that 'the progress of cultures to which Europe traced its own roots stemmed from Ottoman territory' (Shaw 2003: 94).

Islamic art was also a category praised as *authentic* by the West, and the Ottomans, as descendants of the Caliphate, were undoubtedly leaders in this art form too. This encouraged the Ottoman Empire to locate its artistic distinction and superiority on Islamic art. In contrast, modern art had been institutionalized as exclusively Western. Had the museum included modern Turkish paintings in its collection, the Ottomans would have lost face to the West (Shaw 2008).

The Imperial Museum, then, was established as a showcase to promote Ottoman identity as progressive, modern and dominant. Rather than its own public, the museum targeted the foreign gaze. The purpose of the Imperial Museum was the creation of an identity as a survival strategy, not the conveyance of Ottoman identity to its own society for nation-building. In other words, its purpose was the production of affiliations with other nations, and more importantly, the differentiation of Ottoman culture from the rest in order to proclaim its sovereignty. To this end, the Imperial Museum selectively adopted Western museum practices and adapted them to the ideological needs of the Ottoman government.

The art museum in the republic

The Republic of Turkey was established as a modern, secular and Western state in 1923, following the Turkish Revolution. Toynbee described the emergence of the

Turkish nation as 'a total abandonment of one civilization and adoption of another' and this was required for the Turkey that 'sought to be admitted as full members of Western society in order to escape from the terrible position of being its pariahs' (Robbins 1998: 65). The Turkish Revolution was a complete rejection of the Ottoman past on which all of Turkish society's ills could be blamed (Mardin 1971). The new government, by making a complete and immediate break with the past, sought to save Turkish society from the nostalgia of historical romanticism, which might then seek to glorify and mobilize Ottoman expansionism as the basis for remaking the defeated nation (Karpat 2005). A national identity was urgently required in order to invent a tradition for a society, which, supposedly, had no past. The new government's project was to create the nation on ethno-secular lines, which emphasized Turkish identity but rejected its Ottoman components. It also recognized, to some degree, its Muslim identity by embracing personal religious choice (though not Shari'a Law). In the new conjuncture, the society had supposedly completed its transition from dogmatic to secular. This process brought with it the suppression of any historical memory of locality and the promotion of a national identity that rejected the past (old, traditional, ignorant and Ottoman) in favour of the present (new, modern, enlightened and Turkish) (Bozdoğan 2001).

Mustafa Kemal Atatürk (1881–1938), leader of the Turkish Revolution and first president of the Republic, advocated a rational, democratic and civil state (Heper 2006) but did not believe that Turkish society was sufficiently enlightened or civilized to shoulder responsibility for a modern nation. Thus, according to Atatürk, Turkish society had to make *jumps* through the phases of progress to reach the level of Western civilization and become a progressive nation (Bozdoğan 2001). This accelerated agenda required top-down reforms and interventionist state policies developed by a single party, Cumhuriyet Halk Partisi, which operated within a strict bureaucratic system that affected every sphere of society. These bureaucratic officials constituted the bourgeoisie at the time and were unchallenged by a commercial middle class, which simply could not flourish in such a strictly controlled society (Göcek 1996). The ideal Turkish citizen was defined as modern, civilized and loyal to the state; implicitly, the government was all-powerful and knew the benevolence of the nation better than the public.

The Turkish state had to invent new traditions and values, on which to base its new national identity, and was aware of the power of the visual experience – particularly of art and architecture – for a society with a high level of illiteracy as a result of the recent adoption of the Latin alphabet. Modern Turkish art, which had been avoided by the Ottomans, seemed an appropriate tool for propagation of the new ideology. More importantly, modern art was a *tabula rasa* on which the state could build a new modern, Western and secular identity, imbued with references to the West and the concept of modernity, and with no connotations of Islam or Ottoman (Artun 2002). Adopting a populist cultural policy, the state actively supported the Academy of Fine Arts as part of its education policy, and encouraged contemporary artists by buying their work, giving awards, arranging exhibitions,

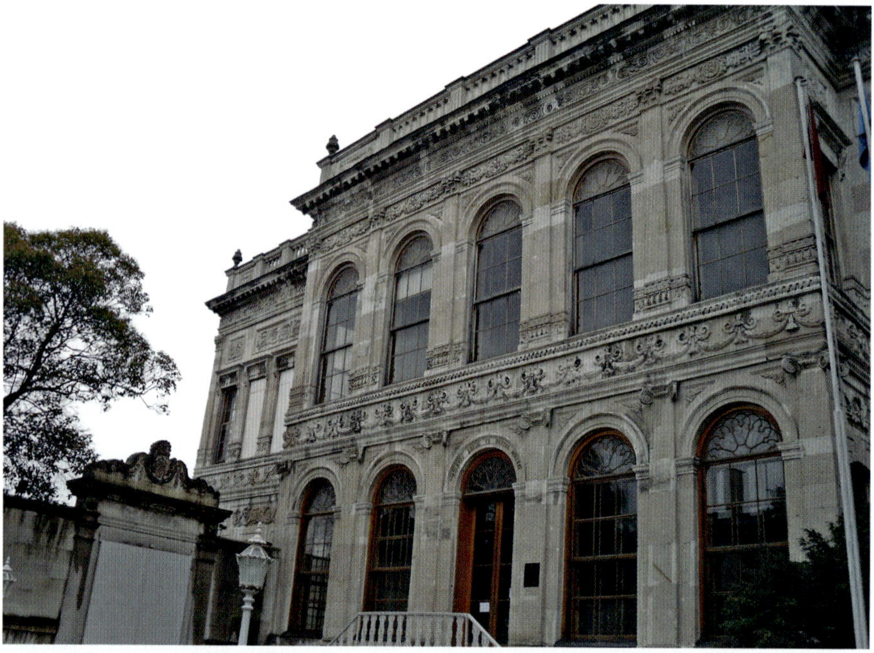

Figure 10.2 The Istanbul Museum of Painting and Sculpture followed the Louvre by being housed in an annex of the Dolmabahçe Palace, the residence of the Sultan's heirs during the Ottoman Empire. Photo: author

sending individuals to Europe for education and organizing artists' visits to rural areas to educate the public through the vehicle of modern Turkish art.[2] During this period, the state officials, being the only bourgeois class, were not only the sole customers and collectors of modern art, but also its exhibitors, audience and critics. The only missing component was a museum for modern art.

The Istanbul Museum of Painting and Sculpture was established in 1937 by direct order of Atatürk, who wisely chose the apartments of the former Ottoman Heir Apparent in Dolmabahçe Palace as its main venue (Figure 10.2).[3] The first Director of the museum, Halil Dikmen (1906–1964), an artist and academic, assembled a staff, organized the museum and immediately began work on expanding the collections.[4] The collection was based on palace treasures and Halil Edhem's Elvah-ı Nakşiye collection, neither of which had ever been on display in the Imperial Museum. The state also ordered its ministries and public offices to send all art works in their possession to the museum.[5] In addition, Dikmen constituted a jury to select for display works from among those 'not damaged and decomposed, which had been thrown into depots and kept in closed rooms for years' (Örik 1937: 477). In order to institutionalize modern art, the museum was affiliated to the Academy of Fine Arts. Devrim Erbil, a former Director of the museum and student of Dikmen, refers to Dikmen's 'charismatic character and the state's generous attitude to modern art which transformed the Istanbul Museum of Painting and Sculpture into a cultural centre where young artists gathered, discussed and produced in the first years of the Republic'.[6]

In contrast to the Imperial Museum, the Istanbul Museum of Painting and Sculpture adopted Western museum practices, a scientific taxonomy and a narrative of progress. Following the Louvre model (McClellan 1996), the rooms where the Sultan's heirs had lived were divided into small galleries so that the visitor followed the narrative programme. Categorization of objects was canonical, displays were chronological with clear, informative labels and the narrative established an evolutionary trajectory, referring to the modern and rational progress of a nation. The Museum possessed 17 chronologically arranged galleries. Galleries 1–6 covered the period 1800–1860 and were devoted to artists from the Military and Engineering Schools. These were called Primitives, based on the criterion that they were the first group to receive a Western education in Paris and were pioneers in producing Western-style art. Galleries 7–9 exhibited the works of the period 1860–1923. These galleries were devoted to works of Academy of Art graduates from the Ottoman period, those who constituted the second wave of contact with the West via a Paris education. Galleries 12–17 contained the most progressive art works, produced from 1923 onwards. These were artists who had reached a certain level of maturity, as they were the third group with a Western education and, at the same time, Republicans. Copies of works by Western painters such as Rembrandt, Rubens, Van Eyk, Dürer and original works by artists such as Utrillo, Bonar, Levy and Preziosi were contained in Galleries 10 and 11. Sculptures, of which there were few, were exhibited in a small gallery and in the garden (Epikman 1938).

Even though the order was chronological, the main emphasis was on the artists' relationship with the West; modern Turkish art history was categorized into three periods – the Primitives, the Intermediate and the Modern – the last of which unquestionably referred to works from the Republican period (Berk 1938). Accordingly the stylistic references were divided as the Primitives-realist, Intermediate-impressionist and the Modern-abstract (Dikmen 1950). This categorization related less to the technical or creative criteria of the art works, and more to their resemblances or contact with Western art; the criterion was Western impact on the artist. Thus, unlike the Louvre, where the narrative of display attributed French art the degree of perfection and claimed the same position in the overall framework of civilization (Duncan and Wallach 2004), the Istanbul Museum of Painting and Sculpture resembled an apprentice seeking the approval of a master, in this case, the West.

It is also worth noting the exclusion of Islamic and traditional art from the collection. The Islamic art of the Ottomans had become the heritage of a distant past, while modern Turkish art was the future of a modern nation. Dikmen, however, was not satisfied with the exclusion of traditional art, remarking, 'If we want to see the Istanbul Museum of Painting and Sculpture as a national art museum we have to include also miniatures and religious paintings in our collection. Otherwise, Turkish painting will never find its character.'[7] However, the museum was the Western face of Turkey and there was no room for any reference to the Ottoman past. Considering the fact that the first art history department was founded in 1943, the museum's inclusion/exclusion of objects and its categorization was accepted as the canon of Turkish painting, due to the lack of any alternative perspective. Thus, the narrative of the Istanbul Museum of Painting and Sculpture that conveyed the state ideology became the representation space where the invented history of the nation overlapped with the art history.

The decision to display modern Turkish art in the reception halls of the former heir to the Sultan was a demonstration of the power of the Republic over the possessions of the old empire. As in the Louvre, the palace was reorganized as a public space; marks of empire were removed and halls and rooms redecorated as showcases for the products of a new nation. Modern art works were here assigned new meanings and as 'public property, they became the means through which a new relationship between the individual as citizen and the state as benefactor could be *symbolically* enacted' (Duncan 1995: 24). The museum became visual proof that the state was committed to the principle of community and was ready to serve its public by providing an equal access to 'liberated' treasures (McClellan 1996: 37).

However, the image of serving the public was ideological rather than a reality. The museum's primary function was to offer the spectacle of a civic state for the purpose of promoting its modern image in the West; it was less interested in addressing its own public. The Istanbul Museum of Painting and Sculpture was not 'an elite experience for everyone' (Zolberg 1994: 49). It provided a place where the elite bourgeois bureaucrats could socialize with each other as an exclusive community and share the codes of cultural capital, secure in their distinction (Bourdieu 1984). Consequently, visitor numbers, which averaged 300–400 per day in 1937[8] had dropped to 70 visitors per day, including school groups by 2006 (Aygündüz 2006). The negligence of the government and the gradual decay of the Istanbul Museum of Painting and Sculpture, which has resulted in temporary closure, never brought about a reaction or protest from the public. Clearly the Istanbul Museum of Painting and Sculpture was never adopted or accepted by Turkish society as a national art museum.

The museum served 'not so much to shore up an existing social order as to provide the script for a new one, and to provide its visitors with new discursive positions within that order' (Bennett 1998: 30). The museum displayed an *idealized,* modern, Western and secular Turkish national identity. Its target was not the Turkish public, but the intellectual elite, who had power in politics, and the West. Unlike Western art museums, which play *within* the league, the Istanbul Museum of Painting and Sculpture campaigned for access *to* the league, in order to prove that Turkey belonged, not on the periphery, but at the centre.

Istanbul Modern and capitalism

The possibility that the Turkish state might gain access to the league became more of a project after the transition to a multi-party political system in 1950. However, the process of accession was not achieved entirely as a result of successful politics. Rather it reflected economic shifts and the growing impact of globalization, which at last gave the commercial bourgeoisie an opportunity to flourish. This new middle class immediately began to access global economic networks and particularly so after the 1980s, when Turkey completed its transition to a capitalist market economy by way of neo-liberal policies, privatization, globalization and adoption of the free market (Keyder 2005). This process changed Turkey and its relationships with Europe and America, opening the possibility of accession to the European Union (EU). The state now enacted economic reforms that permitted the commercial middle class to accumulate wealth, which in turn encouraged them to invest in culture and art.

The old social image of the entrepreneur as somehow corruptive and shameless changed (Buğra 1994). Proud of its wealth, the new bourgeoisie began to claim its right to be visible in every sphere of society, from philanthropic activities to founding new museums. In its search for a way to claim hegemony in the cultural sphere, the middle class created its own shared culture, which included art museums, through which it could also share a global identity (DiMaggio 2004).

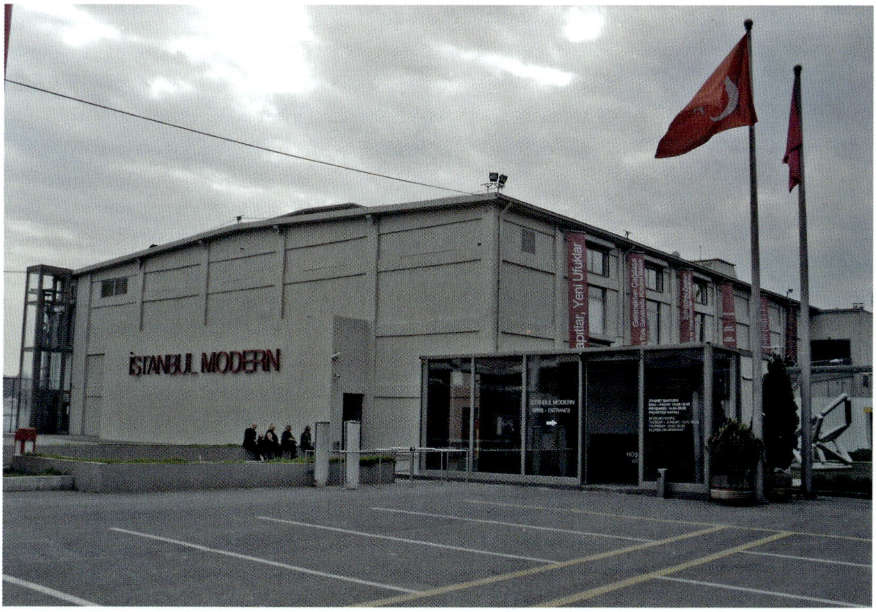

Figure 10.3 Istanbul Modern is located in a converted warehouse. It is a space which enables diverse exhibitions of modern and contemporary Turkish art. Photo: author

The investment boom was centred mostly in Istanbul; the city became the centre of global markets. Increasingly, it became necessary 'to sell' (Keyder 2000) Istanbul through different channels in order to sustain its relationships with the global economy. The positive image of the whole nation became dependent on the city, and art museums became one of the tools helping to boost the overall profile of the city, which was projected as open, globalized and attractive (Aksoy 2008). This attempt produced welcome results; Istanbul became European Capital of Culture for 2010.

Istanbul Modern (Figure 10.3) was founded in 2004 by the most ambitious family among cultural capitalists (DiMaggio 2004), the Eczacıbaşı Holding Group. The Eczacıbaşı family has been sponsoring internationally acclaimed festivals of music (including jazz), theatre, film and biennials for years, but wished to do even more. As noted by Oya Eczacıbaşı, wife of the Chairman of Eczacıbaşı Companies and Board Director of Istanbul Modern (Anon. 2008), the family began collecting works of art as early as 1989 in the hope of fulfilling their long-term desire to found a modern art museum. With an investment of $5 million, they finally reached their goal with the opening of the Istanbul Modern, which housed a collection of 4,000 works.

The site chosen for the Istanbul Modern was a 1950s concrete warehouse in Karaköy, with a floor space of nearly 8,000 square metres. The building was renovated by a group of young Turkish architects who opted for a functional, minimalist approach, installing huge glass cubes, which provided an immense empty space for the exhibitions together with a library, a cinema and an elegant restaurant. 'The architecture is being used to show that Istanbul, that Turkey, has risen. It can't stop. Europe is old news', Melkan Tabanlıoğlu, one of the architects of the project, commented.[9]

The Turkish Prime Minister, Recip Tayyip Erdoğan, intervened to ensure that the museum would be completed in time for the opening ceremony to overlap with the summit meetings to discuss Turkey's accession to the EU.[10] In his opening speech, after emphasizing that Istanbul Modern was proof that 'Turkey is already a part of Europe', Erdoğan continued by saying, 'Today we have gathered here as a people who search for light that will end the darkness. We do not want to rule the world; we want to continue our search for the light' (Anon. 2004: 24).

Being the citizens of a nation in search of enlightenment, people rushed to see the museum: Istanbul Modern hosted 17,500 visitors in its first week (Anon. 2004); 500,000 people visited the museum in its first year (Anon. 2006). This sudden public embracement of the museum reflected good marketing. The museum appeared in more than 2,200 articles in the Turkish press alone in its first year. What was more important, however, was its appearance in the foreign news media. Here, in the *New York Times*, the *Herald Tribune*, *L'Express*, the *Independent* and the *Guardian*, Istanbul Modern was referred to as the modern face of Turkey. Susan Sachs, writing in the *New York Times*, described the courageous attempts by the Eczacıbaşı family to build a museum for contemporary art.[11] Tom Dyckhoff, in the *Times Online*, referred to the architecture of the new museum as one of 'Turkey's New Delights'.[12]

Anna Sussman, also in the *New York Times*, invited everyone to see Istanbul, which now has the Istanbul Modern that

> makes an unmistakable statement about Istanbul's self-perception as a major cultural capital. The gravel courtyard and warehouse-like exterior are reminiscent of P.S. 1 in Long Island City, Queens, and the interior – with its industrial feel and long, fluorescently lighted hallways lined with amply spaced paintings – brings to mind the Pompidou Centre in Paris.[13]

Observation–Interpretation–Multiplicity, the opening exhibition, was an eclectic mix of modern and contemporary Turkish art, from 1900 onwards. While the Eczacıbaşı family collection constituted the basis, Istanbul Modern was supplemented with long-term loans from the Istanbul Museum of Painting and Sculpture, which still had the most diverse collection of modern Turkish art, and also from the collection of a bank, Türkiye İş Bankası. 'The state collection in the private shell' was arranged thematically, in categories such as landscape, still life and interior (Antmen 2007: 25). One of the curators, Çalıkoğlu, asserted that they had no choice but to adopt a thematic display:

> Chronological order is a bequest of the Republican period and it was appropriate for its ideology and its perspective of modernity. But today, this perception seems problematic. We want to convey a better understanding of modern art, providing the socio-cultural context of the objects. There, thematic display works well.
>
> (Yazıcı 2006: 24)

Istanbul Modern followed the trend that Turkey had rediscovered the Ottoman Empire and was no longer afraid of it; it was now accepted as a period of Turkish history (Ortaylı 2008). Interestingly, Istanbul Modern chose to start modern Turkish art from 1900, thus including art works from the Primitives and Intermediate categories in the modern period. By this means, the museum extended the roots of modernity back into the Ottoman Empire, which, ironically, had not considered such art to be part of the national image. Istanbul Modern's new definition of modernity emphasized that Turkish society was modern even before the Turkish Republic came into existence.

The museum also included contemporary art in its collection. It was recognized that contemporary art gave opportunity, visibility and prestige to artists outside of Western countries. Artists, who cannot breach the strict walls of the Western canon and its museums, had already begun to use biennials and contemporary art collections to circumvent the restrictions of the Western establishment (Buddenseig and Wiebel 2007; Vanderlinden and Filipovic 2005). The statement was made even more effective with the most recent organization of display of the permanent collection 'New Works New Horizons' in 2009, which has focused more on acclaimed contemporary Turkish artists and less on the Republican era. Legitimized by discourses of transnationalism, spacelessness and nomadism, Istanbul Modern strove to overcome national limitations. Following this trend, it extended its collection not only to Turkish contemporary art, but also to work by foreign artists. This is undoubtedly a reflection of the Eczacıbaşı family's global aspirations.

The promise of the Istanbul Modern was to demonstrate that Istanbul – and thus Turkey – has entered the global community. The museum re-affirmed that Turkey, with its long tradition of modernity dating back to the days of the Ottoman Empire, deserves entry into the EU. The Eczacıbaşı family's founding of the Istanbul Modern also contained the access code, which connected the city into the global bourgeoisie. Now the state and the private sector were able, hand-in-hand, to promote the museum as a symbol of Turkey leading the country into the light. The enlightenment of Istanbul is now the modern and (trans)national image of the whole Turkish Republic. Thanks to the Istanbul Modern, as former Director David Elliot noted, the time has come to start 'looking to the world from Istanbul'.[14] With its trendy museum practices, postmodern architecture and its collection and display policy, Istanbul Modern emphasizes that it is a global museum, located in a global city in a global country.

Conclusion

The purpose of this chapter has been to contribute to discussions concerning the national museum as an ongoing process of negotiation involving knowledge, ideology and power. The example of Turkey shows how the specific configurations of these elements in time and space can result in radically different conceptions of the national museum as a nation-building institution. Turkey provides a remarkable example of a country which has, in less than a century, undergone major socio-political change that has, with each change of regime, seen the visibility of art in the museums reinvented and redeployed to reflect the political imaginings of the nation possessed by those who sought to control culture: successively, the monarch, the state and a wealthy elite. An overriding political desire of these elites has been to place Turkey on the world stage – to be seen not simply as a neighbour to Europe but as a great civilization in its own right. In order to achieve this, Turkey has repeatedly adopted cultural forms prevalent in Europe. The Imperial Museum was long ago broken up as a symbol of the old order, and while the Istanbul Museum of Painting and Sculpture survives it does so as a relic of earlier values and a regime that has also been surpassed by state progress. In both cases, these museums deployed art to represent the nation to the outside world. Now that nation has again changed, but it is still showing the world an outward-looking face in the form of its art museums. Istanbul Modern, like its predecessors, reflects the contemporary ideology and the controlling interests of elites, but now this city senses that it is no longer on the periphery. It should be noted, finally, that while each of the museums discussed here has been used quite overtly to shape the image of the nation, not one has been labelled 'national museum'.

Notes

1 The collection was on display in the reception hall of Dolmabahçe Palace from 1918 to 1920, only after the Ottoman government had lost power and the Turkish Revolution had begun (Arık 1953).
2 The State Archive of the Prime Minister, 1937/no. 2-6071, 1931/no.10710, 1935/no.6-1276.
3 Archive of the Istanbul Museum of Painting and Sculpture, 1937/no: 358. The statement 'by direct order of Atatürk' confirms that law or legislation did not found the Istanbul Museum of Painting and Sculpture. During the reformation, the state was in a rush to actualize its reforms, and many institutions of the time were founded without any legislation. The Istanbul Museum of Painting and Sculpture finally legislated a museum in 2003.
4 Archive of the Istanbul Museum of Painting and Sculpture, 1937/no: 1763, /1937/no: 25, 1937/no: 42.
5 Archive of the Istanbul Museum of Painting and Sculpture, 1937/no: 4-2837.
6 D. Erbil, pers. comm., 31 May 2007.
7 Archive of the Istanbul Museum of Painting and Sculpture, 1938/no: 68.
8 Archive of Istanbul Museum of Painting and Sculpture, 1937/no: 21
9 Dyckhoff, T. (2006) 'Turkey's new delights', *Times Online*, http://entertainment.timesonline.co.uk/ (accessed 10 July 2009).
10 The government was not involved in funding the institution, but supported the Eczacıbaşı Group by cutting through bureaucratic red tape and facilitating acquisition of the warehouse, which was located in a protected area.
11 Sachs, S. (2005), 'In Istanbul, Creating a Space for Modern Art', *New York Times*, http://www.nytimes.com/ (accessed 11 July 2009).
12 Dyckhoff, op.cit.
13 Sussman, A. (2005), 'Travel To Istanbul', *New York Times*, , http://travel.nytimes.com/ (accessed 10 July 2009).
14 Onat, A. (2007), 'Dünyaya İstanbul'dan Bakmak: David Elliot ile söyleşi', *Milliyet Sanat*, http://sanat.milliyet.com.tr/ (accessed 10 July 2009).

Bibliography

Akay, A. (2008) 'Threads of progress adhering to modern art in Turkey', *Third Text*, 22(1): 99–104.

Aksoy, A. (2008) 'Istanbul's choice', *Third Text*, 22(1): 71–83.

Anon. (2004) 'İstanbul Modern açıldı', *Radikal*, 12 December, 24.

Anon. (2006) 'İstanbul Modern: Bir yılda 500 bin ziyaretçi', *Radikal*, 4 January, 23.

Anon. (2008) 'İstanbul Modern'le daha disiplinli oldum: Oya Eczacıbaşı ile söyleşi 2008', *Milliyet Pazar*, 13 July, 11.

Antmen, A. (2007) 'Özel Kabukta Devlet Koleksiyonu', *Radikal*, 19 September, 23.

Arık, O. (1953) *Türk Müzeciliğine Bir Bakış*, Ankara: Milli Eğitim Bakanlığı.

Artun, A. (2002) 'The museum that cannot be', unpublished paper, presented in 'Exposer L'Art Contemporain du Monde Arabe de Turquie', IISSM–Ecole des Hautes Etudes en Sciences and Sociales.

Aygündüz, F. (2006) '11 bin esere 70 ziyaretçi', *Milliyet Pazar*, 25 June, 10–11.

Bennett, T. (1998) 'Speaking to the eyes', in S. Macdonald (ed.) *The Politics of Display*, London: Routledge, pp. 25–36.

Berk, N. (1938) 'Resim ve Heykel Müzesi I', *Ar Sanat Dergisi*, 4: 10–12.

Bourdieu, P. (1984) *Distinction: A Social Critique of the Judgement of Taste*, Cambridge, MA: Harvard University Press.

Bozdoğan, S. (2001) *Nationalism and Nation Building: Turkish Architectural Culture in the Early Republic*, Seattle: University of Washington Press.

Buddensieg A. and Weibel, P. (eds) (2007) *Contemporary Art and the Museum: A Global Perspective*, Ostfildern: Hatje Cantz Verlag.

Buğra, A. (1994) *State and Business in Modern Turkey: A Comparative Study*, Albany: State University of New York.

Cezar, M. (1971) *Sanatta Batı'ya Açılış ve Osman Hamdi*, İstanbul: Türkiye İş Bankası Yayınları.

Deringil, S. (1993) 'The invention of tradition as public image in the Late Ottoman Empire, 1808–1908', *Comparative Studies in Society and History*, 35(1): 3–29.

Dikmen, H. (1950) 'Resim ve heykel müzesi', in R. E. Koçu (ed.) *İstanbul Ansiklopedisi*, İstanbul: İstanbul Ansiklopedisi ve Neşriyat, pp. 1491–93.

DiMaggio, P. (2004) 'Cultural entrepreneurship in nineteenth-century Boston, part II: The classification and framing of American art', in B. M. Carbonell (ed.) *Museum Studies: An Anthology of Contexts*, Malden: Blackwell, pp. 461–83.

Duncan, C. (1995) *Civilizing Rituals: Inside Public Art Museums*, London: Routledge.

Duncan, C. (1996) 'Art museums and the ritual of citizenship', in S. Pearce (ed.) *Interpreting the Objects and Collections*, London: Routledge, pp. 279–87.

Duncan, C. and Wallach, A. (2004) 'The universal survey museum', in B. M. Carbonell (ed.) *Museum Studies: An Anthology of Contexts*, Malden: Blackwell, pp. 51–71.

Edhem, H. (1970) *Elvah-ı Nakşiye Koleksiyonu*, İstanbul: Milliyet Yayınları.

Epikman, R. (1938) 'Türkiye'de Plastik Sanatlar', *Ar Sanat Dergisi*, 22: 3–9.

Foucault, M. (1980) *Power/Knowledge: Selected Interviews and Other Writings*, New York: Pantheon.

Germaner, S. and İnankur, Z. (1989) *Oryantalizm ve Türkiye*, İstanbul: Türk Kültürüne Hizmet Vakfı Yayınları.

Göcek, M. (1996) *Rise of the Bourgeoisie, Demise of Empire: Ottoman Westernization and Social Change*, New York and Oxford: Oxford University Press.

Güler, S. (1994) 'İkinci Meşrutiyet Ortamında Osmanlı Ressamlar Cemiyeti ve Osmanlı Ressamlar Cemiyeti Gazetesi', unpublished PhD thesis, İstanbul: Mimar Sinan Üniversitesi.

Hallward, P. (2003) 'Interview with Jacques Rancière: Aesthetics and Politics', *Angelaki: Journal of The Theoretical Humanities*, 8(2): 119–21.

Hanioğlu, Ş. (2009) *A Brief History of the Ottoman Empire*, Princeton, NJ : Princeton University Press.

Heper, M. (2006) *Türkiye'de Devlet Geleneği*, İstanbul: Doğubatı Yayınları.

Hobsbawn, E. (1990) *Nations and Nationalism since 1780: Programme, Myth, Reality*, Cambridge: Cambridge University Press.

Karpat, K. (2005) *Osmanlı Geçmişi ve Bugünün Türkiye'si*, İstanbul: İstanbul Bilgi Üniversitesi Yayınları.

Keyder, Ç. (2000) *İstanbul: Küresel ile Yerel Arasında,* İstanbul: Metis Yayınları.

Keyder, Ç. (2005), 'Globalisation and social exclusion in Istanbul', *International Journal of Urban and Regional Research*, 29(1): 124–34.

Kratz, C. and Karp, I. (2006) 'Introduction: Museum frictions: Public cultures/global transformations', in I. Karp, C. Kratz, L. Szwaja and T. Ybarra-Frausto (eds) *Museum Frictions: Public Cultures/Global Transformations*, Durham, NC, and London: Duke University Press, pp. 1–33.

Kreft, L. (2008) 'Sanat ve siyaset: sanatın siyaseti ve siyasetin sanatı', in A. Artun (ed.) *Sanat/Siyaset: Kültür Çağında Sanat ve Kültürel Politika*, İstanbul: İletişim Yayınları, pp. 9–39.

Mansel, A. (1947) 'Halil Edhem ve İstanbul müzeleri', *Halil Edhem Hatıra Kitabı*, Ankara: Türk Tarih Kurumu Yayını, pp. 13–26.

Mardin, Ş. (1971) 'Ideology and religion in the Turkish Revolution', *International Journal of Middle East Studies*, 2(3): 197–211.

McClellan, A. (1996) 'Nationalism and the origins of the museum in France', in G. Wright (ed.) *The Formation of National Collections of Art and Archeology*, Washington: National Gallery of Art.

Ogan, A. (1947) *Türk Müzeciliği'nin Yüzüncü Yıldönümü*, İstanbul: Türkiye Turing ve Otomobil Kurumu.

Örik, N. S. (1937), 'Resimlere ve Ressamlara Dair',*Ülkü*, 9(54): 477–78.

Ortaylı, İ. (2008) *Osmanlı'yı Yeniden Keşfetmek*, İstanbul: Timaş Yayınları.

Rancière, J. (2006) *The Politics of Aesthetics*, Bodmin: MPG Books.

Robbins, K. (1998) 'Interrupting identities: Turkey/Europe', in S. Hall and P. Du Gay (eds) *Questions of Cultural Identity*, London: Sage, pp. 61–87.

Roberts, L. (1997) *From Knowledge to Narrative,* Washington and London: Smithsonian Institution Press.

Şapolyo, E. (1936) *Müzeler Tarihi*, İstanbul: Remzi Kitabevi.

Shaw, W. (2003) *Possessors and Possessed: Museums, Archaeology and the Visualization of History in the Late Ottoman Empire*, Berkeley: University of California Press.

Shaw, W. (2008) 'Museums and narratives of display from the Late Ottoman Empire to the Turkish Republic', *Muqarnas: Journal of Islamic Art and Architecture*, Vol. 24, Leiden: Brill, pp. 233–79.

Şinasi, A. (1934) 'Müzelerimiz ve Hamdi bey', *Ülkü*, 3(14–16): 187–91.

Vanderlinden, B. and Filipovic, E. (eds) (2005) *The Manifesta Decade: Debates on Contemporary Art Exhibitions and Biennials in Post-Wall Europe*, Cambridge, MA: MIT Press.

Yazıcı, M. (2006) 'Modernleşmenin Tarihi', *Radikal*, 17 January, 24.

Zolberg, V. (1994), 'An elite experience for everyone: art museums, the public and cultural literacy', in D. Sherman and I. Rogoff (eds) *Museum Culture: Histories, Discourses, Spectacles*, Minneapolis: University of Minnesota Press, pp. 49–66.

Chapter 11

POLITICAL CHANGE AND THE NATIONAL MUSEUM IN TAIWAN

Chi-Jung Chu

THE DEVELOPMENT OF NARRATIVES in Taiwan's national museums has been influenced by the country's changing political relationships with Japan, China and the US. Taiwan was a colony of Japan for fifty years, ending in 1945, and it was during this period that the country underwent early modernization. The Kuomintang, or the Chinese Nationalist Party (KMT), founded the Republic of China in 1912 and took possession of Taiwan in 1945 following the Japanese defeat. In 1949, the KMT found itself defeated in the Chinese Civil War by the Chinese Communist Party and retreated to Taiwan where it stayed in power until 2000. During this period, the US assisted the development of the KMT as a Cold War ally. The Democratic Progressive Party (DPP) – a home-grown Taiwanese party – came to power in 2000 and governed the country for eight years. The KMT had always supported Taiwan's eventual reunification with China; the DPP now asserted Taiwan's political sovereignty as an independent state.

This chapter will examine how three national narratives – Japanese, Chinese, and Taiwanese – were played out in Taiwan. It will begin with a discussion of the establishment of the Director-General's Museum by the Japanese, and consider colonial governance, modernization, and the museum's role in promoting citizenship. This will be followed by an examination of the period of KMT rule, looking particularly at the National Palace Museum and the Director-General's Museum which was then renamed the Taiwan Provincial Museum. Discussion here will focus on how Chinese political ideology was imposed in these two museums during the Cold War period and how the Taiwanese people were taught to be Chinese through the state's cultural apparatus. The third part of the chapter looks at Taiwan after the political liberation of the late 1980s, when Taiwan's own history and identity emerged after decades of political and military suppression. When the DPP took power, the Taiwan Provincial Museum regained its significance and became the cultural hub through which the DPP mapped its own national narrative. National museums were now heavily exploited by the DPP government to construct Taiwan's cultural and political identity. Evidence for this will be provided by reference to new national museum projects with a strong emphasis on Taiwanese art and culture centring on the National Taiwan Museum.

Since the late 1990s, there has been considerable research interest in Taiwan's museum transformation (Ju 2003, 2007; Lee 2006; Li 2005; Liao 2004). Taiwan's colonial past, its developmental trajectory and its current political circumstances, have made national narratives a controversial issue here.

Japanese rule and the rise of the colonial museum

Before the Japanese took control of the island, Taiwan was a territory of China, and migrants from the mainland had arrived continuously from the seventeenth century.[1] Chinese culture formed the backbone of society and it was China which began the earliest phase of modernization until the island was ceded to Japan in 1895 after the Ch'ing court lost the First Sino-Japanese War.

Wishing to be the imperial power in the East, Japan was then undergoing a nationwide process of Westernization and subsequently implemented modern, Western projects, first in Taiwan and then in its second colony of Korea (Beasley 2000; Myers *et al.* 1984). Japan established Taiwan as its colonial model and made it part of its economic network in East Asia, its aim being to provide and supplement Japan's industrial, agricultural and economic development. It was during this period that Taiwan experienced Western technologies and enterprises, and amongst these was the museum (Chou 1997). By the end of colonial rule, Taiwan's infrastructure was established: public transport, educational system, banking infrastructure, urban plans, census, public health and hospital establishments were all in place. Although people in Taiwan were treated as second-class citizens and were granted limited freedom in social and political activities, colonial rule did set the foundation for Taiwan's later development. Taiwan's experiences of Japanese rule were quite unlike those of the Koreans.

As two popular instruments for the representation of imperial power in the nineteenth century, museums and expositions were Western developments which fascinated Japan. Japanese government officials had visited museums in the US, Canada and Europe in the early 1860s and Japan had established its first museum, the Tokyo National Museum, in 1872 (Shiina 1989). In addition to absorbing museum ideas from the West, the Japanese also developed their own genre of educational museum which emphasized the pedagogical aspect of museums over that of scholarly research (Shiina 2005). This led to the building of Taiwan's first museum, the Tainan Educational Museum, in 1902. Unfortunately this museum was bombed by the US air force during the Second World War and was never rebuilt.

The Director's General's Museum, founded in 1908 and Taiwan's first large-scale museum (Figure 11.1) was established to celebrate the completion of the

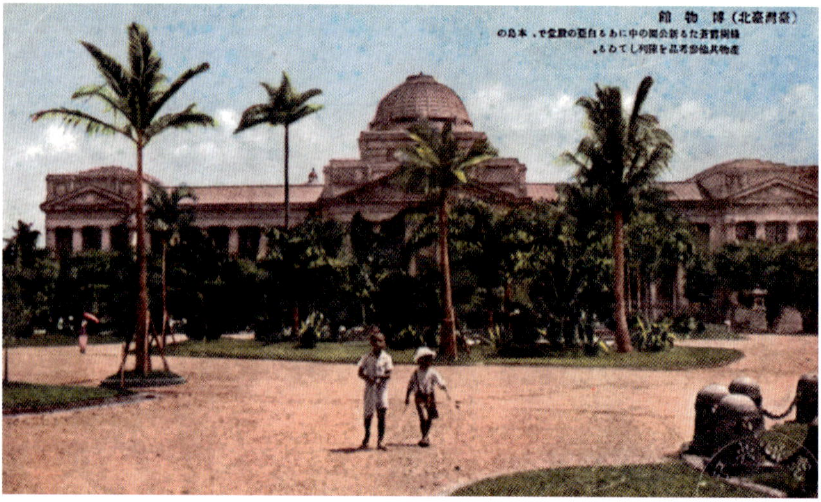

Figure 11.1 The Director-General's Museum, Taipei, in colonial times. Photo: Copyright Institute of Taiwan History, Academia Sinica

North–South railway and the achievement of the two Governors in charge over the period of its building. The architectural components and the interior of this museum – its Doric columns, pediment, dome, grand staircase and galleries – show Japan's ambition and competence in modelling its empire in Western symbols. The early collection was composed of objects formerly displayed at a product display centre under the Director-General's Office, but this was soon supplemented by objects contributed by the Office's committee on the investigation of the culture of indigenous people, set up in 1901 (Lee 2006). By 1915, the museum possessed 23,268 objects, covering geology, botany, zoology, history, agriculture, forestry, marine biology, mineralogy, crafts, trade and the indigenous people and their culture (Lee 2006; T.-N. Li 1999). The collection also drew in objects from outside Taiwan, from areas in South East Asia and southern China; indeed, from wherever Japan's colonial ambitions were directed.

Principally a natural history museum, the Director-General's Museum took a particular interest in the anthropological study of Taiwan's indigenous people and culture, and produced anthropological collections not dissimilar to those in European and American museums. In this manner, the museum was built and developed to show the government's strength and power by replicating the great museums in the West. It was soon fully fledged and professionalized, and participating in both scholarly research and popular education. The museum also housed the Taiwan Museum Association, established in 1929, which published journals promoting professional practices and containing studies of museum objects. The association held island-wide conferences for its members, which included school teachers and principals.

Other museums built by the colonial government were categorized as 'grassroots' museums, 'popular' museums or 'educational' museums (Lee 2006). These museums were built to promote commerce, local industries and agriculture, and sometimes exhibited objects from mainland Japan, Japan's other colonies, or places in which it had an 'interest'. Several 'grassroots' museums resulted from the 1935 Exposition which celebrated the fortieth anniversary of colonial rule (Cheng 2001; Lee 2006); these new small museums inherited exhibits from the Exposition.

Apart from displaying Japan's colonial power and promoting commerce and industries, museums built in Taiwan by the colonial government were positioned as social or informal educational organizations directed at citizens. In 1934, the colonial government published 'Outlines on cultivating society in Taiwan' (Taiwan Governor's Office 1935), which formally promoted social education-related activities. Programmes included the promotion of the Japanese religion and language, youth training, 'social cultivation' and social education networks. 'Social cultivation facilities' included museums, libraries, zoos, botanical gardens, parks, art exhibitions, centres to promote public health and product display centres. Associations were formed, and activities were held, to promote social education. 'Festivals' or 'Days' were created to promote patriotism and discipline and these included the National Day (to promote Japanese nationalism and patriotism), Timekeeping Day, Alcohol Abstinence Day, Library Week and Museum Week (Taiwan Governor's Office 1938). A Western educational system was also introduced into Taiwan by the colonial government, but only a quarter of children had access to an elementary school education; opportunities to enter into higher education were even fewer (Huang *et al.* 2002). For these reasons, social educational organizations, which operated outside of the formal educational system, became important channels for the colonial government to educate citizens and to promote its policies. As a result of efforts made by the Museums Association, Museum Week was launched in 1934, during which the Director-General's Museum had late-night opening and special exhibitions and talks to attract visitors – both the general public and professionals. But, with the coming of the Second World War, the focus of these programmes changed and became centred more on Japanese nationalism and wartime morale. With the end of the war and Taiwan's return to China after a fifty-year separation, many museums disappeared or were merged into local cultural organizations, or became part of the Director-General's Museum collection or, as it would be known, the 'Taiwan Provincial Museum'. [2]

The rise of Chinese museums in Taiwan

In 1925, long before its retreat to Taiwan, the KMT had established the Palace Museum in Beijing, having taken possession of the royal collection of the Ch'ing court and declared that it was owned by the public. The collection moved to a number of sites in China during the turbulent years of the mid-century before being sent from Nan-jing to a remote location in central Taiwan in 1948, during the

Chinese Civil War. It was shortly after that that the KMT also fled to the island where it continued to govern its now much reduced Republic of China (National Palace Museum 2002: 9). Establishing a special committee to oversee the collection and its management, the museum has since then been under the aegis of people closely connected to the inner circle of power of Chiang Kai-shek and subsequent governments.[3]

This, then, was a period in which Taiwanese museums followed the grand Chinese national narrative – a period which coincides with the Cold War, when the US and Soviet Union divided up the globe through separate alliances. Taiwan (or the 'Republic of China', to give it its official title) under the dictatorship of the KMT's leader, Chiang Kai-shek, became allied with the US who considered it 'Free China'. From 1950, Taiwan was in receipt of various aid programmes from the US; then, in 1965, the US changed its foreign policy and befriended Communist China. It was with this assistance that the KMT established its political legitimacy and restored Taiwan's economy after the turmoil of the 1940s. In 1962, the KMT government and the US Department of State signed an agreement to build the National Palace Museum in Taipei and a New York construction company was appointed to undertake the work.[4] The agreement shows that the financing and supervision of the construction project were under US control:

> Whereas, the committee on March 20, 1962, applied for a grant of NT$13,600,000.00 for the Tourism Promotion – Construction of Palace Museum, which project has been approved by the U.S. Agency for International Development Mutual Security Mission to China . . . Whereas, CUSA [Council for US Aid] has retained the J. G. White Engineering Corporation (hereinafter called the Supervisor) to supervise the conduct, execution, and construction of approved projects on behalf of CUSA as defined in the contract between CUSA and the J. G. White Engineering Corporation, and to carry out such engineering and/or consulting services as may be requested by CUSA.

Built in a traditional Chinese palace style, the National Palace Museum (Figure 11.2) was located on the outskirts of Taipei close to highly secure military bases and President Chiang Kai-shek's residence. As the museum was built on the slopes of a hill, visitors had to either climb a long staircase or take a bus to reach it. Upon arrival, they were greeted by statues of Sun Yat-Sen and Chiang Kai-shek and would be in no doubt as to the political purpose of the museum.

The National Palace Museum, together with other cultural and tourist spots, was soon promoted abroad. Although the priority of the aid programmes was to build a sound defence system in Taiwan to serve US military interests, a diverse array of other projects was supported to develop Taiwan's infrastructure and economy, including the Grand Hotel and the Nan-hai 'Cultural Cluster' (which included a broadcasting station, a national art education centre, a science museum, a national

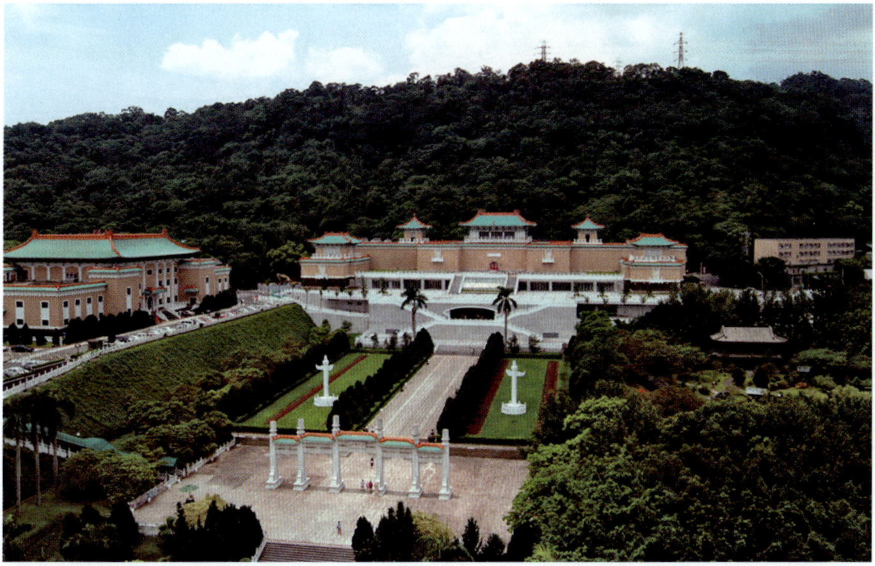

Figure 11.2 National Palace Museum, Taipei. Photo: Copyright National Palace Museum, Taipei

library, and the Museum of National History) (Figure 11.3). The buildings at Nanhai give a strong visual presentation of China and represent the earliest cultural cluster in Taiwan.

During the Cold War, the National Palace Museum was seen politically as the leading cultural organization safeguarding traditional Chinese culture. A showcase to the West, the museum employed young women in traditional Chinese costume, Chi-pao, to conduct museum tours in English for visitors from abroad, expatriates and their families. This engagement was supplemented with bilingual (Chinese and English) museum publications on Chinese art and culture for both general and specialist readerships. The museum also attracted American scholars supported by grants from private foundations and the US government, which led to the production of a number of large, fine-quality publications in Chinese, English and Japanese.

In the early 1970s, with assistance from the Asia Foundation,[5] the museum's director, Chiang Fu-tsung, helped establish the first academic institute focused on Chinese art in Taiwan at National Taiwan University and granted access for graduate students to study the collections in the museum. The Chinese Art History division at the university soon established itself as one of the top such institutes in the world.

However, the National Palace Museum was rather more than an academic institution; it was also an ideological instrument of the KMT which wished to develop Chinese nationalism. The government used coercion to teach the public to be Chinese, and appreciating art and artefacts at the National Palace Museum was one of the state's mechanisms to achieve this. In the 1970s, employees at the museum

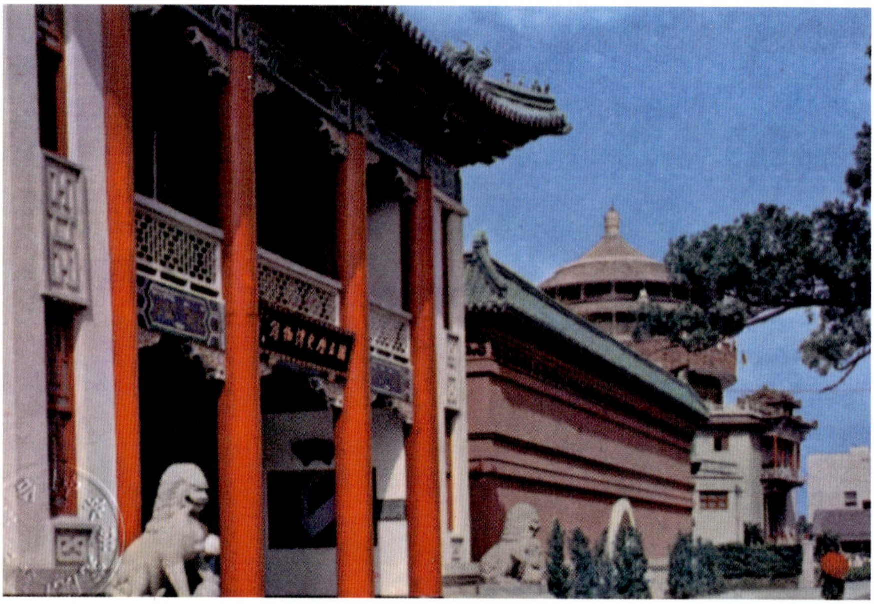

Figure 11.3 A view of the Nan-hai Cultural Cluster, showing the National Museum of History. Photo: Copyright Institute of Taiwan History, Academia Sinica

started each day by gathering in the main auditorium and taking it in turns to read out 'The Sayings of President Chiang Kai-shek' (Chen 2006). All those who grew up in Taipei in the 1970s and 1980s remember having to visit the museum as children. These visits allowed pupils to see the museum's most important objects which were also widely reproduced on stamps, calendars, textbooks, and the state's propaganda. These activities made the museum the number one place to visit by the Taiwanese population.

The museum enjoyed (and continues to enjoy) exceptional institutional status by being governed directly by the Executive Yuan, the highest administrative body of central government. The museum's director is appointed directly by the President and remains closely connected with the political party in power. The museum's annual budget far exceeds that of any other museum or cultural organization and its collection is ranked amongst the world's greatest.

The experiences of the Director-General's Museum, after the resumption of Chinese control, were by contrast fundamentally different: the museum did not serve the KMT's purpose as it could not tell the grand Chinese national narrative. Under its new name, Taiwan Provincial Museum, it was placed under the Educational Department of the Taiwan Provincial Government, thus lowering its institutional status. In this demotion, this museum was no exception; public organizations of Taiwanese culture and art were generally not given 'national' status. Under the new regime the museum became an exhibition venue showing art and propaganda.

Among a total of 23 special exhibitions held in 1965, only one, *The Chinese Cultural History Exhibition*, partly related to Taiwan (Liu 1966). Talks held at the museum were for propaganda or social educational purposes; rarely did they have anything to do with the museum's collection. Many were about American culture and the American way of life. Scholarly research relating to the museum's collection did, however, continue to be published in the museum's annual journal, but this too had also become a vehicle through which to propagate Chinese literature and art.

The rise of Taiwanese identity

With political liberation in the late 1980s, a Taiwanese national sentiment (as opposed to a Chinese national sentiment) became popular and was promoted by the government. The change came gradually in the 1990s when the former President Lee Teng-hui expressed his pro-Taiwanese sentiment while still the KMT party leader. It indicated a sea change in political ideology which did not arrive until the DPP – which, in its party manifesto, proposed building Taiwan into an independent state – won the Presidential election in 2000. The party also won the majority of seats in Congress, and immediately the state's political ideology changed to 'pro-independence' (as against 'pro-unification').

With the intention of promoting Taiwan as an independent state, the new government utilized cultural instruments to convey its political ideas and assert Taiwan as a subject in its own right, separated from the Chinese national narrative. Its first action was to propose the development of a 'National Palace Museum Southern Branch' in Southern Taiwan. This branch museum was to be positioned as 'a museum of Asian culture', a gesture aimed at separating the museum from the Chinese narrative while still keeping the 'National Palace Museum' name.

Under its new director, Tu Cheng-shen, the National Palace Museum was now repositioned in the Taiwanese national narrative, and produced several exhibitions on the subject of Taiwan and imposing a Taiwanese cultural discourse in a museum where the collection was by default about the art and culture of China. Large-scale exhibitions on Taiwan began with *Into the Wilderness: Taiwan During the Seventeenth to Nineteenth Centuries* (2002) and *Ilha Formosa – The Emergence of Taiwan on the World Scene in the Seventeenth Century* (2003). Using loans from various collectors and institutions, the purpose of these exhibitions was to emphasize Taiwan's own history. *Ilha Formosa*, for example, emphasized Taiwan's early connections with other European and Asian countries (other than China). Around the same period, Western-style oil paintings by three Taiwanese artists trained in Japan were exhibited at the museum (Table 11.1). These events created considerable controversy and were rarely repeated after Tu left the museum in 2004. Tu also oversaw a boom in research and publication in Taiwanese arts and culture.

In this new political era, the government found new uses for the National Palace Museum's collections, sending material abroad to major museums in the US, Germany, France and Austria (Table 11.2) as part of its attempts to counter China's

Table 11.1 Taiwan-focused exhibitions held at the National Palace Museum

Exhibitions	*Year held*
Taiwan archives in Ch'ing court: From the NPM collection	1995
Paintings by Yang San-lang	1999
Tracing Taiwan's modernization	2000
The land and her people: Active figures in the history of Taiwan during the Ch'ing Dynasty	2001
Portraits of the land and people: A centennial exhibition in commemoration of Li Mei-shu (1902–1983)	2001
Into the wilderness: Taiwan during the seventeenth to the nineteenth century	2002/2003
Liao Chi-chun's mastery of colours: A centennial exhibition in commemoration of his birth	2002
Ilha Formosa: The emergence of Taiwan on the world scene in the seventeenth century	2003

Source: National Palace Museum *Annual Reports* (2001–2005).

interference in Taiwan's international diplomatic relationships. In return, a number of European countries offered exhibitions of materials from their own national collections for display at the National Palace Museum.

As a hardcore pro-independence politician, Tu argued that the Taiwanese state was the guardian of great Chinese art and artefacts in the manner of many great museums in the West, which displayed foreign treasures for the benefit of all mankind. To further 'de-politicize' the National Palace Museum, Tu removed the statues of Dr Sun Yat-sen and Chiang Kai-shek in 2004 during a major renovation of the main entrance. (In 2008, with the return of KMT control, the museum planned to return Sun's statue to a distinctive location in the museum (Wu 2008)).

In 1999, the Taiwan Provincial Museum became the 'National Taiwan Museum'. It acquired national status and fell directly under the control of central government. After years of neglect, it now found itself politically and financially strengthened with a new management team of high-flying and ambitious individuals. In preparation for its reinvention, a large-scale audit was made of the museum's collections in 2002; the first time this had been done since the end of the colonial period. Despite damage caused by an earthquake in 2003, the museum was soon renovated, with new funds for research and digitization despite budget cuts in other areas of government expenditure. The museum reopened in 2003, and subsequently produced a large temporary exhibition, *Taiwan in Maps*, in 2005. Like the National Palace Museum, the National Taiwan Museum was deployed in cultural diplomacy, presenting exhibitions of Taiwanese art and culture abroad. Although Taiwanese art

Table 11.2 National Palace Museum collections and exhibitions which have travelled abroad

Year	*Country*	*Exhibitions*
1935	UK	735 pieces from the collection were exhibited at the Royal Academy of Art in London.
1940	Soviet Union	Selection from the collection exhibited at the State Museum of Oriental Art, Moscow.
1961	USA	'Chinese Art Treasures' exhibition from the collection travelled to Washington DC (National Gallery of Art), New York City (Metropolitan Museum of Art), Boston (The Museum of Fine Art), Chicago (The Art Institute of Chicago), and San Francisco (The H.M. De Young Memorial Museum) for one year.
1964	USA	50 pieces from the collection were exhibited under the Pavilion of the Republic of China at the World's Fair in New York.
1970	Japan	49 pieces from the collection were exhibited at the World Expo '70 in Osaka.
1973	South Korea	80 pieces from the collection were exhibited at the Chinese Art exhibition in Seoul.
1991	USA	17 pieces from the collection were loaned to the 'Circa 1492: Art in the Age of Exploration' exhibition held at the National Gallery of Art, Washington DC.
1996/1997	USA	'Splendour of Imperial China: Treasures from the National Palace Museum, Taipei' was held at the Metropolitan Museum of Art, the Art Institute of Chicago, the Asian Art Museum of San Francisco, and the National Art Gallery in Washington DC.
1998/1999	France	'Memoire d'Empire: Trésors du Musée National du Palais, Taipei' was held at the Galeries Nationales du Grand Palais, Paris.
2000	Latin America	Around 140 pieces travelled to seven countries (Nicaragua, El Salvador, Guatemala, Costa Rica, Panama, Paraguay, and the Dominican Republic).
2003/2004	Germany	'Treasures of the Sons of Heaven: The Imperial Collection from the National Palace Museum, Taipei' was held at Altes Museum in Berlin, and Bonn.
2008	Austria	'Imperial Treasures: Masterpieces from National Palace Museum, Taiwan' was held at Kunsthistorisches Museum, Vienna.

Source: Na (1993); National Palace Museum (2000).

and culture were not as popular as the Chinese offerings of the National Palace Museum they nevertheless increased Taiwan's exposure on the world stage.

Also in 2005, former president, Chen Shui-bian, announced two national projects centring on the National Taiwan Museum: the 'Taiwan Museum System' and

'Capital Cultural Park'. The projects were to create the National Taiwan Museum as the 'Capital Museum' and the 228 Park – where the museum is located – as the 'Capital Cultural Park'. These were included under the government's major centralized projects, under the objective of 'establishing Taiwan's national identity and cultural focus' (National Taiwan Museum 2006); the National Taiwan Museum was central to both projects.

The Taiwan Museum System aims to form a museum cluster by connecting and converting the National Taiwan Museum's two neighbouring heritage sites into new museums: the Taiwan Museum of Industrial History and the Taiwan Museum of Modernity. Additionally, the current National Taiwan Museum is to be expanded, and further changed to be the 'Taiwan Museum of Natural History'. The Capital Cultural Park is also a network project which aims to connect heritage and cultural sites. The two projects exemplify the DPP's ambition to visualize the Taiwanese national narrative in a distinctive cultural landscape. Like the KMT's Nan-hai cultural cluster built in the 1950s, the DPP sought to create its own political epic.

The DPP saw national museums as a way to construct both cultural and political identity; the new museums established since the late 1990s bear a distinctive Taiwanese focus, and cover industry, the economy, history, natural history and fine arts (Table 11.3). These museums have formed a complete cultural map, embracing major aspects of Taiwanese culture; their 'national' status was not simply a matter of institutional hierarchy or funding, but a statement of political integrity and identity. The older, 'National Museum of History', which essentially told Chinese history through its rich Chinese collections, was now supplemented with a National Museum of Taiwan History opened in 2008. The original National Museum of History ceased to predominantly tell the Chinese story in the late 1990s, when it became a venue for popular and blockbuster exhibitions. These projects show how the DPP's political ideology has been closely connected with constructing Taiwanese culture in national museums.

Table 11.3 New national museum projects in Taiwan since the mid-1990s

Museum	*Year of establishment*
National Taiwan Craft Research Institute	1999
National Centre for Traditional Arts	2001
National Museum of Marine Biology	2000
National Museum of Taiwanese Literature	2003
National Museum of Fine Arts	2004
National Museum of Taiwan History	2008
National Museum of Industrial History	In planning
National Museum of Modernity	In planning

Conclusion

This chapter has demonstrated how three distinctive political ideologies were deployed in museums and museum exhibits in Taiwan. Through its museological activities, Japan exemplified its colonial power and governance; the KMT's emphasis was on representing China; and the DPP constructed the Taiwanese national narrative by collecting and displaying Taiwanese art and culture. The Japanese had also, in their museum projects, introduced Taiwan to Western practices and cultural technologies. At this time, local cultural objects were systematically collected on a large scale, using the colonial government's highly organized machinery. The Director-General's Museum, under Japanese rule, made a significant contribution to Taiwan's cultural history and developed a new social educational function for museums as an important dimension for colonial governance. While the KMT saw a similar power to develop citizens in national museums, with American support it sought to develop and conserve a China that was rapidly disappearing on the Communist mainland. One sees here an agreement between KMT ideology and America's active foreign policy in South East Asia at this time. The National Palace Museum was, for both parties, a means to perpetuate old China. With the arrival of the DPP, national museums were again redeployed, and now those that had been condemned to a 'provincial' (as in a province of China) existence were reinvented and joined a growing number of new national museums. Recently used to establish a cultural conservatism, now national museums were deployed to reimagine the nation. The Taiwanese province was by these and other means to be made a nation.

Notes

1 Holland and Spain had occupied Taiwan briefly for around three decades in the seventeenth century.
2 Lee, W.-I. (2004) 'The construction of national imaginaires in Taiwan's Museums during the decolonization and Cold War era (1945–1971)', paper presented at the Museums, the Construction of Knowledge and Modernity Conference, Taichung, 9–10 September.
3 The committee, National Central Museum and Library Joint Management Committee, also oversaw other museum and library collections removed from China.
4 Institute of Modern History, Academia Sinica, Archive no. 36-06-034-015, 36-06-034-016, 36-06-034-017, and 36-06-034-018.
5 The Asia Foundation was established in the 1950s. The foundation worked closely with the US government to promote democracy and development in Asia. See Chiang (1972).

Bibliography

Anderson, B. R. O. G. (1991) *Imagined Communities: Reflections on the Origin and Spread of Nationalism* (Revised and extended edition), London: Verso.

Beasley, W. G. (2000) *The Rise of Modern Japan: Political, Economic and Social Change since 1850* (Revised edition), London: Weidenfeld & Nicolson.

Chen, P.-C. (2006) 'National Palace Museum in the 1970s', in *Eighty Years of the National Palace Museum*, Taipei: National Palace Museum (in Chinese).

Cheng, C.-H. (2001) 'Study on the 1935 Taiwan Exposition', unpublished MA thesis, National Central University, Chung-Lee (in Chinese).

Chiang, F.-T. (1972) 'Report from National Palace Museum: 1971', in *Annual of Social Education*, Taipei: Chinese Social Education Association (in Chinese).

Chou, W.-Y. (1997) *An Illustrated History of Taiwan* (2nd edition), Taipei: Linking Publishing Company (in Chinese).

Huang, H.-C., Chang, S.-Y. and Wu, W.-H. (2002) *History of Taiwan*, Taipei: Wunan Book Co Ltd (in Chinese).

Jacoby, N. H. (1966) *U.S. Aid to Taiwan: A Study of Foreign Aid, Self-help and Development*, New York: F. A. Praeger.

Ju, J. C. (2003) *'The Palace Museum as Representation of Culture: Exhibitions and Cannons of Chinese Art History'*, in K.-W. Huang (ed.) *When Images Speak: Visual Representation and Cultural Mapping in Modern China,* Taipei: Institute of Modern History, Academia Sinica, pp. 477–507.

Ju, J. C. (2007) 'Chinese art, the National Palace Museum, and Cold War politics', in A. Brzyski (ed.) *Partisan Canons*, Durham, NC: Duke University Press, pp. 115–34.

Lee, (2006) 'The research of natural history in Taiwan during the Japanese occupation and the formation of the preservation of the cultural heritage', unpublished MS thesis, Chung Yuan Christian University, Chung-Lee.

Li, J. (2005) 'The research of Taiwan Governor's Palace Museum', unpublished MA thesis, National Central University, Chung Lee.

Li, T.-N. (1999) *Special Issue on the 90th Anniversary of Taiwan Provincial Museum*, Taipei: Taiwan Provincial Museum.

Liao, C. J. (2004) 'The Taiwan Governor's Palace Museum in Japanese colonialism', unpublished MA thesis, Tainan National University of the Arts, Tainan.

Liu, H. (1966) 'The first year's report from Taiwan Provincial Museum', in *Annual of Social Education*, Taipei: Chinese Social Education Association.

Myers, R. H., Peattie, M. R. and Joint Committee on Japanese Studies (1984) *The Japanese Colonial Empire, 1895–1945*, Princeton, NJ: Princeton University Press.

Na, C.-L. (1993) *Safeguarding National Treasures at the National Palace Museum for Seventy Years*, Taipei: Na, Chi-Liang (in Chinese).

National Palace Museum (2000) *The National Palace Museum across Centuries: a Chronology*, Taipei: National Palace Museum (in Chinese).

National Palace Museum (2002) *A Survey of the National Palace Museum,* Taipei: National Palace Museum (in Chinese).

National Taiwan Museum (2006) *National Taiwan Museum Annual Report 2005*, Taipei: National Taiwan Museum.

Shiina, N. (1989) *Meiji hakubutsukan kotohajime*, Kyōto-shi: Shibunkaku Shuppan.

Shiina, N. (2005) *Nihon hakubutsukan seiritsushi: hakurankai kara hakubutsukan e*, Tokyo: Yūzankaku.

Taiwan Governor's Office (1935) *Social Education in Taiwan: An Overview*, Taipei: Taiwan Governor's Office.

Taiwan Governor's Office (1938) *Social Education in Taiwan*, Taipei: Taiwan Governor's Office.

Wu, Y.-H. (2008) 'National Palace Museum will move back Dr Sun Yat-Sen's statue', *China Times*, 27 May.

Chapter 12

THE BRITISH MUSEUM IN PRINT

From national to universal museum

Sarah A. Hughes

IN 2002, A GROUP OF the world's most powerful museums issued a 'Declaration on the Importance and Value of Universal Museums'[1] in which they claimed the status of universal museums. In doing so, they asked for a public reimagining of their collections as the global heritage of mankind. No longer were these museums to be understood as mere 'national museums', for example. This declaration was a response to growing pressures for the restitution of material culture from these institutions, and the British Museum – the home of so much of the world's great culture – was prominent amongst the signatories. This chapter explores how these claims for universal rather than national status have been made by the British Museum in its various publications. In particular, this chapter will use critical discourse analysis to examine the museum's representation of itself in its catalogues and guides (although these are not the only printed materials the museum publishes). Printed materials of this nature provide a rich and largely overlooked source of data for understanding national museums and their self-portrayal.

Museum books

As complex cultural organisations, museums have offered up various ways to analyse their symbolic representation, including the architecture (Giebelhausen 2003; Forgan 2005), the museum shop (McTavish 1998), museum postcards (Beard 1992) and, more obviously, the design and construction of exhibitions (Ferguson 1996). Duncan and Wallach (1980: 449) take another view; they state that the 'beliefs and values museums communicate' are derived from the museum as an entire experience. They argue that universal museums act 'not in an explicit fashion, but symbolically' (Duncan and Wallach 1980: 456). While it is possible to interpret symbolic representations residing in objects, architecture and exhibition space, historians have long understood the explicit and overt qualities of text as a superior resource from which to distil intentions and actions (Fairclough 2000; Fairclough *et al.* 2002).

Books and collections both attempt to achieve order; they attempt to rationalise and make sensible the chaos of Nature. As organising agents, books and collections have been jointly implicated in the early attempts to impose order in the museum (Impey and MacGregor 1985; Haskell 2000; Swann 2001). While objects and collections have permanence vested in their physicality, this construction of order can be and is reinterpreted and reordered as views of society and culture change. Books, by contrast, present 'fixed' interpretation: 'The book . . .[in] itself becomes an historic document of the state of knowledge and angle of interest at the time of publication' (Borg 1998: ix).

For the major museums in particular, publication is rather more than an act of research dissemination; in the British Museum, for example, it is a major revenue stream (Figure 12.1). This source of income has grown over the last twenty years as museums have sought to control aspects of their finances through commercial enterprises such as museum shops, consultancy, corporate entertainment and partnerships with retail organisations. While financial reporting constraints on these commercial enterprises make it difficult to identify the contribution of publishing to overall income, some indication of its importance is shown by the British Museum Company's reported income of £5.6 million in 2007/8. This was nearly three times greater than income in the previous four years and was attributed to retail sales which 'were higher . . . due to the excellent response to the First Emperor exhibition. The contribution made by the publishing business to the overall profit increased during the year, due to the success of the First Emperor catalogue' (British Museum 2008: 15).

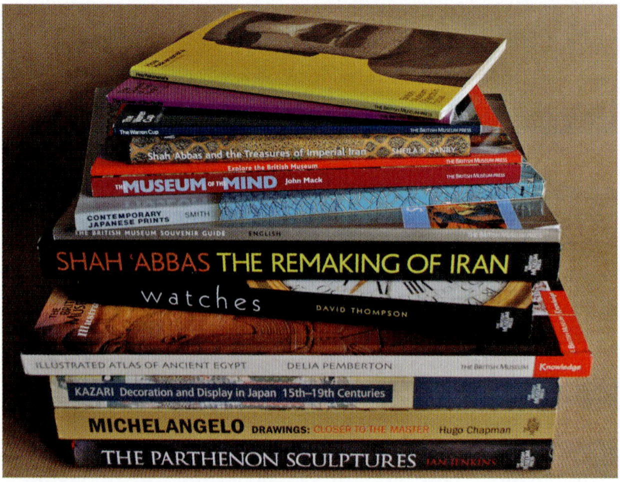

Figure 12.1 A small selection of the over 350 books published by the British Museum Press. Photo: Thomas Lindsay

The commercial considerations of publishing profoundly affect the design, production and sale of museum books. They are complex commodities and as such are open to economic, anthropological and sociological analysis. In the UK, all national museums publish new titles annually and maintain backlists of exhibition and collection catalogues, and visitor guides. While an exact estimate of publishing activity across the nation is difficult to obtain, book catalogues show that national museums are more active in publishing than local, regional and private museums. In these smaller institutions, publishing has long been associated with high risk; in order to achieve a low cover price these museums traditionally published in print runs far greater than needed to address market demand. Publications were sold cheap but always generated an unsold surplus. The dominance of national museums in museum publishing reflects their much greater audience numbers and their capacity for high-profile temporary and permanent exhibitions. All national heritage organisations in Britain publish for general and specialist audiences: these include the British Library, the National Archives, the National Trust and English Heritage.

Reading museum books

This chapter's basis is analysis of four publications produced by the British Museum Press. Since this analysis examines the museum's representation of itself as a universal or world museum to a general audience, the books chosen were examples of guides and exhibition catalogues designed to appeal to a non-technical public. The first book is *The British Museum Souvenir Guide* (English Edition, 2003), a 96-page full-colour publication also available in nine other languages (Figure 12.2). It contains maps of the galleries, information about conservation and research, and an introduction which discusses the history and purposes of the museum. Divided into eighteen illustrated sections, the guide provides a synopsis of the collections and displays under various headings, such as 'The Ancient Near East', 'The Islamic World', and 'Africa'. This book, which will be referred to as the *Souvenir Guide*, was subjected to the most complete analysis, using four major critical discourse analysis categories (see below). The second book is *Explore the British Museum: A Family Souvenir Guide* (2007) by Richard Woff, a short (40-page) volume appealing to a younger audience, which also includes material of use and interest to adults (Figure 12.2). Large titles introduce the various sections of the collection and some pages showcase objects likely to appeal to children, while parts of the text are accentuated with colour panels. This book is here referred to as the *Family Guide*. The third book is the exhibition catalogue, *The First Emperor: China's Terracotta Army* (2007) by Jane Portal, a 240-page volume providing accessible background information regarding the exhibition's subject matter. Finally, the fourth book, which relates to the permanent collection, is *The Parthenon Sculptures in the British Museum*, written by Ian Jenkins and also published in 2007. This heavily illustrated 140-page volume describes the sculptures in detail and discusses their history. In the case of

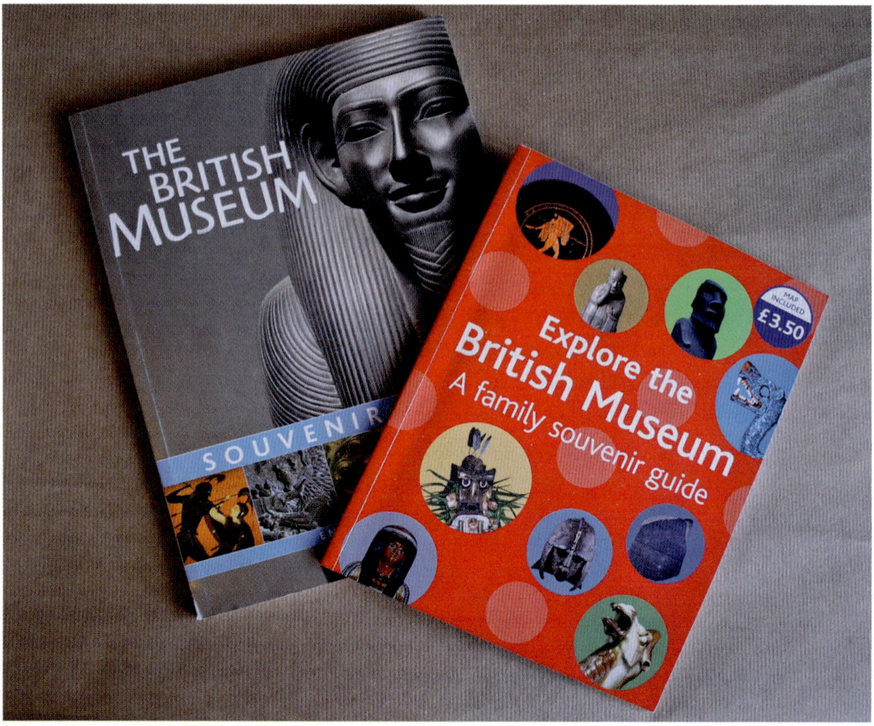

Figure 12.2 Two of the guidebooks available for visitors to the British Museum. Photo: author

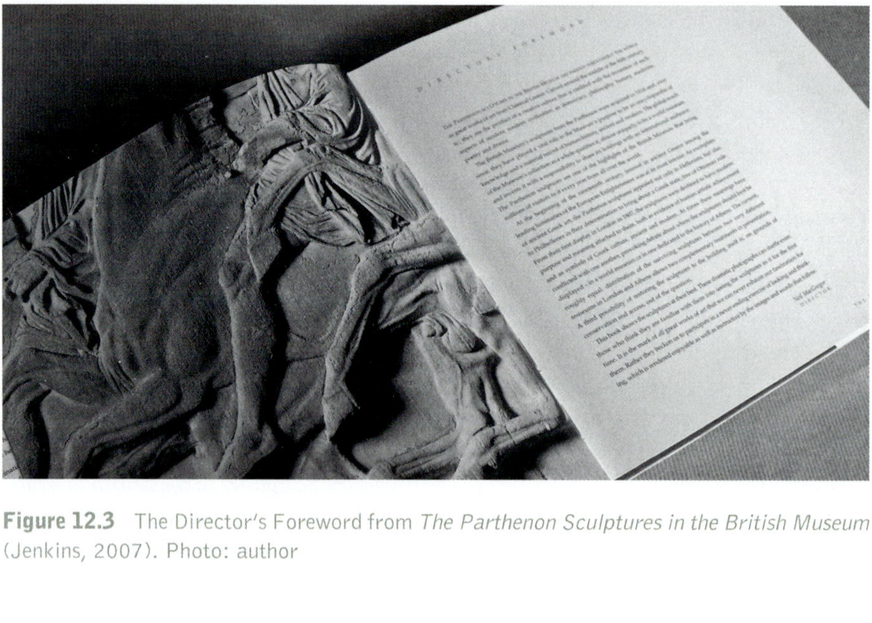

Figure 12.3 The Director's Foreword from *The Parthenon Sculptures in the British Museum* (Jenkins, 2007). Photo: author

these latter two books, only text from the 'Director's Foreword', which provides a global context for the exhibitions, is analysed (Figure 12.3).

These texts were analysed using critical discourse analysis as discussed by Norman Fairclough in his *Analysing Discourse: Textual Analysis for Social Research* (2003). It draws on approaches to theorising the social character of texts found in systemic functional linguistics (Halliday 1978, 1994) and recognises three types of meaning in language present in text: 'action', where a text takes part in an event, for example, as a guidebook within a museum visit; 'ideational', where a text is relevant to ideas and identities, for example, as a description and explanation of a cultural practice or a belief; and 'representational', such as where a text presents the material world, for example, a description of a natural object or thing. By examining these meanings in language, textual analysis recognises that texts have causal effects on, and contribute to changes in, people, activities, social relations and the material world. The analysis provides a number of analytical categories (for which see Fairclough 2003: 191–94).

Text clearly plays a role in the social construction of national museums. Museum texts 'construe (represent, imagine, etc.) the social world' (Fairclough 2003) and analysis of these texts clarifies and makes explicit the contribution of written language to meaning-making. It is important to understand that this analysis of texts is more concerned with the *production* of meaning than with its reception. Critical discourse analysis of text produced by the British Museum gives access to the institution, its intentions and purposes; it is an area little explored. While not a feature of the present study, we also know little of the visitors' use of guidebooks during a museum visit or the use made by purchasers of the exhibition catalogues they take home. Such perspectives are, however, not important to the present study which is concerned with how the institution authors its own identity through the books it produces. Such information can then be triangulated with data from other kinds of analysis.

Of particular interest here is the text, such as the preface or introduction, which presents the author's contextualising statements. Most guidebooks and catalogues include forewords written or approved by a senior staff member (usually the Director) which reflect on the contemporary cultural and political pressures that shape the museum. Questions capable of elucidating the British Museum's representation of its role, purposes and its audience in relation to the nation and the world were formulated in line with four analytical categories of critical discourse analysis. The analysis also recognises the social events and chains of text which provide the context for the use of the books in the museum. Representations of actors, discourses, and difference enable examination of the museum's perceived universal aspect. In short, the analysis aims to elucidate the following questions: Of what social events, network of practices and chains of text are the books a constituent part? How are social actors represented? What discourses are presented? How is difference presented in the text?

Social events and the chain of texts

The British Museum's books contribute to, and are part of, the social event which constitutes a museum visit. The *Souvenir Guide* and the *Family Guide* in particular are conceived and constructed to enhance a visitor's enjoyment and understanding of the institution and the exhibits on display. As a social event, a museum visit is constituted of social actors and social practices such as the use of information desks, museum shops and cafés, and, of course, particular behaviours within the museum galleries. This social event is influenced by the broader discourse shaped by the perceived educational and cultural purposes of museums. As a contributing element to this social visit, guidebooks and catalogues can be examined to build an understanding of how the institution helps to construct this social event. Guidebooks are designed to be used in the museum during the visit. They offer a synopsis of the exhibition(s) and give an overview of the institution. The publications provide context but also aim to orientate the visitor. Catalogues, on the other hand, are rarely used in the galleries. Their purpose is to provide additional in-depth information; they are to be read away from the museum and offer space for longer reflection.

The books are, of course, products of a series of events; the writing, editing, review, design, illustration and production processes. They arise from a number of other social practices. The work can be as painstaking as that which took place to produce the exhibition; indeed, most books have a longevity far exceeding that of their subject exhibitions and consequently greater influence than the displays of real objects.

These guidebooks and catalogues contribute to the chain of texts (Fairclough 2003: 21–38) that 'contextualize objects' (Hooper-Greenhill 1994: 115). Marketing literature, free guides, museum maps, object and wall labels are additional linked texts that interpret collections and which together contribute resources for the social performance of the museum visit. To these are added external reviews in the magazines, newspapers and in other media. Within this text chain, guides and similar publications fulfil the role of explicitly revealing the museum's purposes. In the British Museum, it is here that the universal museum is presented to the general audience and these ideas are then picked up in other media and, not least, in reviews. Journalists quote from these publications as definitive statements of content and intent.

Social actors in the text

The British Museum books portray the universal museum in their deployment or consideration of actors and discourses. When analysing how a text represents the world, Fairclough (2003: 193) suggests asking of the text: Are social actors included or excluded? If excluded, are they absent from the text or can they be inferred? The text's actors may be named and personalised or anonymised or generalised; they can be active (making something happen) or passive (being acted upon).

The *Souvenir Guide* includes several actors in its introductory text. These are not referred to by name but as an impersonal group. The producers of objects in the museum's collection are referred to collectively as 'humankind', a universal term applied to all people without specific reference to location or nation, and intended to carry no intimation of gender. When a location or a culture is specified in the text, the people of that region or culture are inferred rather than named. For example, visitors are informed that the collections of the British Museum are 'vast' so that: 'the Museum does not need to borrow: it can simply dip into its remarkable reserves to treat subjects as diverse as gold from South America, the culture of the Maldives, Rembrandt drawings and Hindu religion' (British Museum 2003: 6).

In this quote, it is the cultures and objects that are the actors, not the constituent people of a nation or the makers of the objects. These people are inferred as participants and contributors to a culture, a religion or a region of the world without any reference to a national affiliation. Rather than acknowledge or celebrate the origins of collections in specific locations, or association with particular nations and peoples, the text places them in the shared, but bland, territory of human culture. Broad categorisations are used, such as continents (South America), religions (Hindu), or cultures (Maldives). Even Rembrandt finds himself no longer a citizen of a country or state, but as an artist whose drawings contribute to the universal category of Western art history.

Further examination finds the British Museum itself presented as an actor in the *Souvenir Guide*, where this state-funded national museum aims 'to be at the centre of international scholarship and to disseminate knowledge for the education . . . of all'. In this way, the British Museum places itself at the centre of a world of broad regions and global cultures. The guide largely omits reference to Britain and nowhere are the words 'United Kingdom' used.

In the *Family Guide* the museum acquires an independent personality: 'The museum . . . celebrated its 250th birthday recently.' It is joined by other actors such as collectors who are presented in generalised roles or positions without reference to nationality or explicit affiliation. The British empire does, however, appear but does not complicate the generality of the narrative: 'The power of the British empire meant that traders, explorers, scientists, generals and admirals brought to the museum all sorts of objects for the scholars to study' (Woff 2007: 2).

These points are repeated in the foreword to the book on the Parthenon sculptures. Here, too, the text tells the reader that the British Museum is not a museum with a responsibility to a nation state, but rather it acts on the wider world stage for citizens of the world. The Director tells the audience: 'the global reach of the Museum's collections qualifies it almost uniquely to be a world museum and invests it with a responsibility to share its holdings with an international audience' (Jenkins 2007: 7). Actors are again impersonalised, almost invisible: 'they [the sculptures] are the product of a creative culture that is credited with the invention of . . . democracy, philosophy, history, medicine and drama' (Jenkins 2007: 7). No individual or group is named and the people who are represented by the objects are passive, that is to say, the collection, display and interpretation of the objects is carried out without direct recourse to the people who created the objects. These British Museum texts distance the objects on display from their creators and seek to weaken the bond both to their precise origins and the ownership claims that might arise from them.

This universalism is carried over to the museum's perceptions of its audience which is conceived to also belong to humankind without further distinction. Texts avoid addressing the British specifically, or indeed any other possible group: the *Souvenir Guide* refers to the 'widest possible audience, not only within the galleries of the Museum, but far beyond them'. The *Family Guide* refers to 'millions of people from all over the world' and explains that the British Museum 'has become a museum of the world for the world'. It also broadens the purposes of these visitors to include those who 'visit or study or . . . help the Museum understand better the objects from their cultures'. At the end of this introduction for the family, the audience is directly addressed as an 'explorer' of the museum:

> We hope that you enjoy exploring the Museum and that the chance you have here to see objects from different countries alongside each other will help you understand better the fascinating variety of world cultures, the connections between them and what human beings have in common.
>
> (Woff 2007: 2)

In the Parthenon book, the audience is described as international, with the Parthenon sculptures bringing 'millions of visitors to it every year from all over the world'. This universal discourse is itself universal in the museum's texts and in everything the museum seems to see, including the objects and the audience. Both are in a harmony and unified by the notion of 'humankind'.

Museum of the world for the world

One recurring discourse present in these British Museum texts is the longevity and consistency of the institution's message. These implied and explicit understandings seek to reinforce the museum's legitimacy as a universal museum. The *Souvenir Guide* establishes that the focus on world cultures is not new and has existed since the museum's inception: 'From its earliest days in the eighteenth century, the Museum collected, displayed, stored and preserved the works of humankind . . . with great earnestness' (British Museum 2003: 6). Using more direct language, a similar message, with the addition of continuous universal access, is present in the *Family Guide*: 'The museum was founded in 1753 and celebrated its 250th birthday recently. It was the first publicly owned museum in the world and was intended to be used by everybody, free of charge' (Woff 2007: 2).

Nowhere could this claim to longevity be more politically useful than in the description of the Parthenon sculptures, where the Director claims the moral high ground of universal knowledge: 'The British Museum's sculptures from the Parthenon were acquired in 1816 and, ever since, they have played a vital role in the Museum's purpose to be an encyclopaedia of knowledge and a record of human history, ancient and modern' (Jenkins 2007: 7). This position permits the museum to fend off local interests; it was now an argument between 'a world museum or . . . one dedicated to the history of Athens'(Jenkins 2007: 7).

This claim to universal representation is also supported by repudiation of the idea that the British Museum might have been built with a national focus in mind. The *Souvenir Guide* states: 'The British Museum has never simply been a museum of antiquities of Britain. In fact, for its first century . . . rather little of British origin was collected. From the very beginning, interests were universal' (British Museum 2003: 6). This universalism was consistently upheld: 'By the turn of the twentieth century, the Museum had established the coverage of the collections just as its founders had hoped it would' (British Museum 2003: 8). This long trajectory of universalism is also projected into the future: the British Museum's active research and fieldwork allowed it to 'fulfil its role as a museum of world cultures past and present'. Indeed, this universalism has real ambition to 'make the British Museum perhaps the best single introduction to world cultures and civilisations that exists today' (British Museum 2003: 9).

Turning now to the catalogue for the First Emperor exhibition, the Director's Foreword situates the exhibition politically. It stressed the uses of culture to forge links between nations and celebrated the global value of the loans which made up the exhibition. It aligned the British Museum with the state of China, presenting them as equal partners in the practice of cultural exchange and understanding:

> The book and exhibition *First Emperor: China's Terracotta Army,* exemplifies the growing exchanges and collaborations between China and the British Museum. Three major exhibitions of British Museum artefacts have toured China's museums during the last two years and many more are planned in both directions to foster better mutual appreciation of the wealth and depth of each nation's cultural history.
>
> (MacGregor 2007: 9)

Within the field of cultural globalisation, museums and nations can be presented as equal partners. Temporary exhibitions and their catalogues provide 'old' museums with the opportunity to reinterpret collections in the light of the 'contemporary challenges associated with economic and political realities of globalisation' (Message 2006: 26), but they also provide a platform where the national museum can claim influence beyond the borders of the state as a global and universal entity.

Yet another discourse which underpins much of the British Museum's writing about itself relates to the gains in understanding which are possible from a museum visit. The First Emperor catalogue states: 'through studying the past it is possible better to understand the present and to prepare for the future' (MacGregor 2007:

9). A suitably child-oriented version of this discourse is presented in the *Family Guide* which draws attention to the opportunity for comparison available in the British Museum and, of course, one of its implicitly universal messages:

> the chance you have here to see objects from different countries alongside each other will help you understand better the fascinating variety of world cultures, the connections between them and what human beings have in common.
>
> (Woff 2007: 2)

These texts show the importance of vast global collections of 'world cultures'; knowledge and understanding were to arise from comparison, though precisely how past and future were united within this possibility for comparison remained unclear. Through comparison, visitors were to locate commonalities and understanding. It was a dominant discourse across a wide range of exhibitions in the museum.

Dialogue and difference in text

Fairclough argues that critical discourse analysis is a means to determine the 'orientation to difference' present in text. He suggests that difference can be present in five 'scenarios': the recognition of difference; the accentuation of difference; the resolution of difference; the accentuation of commonality; and in consensus (Fairclough 2003: 41–42). These forms of difference can be present in various combinations within a text. Difference may include the consideration of intertextuality[2] such as quotations, reported speech, writing and thought, both with and without attribution (Fairclough 2003: 39–40). Linked with intertextuality is the use of assumptions in text: 'What is "said" in a text is "said" against a background of what is "unsaid", but taken as given' (Fairclough 2003: 40). He continues, 'An important contrast between intertextuality and assumption is that the former broadly opens up difference by bringing other "voices" into a text, whereas the latter broadly reduces difference by assuming common ground.' An understanding of the presence of dialogue and difference and the way it is presented can lead to clarification of the agency of the actors in the text. Earlier consideration of social actors has emphasised the role of the museum itself as an active participant and this is further supported in a review of difference and dialogue.

The introductory text in the *Souvenir Guide* considers the historical development of the collections and their housing, and largely avoids any explicit reference to difference or dialogicality.[3] There is one instance of intertextuality which would usually be expected to promote a sense of plurality in a text. However, this text is a quotation from the first museum guidebook of 1761, which through its archaic language actually serves to enhance the sense of distance rather than reduce it: 'Curiosity almost universally prevails . . . Nothing can conduce more to preserve the Learning which this latter Age abounds with, than having Repositories in every Nation to contain its Antiquities, such as is the Museum of Britain' (British Museum 2003: 6).

Less difference and greater dialogue is apparent in the introductory heading for the *Family Guide*. However, the heading text 'Welcome to the world of the British Museum!' maintains some separation and difference of the visitor from the institution. The visitor is entering, is invited into, 'the world' established and controlled by 'the British Museum'. What effect would changing 'the' to 'your' have on the presentation of difference? It would serve to reduce difference and align visitors with the museum. This substitution does not, however, enhance the representation that the British Museum texts promote of an overarching institution in which visitors receive information or explore with museum guidance and supervision. Throughout the introduction, the museum remains an encompassing entity: 'the Museum still contains and collects'; 'objects from the Museum travel the continents' and are 'brought to the Museum'. The assumption within the text is that the museum is a custodian, a cultural agent, laying out treasures for its visitors. Again, in this text, the universal aspects of the institution are stressed.

At the end of the *Family Guide*'s introductory text, the author addresses the reader in a way that, although more personal, continues this difference between museum and visitor: 'We hope that you enjoy exploring the Museum.' The exploration, it is hoped, will lead to understanding but not to a reduction of difference. In fact, one of the discourses of the museum and its texts is a celebration of difference through culture, albeit with an attempt to stress the underlying commonality of cultures.

Conclusion

This chapter has shown how museum texts and critical discourse analysis permit the fine analysis of a museum's representation of itself. In the case of the British Museum as a universal museum, the analysis shows pervasive and ubiquitous attempts to manufacture and legitimise the museum's self-image. While museum books are conceived and produced to interpret collections and to extend national museums' communication with their audiences, these publications also contain subliminal political messages. The visitor, in realising the educational and comparative possibilities of the museum for understanding the past and negotiating the future cannot help but understand that universalism has its role; that museums such as this pursue a moral cause.

Notes

1 Declaration on the Importance and Value of Universal Museums. Online. http://icom.museum/pdf/E_news2004/p4_2004-1.pdf (accessed 5 August 2009).

2 Intertextuality is 'the presence within [the text] of elements of other texts (and therefore potentially other voices than the author's own)' (Fairclough 2003: 218).

3 Dialogicality is the measure of the interplay within text 'between the voice of the author and of other voices, the extent to which these voices are represented . . . excluded or suppressed' (Fairclough 2003: 214).

Bibliography

Beard, M. (1992) 'Souvenirs of culture: deciphering the museum', *Art History*, 5: 505–32.

Borg, A. (1998) 'Preface', in *The Victoria and Albert Museum: A Bibliography and Exhibition Chronology 1852–1996*, compiled by Elizabeth James, London: Fitzroy Dearborn Publishers and Victoria and Albert Museum.

British Museum (2003) *The British Museum Souvenir Guide*, London: British Museum Press.

British Museum (2008) *Annual Report and Accounts*, London: British Museum.

Duncan, C. and Wallach, A. (1980) 'The universal survey museum', *Art History*, 3: 448–69.

Fairclough, N. (2000) *New Labour, New Language?*, London: Routledge.

Fairclough, N. (2003) *Analysing Discourse: Textual Analysis for Social Research*, London and New York: Routledge.

Fairclough, N., Jessop, R. and Sayer, A. (2002) 'Critical realism and semiosis', *Journal of Critical Realism*, 5: 2–10.

Ferguson, B. W. (1996) 'Exhibition rhetorics: material speech and utter sense', in R. Greenberg, B. W. Ferguson and S. Nairne (eds) *Thinking About Exhibitions*, London: Routledge, pp. 175–90.

Forgan, S. (2005) 'Building the museum: knowledge, conflict and the power of place', *Isis*, 96: 572–85.

Giebelhausen, M. (ed.) (2003) *The Architecture of the Museum: Symbolic Structures, Urban Contexts*, Manchester and New York: Manchester University Press.

Halliday, M. (1978) 'The sociosemantic nature of discourse', in *Language as Social Semiotic*, London: Edward Arnold.

Halliday, M. (1994) *An Introduction to Functional Grammar*, Second edition, London: Edward Arnold.

Haskell, F. (2000) *The Ephemeral Museum: Old Master Paintings and the Rise of the Art Exhibition*, New York and London: Yale University Press.

Hooper-Greenhill, E. (1994) *Museums and their Visitors*, London: Routledge.

Hughes, S. (2007) 'Museum publishing: representing the museum', in P. Aronsson and M. Hillström (eds) *NaMu, Making National Museums Program, Setting the Frames*, Linköping: Linköping University Electronic Press.

Impey, O. and MacGregor, A. (eds) (1985) *The Origins of Museums: The Cabinet of Curiosities in Sixteenth- and Seventeenth-century Europe*, Oxford: Oxford University Press.

Jenkins, I. (2007) *The Parthenon Sculptures in the British Museum*, London: British Museum Press.

MacGregor, N. (2007) 'Preface' in J. Portal, *The First Emperor: China's Terracotta Army*, London: British Museum Press.

McTavish, L. (1998) 'Shopping in the museum? Consumer spaces and the redefinition of the Louvre', *Cultural Studies*, 12: 168–92.

Message, K. (2006) *New Museums and the Making of Culture*, London: Berg.

Reeve, J. (2004) *The British Museum Visitor's Guide*, London: British Museum Press.

Swann, M. (2001) *Curiosities and Text: The Culture of Collecting in Early Modern Europe*, Pittsburgh: University of Pennsylvania Press.

Van Dijk, T. (ed.) (1997) *Discourse Studies: A Multidisciplinary Introduction (Vol. 1: Discourse as Structure and Process; Vol. 2: Discourse as Social Interaction)*, London: Sage.

Woff, R. (2007) *Explore the British Museum: A Family Souvenir Guide*, London: British Museum Press.

PART THREE MUSEOLOGY AND PARTICIPATION

Chapter 13

THE NATION DISROBED

Nudity, leisure and class at the Prado

Eugenia Afinoguénova

IN THE YEARS FOLLOWING its nationalization in 1868, Spain's Prado Museum became the subject of humorous stories, poems and sketches in the popular press. Many of these focused on the reactions of uneducated visitors to the many paintings and sculptures of nudes. In 1883, for example, *Madrid Cómico* published a poem about a paterfamilias whose three young children had an unspeakable 'occurrence' in front of a portrait by Ribera, 'the subject of which was a woman unburdened by excessive clothing' (Pérez Zúñiga 1883). In 1887, the same newspaper ran a sketch of two peasants seduced so effectively by a female nude painting that they declare themselves 'a'solutely a'solutist' supporters of the 'Royalist/realist school'[1] (Cilla 1887) (Figure 13.1). A serial novel published in 1888 in *La Risa* depicted a young woman of Madrid guiding her provincial mother-in-law swiftly past paintings of men who were 'naked although well-shaped' (Frontaura 1888: 10). The commoners' reactions to risqué images displayed in a national museum appear to have seized the popular imagination.

What does it take for a museum to be recognized as national by the nation to which it belongs? This chapter will look at nationalization as a process in which the audience's engagement with this institution was equally if not more important than the acts of legislation accompanying its official takeover by the state. From this point of view, humorous anecdotes can lead us to understand how the diffuse entity known as a 'nation' takes possession of its museum. Labanyi (2005) has argued that discussions about art spectatorship in Spain were part of a more general debate about the nature of sovereignty which marked Spain's contested nation-making. Continuing this line of research, this chapter will pay attention to the little-studied comical descriptions of the Prado. These kinds of depictions served to recuperate for the bourgeois the idea of a national museum as a public place allowing the collective viewing of paintings and sculptures by an audience composed of different sexes,

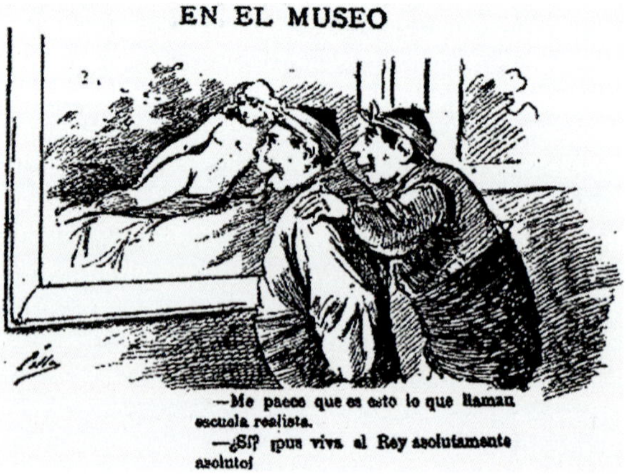

Figure 13.1 In the museum: — I think this is what they call a 'realist (royalist) school' [escuela realista]. — Really? Then long live the a'solutely a'solute King! Photo: Cilla, 'El el museo', *Madrid Cómico*, 18 June 1887: 8

ages and backgrounds; within the space of the museum this social mingling produced a sense of adventure. Such playful representations of uneducated museum behaviour in the popular press and literature reflected a middle-class search for an appropriate type of spectatorship for the 'public sphere' which could be differentiated from religious observance and the consumption of popular entertainment. The chapter begins by examining the dynamics of the attendance at the Prado in the last third of the nineteenth century when it became known as a National Museum of Painting and Sculpture. It will then seek to define the place of the Prado among other leisure activities and institutions before, finally, discussing how middle-class museum spectatorship was negotiated with other forms of leisure, such as bullfights, theatres and musical performances.

Museums and public spectatorship in Madrid

Those who, in the nineteenth century, promoted museums for the moral improvement of the masses were aware of improper behaviour of uneducated visitors (such as eating lunch or sheltering from the rain). This knowledge affected decisions concerning museum regulations, entrance fees, times of public admission and even floor plans (Bennett 1995; Prior 2002: 92–94; Whitehead 2005: 59–69). I will argue that satirical depictions of visitor behaviour not only reflected the concern of the authorities for the education of commoners, but also responded to the bourgeoisie's need to engage with the national art museum by integrating it into the system of modern leisure.

While debates about art spectatorship reflected the development of the 'public sphere', an unusually charged meeting of two quintessentially public activities – museum-going and the reading of the printed press – deserves further attention. Late nineteenth-century depictions of museum-going in Madrid seem to contradict notions of culture, civility and decency associated with museums as sites of 'a discrete activity that involves the communication of ideas and values through looking at and thinking about art' (Lowry 2004: 141). The spectators portrayed by Spanish journalists, writers and caricaturists rarely talk about art. Rather, they talk about morality (the bourgeois 'value') manifested in taste and social status. As we shall see, while Spanish bourgeois men relied on the art museum to validate their identity on the grounds of taste and morals, they also associated the museums with theatres as places of imagination in the 'public sphere'. This tendency could, in part, be explained by the limitations of art education, the belated appearance of nudes in contemporary Spanish art, the unresolved spheres of influence of church and state or the complexities of bourgeois identity in a social order still rooted in the *ancien régime*. Whatever its cause, the linkage between art and commercial entertainment suggests that the national character of a museum was not only officially defined, but also 'produced' from below, in Lefebvre's definition of production as a 'representation' (collective cognitive reflection) and as an experience of a 'lived' space (Lefebvre 1974).

The Prado was nationalized in the public imagination by urban middle-class men who found there a carnivalesque libidinal articulation of their anti-aristocratic spirit. Thus, while museums were enshrined as schools of governance (Bennett 1995), the sites of 'civilizing rituals' (Duncan 1995) and quiet places for the communication of ideas about the world through discussions of art, they were also experienced as playgrounds for meeting, mingling and consumption (Hill 2005). From this point of view, the national museums' bearing on a nation's collective imagination is directly related to their integration into what Bailey has described as the 'leisure zone': a 'new territory of . . . leisure and service industries, a mix of sites, practices and occasions that mediated across the frontiers of the public/private divide' (Bailey 2004: 3; see also Bailey 1987: 68–90). Along with Bennett's (1988) 'exhibitionary complex', Bailey includes in this 'zone' a number of trademark modern sites which time and again emerged side-by-side with the Prado museum in the Spanish press: cafés and restaurants, theatres, music halls, public parks and other places of modern 'adventure' where people from different classes, walks of life and genders could meet (Bailey 2004: 14–15). National art museums had to compete with such establishments to secure a place in nineteenth-century public life.

The depiction of nude-struck laymen in the popular press and in fiction cannot be viewed as a reflection of a lower-class response to museum visiting. Even assuming wide popularity of the sources examined, the visitor books and archives of the Prado feature no complaint left by a shocked soldier or peasant, be it only for the simple reason that by 1900 only one-third of Spain's population could read and write (Carnero 1994: 26). Hence, while undoubtedly telling and symptomatic, these representations were more likely to reflect the literate middle classes' need to project negative behaviour on to citizens viewed as 'imperfect' due to their sex, age, origin or class. As Bourdieu suggests, in this way the middle classes attempted to modify the conventions of taste and in doing so effect their self-identification and hegemony (Bourdieu 1979; Bourdieu *et al.* 1969). Controlling the spectatorship of nude art, which constituted a legacy of the aristocracy, became a key task in the configuration of middle-class culture. In their portrayal of acceptable and unacceptable behaviour, middle-class men could affirm their agency by developing codes for secularized, civic-minded, public – and yet playful – collective spectatorship which was different from the private tastes of the upper classes and the ritualized appreciation associated with religious devotion.

The consumption of nude art was only a part of a much wider problematic of the 'body economy' and social morality that emerged with the growing self-awareness of the middle classes. If morality is considered as an autonomous cultural field, the bourgeoisie defined themselves within it in opposition to both aristocratic libertinism and lower-class promiscuity (Smith 1996: 68–75). According to Smith's (1996: 102–61) analysis of Victorian taste, it was the influence of international exhibitions and French *Salons* which made public displays of nude art acceptable in Britain after 1860.[2] Yet the astonishment of the working-class protagonists of Émile Zola's *L'Assommoir* (1877) at the sight of naked statues in the Louvre (to which they had retreated when their wedding party was rained out) suggests that in France, too, lower-class reactions to art were used to channel bourgeois anxieties about nude bodies in the national museum (Zola 1961 [1877]: 441–46). Such attention to the lower-class museum experience reveals the double standards of bourgeois morality: acceptance of public nakedness in the form of aesthetic or 'disinterested' nudity versus repudiation of both the private enjoyment of nudity by the upper class and the public consumption of nakedness by the commoners.[3]

The nationalization of the Prado and the evolving visitor

The transitional status from royal museum to national museum, which the Prado enjoyed in the latter part of the nineteenth century, makes it a particularly useful site for understanding how the negotiation of taste, morals and middle-class identity is performed as part of a process of nationalization. The Royal Museum[4] collection which had originated in the artistic patronage and the rituals of gift exchange of two royal families[5] had opened to the public in 1819 and became widely known as the Museum of Paintings.[6] It came to occupy a neoclassical building on the Prado Boulevard,[7] which had initially been designed for the Museum of Natural History at the end of the eighteenth century by the architect Juan Villanueva. The first decision to create a public art museum in Madrid had been taken by Joseph Bonaparte during the French occupation of Spain between 1808 and 1814. It was, however, King Ferdinand VII who realized the project, having apparently financed its foundation and maintenance from his own private funds (Madrazo 1843: viii). Between 1835 and 1872, a National Museum of Painting and Sculpture[8] existed in Madrid not far from the Royal Museum. Its collection was composed of artefacts confiscated from the religious communities following the 1835 campaign of 'disentitlement' of the properties of the Spanish church. Lack of funding and space, shared with a government ministry, ensured this national museum of little recognition (Álvarez Lopera 2004).

Decades of discussions concerning the ownership of the collection of the Prado had to pass before it was declared a possession of the Spanish nation. Following Ferdinand's death in 1833, his widow Regent Queen María Cristina initiated several lawsuits[9] with the objective of returning to the collection the status of personal property of the royal family, which could then be divided between herself and her two daughters (Madrazo 1945: 129–32). At the same time, there were discussions as to who owned the museum and proposing the transfer of royal patrimony from the Crown to the nation (Madrazo 1945: 178, 214). On every occasion, however, the museum was declared to be the indivisible property of the Crown. In 1865, Queen Isabella II put an end to these discussions by signing a law which made this status official, protecting the collection from the whims of members of the Bourbon dynasty. Following the Queen's dethronement in September 1868, a Council put in charge of the conservation, custody and administration of the possessions of the Spanish Crown took responsibility for the museum. On 23 February 1872, Amadeo of Savoy, who had been elected King by Spain's Provisional Government after the 1868 revolution, declared the Royal Museum the 'property of the Nation'. The Royal Museum was merged with the National Museum of Painting and Sculpture in 1872, and after King Amadeo's abdication, under the law adopted on 24 July 1873 by the Government of the First Republic regulating the transfer of the former property of the Crown, the administration of the museum was entrusted to the state. The new institution received the name of the National Museum of Painting and Sculpture (*Museo Nacional de Pintura y Escultura*) but also became known as the *Museo del Prado*.

Both before and after its nationalization, the Museum of Paintings was open to anyone at no cost of admission on Sunday mornings and afternoons (10 a.m. to 3 p.m.) and on public holidays (and occasionally on another day of the week) and was accessible to foreigners and visitors to Madrid on any day upon presenting a passport or a document verifying their address. Visitor books signed on the days of non-public admission indicate that even when access was restricted the audience consisted of more than just art connoisseurs. These books feature the signatures of (male) visitors from Spain who could not always spell the names of their professions. These included shoemakers, employees, salesmen, landlords, labourers, ironsmiths, blacksmiths and bakers.[10] The newly nationalized Prado almost immediately began to charge admission, on all days except Sundays and public holidays, to everyone except those carrying special permissions.[11]

If commoners were able to visit the museum long before it was nationalized, they may well have seen the nudes collected by the Spanish Habsburgs and Bourbons, which were displayed there from 1838.[12] These objects had been transferred from the Royal San Fernando Fine Arts Academy to the Museum of Paintings in 1827. Between 1827 and 1838, the nudes were exhibited in a special room on the lower level called 'Reserved Gallery' (*Sala Reservada*) which remained closed on days of public admission. Visitors wishing to see the collection on other days of the week could obtain a permit from the director upon presenting their credentials. Yet even before objects in the Reserved Gallery were dispersed into the galleries representing the different artistic schools, the museum had featured large numbers of male and female nudes (Portús 1998: 240-74). And in addition to paintings, nude sculptures were also a feature of the Sculpture Gallery opened in 1839. It would, then, be incorrect to think of the discussions of middle- and lower-class public spectatorship as a mere reflection of the nationalized museum's growing admission of uneducated visitors. By the same token, the extensive depiction of visitors caught unprepared by the sight of the nudes did not reflect the changes in the availability of these images to the public.

Studies of the Louvre and the British National Gallery suggest that the transformation of royal and aristocratic private collections into national institutions tends to elicit questions of how 'the man in the street' could relate to a museum. When the Louvre was proclaimed a national museum, following the Revolution of 1789, similar questions were discussed by the Museum Commission and the Conservatoire which determined that a display by school and chronology would help explain the idea and function of the museum (McClellan 1994: 107–16). Whitehead (2007) reveals a similar use of art-historical display in Britain. The Spanish public and press took similar steps when the Prado ceased to be a royal museum. In early November 1868 (several weeks after the Revolution), liberal newspapers in Madrid repeatedly advertised a book by the restorer, Vicente Poleró, arguing that the unification of the Prado with the former National Museum would allow for a plausible representation of the Spanish school (Anon. 1868a; Anon. 1868b; Anon. 1868c); this, however, took some time to be accomplished (Afinoguénova 2009). It was only after the two museums merged that the press initiated its discussions of public art spectatorship.

Between 1874 and 1880, the number of paying visitors tripled, while the numbers of non-paying visitors grew only half as much (Table 13.1). Underpinning this growing attendance was the introduction of art history into the university and school curricula following educational reform in 1868. This intensified in 1876 with the foundation of the free school, *Institución Libre de Enseñanza*, which promoted regular museum visits for schoolteachers, schoolchildren and the general public, both male

Table 13.1 Prado visitor numbers, 1874–1880, reported on 19 March 1881[13]

Year	*Paid*	*Permission*	*Public entry*	*Total*
1874	3,700	2,000	20,000	29,900
1875	4,800	1,600	21,400	27,800
1876	5,600	2,020	25,700	33,320
1877	11,700	1,000	26,800	39,500
1878	9,900	1,700	25,000	36,600
1879	9,700	3,000	30,000	42,700
1880	11,400	2,600	29,800	43,800
Total	*57,000*	*13,920*	*178,700*	*249,620*

and female. For interested middle-class citizens, lectures on art history began to be held at the Madrid Athenaeum in the early 1870s. These new opportunities to appreciate art appealed mostly to the more educated inhabitants of Madrid. And, although provision of lectures for the general public at the Prado was announced in 1873, these only began to appear in the following century. Tours of the Prado for the adult and teenage visitors in 1880s led by Manuel Bartolomé de Cossío (one of the leaders of the educational reform and Spain's most prominent art historian) demonstrate that these enlightened attitudes did not explicitly address the spectatorship of nude art or its moral implications. In the tradition of the Spanish Church's acceptance of male nudes and condemnation of the naked female, Cossío did not shy away from Velázquez's *The Forge of Vulcan* (1630). Yet the only female nude featured in his tours was Durer's *Eve* (*c.*1507), suggesting that the topic was perceived as problematic (Cossío 1885, 1889).

Discussions of museum spectatorship in literature and press began in the 1870s and then steadily grew throughout the 1880s and 1890s, when the middle-class audience became well established. It is then that the Madrid newspapers print numerous comical references to visitors, frequently provincials (who were out of reach of the new educational opportunities in the capital) coming to see the museum without knowing why. Their behaviour in front of the nudes occupied the centre stage of these representations, as these paintings offered the maximum challenge for the audiences untrained to relate pictorial and sculptural nudes to ideal beauty.

A museum in the leisure zone

The Prado did not become a part of the Spanish popular imagination until a male middle class of journalists, writers and caricaturists reinvented it as a place of leisure. Situated in the middle of the boulevard, next to the Botanical Gardens where Madrid promenaded, the Prado was a landmark, reference point and location of activities ranging from public protests and demonstrations to public suicides. Throughout the nineteenth and early twentieth centuries, the museum attempted to separate the building from the boulevard by an iron fence and gate but only succeeded in symbolically closing it off at night using an iron chain.[14] As a result, the museum could not help but find itself the location of leisure activities quite unrelated to arts. In the course of the nineteenth century, it became increasingly integrated into the city's tourist attractions and seen as a place of leisure no different from the neighbouring gardens or the city's cafés.

Nineteenth-century novels, tourist guidebooks and etchings depict the boulevard next to the museum as the place of social interaction where Madrid's upper and middle classes were coming to exhibit their dress, carriages and lovers. Richard Ford (1845: 677) called this area the 'mirror of Madrid . . . here all ranks meet, whatever their age, condition or costume'. The Madrid newspapers mocked the provincials rushing to see the celebrations of San Isidro (the patron saint of the city) and then considering it their duty to take a walk through the Prado as well: 'Saw it on Sunday. Ugh, was it hot! There are a lot of paintings and very big rooms. What can I say: no longer shall I die without having seen it.' This imaginary note of a 'conservative old Castilian' coming 'to the yearly celebrations of San Isidro taking advantage of the sale of railway tickets' was advised by the editor of the liberal Madrid newspaper, *Gil Blas*, to keep to less complicated leisure activities: 'You would have had a better time visiting a cold drinks parlour at the Carrera San Gerónimo' (Anon. 1868d). Other newspapers laughed at the middle-class 'folks from outside' coming to attend municipal festivities and bringing their families to the museum, only to meet other families from their own town and discuss food and accommodation (Anon. 1890a).

While local and provincial middle classes mingled in and around the museum, urban déclassé elements joined their celebrations in their own ways, especially after-hours. Memoranda and complaints by the authorities after the museum's nationalization suggest that the Prado was also the centre of lumpen leisure. A water kiosk next to the museum entrance was owned by a woman whose apparently fierce character became a cause of 'scandals sometimes reaching the extremes of shooting firearms and at other times making people hide in the . . . bushes, tricking the guards' vigilance in the darkness of the night'.[15] In the early twentieth century, the museum's immediate surroundings were the gathering point for 'people of bad behaviour, men as well as women' and for 'public women', who were blamed for the disappearance of the museum guards' personal belongings.[16] It is hardly surprising, then, to find a comical 'Nocturne' about a visitor deploying the apparatus of art criticism in front of a pretty woman just outside the Prado, only to make her think she was being taken for a harlot:

> *By the museum of paintings*
> Please allow me
> To admire you *en-face* ecstatically.
> Sir, please step away.
> Keep in mind that, although you see me alone,
> I am a decent girl.
>
> (López Silva 1883)

In 1880, when the Spanish Church requested that, during Holy Week, the authorities cover the nude statue of *Harmony* purchased by the government and installed on the Madrid Puerta del Sol, the liberal press mockingly suggested that the museum statues should also be covered, becoming mannequins for seamstresses and tailors in the process: 'Apollo would look great dressed as a bullfighter. And Venuses, as *manolas*'[17] (Anon. 1880). In the same year, commenting on the Madrid café owners' new habit of exhibiting art, liberal journalist, Miguel Moya, proclaimed that the museum and the café would soon become indistinguishable and there would be only 'one type of café possible. The Museum of Paintings' (Moya 1880). The satirical magazine, *Madrid Cómico*, jokingly announced the installation next to the museum of an exhibit of absurd nationalistic pseudo-events: 'The Discovery of the Balearic Islands by Columbus', 'Charles V taking the Bastille' and 'The Senate's Dream' (de la Vega 1880).

The leisure activities concentrated around the Prado museum promoted modes of spectatorship, consumption and performance mostly concerned with the body. It may be for this reason that common representations of the museum focus on visitors being exposed to nakedness under the respectable cover of nude art. While it is hard to assess the faithfulness of these playful sketches to what visitors actually felt or thought, they undoubtedly projected the authors' own voyeuristic enjoyment at the visitors' shock. This type of humorous exploration of the museum experience formulated in terms of carnivalesque profanity, indecency and the grotesque treatment of the body was a crucial part of the Prado's nationalization 'from below' (Bakhtin 1965). Thus, in 1883, writing for the weekly *América*, Cuban-Spanish writer Tristán de Jesús Medina compared the Prado on a day of public admission to a 'carnival of satyrs abducting virginal nymphs' and stated that visitors who had come 'looking for an Easter of Resurrection' would instead do something 'contrary to the practices of cult' as they became 'by force accomplices of terrible rapes' (Jesús Medina 1883). Such a portrayal of the museum experience as flirtatious and tempting (without necessarily reaching the extremes of abduction and rape) reflected the reconstruction of the Prado as a morally ambiguous yet enjoyable destination. With the attractiveness of the 'leisure zone' also grounded in sexual innuendo and attention to body performance, it was ultimately this playful experience that secured a high profile for the Prado among the affordable bourgeois pleasures involving visual seduction.

This carnivalesque treatment of the nude paintings also suited middle-class political desire for anti-monarchical and anti-aristocratic mockery. The year following the 1868 Revolution, some paintings from the Prado were mockingly imitated in a collection of the pornographic etchings called *The Naked Bourbons* (*Los Borbones en pelota*) (Bécquer and Bécquer 1868–69; see also Portús 1998: 281). The 1880s caricatures of the character of the ignorant soldier or peasant were also used to poke fun at royal morals and aristocratic tastes. These were more printable and further associated the monarchy with pornography. However, it is hard to ignore the suggestion that these caricatures also satirized the effect of art on the museum's lower-class visitors. In 1888 the satirical Madrid weekly, *La Risa*, ran a series of sketches by Ignotus featuring laymen commenting on the immorality of nude art at the museum. In one of these, two soldiers describe Titian's *Venus with a Dog and an Organist* (*c.*1550) as a portrait of King Philip II's lover (Ignotus 1888a) (Figure 13.2). In another, an aged peasant woman accompanied to the Sculpture Gallery by a male compatriot could be seen overwhelmed with shame at the sight of those things 'one can see in Madrid' (Ignotus 1888b) (Figure 13.3). In all these sketches the national museum provided a convenient setting for the bourgeoisie to mock its two class antagonists (aristocrat and peasant), in the process revealing that each possessed base characteristics.[18]

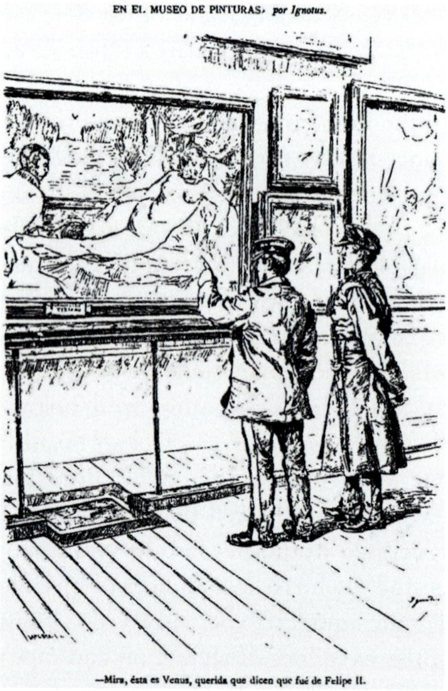

Figure 13.2 In the museum of paintings: — Look, this is Venus; they say she was Philip II's concubine. Photo: *Ignotus*, 'En en Museo de Pinturas', *La Risa*, 4 March 1888: 4

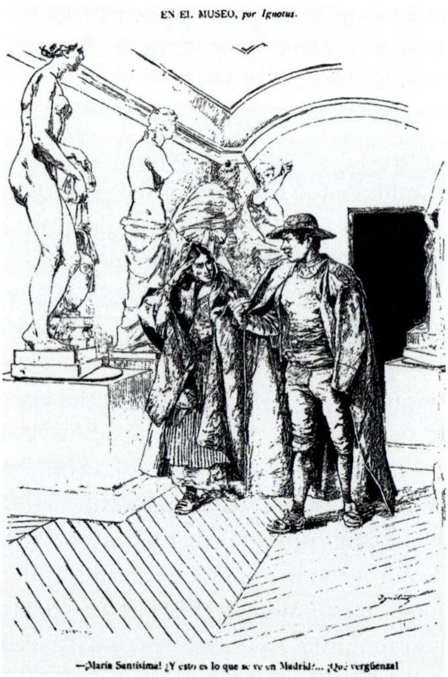

Figure 13.3 In the museum: — Holy Virgin Mary! And this is what one can see in Madrid? . . . What a shame! Photo: *Ignotus*, 'En el Museo', *La Risa*, 1 April 1888: 8

Art, museum and politics

For the liberal middle-class press, the Prado became a politically useful institution because of the social interplay between visitors, and between visitors and art. The earliest comic representations of the museum freely combined politics, art and museum whilst deploying 'naughtiness' for humorous effect. The 'diabolical' Madrid newspaper *Gil Blas*, for example, printed a drawing called 'A projected painting representing the Spanish revolution for the museum of the future', featuring two plump female legs with a Crown still floating on the surface of what might have been the Bay of Cádiz into which the rebellious generals Prim and Serrano had ditched the dethroned Queen (Anon. 1868e) (Figure 13.4). In the following year, the writer and journalist, Eusebio Blasco, waged a war in the pages of the same newspaper against José Aparicio's large-scale patriotic painting, *The Famine of Madrid* (*c.*1818), which was exhibited at the Prado. In the first of this series of articles, Blasco portrayed himself as so shocked by this patriotic depiction of the inhabitants of Madrid (who allegedly preferred to starve and see their children die than surrender to Napoleon's troops) that he decided to seek shelter in other public spaces of the nation. But, having entered into the building of the Senate, he turned away in terror after catching alarming signs of starvation on the deputies' faces and bodies (Blasco 1869; see also Anon. 1871a, 1871b). In mocking tones, he drew terrifying conclusions about what would happen if the canvas constituted a realistic reflection of Spain's political life.

Gil Blas' war against the 'scandalous' presence of this gory picture in the museum was part of a wider struggle concerning the types of spectatorship the Prado should support. Labanyi (2005) has suggested that the popularity of large historical canvases featuring spectacular scenes of madness and assassination at the National Fine Art Exhibitions held in Madrid from 1856 reflected the viewing habits of lower-class visitors whose gaze had been trained through habitual exposure to bullfights, waxworks and freak shows. Since the only art that Spain's lower classes would normally see was religious, a gaze educated in popular entertainment constituted the norm for mass secular spectatorship. Bourdieu (1979) referred to this honest appreciation of art as a 'popular aesthetic' denied legitimacy by the culturally hegemonic bourgeoisie. Some of the battles over paintings exhibited at the Prado reflected a social debate about the validity of the lower-class gaze and, indeed, whether it had any

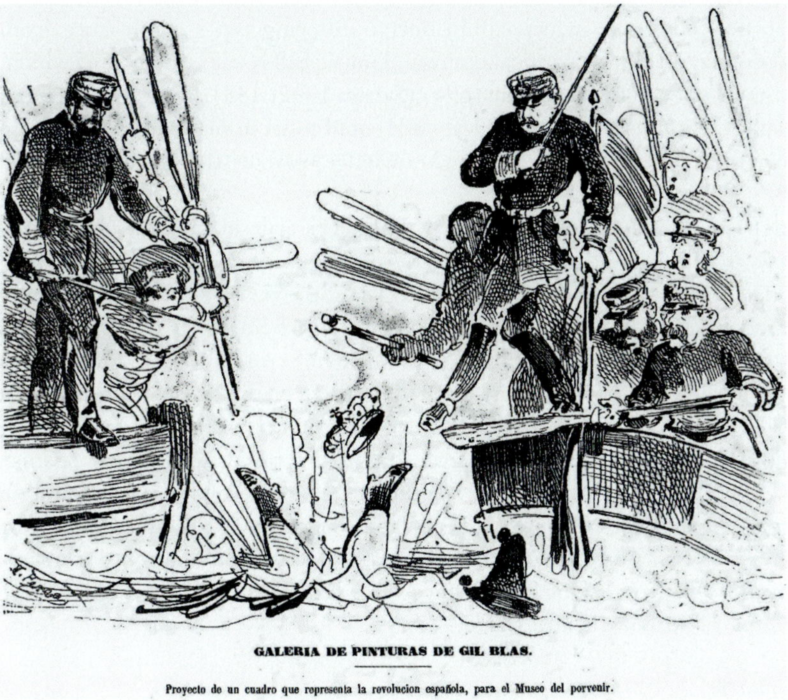

Figure 13.4 *Gil Blas'* Gallery of Paintings. A projected painting representing the Spanish Revolution for the museum of the future. Photo: 'Galería de pinturas de Gil Blas. Proyecto de un cuadro de la revolución española para un Museo del porvenir'. *Gil Blas: Periódico político satírico*, 22 October 1868: 1

place at the national museum. Thus, while the liberal newspapers *Gil Blas* and *El Imparcial* denounced the Prado's spectacular large-scale canvases filled with 'Greeks, Romans and costumed Goths' (Anon. 1890b), the Spanish public demanded that the jury at the 1881 Exhibition grant the Medal of Honour to one of the bloodiest paintings in Spanish history, José Casado de Alisal's *The Legend of the Monk King* (1880). The public defence of Casado's canvas forced the state to purchase it and donate it to the museum together with another large-scale painting of gore, Eduardo Rosales' *Death of Lucretia* (1871). Such debates indicate that, as late as the 1880s, popular taste could be triumphant at the national museum even though middle-class patrons were resistant. There is, however, no evidence that either the bourgeois or the lower-class audience was aware that viewing art at the museum required a special form of seeing, different from the one which was suitable for witnessing the bloody scenes at the bullfighting ring or fairground.

Differentiation was further complicated by the fact that for the middle-class visitors, too, the museum was a performance. The middle-class approach to art differed only in the source of its gaze, since the bourgeoisie contemplated the museum with the eyes of voyeuristic theatre-goers rather than bloodthirsty bullfight *aficionados*. The diversification of theatre genres in the second half of the nineteenth century produced a gaze which could derive differing types of pleasure from historical drama, serious or comical opera and music hall revues (the first of which, the Alhambra, opened in Madrid in 1894) (Salaün 1992: 138). Thus, transplanted to the national art museum, a theatrical gaze could assist in mapping the museum experience to a range of visitors seeking knowledge, voyeuristic pleasure or sexual contact.

Serious visitors educated in history but not necessarily in art experienced the museum as a collection of prototypes for theatrical sets and costumes which they saw on stage in historical opera and drama. In 1865, the reviewer of Verdi's *Hernani* at the *Teatro Real* in Madrid praised the costumes for being copied from Titian's paintings 'which one can admire in our Art Museum' (Anon. 1865). Later in the century the trend of interpreting historical theatre as a continuation of the museum was postulated in positivist terms. Thus, in 1892, Spain's pioneer archaeologist José Ramón Mélida argued that research into the history of costume at the Spanish galleries of the Prado would transform Spanish theatre into a valid tool for education in history (Mélida 1892: 161). Three years later, Diego Luque, the scenic director of the *El Español* drama theatre in Madrid, was reported to have taken his decorators to the Prado to copy Velázquez's paintings, eventually triggering a 'scenographic regeneration' of Spanish theatre (Anon. 1895).

A proper vision of the nude body

In the last decades of the century, the popular and commercial entertainments which had trained the playful gaze of the urban middle classes began to increasingly capitalize on the display and consumption of the female body (Gies 1994; Salaün

1992). Now this gaze had to become civil, educated and mindful of beauty, and differentiate itself from the private avaricious connoisseurship of the aristocracy and lower-class public spectatorship rooted in seasonal celebrations which combined religious masses, bullfights and fairground performances.

While the comparison with historical opera and drama mapped the viewing of art for serious visitors, the parallels between the museum experience and the sights that were open to public eyes during comic opera and other light musical performances gave ample freedom to those seeking voyeuristic pleasures. To these museum-goers, a distanced and disinterested approach to the nudes on exhibit provided a comforting parallel to the female nakedness both on stage and in the audience of light musical performances. Humorous references at the time presented the museum as a place where the gaze was made respectable through a comparison with more frivolous comic opera and operetta. For example, the conservative-leaning journalist and writer, Fernando Martín Redondo, writing for the popular illustrated weekly, *Ilustración Española y Americana*, referred positively to nude art at the Prado while criticizing the consumption of nudity found in light musical shows. The article featured the following dialogue between a bourgeois widow and a friend of the family in the presence of the widow's two young daughters:

> Welcome, don Casimiro, where were you last night?
> Having some fun . . . I went to see some women . . .
> Did you mean 'some ladies'?
> Well . . . sort of . . . I went to see some ladies . . . nude.
> Sir!
> I must say, there was nothing sinful: these ladies were . . .
> . . . painted?
> Exactly.
> So, you were at the Museum; but why at night?
> Oh, no, the ladies I am talking about, although there was a lot of colour on them, were of bone and flesh, all of them young and some were pretty.
> (Martín Redondo 1874: 6)

Martín Redondo's character, in fact, had attended Offenbach's *Genevieve de Brabant* at the popular *Teatro del Circo*.[19] In this comic sketch, both the opera and the museum were places where one could spend time surrounded by women, naked or nude, without committing a sin. However, it is clear that the widow considered one of these venues more respectable.

Given the slow introduction of the nude into Spanish painting in the nineteenth century and the limited possibilities for education in art appreciation in Madrid, it was perhaps inevitable that visitors should bring other forms of spectatorship into the museum. The sketches that appeared in the press can be interpreted both as moral criticism and as testimonies documenting the liberal male authors' recurring excitement at the thought of the national art museum and the paintings in its collection. References to the Prado in contemporary Spanish fiction also show that even the most enlightened members of the middle class were not untouched by the public display of the nudes. Benito Pérez Galdós's novel, *The Disinherited* (1881), shows that even the most canonical Spanish liberal writer, versed in art, was capable of projecting *his* thrill in the presence of nude art on to the female museum visitors (Galdós 1909 [1881]: 64). His protagonist, Isidora, on her visit to the Prado could not help but imagine herself either as a scantily dressed ancient goddess or Venetian courtesan. Another liberal writer, Leopoldo Alas (Clarín), made the young male protagonist of his short story 'Pipá' (1879) imagine that the female paintings on exhibit were looking 'at every century's spectator with a gaze of infinite lust' (Alas 1886: 139). Portús (1998: 277–83) points at over a dozen sensual or frivolous references to the Prado by important Spanish male writers.

Conclusion

While the Prado acquired national status with the stroke of a pen, its 'nationalization' in the eyes of its audiences took much longer and was performed from the bottom up. In the nineteenth century, museums of all kinds were possessed by middle-class men, symbolically and performatively. At the Prado, this was achieved through bourgeois commentators ridiculing incorrect attitudes to nude art as a side-effect or a sign of the visitors' unpreparedness. By denouncing lascivious aristocrats and indignant commoners, they produced an image of the museum as an exciting place where one could rub elbows with people of different social classes and with members of the opposite sex. In these actions, the museum they nationalized in the popular imagination was one with carnivalesque qualities drawn directly from the theatre which had yet to learn distanced aestheticism. Even in the early twentieth century the museum continued to appear in the pages of erotic publications known as '*sicalípticos*' ('naughty'), which were defined as 'halfway between what is artistically moral and what constitutes artlessly frivolous: without being completely the former yet without falling into the latter' (Limendoux *c.*1902: back cover).[20] The collections of such etchings and prints, which mixed the nudes from the Prado and erotic photographs of the so-called 'gallant' women, became customary in illustrated magazines like *Gedeón* and *El Sicalíptico*. In 1904, *El Sicalíptico* advertised a collection of colour prints of nudes advising 'all artists and lovers of fine arts' to purchase them in order to 'have a real museum' of their own (Anon. 1904). Given the national art museum's continued connection with performance and sex, it was only logical that the idea of the museum provided a respectable cover-up for a mélange of erotic images. It took until the 1920s for the museum gaze to break free from the eroticism in Spain.

Notes

1 Escuela realista.
2 Even then, however, an acceptable representation of a nude had to conform to a set of rules which ensured that nudity was properly idealized. The discussions over John Gibson's 'The Tinted Venus' (1851–56), exhibited at the 1862 International Exhibition of Industry and Art in London, demonstrate that the use of colour, for example, had to be restricted (Smith 1996: 121).
3 My use of an opposition between the naked and the nude is based on Clark's oft-cited *The Nude: A Study in Ideal Form* (1956). The confusions in this opposition derive from the social experience of nude art which have been examined by Pointon (1990: 11–16).
4 Museo Real or Museo Fernandino.
5 The Spanish Habsburgs of the sixteenth and seventeenth centuries and the Spanish Bourbons after the eighteenth century.
6 Museo de Pinturas.
7 Paseo del Prado.
8 Commonly called *Museo de la Trinidad* due to its first location at the former Trinity Convent.
9 In 1834–1835, 1845 and 1855. Discussions concerning the conversion of royal patrimony into a national collection took place in 1839, 1842 and 1855.
10 Archivo del Museo Nacional del Prado, Libros-registro de los nombres de las personas que visitan las salas del Real Museo de Pintura y Escultura de S. M., several volumes, 1854–1865.
11 This was announced on 25 September 1871.
12 As Portús has pointed out, the enjoyment of nude art that Spanish kings and high-class patrons and collectors could afford was confined to the space of 'Reserved Rooms' withdrawn from the public eye. In fact, it was the most enlightened Spanish Bourbon King, Charles III, who mandated the destruction of nude art and only Antonio Rafael Mengs's and Marquis of Esquilache's interference saved them from being burned. The paintings were, however, removed from the palace. In 1792, following the request from the Fine Arts Academy, twenty-four (and in 1796, another thirteen) paintings were deposited there as the objects for the study of colouring. They were transferred to the Prado, then the Royal Museum, in 1827.
13 Archivo del Museo Nacional del Prado, Box 1429.
14 1940s typed report, Archivo del Museo Nacional del Prado, Box 1429.
15 Letter to the Mayor of Madrid, 12 September 1873, Archivo del Museo Nacional del Prado, Box 1429.
16 Archivo del Museo Nacional del Prado, Box 369.
17 A name used for lower-class urban females dressed in typical costumes that appeared in nineteenth-century Spanish operettas called *zarzuelas*.
18 While middle-class publications were playfully remaking the museum for the popular leisure imagination, in the properly popular and working-class press the museum remained an alienated and politicized object encapsulating the 'bourgeois artistic sensualism, or art as an aperitif' (*Almanaque de Tierra y Libertad* 1911, quoted in Litvak 2001: 302).
19 Until it fell prey to the 1876 fire, *Teatro del Circo*, situated on Plaza del Rey, was the earliest Madrid theatre staging Italian operas and French operettas.

20 The first series appeared without a date. In 1902, the newspaper *El Liberal* announced the publication of the second series of etchings, suggesting an approximate date of publication of the first series.

Bibliography

Afinoguénova, E. (2009) '"Painted in Spanish": The Prado museum and the naturalization of the "Spanish school" in the nineteenth century', *Journal of Spanish Cultural Studies* 10(3): 319–40.

Alas, L. (*Clarín*) (1879) 'Pipá', *Narrativa Breve*, Madrid: Librería de Fernando Fe, 1886.

Álvarez Lopera, J. (2004) *El Museo de la Trinidad en El Prado: Catálogo de la exposición, 13 de julio–19 de septiembre 1994*, Madrid: Museo Nacional del Prado.

Anon. (1865) 'Revista de teatros', *La Iberia*, 19 February: 1.

Anon. (1868a) *El Imparcial*, 2 November: 3.

Anon. (1868b) *El Imparcial*, 4 November: 1.

Anon. (1868c) *La Iberia*, 5 November: 4.

Anon. (1868d) 'Apuntes de un Castellano Viejo', *Gil Blas*, 21 May: 2.

Anon. (1868e) 'Galería de pinturas de Gil Blas: Proyecto de un cuadro de la revolución española para un Museo del porvenir', *Gil Blas*, 22 October: 1.

Anon. (1871a) 'Cabos sueltos', *Gil Blas*, 11 October: 4.

Anon. (1871b) 'Cabos sueltos', *Gil Blas*, 26 October: 4.

Anon. (1880) 'En broma', *La Iberia*, 24 March: 3.

Anon. (1890a) 'Gente de fuera', *El Imparcial*, 16 June: 6-7.

Anon. (1890b) 'Exposición nacional de Bellas Artes', *El Imparcial*, 15 May: 1.

Anon. (1895) 'Augusto Ferri', *La Gran Vía*, 14 December: 4.

Anon. (1904) 'El desnudo en el arte', *El Sicalíptico*, 29 October: 13.

Bailey, P. (1987) *Leisure and Class in Victorian England: Rational Recreation and the Contest for Control, 1830–1885*, London: Routledge.

Bailey, P. (2004) 'Adventures in space: Victorian railway erotics, or Taking alienation for a ride', *Journal of Victorian Culture* 9 (1): 1–21.

Bakhtin, M. (1965) *Творчество Франсуа Рабле и народная культура Средневековья и Ренессанса* (Tvorchestvo François Rabelais i narodnaya kul′tura Srednevekoviya i Renessansa), Moscow: Khudozhestvennaya Literatura, translated by H. Iswolsky, *Rabelais and His World*, Bloomington: Indiana University Press, 1984.

Bécquer, V. and Bécquer, G. A. (1868–69) (S.E.M.) *Los Borbones en pelota*, ed. R. Pageard, L. Fontanella and D. Cabra Loredo, Madrid: El Museo Universal, 1991.

Bennett, T. (1988) 'The exhibitionary complex', *New Formations* 4: 73–102.

Bennett, T. (1995) *The Birth of the Museum: History, Theory, Politics*, London: Routledge.

Blasco, E. (1869) 'La semana del hambre', *Gil Blas*, 8 July: 2.

Bourdieu, P. (1979) *Distinction: A Social Critique of the Judgement of Taste*, translated by R. Nice, Cambridge, MA: Harvard University Press, 1984.

Bourdieu, P., Darbel, A. and Schnapper, D. (1969) *The Love of Art: European Art Museums and their Public*, translated by C. Beattie and N. Merriman, Stanford, CA: Stanford University Press, 1991.

Carnero, G. (1994) 'Cultura y literatura en la vida española del siglo XIX', *El mundo literario en la pintura del Museo del Prado,* Madrid: Ministerio de Cultura, pp. 19–66.

Cilla (1887) 'El en museo', *Madrid Cómico*, 18 June: 8.
Clark, K. (1956) *The Nude: A Study in Ideal Form*, New York: Pantheon Books.
Cossío, M. B. de (1885) 'Los cuadros más importantes del Museo del Prado', *Boletín de la Institución Libre de Enseñanza* 9: 301–03.
Cossío, M. B. de (1889) 'Los alumnos de la escuela de Madrid en el Museo de Pintura: Consejos prácticos para hacer una excursión', *Boletín de la Institución Libre de Enseñanza* 13: 5–6, 22–25.
Duncan, C. (1995) *Civilizing Rituals: Inside Public Art Museums,* London: Routledge.
Ford, R. (1845) *A Handbook for Travelers in Spain and Readers at Home*, Part II, 3rd edition, London: John Murray, 1855.
Frontaura, C. (1888) 'La viuda de Zaragata', II, *La Risa*, 12 August: 10–11.
Gies, D. T. (1994) *The Theatre in Nineteenth-century Spain*, Cambridge: Cambridge University Press.
Hill, K. (2005) *Culture and Class in English Public Museums, 1850–1914,* Aldershot: Ashgate.
Ignotus (1888a) 'En en Museo de Pinturas', *La Risa*, 4 March: 4.
Ignotus (1888b) 'En el Museo', *La Risa*, 1 April: 8.
Jesús Medina, T de (1883) 'Visita a los muertos', *América*, 28 November: 11.
Labanyi, J. (2005) 'Horror, spectacle and nation-formation: Historical paintings in late-nineteenth-century Spain', in S. Larson and E. Woods (eds) *Visualizing Spanish Modernity*, London: Berg, pp. 64–80.
Lefebvre, H. (1974) *La Production de l'espace,* Paris: Éditions Anthropos.
Limendoux, F. (*c.* 1902) *Las mujeres galantes*, No. 1, Barcelona: Ramón Sopena.
Litvak, L. (2001) *Musa libertaria: Arte, literatura y vida cultural del anarquismo español (1880–1913)*, Barcelona: Antoni Bosch.
López Silva, J. (1883) 'Nocturnos', *Madrid Cómico*, 9 September: 7.
Lowry, G. (2004) 'A deontological approach to art museum and public trust', in J. B. Cuno, N. MacGregor, P. de Montebello, G. D. Lowry and J. N. Wood, *Whose Muse? Art Museum and the Public Trust*, Princeton, NJ: Princeton University Press, pp. 129–50.
Madrazo, M. de (1945) *Historia del Museo del Prado, 1818–1868,* Madrid: Ministerio de Asuntos Exteriores, Junta de Relaciones Culturales.
Madrazo, P. de (1843) *Catálogo de los cuadros del Real Museo de Pintura y Escultura de S. M.,* Madrid: Aguado.
Martín Redondo, F. (1874) 'Manufacturas poéticas', *Ilustración Española y Americana*, 22 August: 6.
McClellan, A. (1994) *Inventing the Louvre*, Cambridge: Cambridge University Press.
Mélida, J. R. (1892) 'La arqueología en el teatro', *La España Moderna*, 15 March: 161–63.
Moya, M. (1880) 'El Café', *La América*, 8 June: 14.
Pérez Galdós, B. (1881) *La desheredada*, Serie Contemporánea I, Madrid: Librería de Perlado, Páez y Cª, 1909.
Pérez Zúñiga, J. (1883) '¡Qué ocurrencia!', *Madrid Cómico*, 22 July: 6.
Pointon, M. (1990) *Naked Authority: The Body in Western Painting 1830–1908,* Cambridge: Cambridge University Press.
Portús, J. (1998) *La Sala Reservada del Museo del Prado y el coleccionismo de la pintura de desnudo en la corte española, 1554–1838,* Madrid: Museo del Prado.
Prior, N. (2002) *Museums and Modernity: Art Galleries and the Making of Modern Culture*, Oxford: Berg.

Salaün, S. (1992) 'Apogeo y decadencia de la sicálipsis', in M. Díaz-Diocaretz e I. Mª. Zavala (eds) *Discurso erótico y discurso trasgresor en la cultura peninsular. Siglos XI al XX,* Madrid: Ediciones Tuero, pp. 129–53.

Smith, A. (1996) *The Victorian Nude: Sexuality, Morality and Art,* Manchester: Manchester University Press.

Vega, R. de la (1880) 'De todo un poco', *Madrid Cómico*, 19 September: 1–2.

Whitehead, C. (2005) *The Public Art Museum in Nineteenth-century Britain,* Aldershot: Ashgate.

Whitehead, C. (2007) 'Establishing the manifesto: Art histories in the nineteenth-century museum', in S. J. Knell, S. MacLeod and S. Watson (eds) *Museum Revolutions*, London and New York: Routledge, pp. 48–60.

Zola, É. (1961 [1877]) *L'Assommoir, Les Rougon-Macquart* II, Paris: Gallimard.

Chapter 14

TAKING PART

Performance, participation and national art museums

Stuart Burch

On Saturday, 28 March 2009, hundreds of demonstrators gathered outside Nasjonalgalleriet, Norway's national museum of fine art. At 2 o'clock they joined hands to form a symbolic chain around the building (Figure 14.1). The reasons for this unusual action were explained in a website entitled *Bevar Nasjonalgalleriet* (Save the National Gallery). It called upon the Norwegian government to reverse its decision to close the museum and move it to a new building that it planned to erect in another part of Oslo.[1]

A protest such as this refutes a still widespread suspicion that national museums are, at best, solid and unchanging and, at worst, dull and irrelevant. Recent events in Norway confirm that national museums are in fact contested places in a state of constant change. Often, these permutations take the form of scarcely perceptible refurbishments or seemingly unremarkable redisplays. Occasionally they undergo more radical transformations such as the provision of an additional wing or construction of a brand new building. This is well illustrated by Norway's Nasjonalgalleriet. Its present home gradually took shape in three distinct architectural phases from 1879 until 1924. And, despite the best efforts of *Bevar Nasjonalgalleriet*, it might well end up experiencing a complete metamorphosis in the near future.

The demonstration that took place in March 2009 served as a reminder that national museums are emphatically performative spaces. The exhibitions they mount are not the only form of display. On show too is the public (Solkin 2001: ix). Those people that chose to link arms outside the walls of Nasjonalgalleriet were placing themselves, as it were, in the frame. Their carefully choreographed actions formed the prelude to the theatrical submission of a petition signed by over 11,000 people opposed to the relocation of the museum. This was captured on camera at the moment it was placed into the hands of Norway's then minister of culture, Trond

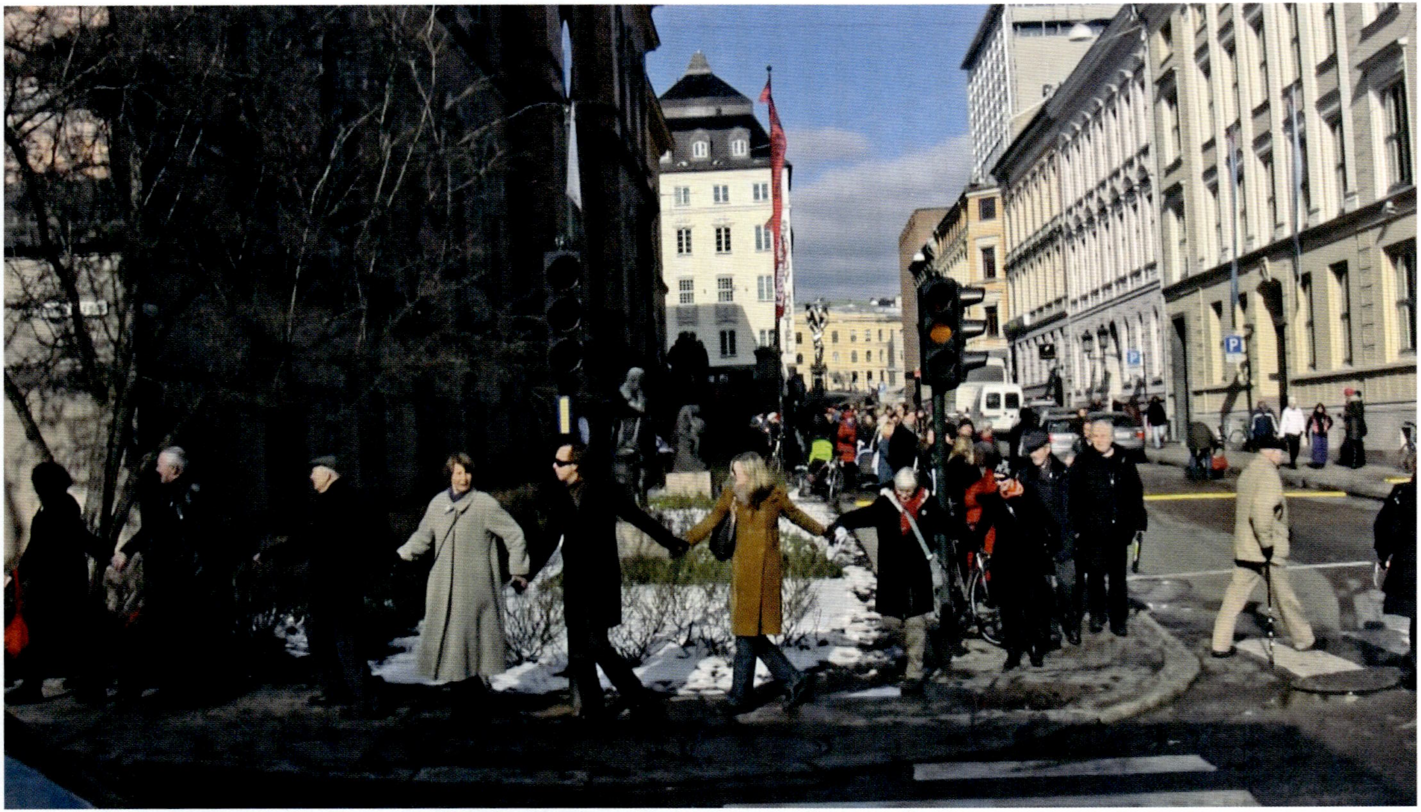

Figure 14.1 Bevar Nasjonalgalleriet (Save the National Gallery) demonstration, Oslo, 28 March 2009. Photo: Terje Lerberg

alised in favour of the children. The children participated for free; only adults had to pay. So many wanted to attend *Modellen* that numbers had to be restricted and an additional play area built outside the museum. This meant that the museum had literally as well as metaphorically broken out of its architectural frame and become 'a model for experiencing life outside its walls' (to recall Kirshenblatt-Gimblett).

An important aim of *Modellen* was to assess how the children would react to this stimulating environment, one where 'everything was, without reservation given over to the visitors – the children' (Granath and Nieckels 1983: 190). This was participation of the most comprehensive kind; a variety that did not just modify the orthodox authority structure – it completely reworked it (Pateman 1970: 106). It was motivated by dissatisfaction with the status of children in society and an education system that failed to value and nurture a child's artistic and creative potential (Hultén, in Nielsen 1968: 32). *Modellen* was also a 'protest' against a society that was seen as prioritising commercial interests over giving full rein to children's fantasy. It was anticipated that *Modellen* would elicit 'constructive' play; constructive in the sense that the children would learn to communicate and collaborate with their peer groups and because these activities were to be filmed and studied by psychologists (Hultén 1983: 43–44).

Nielsen used *Modellen* to hit out at what he perceived to be a sense of anxiety and powerlessness felt by many adults. He did not want adults to project their fear of society onto their children. He also hoped that *Modellen* would serve to release 'the child in the adult' (Tellgren 2008: 265). This was embodied in a series of oft-reproduced images of a man whose face is still widely recognised today, even though well over twenty years have passed since he was tragically gunned down in a Stockholm street. It is the Swedish politician and statesman, Olof Palme (1927–1986), who at the time was the Swedish Minister of Education.

Palme was an advocate of an active, participatory politics. Whilst Nielsen was doubtful about the benefits of street protests – preferring instead the 'direct action' of building playgrounds – Palme saw no such distinction. His photographed leap into *Modellen*'s foam blocks is a performative action that is on a par with his extremely controversial decision to participate in the anti-Vietnam war demonstration that took place in Stockholm on 21 February 1968. An image of him carrying a flaming torch appeared on the front page of *Dagens Nyheter*, eight months before the same national newspaper depicted him leaping into the air of Moderna Museet.

Palme's performative actions on the streets and in the museum literally embody the notion of participation when it comes to democracy. Politics is everywhere and happens all the time. It is not restricted to political representatives and periodic elections. The museum as an essential 'element of cultural politics' played a key role in this (Palme 1971). Carole Pateman's aforementioned book *Participation and Democratic Theory* (1970) came out a few years after *Modellen*, and her focus was elsewhere rather than museums. It nevertheless encapsulates the notion of establishing 'local political institutions' as 'training grounds for democracy' (John Stuart Mill, cited in Pateman 1970: 47). By allowing Nielsen to construct *Modellen*, Sweden's national museum of modern and contemporary art was aligning itself with a radical political project: an unambiguous attempt to establish a model not only for *experiencing* life but for *changing* it. That was the explicit objective of Pontus Hultén, the Director of Moderna Museet from 1960 until 1973 and the man who had sanctioned Nielsen's 'experiment'. In the catalogue to *Modellen*, Hultén declared boldly that the exhibition provided nothing less than 'a model of a new form of society'. He used this to advocate a 'new role for museums', one whose most 'pressing task' was to provide space for precisely these kinds of 'experimental models' (Hultén, cited in Nielsen 1968: 32).

Modellen was the most uncompromising endeavour to date of an institution that was only a decade old. Since its founding in 1958, Moderna Museet had mounted a series of bold experiments in art: exhibitions such as *Movement in Art* (1961); the *5 New York Evenings* of 'happenings' of 1964 including Robert Rauschenberg's legendary performance alongside a live cow in slippers; and 'interactive' works such as Niki de Saint Phalle's *She – A Cathedral* (1966), an enormous female figure that contained the exhibition and which was accessible through her 'vagina'. This programme established Pontus Hultén's reputation and left an indelible impression on Moderna Museet. Hultén may have departed the museum to become the Pompidou Centre's first Director in 1973, but his presence at the museum remains there to this day – as we shall soon see.

One artist who featured at Moderna Museet during the Hultén era was Allan Kaprow (1927–2006), a pioneer of 'happenings'. These were overtly participative artistic events that invited the audience to 'come out and play' in marked contrast to the more usual museal injunction: 'do not touch' (Harding 2005: 10). The Swedish artist Öyvind Fahlström (1983: 169) was therefore right to compare *Modellen* to a Kaprowian 'happening'. Kaprow's work was all about 'the blurring of art and life'. This is the title given to a book of his collective writings. In the introduction its editor noted that 'actual participation [or 'full participation'] in a work of art courts anarchy' (Kelley 2003: xviii). And it was anxieties over the anarchic nature of *Modellen* that led to criticism in the media, not least when injuries resulted from youthful dives into the sea of foam. The exhibition was forced to close when the fire chief of Stockholm declared the exhibition to be a safety hazard. This necessitated the replacement of the foam blocks with slides before the show could reopen. Nielsen interpreted this decision as being politically motivated. Blurring art and life in the way he had was, for some at least, a participatory step too far.

Modellen is a landmark exhibition. It constitutes a vitally important rider to the evolutionary claims that museums are progressively becoming more engaging, inclusive and participatory. Take, for example, this statement in the opening chapter of *The Responsive Museum*:

> Over the past 50 years there has been a major shift in the relationship between museums and their audiences. In the 1960s the relationship could

> have been considered simple and one-dimensional; the museum was all-powerful and the uncontested authority.
>
> (Reeve and Woollard 2006: 5)

It must be stressed that the authors here were referring to the history of British museums, so they can be forgiven for not qualifying their line of argument by referring to *Modellen*. But their argument overlooks an analogous example that *did* take place in Britain, namely Robert Morris's infamous 1971 exhibition at the Tate Gallery in London. Motivated by a desire to establish 'a new role for the artist in relation to society', Morris oversaw the construction of a series of large-scale structures which 'invite[d] the physical participation of the public' (Compton and Sylvester 1971). Akin to an adult version of *Modellen*, it closed after only four days due to the 'wild' behaviour of the participants (Kennedy 2009).

In 2005, the foremost writer on *Modellen*, Lars Bang Larsen pondered whether an exhibition such as that devised by Palle Nielsen would nowadays be possible. He quite reasonably concluded that 'no public art gallery would today allow any activist to take it over for the purpose of uncontrolled social experiment' (Larsen 2005: 217). This was affirmed indirectly in 2009 when Tate Modern decided to resurrect Robert Morris's abortive exhibition of 1971. Entitled *Bodyspacemotionthings*, it would be more accurate to describe it as a 'tame' reworking of the original.[5] Health and safety concerns led to a number of changes. The first thing that visitors encountered was a rope suspended over a huge ball (Figure 14.2). In 1971, participants could

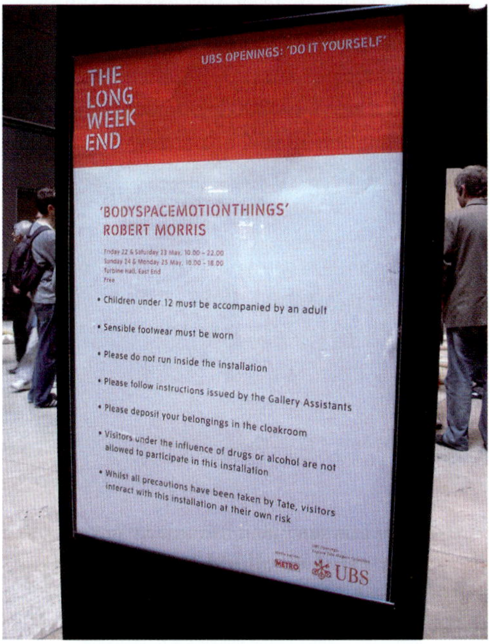

Figure 14.2 Robert Morris, *Bodyspacemotionthings*, Turbine Hall, Tate Modern, 22–25 May 2009. Photo: author

hang from it and manipulate the sphere with their feet. The 2009 version retained this as a centrepiece for the display. But the rope was tied into a knot to keep it just out of reach – a poignant metaphor in other words for this decidedly *controlled* social experiment, one that was given over to the visitors *with* reservation.

An abortive model

The 1968 catalogue to Moderna Museet's *Modellen* featured quotations from a diverse range of writers and thinkers, including Sören Kierkegaard, Mao Tse-tung and a boy aged six-and-a-half called Mats. Characteristically it was the boy who was given the first word. His thoughts and sketches doubled as preparatory designs for what was actually built in the museum. Mats revealed that he, like other children, found it boring to be stuck on the ground. He dreamed of escape via an enormous ladder reaching straight up in the sky – rather like (not) climbing that elusive rope in Tate Modern's *Bodyspacemotionthings*.

That it is not just children who find life on the ground mundane was confirmed by Yoko Ono. In 1968 she conceived a conceptual work entitled *Sky Event for John Lennon*. This consisted of asking as many participants as possible to get hold of a telescope or pair of binoculars, dress in their smartest clothes and gather together to 'check the sky'. In addition, 'ladders of great height should also be prepared for people who wish to climb up high to check'.[6] It would take exactly forty years for this surreal event to be realised. Renamed *Liverpool Skyladders* it took place in the ruined church of St Luke's as part of Liverpool's year as European City of Culture. When I visited the exhibition I was rather surprised to find that none of the ladders had any occupants. I soon realised why: health and safety concerns had again intruded. It was strictly forbidden to climb up high to check the sky.

I was reminded of this a few days later as I stood in the Duveen Gallery of Tate Britain and watched as every thirty seconds a lone runner sprinted down the empty hall. This was the artist Martin Creed's *Work No. 850* (The Duveen Galleries Commission 2008, 1 July–16 November 2008). My hopes of joining in this dash through a national museum were thwarted by a sign warning that: 'For reasons of safety, we ask the public not to run or obstruct the runners.' This was ironic given that it was on the very site where Robert Morris's exhibition took place in 1971. Perhaps anxieties about physical involvement are part of the museum's institutional memory? If so, its sibling Tate Modern is slightly less encumbered: as demonstrated by its ongoing 'Unilever Series' in the museum's Turbine Hall, which has included such participative works as Carsten Höller's *Test Site* (2006) and the (albeit less anarchic) reworking of Robert Morris's project.

Nevertheless, *Work No. 850* and *Liverpool Skyladders* marked a morose fortieth anniversary of Moderna Museet's *Modellen*. This despondency was compounded by another temporary exhibition that ran concurrently at Tate Britain. Forming part of the museum's 'Art Now' scheme it was entitled *The way in which it landed* (2 August–26 October 2008). This was a group show curated by the British artist Ryan

Gander (born 1976). For his own contribution, Gander opted to randomly select two metal storage racks or screens from Tate Britain's store and, maintaining the same configuration of the paintings hung on them, relocate them to the 'white cube' of the gallery. This subverted normal curatorial conventions and had the potential to generate a plethora of questions concerning the museum's collecting practices and exhibition policy. Yet nothing of any consequence was forthcoming beyond the inane transcript of an interview conducted by Gander *with himself*. The audience had absolutely no influence in determining 'The way in which it landed' and were denied the opportunity to access the museum store. Similarly, visitors could not take part in Martin Creed's *Work No. 850*, not unless they had made prior contact with the museum and been 'trained' how to comport themselves. This is all a very far cry from the liberating autonomy of *Modellen*.

This is not to deny that there was a participatory dimension, however limited, to the work of both Gander and Creed. That much is clear from the museum's decision to invite these artists into the institution and allow them to comment on its collections and enact a performance in its galleries. There is a very long line of such actions, many of which come under the loose category of 'institutional critique'.[7] At its best this phenomenon has the potential to activate the sorts of insights identified in Pateman's study of democracy:

> The major contribution to democratic theory . . . of . . . theorists of participatory democracy is to focus our attention on the interrelationship between individuals and the authority structure of institutions within which they interact.
>
> (Pateman 1970: 103)

Exploring the limits of what can be tolerated reveals a great deal about museum attitudes. That much became abundantly clear at another exhibition held in 2008, this time at Nationalmuseum in Stockholm (the institution from which Moderna Museet disaggregated in 1958). Entitled *Examining Form*, it consisted of a number of interventions in the museum by a series of artists, designers and a choreographer. An introductory panel related that they had been 'invited to give their own personal interpretations of the museum and its collection of applied art and modern design'. In so doing Nationalmuseum posed itself a question: 'What happens if we change the rules and let others interpret the museum?'

The answer to this query by the designer and writer Zandra Ahl (born 1975) did not make for comfortable viewing for the museum authorities. Ahl had produced a video entitled 'Nationalmuseum and me' (*Nationalmuseum och jag*). She had, as the title suggests, responded to the institution in a very personal, direct manner. She chose to address 'the delicate situation that exists between practitioner and institution when it comes to taste, power, gender, influence and hierarchies of interpretation' (Ahl, cited in Zetterlund and Plöjel 2008: 29). Ahl recorded interviews with members of staff and edited what they had to say to form the material for her video piece.

Ahl was explicit about the fact that, for once, it was she who held the upper hand. The roles had shifted: from being the *object* of analysis by curators, Ahl was now the *subject* with the power of interpretation over the museum and its staff (Zetterlund and Plöjel 2008: 31). Her critique of the institution and its collecting practices proved too much for some of the museum's personnel. They objected to 'the malevolent tone' ('*den illvilliga tonen*') of the video and, because it was perceived to have a negative impact on the working environment, ordered its withdrawal from display.[8]

When I visited the *Examining Form* exhibition just as it was about to close in August 2008 I got the distinct impression that the museum could not wait for it to be over. The supply of leaflets guiding the visitor around the various works had run out. And no explanation was given in the gallery concerning the Ahl piece: the video was simply switched off and I was unable to view it. It was only by reading about what had transpired in the media and speaking subsequently to members of staff at an unconnected museum seminar that I learned what had happened.

In this example we can see the pernicious effect of the rhetoric of engagement. Nationalmuseum's act of self-censorship utterly contradicted the museum's own professed aims. In the pamphlet produced in connection with the exhibition the museum's Director, Solfrid Söderlind, had argued persuasively that because there is not one, authoritative interpretation of either the past or the present, it is 'important that we have a lively discussion about the museum, our collections and exhibitions' – and that this was all part of the museum's goal of initiating an 'active dialogue with the contemporary art and design world' (Söderlind, cited in Zetterlund and Plöjel 2008: 3). But – to recall the words of Carole Pateman – it would be a mistake to describe this 'friendly approach' as 'democratic', ignoring as it did 'the authority structure within which this approach occurs' (Pateman 1970: 106).

This affair reveals the abortive consequences of a museum buying the illusion that it can unconditionally 'change the rules and let others interpret the museum' (Soren 2009: 235). Ahl's video was later purchased by the Röhsska Museum of Fashion, Design and Decorative Arts in Gothenburg (Öqvist 2008: 66). This means that Nationalmuseum is the real victim here. A victim of the omnipresent nature of the literature on engagement and a denial of any frank understanding or assessment of participation and the real implications it has for museums.

A promising model

A pre-eminent example of a national museum that still likes to promote itself as an open, participative institution is Moderna Museet. Lars Nittve, its Director since 2001, repeats again and again that the institution over which he presides strives to remove physical and intellectual barriers: 'all unnecessary obstacles, anything that makes the visitor feel excluded' should be set aside in order to ensure 'a closer contact and more encounters between artists and the public'.[9] Such pronouncements from the leader of an organisation are important in establishing a participatory 'climate' (Pateman 1970: 71).

One current initiative that comes close to realising *Modellen*'s goal of empowering children and young adults is the 'art project', *Zon Moderna*. Positioned on one of the museum's lower floors it is described as a 'zone of freedom . . . where there is loads of material, loads of freedom and helpful adults' (Nittve, cited in Malmquist 2006: 120–21). Participants work with leading artists, undertake guided tours of the gallery spaces and produce their own work. Its proponents claim that schoolchildren have 'made themselves at home [in the museum] and created fantastic things'. This has clearly made a lasting impression on some of those involved, given that a number of former participants have 'come back . . . with their own suggestions and initiatives' (Malmquist 2006: 127).

Those behind *Zon Moderna* argue that it 'has affected the entire museum' (Malmquist 2006: 127). This is key if a participatory ethos is to become embedded in an organisation. Zones of participation need to go beyond 'outreach' projects, temporary initiatives and activity rooms if an organisation's commitment to it is to be credible, sustainable and pervasive.

A further avenue for novel participative ventures presented itself in December 2009, when Moderna Museet inaugurated a new branch museum in the southern city of Malmö. This development draws attention to Moderna Museet's *national* role (it is no longer simply a 'Stockholm institution'). Pateman, as we have seen, paid particular attention to the role of 'local political institutions' as 'training grounds for democracy'. Indeed, she noted that people are generally more interested in matters pertaining to their own region than they are in national concerns. This led Pateman (1970: 46) to argue that the building of political confidence and competence was more pronounced at local level. This represents a critical challenge to all 'national' museums: how to bolster their relevance to the whole population, not just those that live in the capital city.[10]

The opening of 'Moderna Museet Malmö' inevitably prompts comparisons with its dominant partner in the Swedish capital. How might the participatory 'climates' of the two differ? What 'model' might Malmö follow, assuming that an 'uncontrolled social experiment' such as *Modellen* will be unfeasible? Well, one exemplar that would certainly be worth considering when it comes to participation is the Pontus Hultén Study Gallery (Burch 2008). Designed by the Renzo Piano Building Workshop, it opened in May 2008 in a former library space of the main Moderna Museet building (Figures 14.3 and 14.4). Its principal purpose is to house the 700 works from Hultén's private collection, which were bequeathed to the museum shortly before his death in 2006. Hultén stipulated that these holdings were to be always accessible, even when not on display in the main galleries. This has been realised thanks in part to the support of 'the Access project', a state-funded initiative 'to care for, preserve and make available collections, objects and documents'.[11] Such undertakings form the bedrock of participation, as is shown by what they facilitate at the Pontus Hultén Study Gallery. Utilising the sort of machinery normally found on a factory production-line, visitors can, with the help of a gallery assistant, access works of their choice via a touchscreen computer and literally download one of

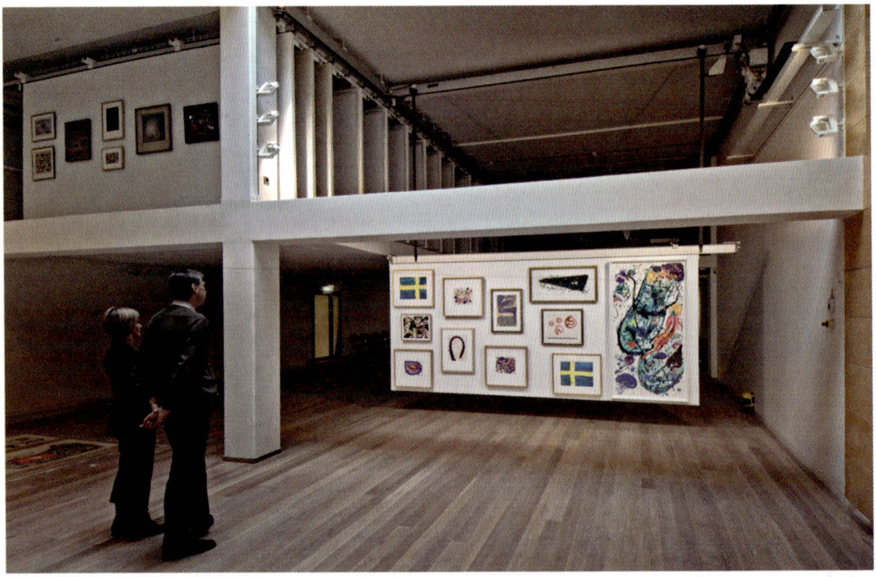

Figure 14.3 Pontus Hultén Study Gallery, Moderna Museet. Photo: Copyright Moderna Museet

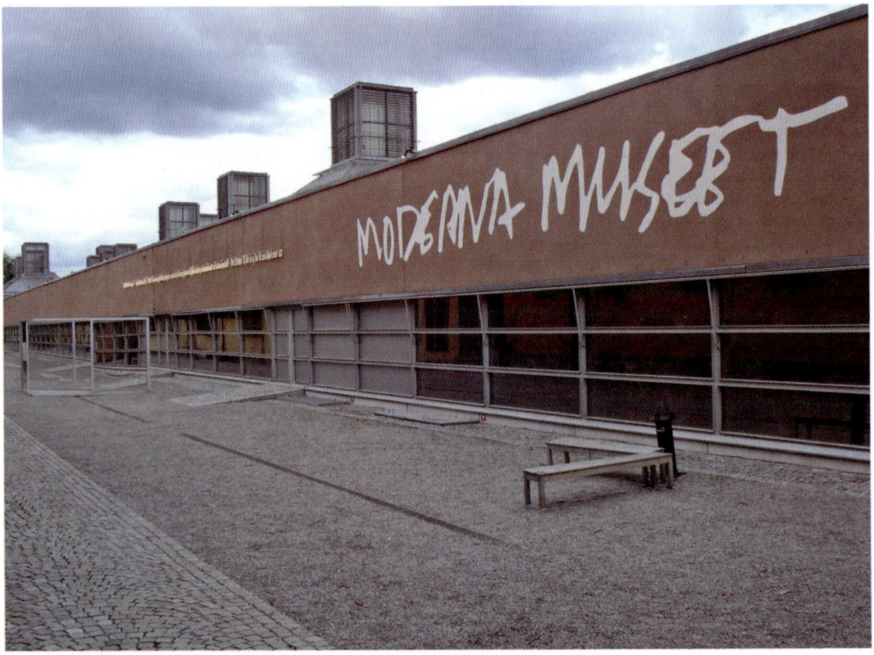

Figure 14.4 Moderna Museet, Stockholm. Photo: author

thirty panels or screens that are stacked in the air above. The ways in which the paintings 'land' in the gallery are determined by *all* visitors – not just 'special' (*sic*) mediators like Ryan Gander, the self-indulgent artist responsible for *The way in which it landed* at Tate Britain.

A visit to the Hultén Gallery is a decidedly performative experience for all concerned. The public is as much on show as the art: both are equally visible through a glass wall and a series of windows that look down from the museum's entrance hall. The Pontus Hultén Study Gallery takes the idea of 'open storage' a stage further. Thus the Luce Foundation visible storage areas at Brooklyn Museum and the Smithsonian's American Art Museum in Washington might be far larger than the Hultén Gallery, but the latter is superior in two important respects: the paintings are not trapped behind glass walls, and it is the visitor who actively decides what to look at from the available selection.

With this in mind it will be interesting to see how the Pontus Hultén Study Gallery is utilised. Should it be a 'curated' space? It has the potential to complicate traditional divisions between curator and visitor in much the same way as Web 2.0 blurs the status of reader and writer.[12] The computer terminal used to access the holdings could be programmed to log the number of times individual works are selected and in so doing help the museum to identify its 'bestsellers'. In a similar vein, profiles of the museum's 'customers' could be established and, by acting like an online retailer, the information gleaned would make it possible to flag up related works: 'other people who looked at this painting were also interested in . . .'. The Study Gallery might, by extension, be used to allow groups or individuals to curate 'their' screen on a topic of their choosing. It could moreover act as an adjunct to exhibitions in the main galleries. It has the potential to provide an opportunity for 'hot interpretation': rapid responses to current crises or topical news stories. This would serve to embed the museum in the politics of the present – a vital objective if national museums really are to become 'training grounds for democracy'.

Yet all this will only ever remain instances of 'partial-participation' so long as the public's involvement is limited to responding to a predetermined collection. Full-participation will only come if they can actually shape it in the first place. Acquisition constitutes the acid test of a museum's commitment to participation. The sorts of challenges it raises are conveyed vividly in Bruce Altshuler's edited volume, *Collecting the New* (2005). In it, Howard N. Fox of Los Angeles County Museum of Art considers the widespread notion that museum curators are 'astute experts' whose 'innate insightfulness, cultivated sensibilities, and special training . . . allows them to make . . . judgements on everyone's behalf' (Fox 2005: 15). Working on the hypothesis that the authority of museums is all too often left unquestioned by society at large, Fox wonders whether those employed in museums acknowledge properly the authoritarianism of their institutions and, by extension, their own powerful role. This is taken up by Robert Storr (2005: 40), who points out that museum collecting, despite appearances to the contrary, is very far from 'being a matter of serenely disinterested scholarship or discernment'. This is particularly evident when it comes to 'collecting the new', given that, unlike older works of art, recently produced pieces have not been validated by time (Altshuler 2005: 1, 8). In a striking assertion, Altshuler (2005: 2) claims that the future value of art – be it financial or art historical – is 'a matter of guesswork'.

This is absolutely *not* the impression conveyed by Moderna Museet when it exhibited some of its most recent acquisitions in late 2008. The works were simply presented without comment, with no indication of their cost or the process by which they were acquired. The museum has been slightly more forthcoming when it came to its recent call for additional funds to address the gender imbalance of its collection. An interior wall in one of the service areas of the museum was devoted to a large text panel explaining the reasons for this initiative and a list of the names of those female artists that Moderna Museet wished to acquire (Figure 14.5). The same

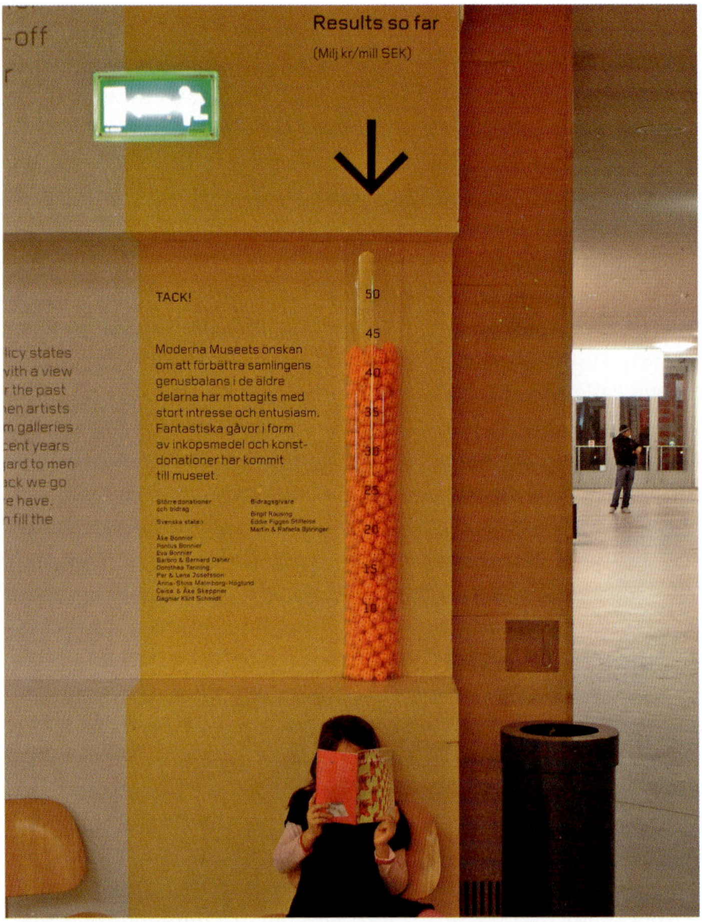

Figure 14.5 Graphic panel charting the progress of Moderna Museet's campaign to acquire more works by female artists. Photo: author

information was mirrored on the museum's website in a text signed by Nittve in which he set out why these female 'masterpieces' were necessary.[13]

Yet who is to say that the 'right' women (or works) are being sought? How was the list drawn up? Is there a hierarchy of importance? Which names have been left off? Has there been universal agreement either within the museum or outside? None of this was addressed. The public were invited to accept a decision that had already been reached. Indeed, it was expressed in such a manner as to obscure the fact that there was anything to decide in the first place. The museum's practice was naturalised. Its cool, measured demeanour exuded authority: Nittve and the other (unnamed) connoisseurs have decided and we can relax, secure in the knowledge that the museum is safe in their hands. Yet how can the experts (*sic*) of yesteryear have presided over a collection with such glaring 'gaps'? Their oversight dictates the collecting practices of the museum. Lacunae in the collection as well as the areas of coverage provide the rationale for the acquisition policy of the present (Tellgren 2008: 20).

It was precisely this issue that Zandra Ahl raised when she participated in Nationalmuseum's 'Examining Form' exhibition (Zetterlund and Plöjel 2008: 31). 'Museum collecting today is weighed down by tradition', Ahl complained. This means that museums 'continue to build on their history', to the detriment of those whose value systems are based on an alternative frame of reference. Perhaps that's the real reason why Ahl's video sat so awkwardly in the hallowed halls of Nationalmuseum? But, of course, I don't know the answer to that question, given that the museum had switched off the monitor, thus denying me the opportunity to make up my own mind.

A mute model

Detecting signs of participation is relatively easy – the squeals of children in *Modellen*, the splattering of paint at Zon Moderna, the mechanical clicks and whirls as paintings descend in the Pontus Hultén Study Gallery. Seeing evidence of non-participation is significantly more challenging. Is that Zandra Ahl video broken, or has it been censored? These sorts of thoughts surfaced in my mind during my many research trips to Moderna Museet in 2008. I spent a considerable amount of time in one particular gallery and watched as people moved about the space reading the little labels and glancing momentarily at the accompanying paintings. What, I wondered, were they thinking when they read this:

> An intensely red, orange and blue meadow of flowers appears from out of the thick layers of paint. Direct painting – no preliminary sketches were made. The lush garden, part of the artist's farm in Utenwarf on the German west coast, was painted in the midst of the First World War, in a time of schism and violence. To Nolde flowers incarnated the eternal cycle of birth–life–death. He continuously found new motifs in his garden. Flowers

were also emblematic of art – Nolde saw both works of art and flowers to be the products of natural forces. For a short period he was a member of the expressionist artist group Die Brücke.

This classic piece of art-speak referred to *Garden (Utenwarf)* painted in 1917 by Emil Nolde (1867–1956). Similar material was repeated in the museum's online catalogue, along with basic information about its provenance. This indicates that the work was in the possession of one Otto N. Deutsch of Frankfurt am Main prior to 1923. What happened to it between then and its sale at auction in the 1960s was not mentioned. Moderna Museet's catalogue of 2004 states that it was purchased by the museum in 1967. It uses much the same language as that seen on the text label but adds one additional piece of information: Nolde was viewed as a so-called 'degenerate' artist by the Nazi regime and over one thousand of his works were confiscated from museums. Twenty-nine of them appeared in the infamous Entartete Kunst exhibition held in Munich in 1937 (Müller-Westermann 2004).

In 2002, the heirs of Otto Nathan Deutsch contacted the museum to notify them that their forebear had been forced to part with the painting during the Nazi period. It took a further seven years before a mutually acceptable decision was reached – one that will ultimately see Moderna Museet relinquish ownership of the painting (Eriksson 2008).[14] Yet when I conducted research for this paper in 2008, *Garden (Utenwarf)* looked just like any other artwork on the walls of the museum. Those visiting the museum at that time would have learnt nothing about this affair, one that raised absolutely fundamental questions about what Moderna Museet is, how it is managed and what role it plays in society.

I felt this particularly acutely when I saw middle-aged Swedish men looking at Nolde's wonderful painting hanging mute on the wall. Was one of them called Mats and did he as a six-and-a-half-year-old visit *Modellen* to dive into its sea of foam just like Olof Palme? If so, did it prepare him for a life of participation? Would he have welcomed the chance to have a say in whether the painting he was looking at should stay in the national art collection of Sweden or be returned to the heirs of Otto Nathan Deutsch?

None of this mattered, however, because the museum opted to remain silent on the subject until a decision had been reached. This is reminiscent of the process by which the canon of female artists was drawn up and announced. Only once a judgement had been made on the part of the museum leadership were the public invited to enter into the process by attending an ongoing series of debates focusing on some of the artists that had been pre-selected.

The imminent deaccessioning of the Nolde painting and the choice or appropriateness of seeking to 'fill' gaps in the collection get to the heart of the form, function and financing of this national museum. Nevertheless, these two matters, like the Ahl incident at the nearby Nationalmuseum, reveal the limits of participation rather than the scope of critical enquiry. They can be seen as indicative of the constrained nature of today's national museums and affirmations that they fail when it comes to 'encouraging genuine participation' (to misquote David Carrier). Because simply initiating a debate and celebrating the fact that 'everybody talks about the museum' does not equate to full or even partial participation. Yes, the pronouncements of museum Directors like Sune Nordgren, Lars Nittve and Solfrid Söderlind are important in establishing a participatory climate – but the end results constitute instances of pseudo-participation at best. Only the case of the Pontus Hultén Study Gallery might be deemed an example of partial participation.

But perhaps this is the most that national museums can manage. At the outset of this chapter it was asserted that participation is always a matter of degree. This is easily missed when being seduced by the museological literature and soothed by reassurances that 'all unnecessary obstacles' have been removed from today's museums. Yet that phrase – 'all unnecessary obstacles' – is more revealing than its author, Lars Nittve, perhaps intended. Obstacles come in many guises. Yes, some can be abolished. But others are *necessary*. Necessary for reasons of health and safety. Necessary so that certain vested interests retain power. Necessary to ensure the practical functioning of the institution. Necessary to safeguard the collection. The injunction 'do not touch', for example, might well be a necessary obstacle. But what about a ban on taking photographs? The simple point that photography is forbidden in Sweden's Moderna Museet but not down the road at Nationalmuseum reveals that determinations over whether an obstacle is necessary or not are far from fixed. A video by Zandra Ahl is not inherently controversial: its acceptability or otherwise is determined by those who wield the greatest power in the museum. It is they who adjudicate which obstacles are incontrovertible.

A clearer attention to the nature and extent of participation – both physical and intellectual – will help us identify, locate and define the framework of obstacles that of necessity constitute the parameters of any given museum. Who has determined which 'obstacles' are necessary or not? Do they apply to everyone, or are some 'special' groups or individuals exempt? In what ways have these strictures changed and how might the boundaries shift in the future? If opportunities for partial participation are accorded to visitors, then perhaps museums – those 'training grounds for democracy' – could scale back a previously necessary obstacle, opening up possibilities for greater participation?

That such modifications *do* occur confirms that museums are indeed contested places in a state of constant change and that participation, rather than being a fixed entity, is instead a dynamic, contingent phenomenon that merits interrogating in a far more nuanced, critically engaged manner. Yes, audiences participate in national museums. But we would do well to remember the rope that remained just out of reach at Tate Modern in 2009. For it is obstacles – whether necessary or not – that set limits on everything that happens in our national museums. It is up to us to decide if these impediments are justifiable and in the right places.

Notes

1 Collett, E. *et al.* 'Save the National Gallery of Norway'. Online. <http://www.bevarnasjonalgalleriet.org/index-filer/Page292.htm> (accessed 3 October 2009).

2 O'Reilly, T. (2005) 'Web 2.0: compact definition?' Online. <http://radar.oreilly.com/archives/2005/10/web-20-compact-definition.html> (accessed 28 February 2009).

3 IMC (2009) 'Enabling diversity sustaining development', second 'Inclusive museum conference'. Online. <http://www.museum-conference.com> (accessed 28 February 2009).

4 Simon, N. (2007) 'Museum 2.0 blogspot 20/03/2007: hierarchy of social participation'. Online. <http://museumtwo.blogspot.com/2007/03/hierarchy-of-social-participation.html> (accessed 28 February 2009).

5 Rees Leahy, H. (2009) 'Making an exhibition of oneself: reconstruction and reflexivity in the art museum', unpublished conference paper, *Museums and Biographies*, National Gallery, London, 11–12 September.

6 Liverpool Biennial. Online. <http://www.biennial.com/content/LiverpoolBiennial2008MADEUP/YokoOno/Overview.aspx> (accessed 13 October 2008).

7 Burch, S. (2007) 'Museum landscapes: Zoning in on a complex cultural field', *NaMu I, Making National Museums: Setting the Frames*, Linköping: Linköping University, 49–67, p. 50. Online. <http://www.ep.liu.se/ecp/022/005> (accessed 27 October 2008).

8 Josephi, N. (2008) 'Kritisk film stoppad av Nationalmuseum', Sveriges radio, P1, 2 April. Online. <http://www.sr.se/cgibin/P1/program/amnessida.asp?programID=478&Nyheter=1&grupp=2177&artikel=1987064> (accessed 28 February 2009).

9 Nittve, cited in Burch (2007: 55, note 7).

10 Ibid.: 54.

11 Ibid.: 55.

12 O'Reilly, T. (2005) 'Web 2.0: Compact definition?' Online. <http://radar.oreilly.com/archives/2005/10/web-20-compact-definition.html> (accessed 28 February 2009).

13 Nittve, L. (2006) 'The second museum of our wishes'. Online. <http://www.modernamuseet.se/v4/templates/template3.asp?lang=Eng&id=3106> (accessed 28 February 2009).

14 Nittve, L. and Rowland, D. (2009) 'Press release: Blumengarten settlement'. Online. <http://www.modernamuseet.se/v4/templates/template3.asp?lang=Eng&id=4300> (accessed 9 September 2009).

Bibliography

Altshuler, B. (ed.) (2005) *Collecting the New: Museums and Contemporary Art*, Princeton, NJ: Princeton University Press.

Berg, K. (2005) 'Nasjonalgalleriets endelikt', *Aftenposten*, 7 April, Kultur, 12.

Black, G. (2005) *The Engaging Museum: Developing Museums for Visitor Involvement*, London: Routledge.

Bloom, B. (2007) 'A magical land of roving Santa Claus armies, pirated energy drinks and a giant squatted urban village: Political art activities in Denmark', in J. MacPhee and E. Reuland (eds) *Realizing the Impossible: Art against Authority*, Oakland: AK Press, pp. 131–51.

Bringager, F. (ed.) (2005) *Guidebook*, Oslo: National Museum of Art, Architecture and Design.

Burch, S. (2008) 'Sliding doors', *Museum Practice*, 43: Autumn, 18–23.

Carrier, D. (2006) *Museum Skepticism: A History of the Display of Art in Public Galleries*, Durham, NC, and London: Duke University Press.

Ciolfi, L., Bannon, L. J. and Fernström, M. (2008) 'Including visitor contributions in cultural heritage installations: Designing for participation', *Museum Management and Curatorship*, 23(4): 353–65.

Compton, M. and Sylvester, D. (1971) *Robert Morris*, London: Tate Gallery.

Corsane, G. (ed.) (2005) *Heritage, Museums and Galleries: An Introductory Reader*, London: Routledge.

Duncan, C. (1995) *Civilizing Rituals: Inside Public Art Museums*, London and New York: Routledge.

Eriksson, T. (2008) 'Moderna anklagas för utpressning', *Dagens Nyheter*, 9 June. Online. <http://www.dn.se/DNet/jsp/polopoly.jsp?d=2206&a=777698> (accessed 28 February 2009).

Fahlström, Ö. (1983) 'Moderna Museet – från pompa och ståt, till informationscentrum', in Granath and Nieckels (eds), *op. cit.*, pp.169–71.

Fox, H. (2005) 'The right to be wrong', in B. Altshuler (ed.), *op. cit.*, pp. 15–27.

Frivik, K. R. (2009) 'Slo ring runt Nasjonalgalleriet', *NRK*, 30 March. Online. <http://www.nrk.no/nyheter/kultur/1.6547561> (accessed 9 October 2009).

Granath, O. and Nieckels, M. (eds) (1983) *Moderna Museet 1958–1983*, Stockholm: Moderna Museet.

Harding, A. (ed.) (2005) *Magic Moments: Collaboration Between Artists and Young People*, London: Black Dog Publishing.

Hultén, P. (1983) 'Det lilla museets stora roll', in Granath and Nieckels (eds), *op. cit.*, pp. 38–46.

Hyman, J., Thompson, P. and Harley, B. (eds) (2005) *Participation and Democracy at Work: Essays in Honour of Harvie Ramsey*, Basingstoke: Palgrave Macmillan.

Kelley, J. (ed.) (2003) *Essays on the Blurring of Art and Life*, Berkeley: University of California Press.

Kennedy, M. (2009) 'Art to ache for: Tate Modern's bruising interactive exhibition reopens', *Guardian*, 23 May, 9.

Kirshenblatt-Gimblett, B. (1991) 'Objects of ethnography' in I. Karp and S. Lavine (eds) *Exhibiting Cultures: The Poetics and Politics of Museum Display*, Washington and London: Smithsonian Institution Press, pp. 386–443.

Lagerkvist, C. (2006) 'Empowerment and anger: Learning how to share ownership of the museum', *Museum & Society*, 4: 52–68.

Lang, C., Reeve, J. and Woollard, V. (eds) (2006) *The Responsive Museum: Working with Audiences in the Twenty-first Century*, Aldershot: Ashgate.

Larsen, L. (1999) 'Sometimes I'm up, sometimes I'm down, sometimes I'm underground: Making social aesthetics operative', in D. Birnbaum and J.P. Nilsson (eds) *Like Virginity Once Lost: Five Views on Nordic Art Now*, Malmö: Propexus, pp. 37–59.

Larsen, L. (2005) 'Play and nothingness: The Model for a Qualitative Society, an activist project at the Moderna Museet, Stockholm 1968', in A. Harding (ed.), *op. cit.*, pp. 206–17.

Malmquist, K. (ed.) (2006) *Zon Moderna*, Stockholm: Moderna Museet.

Message, K. (2007) 'Museums and the utility of culture: The politics of liberal democracy and cultural well-being', *Social Identities*, 13(2): 235–56.

Müller-Westermann, I. (2004) 'Emil Nolde: Garden (Utenwarf)', in Widenheim, C. (ed.) *Moderna Museet – the Book*, Stockholm: Moderna Museet.

Nielsen, P. (1968) *Modellen – En modell för ett kvalitativt samhälle*, Stockholm: Bonniers.

Öqvist, P. (2008) 'Argast i designvärlden: Zandra Ahl', *Konstvärlden & disajn*, 8: 62–68.

Owe, T. and Hverven, M. (2009) '11,500 vil bevare Nasjonalgalleriet', *NRK*, 30 April. Online. <http://www.nrk.no/nyheter/kultur/1.6589984> (accessed 9 October 2009).

Palme, O. (1971) 'Museet – en del av kulturpolitiken', *Svenska Museer*, 1: 27–34.

Pateman, C. (1970) *Participation and Democratic Theory*, London: Cambridge University Press.

Reeve, J. and Woollard, V. (2006) 'Influences on museum practice', in C. Lang, J. Reeve and V. Woollard (eds) *The Responsive Museum: Working with Audiences in the Twenty-first Century*, Aldershot: Ashgate, pp. 5–17.

Russell-Ciardi (2008) 'The museum as a democracy-building institution: Reflections on the Shared Journeys program at the Lower East Side Tenement Museum', *The Public Historian*, 30(1): 39–52.

Söderling, F. (2006) 'Kampanj fällde Sune Nordgren', *Dagens Nyheter*, 5 September. Online. <http://www.dn.se/DNet/jsp/polopoly.jsp?d=2206&a=570347> (accessed 24 September 2006).

Solkin, D. (ed.) (2001) *Art on the Line: The Royal Academy Exhibitions at Somerset House 1780–1836*, New Haven, CT, and London: Yale University Press.

Solomon, R., Heen, H. and Grimsmo, A. (2008) *Arbeidsmiljø og Arbeidsorganisering ved Nasjonalmuseet for Kunst*, 30 June, Oslo: Arbeidsforskningsinstituttet.

Soren, B. (2009) 'Museum experiences that change visitors', *Museum Management and Curatorship*, 24(3): 233–51.

Storr, R. (2005) 'To have and to hold', in B. Altshuler (ed.), *op. cit.*, pp. 29–40.

Strömberg, H. (2005) 'Nordgren hängde om norsk konst', *Göteborgs-Posten*, 5 March.

Tellgren, A. (ed.) (2008) *The History Book: On Moderna Museet 1958–2008*, Göttingen: Steidl.

Zetterlund, C. and Plöjel, M. (eds) (2008) *Undersöka Form: Dialoger*, Stockholm: Nationalmuseum.

Chapter 15

REPRESENTING WALES AT THE MUSEUM OF WELSH LIFE

Rhiannon Mason

> The very notion of a nation with a fixed 'given' cultural identity is a sign of the success of a whole array of practices in naturalizing that identity. It is also the sign of the success of a particular construction or version of national culture, for all definitions of the national essence selectively ignore competing definitions.
>
> (Foster 1991: 238)

> [I]f St Fagans is the Museum of Welsh Life should it also seek to represent Somali, Italian, and Asian heritage and culture as it exists in Wales? Or should it remain an icon and expression of a principally Welsh speaking, white, rural based culture? St Fagans stands at an important crossroads.
>
> (Houlihan 2005: 6)

HERITAGE AND THE PAST provide an important means of claiming and securing identity positions. Heritage supplies identities with precedent and legitimacy through the invocation of 'tradition' and, in the case of established cultural institutions like museums or heritage sites, by lending those identity-claims the authority which public institutions command. The act of creating an institution like a national museum has been and continues to be an act of assertion: a gesture designed to claim recognition for a given identity and an attempt to translate a set of intangible beliefs about the special quality of a certain cultural group into an identifiable, material and visible presence.

A national museum can be therefore understood performatively in that the nation's citizens are presented with the national story as celebration and affirmation. The museum instructs citizens how to locate themselves in relation to the 'national story'. However, complexities quickly arise when museums try to define the national identities they represent and when they endeavour to construct displays and collecting policies to present and capture such identities. This is particularly acute when they are obliged to deal with the changes to 'the national story' that occur over time, or indeed to acknowledge multiple and competing versions of those national stories. Museums have the added challenge of dealing with the material legacies – in terms of buildings, collections, displays – bequeathed to them by earlier generations who may have held different ideas about what constituted appropriate national culture for display in museums.

The second branch of Wales's national museum, St Fagans: National History Museum, which is the subject of this chapter, clearly illustrates all of these difficulties and offers another opportunity to examine the processes involved in the representation of national identities within national museums in Wales. This museum started life as the 'Welsh Folk Museum' but in 1995 became the 'Museum of Welsh Life'. At the time of conducting the research for this chapter the museum was still known by that name; it only received its newest name in late 2005. For the purposes of consistency, this chapter will refer to it with its penultimate name and will restrict the discussion of its new name and future development to the concluding section.

This chapter examines how and why the previous name-change occurred and the challenges this shift in remit and redrawing of representational parameters brought to the museum and its curators. What emerges from this case study is not only the extent to which museums are palimpsests but also the way that any subsequent changes are circumscribed by, and sometimes at odds with, those earlier legacies. The chapter also examines the claim that folk museums are inherently conservative in their politics and that they suffer from a form of amnesia over divisions within national stories. Finally, the case study offers two museological insights. First, it illustrates the need to consider not only the historical, macrocosmic reasons for why a museum representation has come to be as it is, but also the effects of practical current factors like marketing, audience development, visitor profiles and visitor surveys. Most studies on national museums tend to privilege the historical and ideological context at the expense of the more diffuse, but equally important, day-to-day practical elements of museum work.

Secondly, while it is certainly possible to identify dominant discourses at work in museum representations, there is a danger of reading museums as too internally coherent, too unitary in their meanings. By contrast, this analysis of the Museum of Welsh Life leads me to concur with Andreas Huyssen's (1995: 15) comment that: 'No matter how much the museum, consciously or unconsciously, produces and affirms the symbolic order, there is always a surplus of meaning that exceeds set ideological boundaries, opening spaces for reflection and counter-hegemonic memory.' Tony Bennett makes a similar point when he observes that museums which make universalising statements – that is to say those which claim to represent a whole way of life or a whole people – always open themselves up for criticism and accusations of misrepresentation. They are always already internally flawed because of the inevitable inadequacy of their overreaching representational claims (Bennett

1995: 102–03). This means that museums can function as both a catalyst for debate and a public forum within which debates over the accepted nature of national identities and histories can occur. The Museum of Welsh Life enables us to see the above in action because it accommodates competing discourses of Welsh identity and because it makes visible the processes by which dominant representations are created but equally challenged and revised. This chapter also enables us to see how the museological approach adopted has shaped the way that national identities, cultures and histories are represented. Whereas the very first National Museum of Wales followed the 'universal-survey museum' approach, this second branch is much closer to the idea of the museum as 'nation-in-miniature' (Mason 2007). The original intention underpinning the creation of the folk museum was to capture a holistic view of the nation and be explicitly representative of Wales's vernacular national culture as recognised at that time.

The creation of the Welsh Folk Museum

The first National Museum of Wales based in Cardiff is reminiscent of the archetypal nineteenth-century museum, with its neo-classical, temple-like architecture and European 'high art' collections (Figure 15.1). The Museum of Welsh Life, or Amgueddfa Werin Cymru to give it its Welsh name, is its opposite – museologically speaking. The first National Museum of Wales was originally intended to show that Wales was a fully fledged nation with the requisite institutions: a national museum, a national library and a national university. By contrast, the Museum of Welsh Life – or the Welsh Folk Museum as it was first called – was, and still remains, about valuing the vernacular culture of Wales. Its 2001 guidebook stated that 'The Museum shows how the people of Wales lived, worked and spent their leisure time over the last five hundred years' (NMGW 2001: 3). The emphasis at the Museum of Welsh Life is thus intended to be on the 'people of Wales' and their lives. The complex questions of who these people might be, what concept of 'Wales' such a mission statement invokes, and how all this comes to be represented is considered below.

The Museum of Welsh Life today is based in the village of St Fagans situated on the periphery of the city of Cardiff. The museum grew out of a 'Bygones' collection of folk material exhibited at the National Museum in Cardiff first in 1913 and then in 1926 and the subsequent establishment of a Department of Folk Culture and Industries there in 1936 (Figures 15.2 and 15.3) (National Museum of Wales 1938: 71). It is worth noting that the use of the term 'industries' at this point is somewhat misleading in that it means rural crafts: 'The department deals, in short, with the life of Man in Wales during the last four hundred years, excluding only modern industrial developments' (National Museum of Wales 1938: 71). The collections were relocated into the Welsh Folk Museum at its current St Fagans site following the Earl of Plymouth's 1946 donation of St Fagans Castle with its 20-acre grounds and the collection of funds from a public appeal. The museum was the first of its

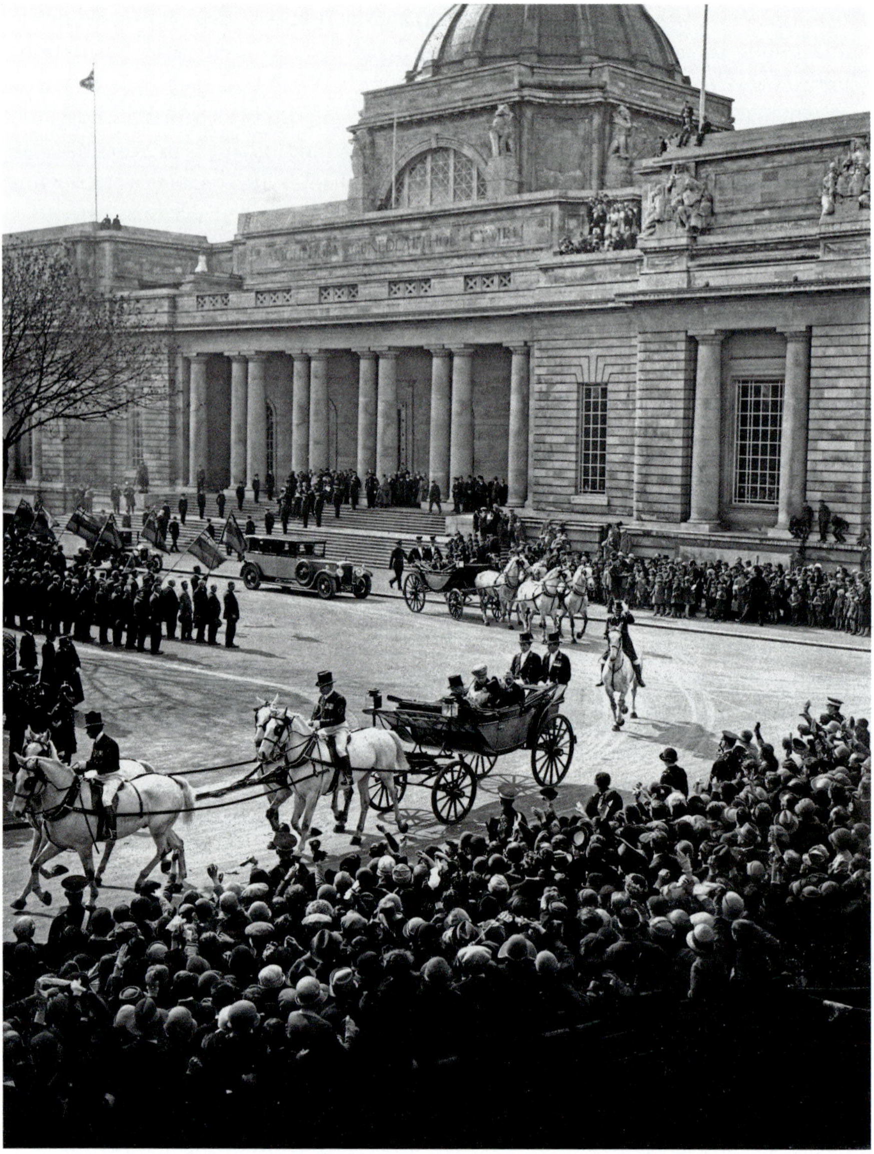

Figure 15.1 The formal opening of the first branch of the National Museum, 21 April 1927. Photo: Copyright National Museum Wales

type to be built in Britain, although smaller folk museums did exist in the Isle of Man and the Highlands (Kavanagh 1993). It was the first national open-air museum in Britain, and museum documentation from the time of its creation states that: 'The aim will be to form in the Folk Museum as complete a picture of the Welsh

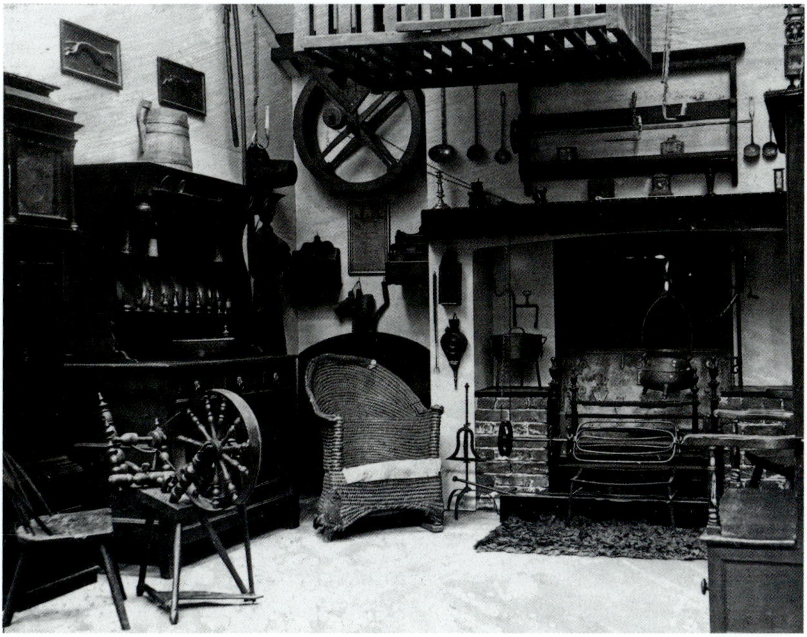

Figure 15.2 Welsh Kitchen Display at the National Museum of Wales, Cardiff. Photo: Copyright National Museum Wales

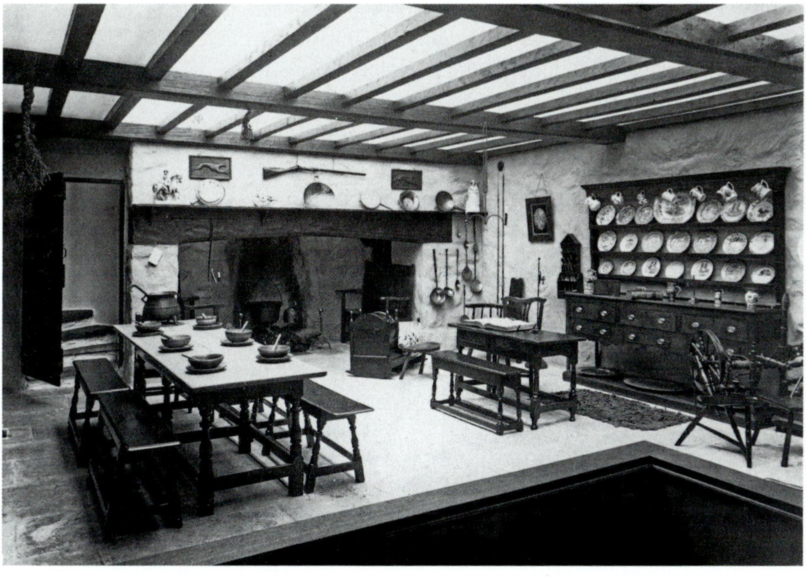

Figure 15.3 Welsh Kitchen Display at the National Museum of Wales, Cardiff. Photo: Copyright National Museum Wales

past as is possible, to create a "Wales in miniature"' (National Museum of Wales 1946: 6). The first open-air museum in the UK was on the Isle of Man at Cregneash in 1938 but this does not appear to have been termed 'national' (Kavanagh 1993).

Since its official opening in 1948, over thirty buildings have been taken from around Wales to be rebuilt in a village setting of an extra 45 acres surrounded by a woodland area of about 40 acres. The buildings include:

- farm houses, cottages, barns, hayshed, pigsty
- bakehouse, pottery, tannery, gorse mill, sawmill, smithy, saddler, woollen mill, cider mill, coach house
- Oakdale Working Men's Institute, Rhyd-y-car ironworkers' cottages decorated in the styles of 1805, 1855, 1895, 1925, 1955 and 1985
- tollhouse, chapel, school, local stores, cockpit, post office, a cenotaph, church
- Celtic village, St Fagans Castle and gardens, boathouse, summer house, a post-Second World War prefab, and an ecological experiment: the House for the Future.

The museum also includes a visitor centre with a temporary exhibition space, large permanent galleries of material culture, agriculture and costume, restaurants and a shop. Attractions at the site include storytelling, rides in a horse and cart and a photography studio at which visitors can be photographed in period or national costume. The craftsmen – wood turner, cooper, clog maker, tailor, smithy – offer demonstrations of traditional skills, and visitors can purchase the finished products. The museum houses an extensive oral testimony archive created in 1958 of folk tales, folk music, folk customs, and the definitive national archive of Welsh language and dialect. It also links to local intangible heritage, for example, by commemorating events like the Battle of St Fagans: the last major battle of the 1648 Civil War (NMGW 2001: 39). Such events – and other activities like the celebration of St David's Day – constitute a year-round programme of events and are an equally significant part of the museum's representational work.

Between April 2005 and March 2006, the Museum of Welsh Life received more than 580,000 visits, making it the most visited of all National Museum Wales's sites. The National Museum Cardiff, located in the capital's city centre, received over 300,000 visitors for the same period, making it the second most visited site.[1] The Museum of Welsh Life also saw the biggest increase in visitors following the introduction of free admission in 2001. In 2000–2001 it received more than 320,000 visits, while in 2001–2002 visits soared to nearly 700,000 (NMGW 2003: 26).

Existing literature

While there is not a vast literature on the Museum of Welsh Life specifically, a small number of critics, mainly historians but also sociologists and curators, have written about it as part of a larger commentary on heritage or museums within Wales. The one exception is Bassett (1982, 1983, 1984, 1990; Bassett and Bassett 1993) who has written the definitive account of the national museums of Wales to date.[2] Regarding the Museum of Welsh Life specifically, Bassett reports the recorded motivations and factors leading up to its creation but gives less discussion to the content or museological approach.

Peter Lord (1992b) also considered the Museum of Welsh Life as part of his much larger polemic on the treatment of visual culture in Wales and the role he claimed that the National Museum Wales had played in marginalising indigenous visual culture in favour of an Anglicised, Europeanised aesthetic canon. For Lord, the Welsh Folk Museum (as it was still called at the time of his writing) was complicit in this process because its self-identification as the supposedly most Welsh of all Wales's museums had abnegated the National Museum and Galleries, Cardiff from its responsibility to represent Wales in all its facets and, especially, in the realm of visual culture. Criticism also came from within the museum itself in the form of one of the curators, Geraint Jenkins, who was appointed in 1987 and argued that the collecting policies had been driven by a romantic and highly selective view of Wales (Dicks 2000: 91).

This criticism of the Welsh Folk Museum's tendency to prioritise a single aspect of Welsh society – that of rural culture – above all others was voiced by a number of other Welsh academics from at least the 1960s but increasingly during the1980s. In her book, *Heritage, Place and Community*, Bella Dicks provides a useful summary of this episode in the museum's history and contextualises the critiques in terms of wider movements relating to the development of urban history in museums and community studies (Dicks 2000: 78–102). Dicks also discusses the accusation made by a number of critics that the folk museum presented a past that was too static, too unified and simplified:

> Dai Smith, for example, whilst welcoming the new direction taken by the museum under Jenkins's curatorship, identifies in it a reluctance to countenance the display of conflict or dissension. Instead, he sees a continuing reliance on visions of 'wholeness, and harmony, and community, and distinctiveness and togetherness, and uniqueness'.
>
> (Dicks 2000: 94)

Lord made a similar criticism of the Welsh Folk Museum as devoid of politics and as a static, closed version of history. Writing about the presentation of Wales in the museum, Lord commented: 'Wales has come to an end at some indeterminate point in the nineteenth century, a passive nation existing in a time warp. It is a concluded story' (1992b: 40). This criticism is not unique to the Museum of Welsh Life; it is commonly levelled at folk and open-air museums as a genre and is echoed in Tony Bennett's (1995: 111–12) criticism of an 'institutionalised mode of amnesia' at Beamish Open-air Museum in the North of England. A similar accusation of political amnesia can also be found in Richard Kirkland's (1996) discussion of the Ulster Folk and Transport Museum. Kirkland is critical of this museum, founded not long after the Welsh Folk Museum in 1958, because it avoids any discussion of the histories of post-partition Ireland. The extent to which such criticism holds true for the Museum of Welsh Life will be discussed below.

A key point here is that much of the literature – while extremely informative – does tend to attribute the causes of representational choices and changes within the museum to predominantly historical, macrocosmic change or to shifts in ideological positions held by curators and academics. A particularly clear example of this can be found in Adamson (1999) who discusses the Welsh Folk Museum as a vehicle for the dissemination of the ideas and values of Welsh intellectuals. This chapter argues that there were equally important internal, microcosmic reasons why the 'text' of this museum shifted as, and when, it did. Contrary to Lord, Bennett and Kirkland's critique of folk museums as frozen in time or as presenting history as a 'story concluded', I want to argue that the text of this museum is far more organic, open-ended and internally contradictory. My argument is that it operates as a space in which it is possible to identify competing definitions of Welshness and that close examination reveals evidence of the ongoing process of remembering and the remaking of cultural memory in response to the demands of the present.

Formative factors

Accounts of the history of the Museum of Welsh Life often attribute the form adopted at the time of its establishment in the 1940s to the personal experiences of the museum's first curator, Dr Iorwerth Peate (NMGW 1998; *Western Mail* 1998: 8). Peate joined the National Museum of Wales in 1927 and was head of the Welsh Folk Museum from 1948 to 1971. Brought up in Montgomeryshire, mid-Wales, he was a well-known Welsh poet and literary figure, a member of Plaid Cymru (the Welsh nationalist party) and a conscientious objector in the Second World War. While Peate did clearly exercise a formative influence over the development of the Museum of Welsh Life and its public profile, other individuals were equally instrumental, as were wider cultural factors. National Museum Director Cyril Fox was extremely supportive of the idea of a Welsh folk museum, particularly after visiting open-air museums in Scandinavia in 1930 with two members of the museum's Council. Evans Hoyle, a previous director, had also declared support for the idea of a Welsh Folk Museum (Bassett 1984: 26). Other curators similarly influenced the museum's development over the years: it would be impossible and inappropriate to list all of these. However, Haycock's (2004) article on Ffransis Payne illustrates his role in the early years of the museum.

A major influence on the Welsh Folk Museum was the open-air folk museum tradition pioneered at Skansen, near Stockholm in 1891, and subsequently repeated across Europe during the late nineteenth and early twentieth centuries (Bennett 1995: 115; Bassett 1982, 1983; Lord 1992b: 37).[3] In its own time, this folk museum movement was considered a radical move in museological terms because of its valorisation of ordinary people and their lives and also, according to Dicks, because of its return to spectacle; something which other museums were consciously rejecting as unscientific and outmoded (2000: 86). The debt to Skansen is made explicit by Peate in various early texts about the Welsh Folk Museum:

> A folk museum represents the life and culture of a nation, illustrating the arts and the crafts, and in particular the building crafts, of the complete community, and including in its illustration the activities of the mind and spirit – ceremonial, drama, dance and music – as well as of the hand. . . . This is no imaginary picture: it has been fully achieved on several sites in the Scandinavian countries, where the influence of the folk museum in improving the standard of taste and maintaining the pride of the people in the best of traditions of their past has been remarkable.
>
> (National Museum of Wales 1958: 5)

As indicated by the language of this passage, writers such as Peate conceived of folk culture as a holistic and authentic representation of the 'true' character of a nation. Bjarne Stoklund (1983: 8) writes: 'while the culture of the higher classes was subject to changing foreign fashions, the popular culture, with its deep roots, was supposed to represent the true national culture.' This interest in European folk culture was a response to the dramatic socio-economic shifts produced by industrialisation (Morgan 1982: 126; Smith 1984: 14–15). As elsewhere in Britain, the effects of industrialisation were extremely far-reaching within Wales during the late nineteenth and early twentieth centuries.

The transition from rural crafts to mechanised industrial and urban life equally prompted a shift in how culture was conceptualised and also supported the romantic idea of folk culture. The correlation between support for the growth of folk culture and anxiety over the growth of urban industry underpins both Iorwerth Peate's and Cyril Fox's writings on the need for a Museum of Folk Life in Wales. For example, in his introduction to the 1929 collection of Welsh Bygones, Fox wrote:

> It may be urged that the change in the organization of society to which I have referred is a continuous and orderly process, and that our case to-day differs only in degree from that manifest in past centuries. This is, I venture to think, a mistaken view of the situation. . . . The fruitful interaction thus developed, century after century, has now, however, come to an end. The growth of centralized industry and of rapid transport is flooding the Welsh countryside with machine-made goods, of a class hitherto produced locally, exported from the great towns.
>
> (National Museum of Wales 1929: xvi)

Peate similarly expressed alarm over the social impact of the demise in the rural crafts industry and the loss of 'real Wales':

> Anyone who knows the real Wales well can estimate the importance of these craftsmen in the life of their communities, and with the decline of the demand for their services comes the disintegration of small societies of folk which are of real value in a civilised state. . . . The mass-production resulting from the Industrial Revolution in squeezing the rural craftsman out of existence has also, it is more than probable, impoverished the spiritual life of the people.
>
> (National Museum of Wales 1929: 1, 2)

This notion of the craftsman being free from the curse of alienation from the fruits of his labour has a long precedent in Pugin, Marx, Ruskin, Morris and the Arts and Crafts movement. Indeed, Peate often echoed or directly quoted from Morris in his writings. However, as Tony Bennett (1995: 115) has observed, support for folk culture contained both progressive and reactionary elements and was bound up with class relations. Bennett cites Michael Wallace on Skansen:

> The Skansen movement blended romantic nostalgia with dismay at the emergence of capitalist social relations. As the new order had introduced mechanised mass production, a burgeoning working class, and class conflict, these museums, often organised by aristocrats and professionals, set out to preserve and celebrate fast disappearing craft and rural traditions. What they commemorated, and in some degree fabricated, was the life 'of the folk', visualised as a harmonious population of peasants and craft workers.
>
> (Bennett 1995: 115)

This concurs with what Antony Smith (1999: 18) calls 'a route of vernacular mobilization whereby an indigenous intelligentsia uses folk culture to mobilize middle and lower strata and create ethnic nations'. This can be seen in Peate's writings:

> It is, therefore, natural to find that Welsh social life had the village as the centre of a self-sufficing community where work and leisure, individual enterprise and mutual co-operation were combined to produce a rural polity where poverty was never extreme nor wealth out of proportion to the needs of those who enjoyed it . . . the outlook of the people of the Welsh countryside has always postulated a natural courtesy and a willing kindness engendered and fostered by the combination of labour and leisure for the benefit of the community.
>
> (National Museum of Wales 1929: 2)

The important point here, as far as the current Museum of Welsh Life is concerned, is that while the Welsh Folk Museum did not open its gates to the public until 1948, its roots – both in the ideological and material sense – extend back to the nineteenth century and were closely entwined with, and responding to, the form adopted by the first National Museum in Cardiff from 1927.

Repeated calls were made by museum staff and other interested museum professionals from the 1930s onwards for the relocation of the early Bygones collections into a self-contained, Swedish-style, open-air folk museum (Bassett 1984: 29). However, it was not until after the Second World War that various factors combined to boost and finally realise the museum's cause (*Western Mail* 1998: 3; Bassett 1982, 1983). At the UK level, the museum was undoubtedly helped by the post-war interest in the 'daily lives, customs, rituals, and traditions of non-elite social strata' which prompted a 'flurry of new museum initiatives' (Bennett 1995: 109). In Wales, the display of folk culture was thought of by museum professionals as a powerful tool for rebuilding cultural pride and a cohesive national identity (National Museum of Wales 1958). A 1998 *Western Mail* newspaper supplement – '50 years at St Fagans' – noted that: 'In 1943, during some of the darkest days of World War II, the museum council reiterated its belief that part of the post-war reconstruction of facilities in Wales should be the establishment of such a museum' (*Western Mail* 1998: 3; see also Kavanagh 1993). This concurred with the reopening of museums generally across post-war Europe and the symbolic importance attached to this gesture of post-war restoration. Most importantly, the Earl of Plymouth's 1946 gift of St Fagans Castle and gardens provided a physical home for the project, while a public appeal for £100,000 and a Treasury contribution towards the maintenance cost for the new site, albeit not for its capital development, gave the necessary financial support (Bassett 1984: 47).

Collecting Wales

From its outset the Welsh Folk Museum set out to collect predominantly rural rather than modern industrial material culture. Although curators at the Welsh Folk Museum had been actively collecting oral testimony from the industrial areas since the late 1960s – particularly regarding domestic life, coal-mining vocabulary and folklore, as well as information about lead mining in mid-Wales and the tinplate industry – industrial material culture remained seriously underrepresented. It was not until the 1980s that curators decided to 'yield to urban and industrial Wales' and to represent the area where, as Loveluck puts it, 'the majority of Welsh people had lived since 1841'.[4] This reorientation entailed taking in first a row of ironworkers' cottages (Rhyd-y-car) in 1987, the Gwalia Stores in 1991, the Oakdale Miners' Institute in 1995, and most recently, a post-war prefab from a suburb of Cardiff (2001). Ironically, it is the ironworkers' cottages which are reportedly the most popular aspect of the site today possibly because they relate to recent memory and the local knowledge of many visitors who come from the South Wales area (Figure 15.4).

The museum's initial tendency to privilege rural culture as more 'authentically' Welsh than its supposedly alien and Anglicised industrial counterpart follows an extremely well-established precedent within Welsh culture and Welsh nationalism. As I have discussed elsewhere, it was not until the 1980s that the balance of national

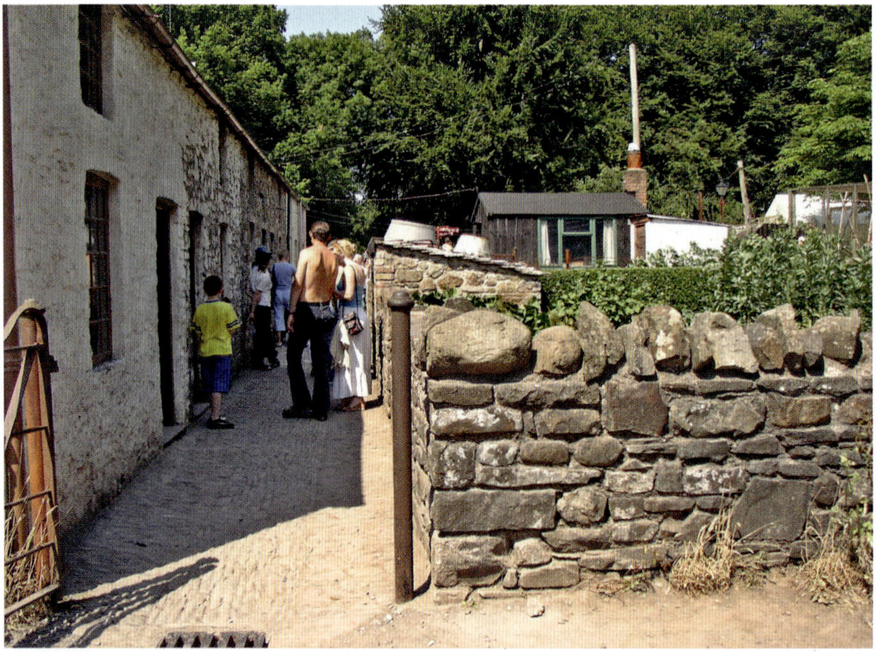

Figure 15.4 Rhyd-y-car ironworkers' houses (Mid-Glamorgan, built about 1800, re-erected 1987). Photo: author

discourse as represented in museums began to shift decisively away from the predominance of rural, Welsh-speaking Wales towards an increasing recognition of the contribution and legitimacy of industrial, Anglophone Wales and an acceptance of a more pluralistic understanding of national identities within Wales (Mason 2007; Dicks 2000: 78–102).

While these broader discursive shifts were evidently significant in terms of the context within which the Welsh Folk Museum was operating during the 1980s, the timing of its decision to reorient its collecting policies and representational remit is also bound up with earlier internal museum issues and indicative of the ways in which it responds to external events. In terms of date for the inclusion of the first industrial exhibits, the Rhyd-y-car cottages, it is crucial to know that the Museum of Welsh Life has a policy of waiting until buildings are offered to it.[5] This is because the museum follows a principle of conservation *in situ* and would rather see the properties remain in their original place wherever possible. They will therefore only take those which are threatened with demolition – the Rhyd-y-car cottages being a case in point. Following a heavy flood in the Merthyr area of South Wales in 1979, the cottages had fallen into serious disrepair and in 1980 the Merthyr Tydfil Borough Council offered some of the houses to the Museum of Welsh Life (Wiliam 1987: 2). Although it is hard to pinpoint evidence within the museum's own records for the exact decision to begin collecting industrial heritage at this time, the wider social context of the 1980s in South Wales, and in particular, Britain's miners' strike of 1984–1985 must have been another significant contributing factor. The difficulty is that the Annual Reports have a ten-year gap which effectively spans most of the 1980s. It may be that the Welsh Folk Museum was taking account of more interest generally in the UK in industrial history. According to Eurwyn Wiliam the use of the Rhyd-y-car cottages to show a chronological narrative of change in living styles was a first among European folk museums.[6]

Another reason for the exclusion of the influence of industries, such as iron and coal, for over a hundred years right up until the 1980s is quite possibly because these industries were perceived to be still so modern and culturally dominant as to not require representation within a museum. It was only when overtaken by newer technological industries that an anxiety that they would disappear from the present sphere of work, and from culture in general, prompted the reappraisal and revaluation of such industrial history throughout Britain. This kind of rescue-mission mentality is common to museums and especially history museums. As Huyssen (1995: 15) puts it, the museum is 'the paradigmatic institution that collects, salvages, and preserves that which has fallen to the ravages of modernization'. In this respect, this museum has been engaged not so much in the representation of what constitutes Welsh life, but more in the representation of what has disappeared from it. There may be another reason which links specifically to Celtic cultures. Hale (2001), for example, identifies the same reluctance to recognise industrial culture within representations of the Cornish past until the 1980s, and Cooke and McLean (2002) observe a longstanding tendency to align Celtic culture with the 'natural', the 'feminine' and the 'ancient' rather than with the industrial and the modern.

Some of the above can only remain at the level of speculation because of the absence of accessible documentation detailing the wider discussions informing collecting policy decisions. However, in the libraries and archives of National Museum Wales there is evidence for another reason for the realignment of its representational strategies. This comes in the form of marketing and audience development reports.

Markets and audiences

During the 1980s the predominance of rural culture over and above representations of the industrial in the Welsh Folk Museum became increasingly recognised within the museum itself. This problem was not simply one of representation but was caught up with wider issues of funding, overall visitor trends, and public and media profile. In 1986 the local press reported that the National Museums & Galleries of Wales (as it was then) as a whole organisation was experiencing a number of difficulties (Underwood 1986: 10). The then Director of the National Museums & Galleries of Wales, Dr David Dykes, was quoted as saying, 'We have problems: problems with our buildings, a shortage of money, and we suffer from staffing problems. I think there has obviously been a loss of morale' (Underwood 1986: 10). In addition, the organisation's public profile had been damaged by a highly publicised and unresolved dispute over the authenticity of Rubens cartoons purchased by the National Museum & Gallery, Cardiff in 1979 (Underwood 1986: 10).

The Welsh Folk Museum (as it was still called) was equally affected at this time. In response to the funding shortage, the National Museums & Galleries of Wales decided to increase its admission charge at all its sites. A consultancy report by John Brown Tourism Services observed that in June 1985 admission prices were increased dramatically in line with findings published in a national report into visitor attitudes at thirty-six National Trust and Department of the Environment heritage sites.[7] This national report had suggested that 'demand was inelastic and relatively constant and would not diminish to any great extent if prices were increased'.[8] Admission charges at the Museum of Welsh Life were thus increased dramatically (according to Brown's calculations by 1,900 per cent in five years, from 10p for an adult admission in 1980 to £2 in 1985). However, contrary to expectations, visitor figures did fall sharply: 'Prior to 1985, visitor figure decreases were in line with other similar attractions but in 1985 they bucked the general trend and dropped by 19.0% while other attractions in South Wales rose by 9.7% aggregate figure.' The Brown report attributes this to the excessively dramatic price increase which deterred casual, impulse visitors and local visitors, and a public perception that the museum had decided to pursue a tourist market at the expense of the local one.

By 1991, the situation had improved for the National Museums & Galleries of Wales. Their visitor figures overall were improving again, as were the numbers for the Welsh Folk Museum (Betts 1991: 7). This was attributed in some quarters to a 'redirection of the museum's entire policies and the introduction of a marketing strategy' (Betts 1991: 7). As part of this marketing strategy the Welsh Folk Museum was renamed the Museum of Welsh Life in 1995 and all the National Museums and Galleries of Wales were given a standardised corporate image, with a shared logo, colour scheme, and single website. Part of this marketing strategy involved responding to findings regarding the Welsh Folk Museum's audience profile. The Brown report of 1986 had indicated that 55 per cent of the museum's visitors at this time were day visitors to the local area and that 85 per cent of these day visitors came from within Wales. Furthermore, the report found that a high proportion of the museum's visitors lived within a 30- to 60-minute radius of the site.[9] Consequently, it reported that: 'The main market for the Museum will nevertheless remain people living within South Wales. Policies must at all times recognise the overriding need for the Museum to maintain, and if possible, increase its appeal within this area.'

Brown's analysis of the audience profile for the Museum of Welsh Life continues to be true, according to a 2003 survey by different consultants showing that in 2003, 61 per cent of visitors lived within 30 minutes of the museum, 29 per cent within 30–60 minutes' radius, 8 per cent at 1–2 hours and only 1 per cent at more than 2 hours' radius.[10] In 2003, 71 per cent of visitors were from Wales, 23 per cent from 'other UK' and only 6 per cent from overseas. This means that the majority of the Museum of Welsh Life's potential audience originate precisely from the Anglophone, post-industrial area of South Wales which had for so long been excluded from the museum. Ironically, this means that for reasons of pure geography and visiting behaviour, a tension existed between the representational ideals envisaged by Peate and the other founders – to focus on rural, Welsh-speaking, 'Y Fro Gymraeg' – and the contemporary need for the museum to cater for, and engage with its local audience – South, industrial 'Welsh Wales' (Balsom 1985; see Mason 2007). The National Museums & Galleries of Wales has recognised this in the reorientation of its remit since the 1980s to include industrial history and, in particular, with its development of a network-wide 'industrial strategy' since 1998. The development of the strategy involved a considerable degree of public consultation and was also arguably prompted by the highly contentious closure of the Welsh Industrial and Maritime Museum in Cardiff Bay in 1998. What this clearly demonstrates is that the museum's representations of Welsh national identities are not driven solely by changes in academic and curatorial thought but result from the complex interplay of concrete issues of location, marketing, and audience development with theoretical discourses and debates surrounding national identities.

Displaying Wales

As the name-change implies, the overall collecting remit was explicitly amended in 1995. This led to an increased interest in acquiring further industrial artefacts and more recent ventures like the House for the Future and the post- Second World War prefabricated house. The result today is a diverse but ultimately quite confusing mix of objects and chronologies. Buildings from different time periods appear to be grouped together according to their functions rather than by date. Agricultural buildings, covering the period 1508–1850s, such as farms, barns and a cider mill, are scattered around the periphery of the site in little glades only accessible by individual paths, thus creating the illusion of rural seclusion and isolation (Figures 15.5 and 15.6).

By contrast, although they differ radically in time period, the more institutional and social buildings such as the tollhouse (1771), the tailor's workshop (1896), the bakery (1900), and the Post Office (1936), are all closely grouped into a kind of village centre connected by roads. The row of ironworkers' cottages (1805–1980), the school (1880), the Gwalia stores (built in 1880 but displayed in the style of the 1920s), and the Miners' Institute (1916), are similarly clustered in the vicinity. Even the Cockpit, which dates from the 1700s, is placed relatively close to this apparently urban social centre, nestling somewhat incongruously behind the Post Office with its Second World War memorabilia (Figure 15.7). This layout therefore takes as its organising principle not the original date of the construction of the building, nor the date of its relocation to the museum, but rather the exhibit's social purpose. This aspect is foregrounded to create a recognisable pattern, albeit with trans-historical continuity. In terms of narrative this arrangement could be read as suggesting that towns and rural areas have always existed in this kind of harmonious proximity.

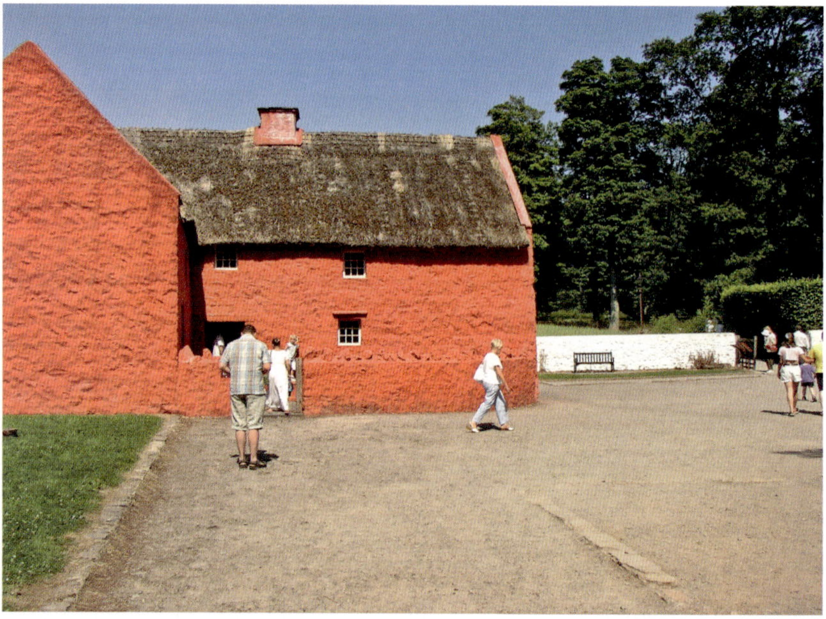

Figure 15.5 Kennixton Farmhouse (West Glamorgan, built 1610, re-erected 1953). Photo: author

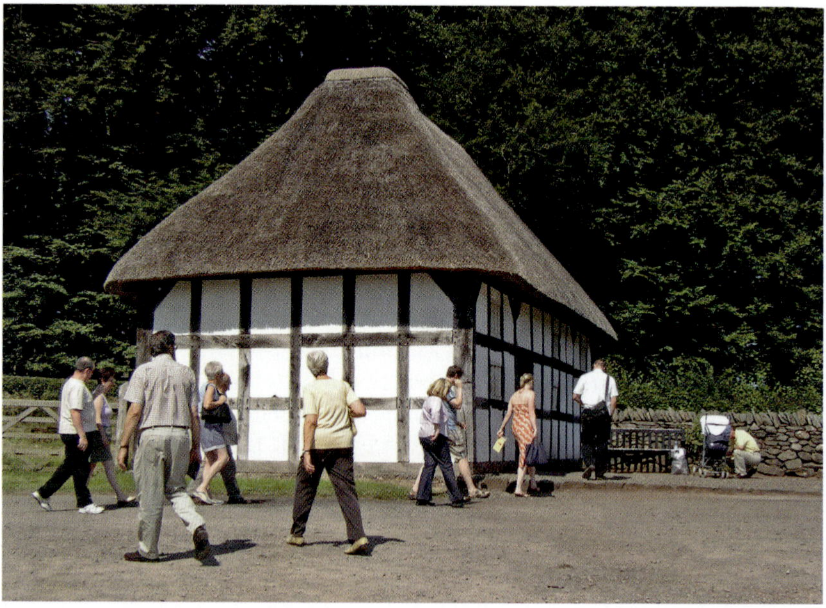

Figure 15.6 Aber-Nodwydd Farmhouse (Powys, built 1678, re-erected 1955). Photo: author

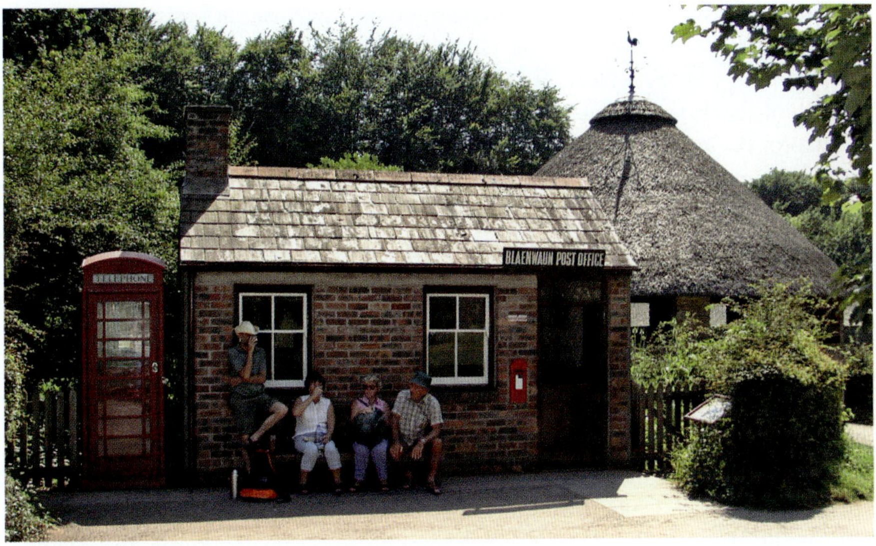

Figure 15.7 Post Office (Dyfed, built 1936, re-erected 1993), Cockpit (Clwyd, seventeenth century, re-erected 1970). Photo: author

A practical explanation for this arrangement is that the collection and siting process has been a gradual one carried out over the last fifty years and has been variously influenced by considerations of space and resources. For example, it was self-evident to site the mill near the river in order to provide it with a water source, while the tollhouse was sited at the crossroads because it was originally situated at the intersection of five roads in Aberystwyth (Jenkins 1990: 167). The Museum of Welsh Life also includes a protected forested area which cannot be altered and this presents another set of constraints to the possibilities for siting and displaying artefacts.

Another pragmatic reason for the museum's current appearance relates to its open-air quality; the Museum of Welsh Life has been useful to the overall National Museums & Galleries of Wales network because it provides room for parts of the national collection which cannot be accommodated elsewhere. A pertinent example here is the Celtic Village which was originally part of an exhibition at the National Museum & Gallery, Cardiff, in May 1991. This exhibit is, in one respect, out of place in the Museum of Welsh Life because it is a reconstruction, not an actual rebuild like the rest of the buildings. It has been 're-created' from excavated remains of buildings, not just from modern-day Wales, but also from Warwickshire. This is a joint archaeology/education initiative and both of these departments contribute to its interpretation. The Department of Social and Cultural History which looks after the majority of the rest of the Museum of Welsh Life provides advice and assistance with the maintenance of the buildings.[11] What is significant about this exhibit is that it came to be sited at the Museum of Welsh Life for a number of reasons – some practical and some intellectual. Thomas suggests that it was felt that it would be better developed and interpreted on a permanent basis in an open-air site. Eurwyn Wiliam, who was the curator of the folk museum at the time and had an archaeology background, also remembers that he 'had always been interested in pushing back the timeframe represented at St Fagans and in experimental archaeology'. It also fitted well with plans afoot at the time to introduce other medieval reconstructions, which did not eventually come to fruition.[12] The reasons why the Celtic Village has come to be in the Museum of Welsh Life illustrate the inter-textual nature of networks of museums like National Museum Wales. However, the complex life-history of the Celtic Village is not apparent to the casual visitor to whom it will appear as simply another part of the 'story of Wales'.

In terms of how it fits within the overall narrative of the Museum of Welsh Life, the Celtic Village presents both problems and opportunities. Its addition creates a considerable temporal gap in the museum's timeline which now jumps from the Celts to the medieval period. It also causes a conflict in the Museum of Welsh Life mission statement, which is variously described on the current website as: 'A walk around Wales from Celtic times to the present day' and 'to show how the people of Wales lived, worked and spent their leisure time over the last five hundred years'. At the same time, the Celtic Village is useful to the Museum of Welsh Life in terms of telling the national story because it offers an historical point of origin for the Welsh, fulfils visitor expectations that a story about Wales will include something about the Celts, and links to the national curriculum.

The complexity and idiosyncrasies of the Museum of Welsh Life and its uneven mix of rural and industrial Wales has been recognised by the museum staff and is discussed in various reports commissioned into visitor responses to the site. Indeed, in 1986, the John Brown report recommended a return to the earlier remit stating that:

> The National Museum of Wales should reconsider and redefine the objectives of the Welsh Folk Museum. It should be accepted that it is impracticable for it to attempt to cover all aspects of the history of the Welsh people, their way of life and their culture. The aim should perhaps become more limited: to be a museum of traditional Welsh rural life.[13]

Curators and consultants also identify part of the problem as being linked to a lack of orientation for the visitor upon arrival. At present, most visitors are unaware that the museum began life as a folk museum, and there is no room to discuss issues of identity within Wales which would contextualise what they will see outside. John Brown's survey found that non-Welsh visitors found it especially difficult to gain a sense of what constituted Welsh culture.

> There was an expectation that the Museum would be 'Welsh' and this was sometimes followed by a feeling of disappointment that the 'Welshness' in a cultural sense did not really come through . . . it was generally more difficult for overseas visitors to form an impression of Welsh culture and Welsh life.[14]

Moreover, although staff recognise that the museums belong to a network, recent audience research suggested that visitors were not aware of NMGW as a brand. A 2002 'Site Audit Debrief' report suggested that individual sites possessed strong brand identities of their own but in general visitors were unaware of individual museums' relationships to other sites in the network. It was concluded that the National Museums & Galleries of Wales 'brand' is 'remarkable yet confusing: it's difficult to put your finger on what NMGW is as a group' (Golley Slater & Partners 2002). Indeed, many visitors continue to call the Museum of Welsh Life simply 'St Fagans' after the village in which it is situated.

Multi-sited displays of identity

It is crucial to remember that the Museum of Welsh Life is only one part of the network of National Museum Wales. In this respect, it is perhaps inappropriate and unrealistic to criticise the Museum of Welsh Life for not representing all of Wales in one site. As its President, Paul Loveluck (2003: 9) notes, industrial history has been collected within the overall National Museum network since the creation of a Department of Industry in the city-centre Cathays Park site in 1959. This collection was transferred in 1977 to the Welsh Industrial and Maritime Museum in Cardiff Docks until its closure in 1998. Since this time, the collection was held in the new Collections Centre in Nantgarw (north of Cardiff) until the creation of the new National Waterfront Museum, Swansea. Industry has also been represented within National Museum Wales's network at a number of sites elsewhere in the country, including the National Slate Museum in the North of Wales (since 1972), the National Wool Museum in the West (since 1976) in the West of Wales, and most recently at Big Pit National Coal Museum (since 2001) in South Wales. The dates here indicate the adoption of the sites into the NMW network. However, in the case of both the Wool Museum and Big Pit, the NMGW had some degree of involvement before formally taking over full control of the sites.

Such large-scale industry is arguably best displayed *in situ*. However, the multi-sited nature of National Museum Wales network as it has evolved through the twentieth and twenty-first centuries raises representational problems concerning demarcation between sites, disciplines and collections. While on paper the range of museums might cumulatively represent a national identity, visitors on the ground may only visit one or two branches because of geography, cost, and perhaps habit. For the inhabitants of Cardiff, for example, the National Museum Cardiff and the Museum of Welsh Life are their most immediately accessible sites. There are practical difficulties too – what happens when one branch is closed for over seven years as was the case with the Welsh Industrial and Maritime Museum? Moreover, with specific titles come specific expectations. In the case of the Museum of Welsh Life which claims to 'illustrate and interpret the daily life and work of the people of Wales from the Middle Ages to the present day' its stated remit sets the parameters against which it is measured.

Conclusions

To summarise, there are two key issues which the Museum of Welsh Life has not addressed. The first is an explicit discussion of the nature of Welsh identities – be they based around class, place, gender, generation, ethnicity, language, or work. The second is a self-reflexive history of the museum itself. The two things are interrelated; unless one is aware of the issues surrounding national identities within Wales the actions of the early museum founders do not make sense. These suggestions echo in part those proposed by the Welsh Assembly and the Wales Tourist Board. In the priority action plan section of its 2002 cultural strategy document, *Creative Future*, the Welsh Assembly Government nominated NMGW to 'review the options for providing a one-stop venue for an overview of Welsh history (NMGW 2004–05)' (WAG 2002: 4). The Wales Tourist Board similarly identified the need to provide more interpretation and orientation particularly for overseas visitors unfamiliar with Wales. These calls converged initially in the proposal for a new 'Gallery of Welsh Histories' although the challenge of addressing such issues in a limited space would be considerable. It is worth remembering that the (1998) Museum of Scotland deploys six floors to 'present Scotland'.

Such changes also fall in line with new museological thinking which argues for a greater self-reflexivity about museum practice and with the broader trend towards visitor-centred displays (Anderson 2004). This is not to imply that the museum should either abandon its duty of care to its objects, nor to seek to didactically tell visitors what their identities should be. As James Braeburne (2000) has argued, people's identities are too complex and too various for a museum to simply impose a top-down definition upon their visitors. For Braeburne the answer is to move towards a more participatory, 'bottom-up' form of museum work; he cites the Twentieth Century Gallery in the Museum of Scotland as exemplary. However, what is gained in terms of inclusiveness and populism may sometimes be at the expense of a coherent framework and the broader historical perspectives which curators can provide. Instead, the Museum of Welsh Life might function as a jumping-off point to encourage visitors to critically review their ideas of identity and 'Welshness'. This could prompt them to examine what exactly they bring to the museum – in the way of assumptions, beliefs, and ideas about the past. It might begin with a consideration of the nature of national identities. Indeed, the museum might pose Day and Suggett's (1985: 96) questions: 'How many Wales [are there]?' or 'How many ways [are there] of being Welsh?' One way to approach this task might be through a discussion of immigration and emigration and its effects on Welsh culture and identity. Given the considerable cross-border traffic between Wales and England this would have the advantage of deconstructing the conventional binaries between Welsh and English and would encourage recognition of the multicultural nature – historic and contemporary – of Welshness which Charlotte Williams so compellingly describes in her work on Welshness and multiculturalism (Williams *et al.* 2003). This kind of interpretive activity is no small task and placing this amount of emphasis on museum communication can, as Andrea Witcomb (2003) has argued, require the rethinking of traditional museum staff responsibilities and remits. However, this would allow the Museum of Welsh Life, and by implication National Museum Wales, to fulfil both its current aim to 'tell the world about Wales, and Wales about the world' and its other founding aim which was to 'tell the Welsh about themselves'.

The above discussion demonstrates that it is impossible to identify one single reading of national identity at the Museum of Welsh Life. There are elements which certainly dominate – for example, rural culture – but there are alternative representations. Visitor attitudes, the degree of their prior knowledge, and their visiting habits complicate the picture further. I would suggest that it is therefore more accurate to say that the Museum of Welsh Life represents a meeting point for competing ideas about national identities. This has resulted from shifts in the museum's working definition of what counts as 'Welsh'. The change in collecting policy in 1995 is a clear example of this process. In response to Bennett's and Kirkland's charges that such museums exhibit a form of 'institutional amnesia', I would counter that there remain problematic gaps in the account of life in Wales which it presents. Its representations are certainly less politicised and less explicit about the inequalities of social and industrial relations than, for example, at Big Pit: National Coal Museum. However, close examination of the Museum of Welsh Life reveals evidence of the ongoing process of remembering and the remaking of public memory in response to the changing demands of the present. It demonstrates how museums function as palimpsests of national identities where the residue of what has gone before conditions what follows.

Postscript

Since the initial writing of this chapter, there have been a number of significant developments at the Museum of Welsh Life. The overall network of museums has changed its name from the National Museums & Galleries of Wales to National Museum Wales. Branches have also been renamed. The Museum of Welsh Life has now been retitled Sain Ffagan Amgueddfa Werin Cymru/St Fagans: National History Museum. The Welsh version retains its older name 'Amgueddfa Werin Cymru' which corresponds with the previous 'Welsh Folk Museum'. Following a future reorganisation of National Museum Cardiff, St Fagans will eventually house the archaeological collections which are to be moved there from the branch in the centre of Cardiff (Heal 2005: 11). The biggest single change will be the creation of a new gallery – Oriel 1 – which will focus on the theme of 'Belonging', or 'Perthyn' in Welsh. At present it is envisaged that this gallery will focus on themes such as voices, beliefs, family and nation, with a clear emphasis on the multiplicity and diversity of identities and ways of 'belonging' present within Wales today. The new vision for St Fagans has been publicised as follows:

> With over £3.5 million being invested at the museum over the next few years, St Fagans: National History Museum, will be developed to tell the stories

> of the peoples of Wales through the ages. The first step in this work is the exciting new exhibition, *Perthyn: Belonging,* in Oriel 1, one of the museum's indoor galleries. Opening in March 2007, *Perthyn: Belonging* is a totally new way of using Amgueddfa Cymru – National Museum Wales's social history collections. Oriel 1 is an exciting and experimental exhibition space and a chance for the museum to work hand in hand with visitors. These developments also respond to one of Amgueddfa Cymru – National Museum Wales's vision for the next ten years, developing ideas for a national history museum for Wales in the future, and it is also an integral part of the development of Amgueddfa Cymru – National Museum Wales and a world-class museum of learning. Beth Thomas, Keeper of Social & Cultural History at St Fagans: National History explains:
>
> '*Perthyn: Belonging* is the first opportunity we've had in years at St Fagans to experiment with new ways of interpreting our collections. Curatorial ideas and visitor expectations have changed considerably, and we're looking forward to being able to interpret Wales's social history in a different way.
>
> 'Through Oriel 1 we hope to work with communities from all parts of Wales to create displays which will challenge stereotypes and encourage people to respect each other's languages, beliefs and customs.'
>
> (National Museum Wales 2006)

Some changes have already begun. For example, the museum's café has been redeveloped along the theme of a Welsh–Italian café. 'Bwyty Bardi Café' opened in July 2006 and is so named in recognition of the Italian–Welsh communities who emigrated to South Wales from the town of Bardi in Italy. Plans are also underway to link the new indoor displays at Oriel 1 with the external buildings. This is crucial as visitors are largely drawn to the open-air aspect of the site rather than specifically to its interior galleries. However, as the comments above make clear, the planned changes are about much more than reorganising the collections. The changes will involve a reinterpretation of the museum's representations of Wales's national cultures, histories and identities. With its chosen theme of 'belonging', Oriel 1 will explicitly encourage visitors to reflect upon issues of identity and what people think it means to be Welsh. These latest developments in this museum's history clearly demonstrate the processes of reimagining national identities in which many national museums are engaged.

Acknowledgements

An earlier version of this paper was published in the online journal *Museum and Society* (Mason 2005) and this version was published in Mason (2007) *Museums, Nations, Identities: Wales and its National Museums*, Cardiff: University of Wales Press. I am grateful to both *Museum and Society* and University of Wales Press for allowing me to republish this material. My thanks also to all those who assisted with this research – see my book (Mason 2007) for full details.

Notes

1 National Museum Wales website, http://www.museumwales.ac.uk/ (accessed 1 March 2007).

2 Bassett, D. A. (unpublished notes, no date given) 'The Museum Movement: 1890–1905: A Selection of Items From the Museum Diary; Concurrent Events in Cathays Park' (held in the National Museum Cardiff Library), pp. 1–6.

3 Loveluck, P. (2003) 'Museums and national identity in post-devolution Wales'. Unpublished transcript of the paper given to the Society of Antiquaries of London and the Cambrian Archaeological Association at the National Museums & Galleries of Wales, 20 March 2003, Cardiff: Society of Antiquaries of London.

4 Museum of Welsh Life exhibition text, *50 Years at St Fagans* (1998). P. Loveluck, 'Museums and national identity in post-devolution Wales', a paper given to the Society of Antiquaries of London and the Cambrian Archaeological Association at the National Museums and Galleries of Wales, Cardiff, 20 March 2003.

5 Interview with Beth Thomas, Keeper of the Department of Social and Cultural History, National History Museum of Wales (formerly known as the Museum of Welsh Life, and the Welsh Folk Museum), personal communication, 2004.

6 Correspondence with Eurwyn Wiliam, Deputy Director General and former Curator of the National History Museum of Wales (formerly known as the Museum of Welsh Life, and the Welsh Folk Museum), personal communication, 2004.

7 Brown, J. (1986) 'Increasing the visitor appeal of the Welsh Folk Museum: A strategy for presentation, interpretation, management and marketing, prepared for the NMW and the Welsh Tourist Board by John Brown Tourism Services'.

8 Brown 1986: 92.

9 Ibid.: 78.

10 Beaufort Research Ltd (2003) 'National Museums & Galleries of Wales: Visitor survey 2003', (unpublished presentation).

11 Thomas, personal communication, 2004.

12 Wiliam, personal communication, 2004.

13 Brown 1986: 13.

14 Ibid.: 63.

Bibliography

Adamson, D. (1999) 'The intellectual and the national movement in Wales', in R. Fevre and A. Thompson (eds) *Nation, Identity, and Social Theory: Perspectives from Wales*, Cardiff: University of Wales Press, pp. 48–68.

Anderson, G. (2004) *Reinventing the Museum: Historical and Contemporary Perspectives on the Paradigm Shift*, Walnut Creek, CA, and Oxford: Altamira Press.

Balsom, D. (1985) 'The three Wales model', in J. Osmond (ed.) *The National Question Again*, Llandysul: IKGomer, pp. 1–17.

Bassett, D. A. (1982) 'The making of a national museum: Part I', *Transactions of the Honourable Society of Cymmrodorion*, pp. 153–85.

Bassett, D. A. (1983) 'The making of a national museum: Part II', *Transactions of the Honourable Society of Cymmrodorion*, pp. 187–220.

Bassett, D. A. (1984) 'The making of a national museum: Part III', *Transactions of the Honourable Society of Cymmrodorion,* pp. 1–100.

Bassett, D. A. (1990) 'The making of a national museum: Part IV', *Transactions of the Honourable Society of Cymmrodorion*, pp. 193–260.

Bassett, D. A. and Bassett, M. (1993) *Wales in Miniature: Amgueddfa* (Special Edition), Cardiff: National Museum of Wales.

Bennett, T. (1995) *The Birth of the Museum: History, Theory, Politics*, London and New York: Routledge.

Betts, C. (1991) 'Museum turns corner', *Western Mail,* 23 February, 7.

Braeburne, J. (2000) 'The poverty of nations: Should museums create identity?', in J. M. Fladmark (ed.) *Heritage and Museums: Shaping National Identity*, Oxford: Donhead Publishing Ltd, pp. 379–93.

Cooke, S. and McLean, F. (2002) 'Picturing the nation: The Celtic periphery as discursive other in the archaeological displays of the Museum of Scotland', *Scottish Geographical Journal,* 118(4): 283–98.

Day, G. and Suggett, R. (1985) 'Conceptions of Wales and Welshness: Aspects of nationalism in nineteenth-century Wales', in G. Rees, J. Bujra, P. Littlewood, H. Newby and T. Rees (eds) *Political Action and Social Identity: Class, Locality and Ideology,* Basingstoke and New York: Macmillan, pp. 91–116.

Dicks, B. (2000) *Heritage, Place and Community,* Cardiff: University of Wales Press.

Foster, R. J. (1991) 'Making national cultures in the global Ecumene', *Annual Review of Anthropology,* 20: 235–60.

Golley Slater & Partners (2002) 'National Museums and Galleries of Wales: Site Audit Debrief', unpublished presentation.

Hale, A. (2001) 'Whose Celtic Cornwall? The ethnic Cornish meet Celtic spirituality', in D. Harvey, R. Jones, N. McInroy and C. Milligan (eds) *Celtic Geographies,* London: Routledge, pp. 157–72.

Haycock, M. (2004) 'The scholarship and creativity of Ffransis G. Payne', *Trans Radnorshire Society*, 74: 24–49.

Heal, S. (2005) 'News', *Museums Journal,* April: 11.

Houlihan, M. (2005) 'Social history: Closed for reconstruction', *Social History in Museums: Journal of the Social History Curators Group,* 30: 5–8.

Huyssen, A. (1995) *Twilight Memories*, London and New York: Routledge.

Jenkins, G. (1990) *Exploring Museums: Wales, a Museums Association Guide*, London: HMSO.

Kavanagh, G. (1993) 'History in museums in Britain: A brief survey of trends and ideas', in D. Fleming, C. Paine and J. G. Rhodes (eds) *Social History in Museums: A Handbook for Professionals,* London: HMSO, pp. 13–26.

Kirkland, R. (1996) *Literature and Culture in Northern Ireland since 1965: Moments of Danger,* London and New York: Longman.

Lord, P. (1992a) 'Life before Wilson: Reconstructing Welsh art', *Planet: The Welsh Internationalist,* 95: 42–48.

Lord, P. (1992b) *The Aesthetics of Relevance,* Dyfed: Gomer.

Mason, R. (2005) 'Nation building at the Museum of Welsh Life', *Museum and Society,* 3: 18–34.

Mason, R. (2007) *Museums, Nations, Identities: Wales and its National Museums,* Cardiff: University of Wales Press.

Morgan, K. O. (1982) *Rebirth of a Nation: Wales 1880–1980,* Oxford: Oxford University Press and Cardiff: University of Wales Press.

National Museum Wales (1929) *Guide to the Collection of Welsh Bygones: A Descriptive Account of Old-fashioned Life in Wales: Preface,* Cardiff: National Museum of Wales and the Press Board of the University of Wales.

National Museum Wales (1938) *The Museum and its Contents: A Short Guide,* Cardiff: National Museum of Wales and the Press Board of the University of Wales.

National Museum Wales (1946) *St Fagans Castle: A Folk Museum for the Welsh Nation*, Cardiff: National Museum of Wales.

National Museum Wales (1958) *Welsh Folk Museum St Fagans: Handbook,* Cardiff: National Museum of Wales.

National Museums & Galleries of Wales (NMGW) (1998) *Amgueddfa 1997,* Cardiff: NMGW.

National Museums & Galleries of Wales (NMGW) (2001) *Museum of Welsh Life: Visitor Guide,* Cardiff: NMGW.

National Museums & Galleries of Wales (NMGW) (2003) *Annual Report 2002–03: National Delivery,* Cardiff: NMGW.

National Museums & Galleries of Wales (NMGW) (2004–05) *Annual Report,* Cardiff: NMGW, pp. 1–48.

National Museum Wales (2006) *Museum News Release,* 29 June.

Smith, A. D. (1999), *Myths and Memories of the Nation*, Oxford and New York: Oxford University Press.

Smith, D. (1984) *Wales! Wales?,* London: Allen & Unwin.

Stoklund, B. (1983) *Folk Life Research: Between History and Anthropology: A 75th Anniversary Lecture delivered at the Welsh Folk Museum, St Fagans,* Cardiff: National Museum of Wales.

Underwood, P. (1986) 'Looking Ahead: The Man Who Must Preserve Our Past', *South Wales Echo,* 11 March, 10.

Welsh Assembly Government (WAG) (2002) *Creative Future,* Cardiff: WAG.

Western Mail (1998) '50 years at St Fagans: Special Supplement'.

Wiliam, E. (1987) *Rhyd-y-car: A Welsh Mining Community: The Life, Death and Re-creation of a Welsh Mining Community 1800–1980,* Cardiff: National Museums and Galleries of Wales.

Williams, C., Evans, N. and O'Leary, P. (eds) (2003) *A Tolerant Nation? Exploring Ethnic Diversity in Wales,* Cardiff: University of Wales Press.

Witcomb, A. (2003) *Rethinking the Museum: Beyond the Mausoleum,* London: Routledge.

Chapter 16

SAME EXHIBITIONS, DIFFERENT LABELS?

Romanian national museums and the fall of communism

Simina Bădică

ROMANIA UNDERWENT major political change in 1989 but it soon became apparent that this revolution did not extend to other domains of the country's political life. Over the following two decades a strong anti-communist discourse appeared in museums but museologically these institutions changed little. In the National History Museum, for example, exhibitions switched from praise to accusation but essentially it was the same style of political discourse. This chapter enquires into the differences and similarities between communist and anti-communist exhibitions in Romania, before and after 1989. Were these really the same exhibitions with the labels changed?[1] What was 'communist museography' and how did the new discourse seek to address it?

Museums of communism

Much research has concerned those museums of communism which first appeared after 1989 (Apor and Sarkisova 2007; Rév 2005; James 1999; Boym 2001). One could say these museums sprang up like mushrooms all over Eastern Europe, in the form of 'statue parks' like Memento Park in Budapest and the Muzeon Park in Moscow, and commercial museums for Western tourists, like the Museum of Communism in Prague, to the memorial museums and museums of terror, such as Terror Haza in Budapest and the Sighet Memorial in Northern Romania. What is forgotten is that museums of communism existed all over Eastern Europe even before 1989. Socialist regimes used museums to enforce their view of history on the present; museology was just another form of propaganda. The sudden disappearance of these museums in 1989, or shortly afterwards, is barely documented. They simply vanished, leaving behind empty exhibition galleries and archives. And while the destruction or movement of statues and mausoleums was well documented, the emptying of the museums occurred behind thick walls and closed doors.

Three of the most important national museums exhibiting communism before 1989 are the subject of this chapter. They are: the History Museum of the Romanian Communist Party, a museum that underwent changes of name, exhibition and ideology from the 1950s to the 1980s; the Doftana Museum, which was a former prison of the interwar period for political inmates (among them communists) and was transformed into a museum in 1949; and the History Museum of the Romanian Socialist Republic, which was inaugurated in 1972.

The idea of a museum exhibiting communist ideology and achievements soon became established in the new Popular Republic of Romania. The monarchy had been abolished at the end of 1947, and by July 1948 the new Moments from the People's Struggle Museum had appeared. In 1950, this museum was reorganized and renamed the Revolutionary Struggle of the People Museum (shortened to the Revolutionary Museum). Re-baptized again in 1954 , it became the History Museum of the Romanian Workers' Party. It then closed in November 1957 only to be reopened in July 1958 in a more sumptuous location: the Neo-Romanesque palace on Kiseleff Boulevard which had been expropriated from the Museum of National Art in 1952. At the time it shared the building with another museum of communism: the V. I. Lenin–I. V. Stalin Museum (which was later renamed the Marx–Engels–Lenin Museum). In 1966 it acquired its last and longest name, the History Museum of the Communist Party, of the Revolutionary and Democratic Movement of Romania. This museum was informally called 'the Party Museum' and despite these several changes of name and location it continued to be, throughout the communist period, a true museum of communism: of the communist movement, ideology and party.

The Doftana Museum was a section of the Party Museum. It had functioned as a prison from 1895 to 1940 and was turned into a museum by the communist regime in 1949. Doftana became the ceremonial place where thousands of children (myself among them) became pioneers. Today, the building is abandoned but the ruin remains and is listed as a historical monument. From time to time, private firms organize paintball games there.

The last of the three museums of communism considered in this analysis is the History Museum of the Romanian Socialist Republic inaugurated by Nicolae Ceauşescu in 1972, following a decision of the Central Committee, in 1968. Romania had not possessed such an institution. This new History Museum claimed to be the proud heir to the Antiquities Museum, founded in the middle of the nineteenth century (Georgescu 1975: 5). It aspired to be *the* national museum Romania lacked; 'the most representative museal institution in our socialist country: THE NATIONAL MUSEUM' (Georgescu 1974: 2). Unsurprisingly, then, the annual review of this museum was boldly entitled *The National Museum*.

In addition to examining these pre-1989 communist museums, this chapter will also discuss three major museums exhibiting communism after 1989. All three of these latter institutions possess interesting similarities and each is in some way deeply connected to those pre-1989 museums. The Romanian Peasant National Museum occupies the former building of the Party Museum on Kiseleff Boulevard in the centre of Bucharest. The Peasant Museum hosts the only permanent exhibition in Bucharest on the communist past of the country. In 1996, six years after its establishment, it received the prestigious European Museum of the Year Award. The Sighet Memorial is a former prison where, in the 1950s, the political elite of the interwar period were imprisoned and exterminated by the communist regime. In 1993, it was transformed into a museum and memorial for the victims of, and resistance against, communism. The final museum is the National History Museum which during its reconstruction, ongoing since 2002, presented temporary exhibitions on the communist regime. This chapter will examine how these different museums have told their often contradictory stories.

The antidote museum

In the enthusiasm that followed the 1989 change of regime, everything had to be reinvented. Or so these enthusiasts thought. Others simply waited for the storm to pass and the return to normality. Museums and museum professionals were similarly minded: there was no middle way; it was either total change or no change at all.

Little has been written about those museums that survived the early 1990s, probably because, in the excitement of transition, museums simply were not a hot topic. However, Irina Nicolau's (1996) *Me and the Museums of the World: The History of a Museum Experience in an Eastern European Country* (published in French) provides important insights into this moment of change and the high expectations placed on museums by the enthusiasts. Nicolau was a writer, an ethnologist and a museum curator; she could be called a museum *creator*. Together with Horia Bernea, she was responsible for creating the most well-known of Romanian museums, the Romanian Peasant Museum.

In 1994, Nicolau developed the oppositional concepts of the 'mother-museum' and the 'father-museum'. Nicolau explains these terms as follows:

> As opposed to mother-museums, where you meet unknown objects – appropriated although remaining unknown – the father-museums give explanations, produce reasoning, educate. . . . The mother-style is an antidote to the hyper-amnesia towards which the father-museums push us, together with all our society.
>
> (Nicolau 1996: 18)

The mother-museum is the prototype of the museum as 'A place where you see objects that you like' (Nicolau 1996: 18).

Nicolau also developed the notion of the 'antidote-museum'. The antidote-museum responded to the double crisis faced by Romanian museums in the post-communist period. 'The Romanian museum is in a double crisis, provoked by the consequences of communist ideology and by the danger of badly appropriated occidental museology' (Nicolau 1996: 37). She does not give a clear definition but rather composes a 'Decalogue of the antidote-museum' that she thought necessary in 'periods of cultural, social and political convalescence', including:

> 3. One doesn't go to the antidote-museum as one would go to a church, neither to a school, a tribunal, nor a hospital or a cemetery.
> 4. The antidote-museum is the museum of 'Look at that!' Its exhibitions free the object of any stereotyped interpretations.
> 5. One comes to the antidote-museum to see the objects.
> 8. The antidote-museum shows, but also hides. It is for people willing to invest imagination and time.
>
> (Nicolau 1996: 38)

There is vehemence behind the idea: the antidote-museum is to be the antonym and mortal enemy of the apathetic or blasé museum. Nicolau is here arguing against a particular kind of museum practice rather than against the political ideologies that museums communicate. Were communist museums museologically of this blasé type? Was communist museum practice the reason for their deplorable state in 1989? Did this result from the lack of personal commitment to the ideas they were exhibiting?

The argument goes both ways. Leaving aside the communist ideology, in which few people believed anymore, museum practice in the communist era over-emphasized propaganda and the place of a single master narrative. It led museums to a museological dead-end, where many of them remain today. There was to be just one way, one story:

> As for the period after August 23rd 1944, there is only one solution for museum representation, the chronological one, as events unfold only in one direction: the fight of the people led by the Communists for the construction of a new society lacking exploitation and the contradictions that characterized former epochs.
>
> (Cloşcă 1972: 36)

Evgeny Dobrenko has remarked on Stalinist museums which adopted this singular voice: 'Museumification acts as one form of destruction, in so far as the only thing that is not acceptable to Stalinism is the "living past"' (Dobrenko 2008: 9).

Communist museology

In the 1990s, Romanian museums were challenged to reinvent themselves, to renounce the blasé museum and become antidote-museums. Few people or museums took the challenge seriously. Most museums simply cleansed their exhibitions of any reference to the communist past and considered that the rest of the exhibitions were acceptable. But, were they really? What exactly was the basis of communist museology, if such a thing existed?

It is important to start this genealogy (Foucault 1977) of museology in communist ideology and in communist reality by acknowledging that Soviet museums were threatened with disappearance after the Bolshevik Revolution of 1917. The radical artistic movement emerging in the young state of the Soviets wanted to get rid of these symbols and places of bourgeois decadent art. It was the bureaucrats, the political figures that saved Soviet museums and, at the same time, sealed their fate as institutions subordinated to bureaucracy rather than to creativity.

The story of 'the ideological struggle against the museum and museum art conducted by the Russian avant-garde' is convincingly told by the Russian professor of philosophy and émigré, Boris Groys (1994: 145). He begins by acknowledging that the anti-museum discourse (see also Sherman 1994) was part of the avant-garde movement in the first decades of the twentieth century. However, it was only in revolutionary Russia that the discourse came so close to producing a museum-free reality. The painter Kasimir Malevich wrote: 'Do we need Rubens or the Cheops Pyramid? Is the depraved Venus necessary to the pilot in the heights of our new comprehension? Do we need old copies of clay towns, supported on the crutches of Greek columns?' (Groys 1994: 147). Another of the Russian avant-garde, Nikolai Tarabukin, compared museums to graves and art historians to gravediggers:

> Everything created by the 'left' wing of contemporary art will find its vindication nowhere else but within the museum's walls, and the whole revolutionary tempest will find its peace in the silence of that cemetery. And 'art historians', those indefatigable grave robbers, for their part have the task of composing the explanatory texts for these mortuary crypts.
>
> (Groys 1994: 150)

The revolutionary poet, Vladimir Mayakovsky, declared, 'it's time bullets rang out against museum walls' (Dobrenko 2008: 8).

'One topos of Soviet avant-garde criticism', Groys explains,

> was the conviction that the very institution of the museum was connected with *vita contemplativa,* understood as a purely consumerist attitude toward life deserving no place in a society in which he who does not work does not eat. . . . For a proletarian to go to a museum to gaze on the beautiful meant that his labours toward the aesthetic transformation of life itself had failed.

> The avant-garde aesthetic understood the museum-as-temple-of-beauty exclusively as compensation for an absence of beauty in life itself.
>
> (Groys 1994: 145)

The Soviet leadership seemed for once less radical than 'the people'. It was Soviet bureaucracy that saved the museum, if only for the reason that the immense number of artworks inherited from the former regime were not to be easily disposed of. 'From the very beginning, this radical point of view failed to satisfy the Soviet authorities, if only because it failed to explain what was to be done with the mass of artworks they had inherited' (Groys 1994: 145). The propaganda potential of the museum must also have appealed to the newly established Soviet government. Consequently, it is in the form of propaganda and didactic material that the museum was reborn from the ashes of the bourgeois museum:

> The goal of this newly conceived museum became not to present objects and artifacts that might be considered original, characteristic, and specific in the historical development of art; rather, it was to present only those elements that appeared useful from a didactic point of view.
>
> (Groys 1994: 145)

What was problematic about bourgeois museums, from a communist point of view, was that they did not reflect the truthfulness and beauty of life. Yet, by transforming museums into propaganda, by commissioning art specifically for the museum, the Soviets managed to estrange the museum still further from real life. The solution to this dilemma came not by another transformation of the museum but by a transformation of reality itself. As Boris Groys argues, 'Soviet communism itself came to look more like a museum exhibit than like the society of the future' (Groys 1994: 145).

> In essence the Stalinist aesthetic preserved the same goal of overcoming the division between art and life. However, it undertook to solve the problem not by destroying the museum but by a sort of equalization of museum exhibits and their surrounding milieu, accomplished by physically filling the milieu with art indistinguishable from that in museums. It is this strategy of equalizing what is in museums with what lies outside them that creates the very specific aesthetic atmosphere found in totalitarian societies of the Stalinist type.
>
> (Groys 1994: 157)

By the time communism and communist museums became established in Romania, after the Second World War, these debates were long over and their advocates long dead. Socialist Realism and Stalinist aesthetics were then the mandatory fashion and only mere traces of other approaches reached museum specialists. The manual, *Bases of Soviet Museology* (Galkina *et al.* 1957) was translated from Russian and printed for internal use by the Romanian Ministry of Culture in 1957. It mentions and simultaneously denigrates various past trends in museum exhibition. The sociological school of M. N. Pokrovsky and the first Soviet Museological Congress of 1930 is mentioned only to be aggressively refuted. Pokrovsky is accused of 'vulgar sociologism'[2] and 'anti-scientifism' because he promoted the display of interior dioramas in order to represent historical eras.

> These exhibitions were based on the idea of presenting 'pieces' of real life in order to mirror characteristic traits of different historical periods. According to the authors of these exhibitions, these were the interiors, that is, they were the interior aspects of the life of people belonging to different classes of society. In reality, such interiors were only presenting exterior aspects of life, life-conditions of different classes, but gave no notions of the production modes. They could not unveil the essence of main historical periods, of socio-economical formations, of correlations between classes, of the class struggle.
>
> (Galkina *et al.* 1957: 175)

The authors equally criticize 'vulgar sociologism' for refusing to place original objects in exhibitions and using 'pure formalist procedures' and 'colour and construction symbolism'. During the 1920s, sociologism had been the official trend not only in museums but also in art and literary theory. By the 1930s, the assessment of an artist, writer or work of art according to their social origin came to be considered 'vulgar sociologism' and thus harmful to the Soviet state.

> If, for example, in the early 1920s A. Fedorov-Davydov, one of the founding theoreticians of Soviet museum administration, was still able to boast that both permanent displays and large exhibitions had been organized according to sociological criteria, before long the sociological method had been subjected to harsh criticism and was open to serious accusations like the following: 'The role of the folk as the chief creator of artistic values was negated [by the sociological method], and all the achievements of the past were given away to the dominant exploiting classes'.
>
> (Groys 1994: 155)

This debate had never entered Romanian circles but it was used to raise awareness of the possibly devastating effects that might result from such apparently benign artefacts as dioramas. Museum work in the new Romanian Socialist Republic was to be done according to strict rules and advice. The exhibition discourse had to be of Soviet origin and the techniques used should have been tested and approved beforehand by Romania's powerful Eastern neighbour. *Bases of Soviet Museology* was not meant to be an inspirational book. It was actually a manual of museology that

Romanian museum professionals had to understand, internalize and apply, leaving practically no place for personal creativity or interpretation. The strict rules of museum display in 1950s may account for the deathly boredom that was to be found in Romanian museums at the end of the 1980s and even after the political change of 1989.

Communist museums were, to use Nicolau's terminology, father-museums (Nicolau 1996); they were there to give answers and explanations. Their mission was to educate the public according to the ideology of the day. Indeed, *Bases of Soviet Museology* opens with the following statement: 'Soviet Museums are an efficient propaganda tool for communist ideas, for the communist conception of the world and thus for communist education' (Galkina *et al.* 1957: 3). If one reads the manual, the message is obvious: the communist museum is a museum where people find 'The Answers'.

Words versus objects

A museum of answers, a father-museum, is actually afraid of the object. The object appears silent and dangerously ambiguous; it is difficult to control. The solution was to get rid of the object or surround it with words and explanations. Soviet guidance, while recommending the use of original objects, points to the dangers of leaving the objects uninterpreted. The ambition of the Soviet museum was to turn visitors into eyewitnesses, to make them feel that, after leaving the exhibition, they could testify to the things they saw with their own eyes. 'In the exhibition, the visitor will see the concrete materials that he would have seen were he the eye witness of the strikes at the beginning of the 20th century' (Galkina *et al.* 1957: 176). However, the curator should not trust the eyes of the visitor for they can lead him or her away from the orthodox interpretation of events. 'In order to help the visitor examine the exhibits in a correct manner and to draw the correct conclusions, texts-quotations and texts by the curators are introduced into the exhibition' (Galkina *et al.* 1957: 177). Objects in exhibitions have to be surrounded by words and explanations; this is the recommendation of Soviet experts. It seems Romanian museum professionals took this recommendation very seriously and focused their creative energies on countless exhibitions where the text, the slogan and the explanation displaced the object. Even so, many professionals saw the text as auxiliary. As a Romanian museum manual suggested in 1979, 'The primary quality of every helping text is discretion. They will be clear, readable and visible without competing with or covering the object' (Nicolescu 1979: 173). However, the text was then still gaining prominence (and not just in Eastern Europe). It became an exhibit in itself. It, in effect, became the object.

No object or photograph was allowed to remain uninterpreted, and museum employees had to make sure no double meaning could be extracted from their exhibitions. As a result, exhibition titles became longer and longer. In 1978, an exhibition opened at the History Museum of the Romanian Socialist Republic entitled *Proofs of love, appreciation and deep gratitude manifested by the entire party and people towards comrade Nicolae Ceauşescu, Secretary General of the Romanian Communist Party, president of the Romanian Socialist Republic, on the occasion of his 60th anniversary and over 45 years of revolutionary activity*. Any review of the exhibition (and there were many) had to cite the full title (Ştefănescu 1984: 33–35). A major part of the exhibition featured gifts received by Ceauşescu on special occasions when he met teams of Romanian workers. In 1979, the History Museum of the Romanian Communist Party opened an exhibition called *90 years since the declaration of May 1st as International Worker's Day (1889). 40 years since the big Antifascist and Antiwar demonstration in our country (May 1st 1939). 35 years since the establishment of the Worker's United Front*. It contained panels, photographs, newspapers, drawings, texts and a few showcases with objects (such as flags). Another exhibition, opened in December 1981, hosted by the History Museum of the Romanian Socialist Republic, with a shorter but equally eloquent title, was called *President Nicolae Ceauşescu and World Peace* (Figure 16.1). There were no objects in this exhibition. There were only interpretive panels showing images and slogans, quotes and explanations (Lungu and Leahu 1982). It was a typical father-museum; a museum that seeks to explain, to convince, to transmit precise knowledge; a museum that knows exactly what the visitors should retain in their minds when they exit the exhibition halls.

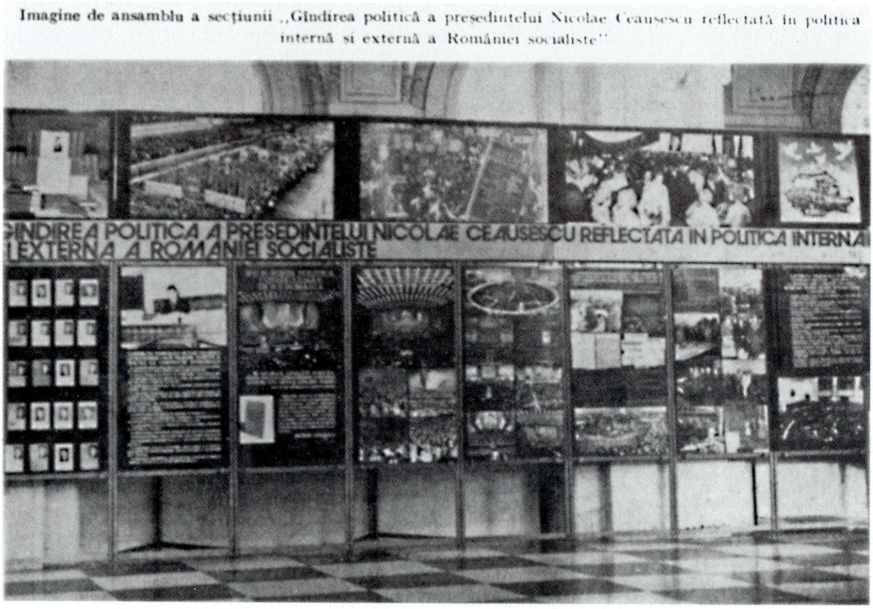

Figure 16.1 Image from the exhibition, *President Nicolae Ceauşescu and World Peace*, at the National History Museum of the Romanian Socialist Republic in 1981–82. This section of the exhibition, entitled 'The political thinking of President Nicolae Ceauşescu', reflected in the internal and foreign affairs of Socialist Romania. Photo: Revista Muzeelor 1982: 3

The situation was no different for the permanent exhibitions in communist museums. Those dealing with the modern and contemporary eras were most keen on using lots of photographs, newspapers, photographs of newspapers and extensive texts. These exhibition halls were crowded with words, explanations and labels. No detail could be left to the free interpretation of the observer. However, the exhibits that these words surrounded were not the objects usually found in historical museums. Ioana Popescu, a visual anthropologist and exhibition curator, once remarked that the History Museum of the Communist Party was in fact a museum of photography – of fake photography.[3]

These interpretive rules prevented the production of diverse exhibits. But there was also a dearth of material from which to construct historical interpretations of the Romanian Communist Party and particularly so for the period before 1944. This history had to be produced after 1945. Consequently the History Museum of the Communist Party was full of paintings (most of them reproductions of famous Soviet paintings), drawings and statues re-creating events. For the period after 1945, museums favoured charts, maps and photograph (Figure 16.2). Museum debates of the time explain why paper was 'The Exhibit' and why museums should always prefer paper to objects.

> No matter how much we would argue and want to exhibit three-dimensional pieces, the efficient solution to the problem can only come from paper (meaning documents, printed press and other prints, photographs). . . . Compared to the object of a politician or of a leader of proletarian movement, we prefer the paper on which a law was written, a manifesto, a calling. . . . The big social, economic, political problems that tormented the epoch in question are truly and fully mirrored on paper.
>
> (Cloşcă 1972: 35)

This view was strongly supported in this publication *Revista Muzeelor* (*Museums' Review*): 'Admitting that the paper-document, in its multiple forms, constitutes the main exhibit in illustrating the contemporary era, we have the obligation, as specialist museographers, to valorize it in the most diverse ways' (Pavel 1972: 308).

The situation was broadly the same in neighbouring countries. In Bulgaria, for example: 'In the majority of cases they were standard photographic exhibitions with an emphasis on their documentary value and clear propaganda overtones' (Voukov 2007: 316). The Bulgarian Museum of the Revolutionary Movement held '3000 photos, objects, weapons, documents, maps and works of art that recreated "the struggle of the working people under the guidance of the BKP against Capitalism and Fascism"' (Voukov 2007: 314). These objects supposedly belonged to the partisan fighters. In the Romanian case, where the Party propaganda worked hard to invent traces of partisan movement during the Second World War, these objects were quite hard to find, as is made clear in the inventory of the Romanian Lenin–Stalin Museum. The museum had on display, in ten rooms, 'copies of manuscripts

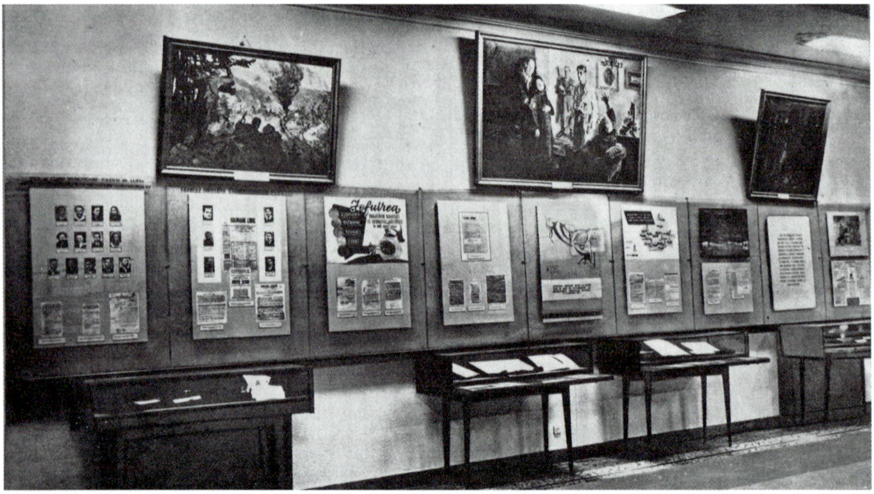

Figure 16.2 The Romanian Workers' Party Museum in the late 1950s. This room exhibits 'Aspects of the workers' heroic fight led by the Communist Party against the military-fascist regime and the criminal anti-Soviet war. Portraits and objects belonging to working-class heroes who fell in the fight against fascism, in our country and abroad.' Photo: History Museum of the Romanian Workers' Party 1950s: 21

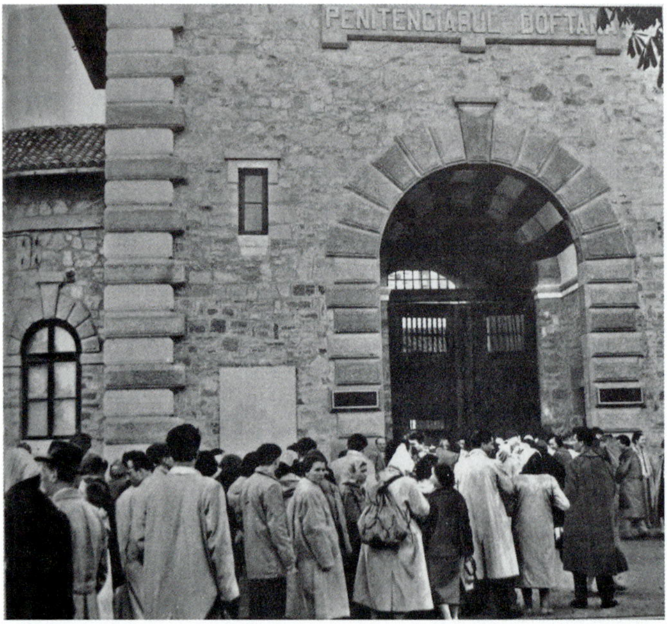

Figure 16.3 'Working people visit the Doftana Museum.' Visits to communist museums were mandatory. Photo: History Museum of the Romanian Workers' Party 1960: 39

and printed texts, original newspapers and brochures, photographs, maps, graphics, paintings, sculptures and small-size models of important buildings *(machete)*' (National Office for Tourism n.d.: 32). The impact these new museums had in the 1950s and 1960s is difficult to assess. However, in the 1970s and 1980s most people entered these exhibition halls as part of a mandatory organized visit, through the school system or the workplace (Figure 16.3).

Exhibiting communism after 1989

After 1989, it is not hard to imagine how museums were perceived within Romanian culture. The situation was the same across all the former socialist bloc: 'It is not surprising that, given this overexploited encounter between museums and the population in the socialist period, a critical lack of interest in museums with historical collections became dominant after the change of regime' (Voukov 2007: 319). These museums were simply boring. It is not only that the story they told became increasingly surreal – not believed by those who produced the exhibitions or those who consumed them; it was also that the exhibitions themselves had ceased to be anything more than an oversized photograph album. It is easy to understand why Irina Nicolau's work has focused so closely on the reintroduction of the unmediated object (Nicolau 1996: 37).

So, what changed after 1989? For some time communism, both as regime and ideology, disappeared from museums. Without an official version of the country's communist past, museums kept silent. History museums ended their chronological enfolding of Romanian history in 1945. It was only in 2007 that the National History Museum began to organize temporary exhibitions on the communist era, using material from the collections gathered during state socialism. Rather more problematic was the continuation of museum practices developed for the production of exhibitions during state socialism. The only obvious change seemed to be in the words surrounding the exhibits, and not in the exhibition style.

The exhibition, *Communism in Romania (1945–1989)*, opened on the ground floor of the National History Museum in July 2007 (Figure 16.4). It gave a grand overview in one large room using panels, propaganda objects and didactic explanations on the meaning of the exhibits. The exhibition, in its press release, claimed to be impartial and based only on facts. This emphasis on 'facts and figures' is another legacy of communist museum practice, even if this exhibition was deeply anti-communist and exposed the horrors and illegitimacy of state socialism in Romania. The display technique of didactic panels, crowded photographs and relics of communist propaganda produced that same intellectual dead-end known to visitors of communist museums.

The National Museum of the Romanian Peasant or 'the Peasant Museum' (Muzeul Ţăranului), as it is known, was the first museum to attempt to break with these old ways. The museum was re-established in 1990 in the building occupied for forty years by the Communist Party Museum. Only six years after its establishment, the novelty and courage of its new displays was internationally acknowledged with the

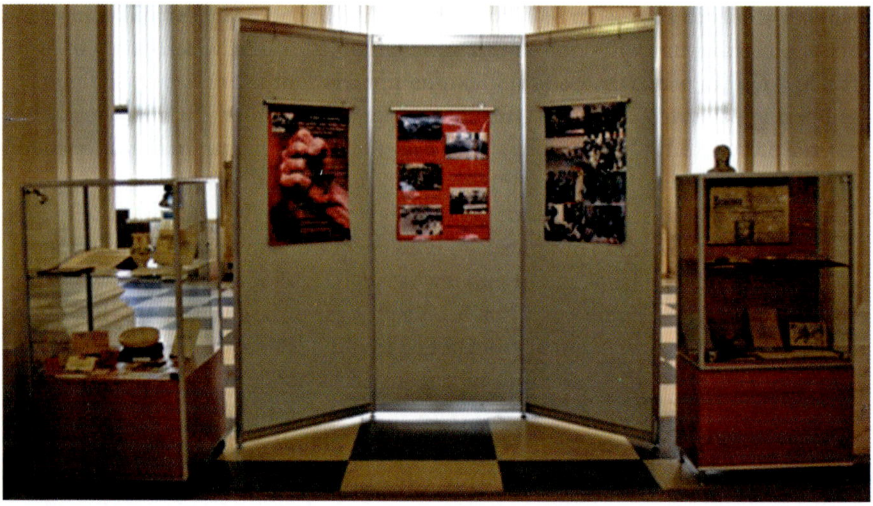

Figure 16.4 *Communism in Romania (1945–1989)*, exhibition at the National History Museum in 2007. This section exhibits the Sovietization of Romania. Photo: Muzeul National de Istorie a României

European Museum of the Year Award. I provide a more detailed discussion of the ways this museum has addressed the heritage of the communist museum in 'Raising the cross: Exorcising Romania's communist past in museums, memorials and monuments' (Cristea and Radu-Bucurenci 2007).

The Peasant Museum is a museum without labels, a museum of objects and a mother-museum; a museum of questions. Although Horia Bernea, the creator of this museum, admitted he was a keen reader of labels in all museums, there are few in his museum. 'The objects will be connected to one another through relations that the labels might disrupt' (Bernea 2003: 19). The visitor is informed that this lack of labels is not an omission on the part of museum professionals:

> We usually present collection pieces by indicating the name, the age, the author (when it is known), the place of origin, the way it was made and the way it functions. By doing this, we impoverish the testimony of the object which might have much more to say. The curiosity of the visitor towards the object is thus reduced to the same questions: what's the name of the object, how old is it, where does is come from. . . . Is it not possible that you come to the museum to freely confront the objects and let the spontaneous connections arise?
>
> (Bernea 2003: 89)

The Peasant Museum is the only museum in Bucharest to host a permanent exhibition on the communist past. In 1997, curators Horia Bernea and Irina Nicolau transformed a small room in the basement near the toilets into an exhibition room called *The Plague – Political Installation* (Figure 16.5). The only explanation given to the visitor is on the small notice at the entrance: 'A memorial of the pain and suffering collectivization caused to the peasant world.' It is an exhibition that appeals to your senses, very much unlike the exhibitions presented above. The upper part of the walls of the room are painted with red hammers and sickles, 'painted in oil on a strip of blue, they still look like blood drops' (Nicolau and Huluţă 2003: 132) while the lower part is covered with issues of the communist newspaper *Scânteia* (*The Spark*), bearing lists of peasants imprisoned and killed for resisting collectivization. Numerous busts of Lenin and Stalin facing each other or facing the walls, large pictures of Gheorghe Gheorghiu-Dej are crowded together on the same red-painted wall. The centre of the room is occupied by a huge, communist-style vase

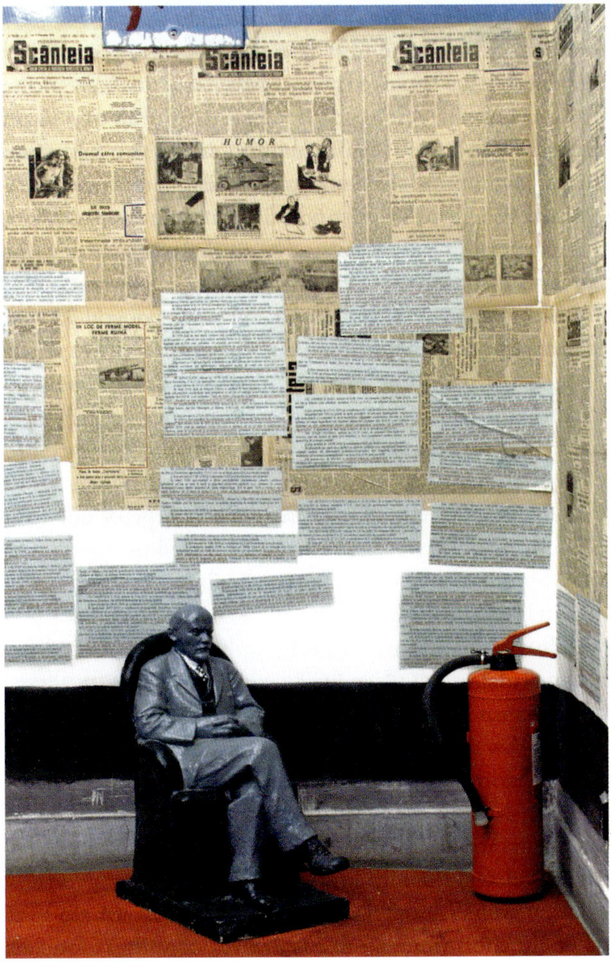

Figure 16.5 Image from *The Plague*, a permanent exhibition in the Romanian Peasant National Museum since 1997. Photo: Copyright Roald Aron and the Romanian Peasant National Museum

with the inscription: 'To comrade I. V. Stalin, a sign of love and gratitude from the Romanian Association for Tightening Relations with the Soviet Union.' To be able to read the inscription, one has to tour the room four times. Four huge ashtrays support the cord that surrounds the vase.

There are sharp discrepancies between this room and the rest of the museum. One is the special status of the objects in this exhibition. One of the strongest points of the new 'testifying museography' (Bernea 2003: 11) proposed by the Peasant Museum was dialogue with the objects: letting the objects speak for themselves, letting them conquer the space and find their most appropriate place in the display. Horia Bernea claims 'The Romanian Peasant Museum was born out of dialogue with the objects, an accepted, provoked, always attentive dialogue, without any preconceptions' (Bernea and Nicolau 1998: 225). In the case of *The Plague* exhibition, he confessed, they totally ignored this freedom of the object. The objects in this exhibition used to be exhibited in the Communist Party Museum. They were thus considered to be objects of a fake past, of an unreal reality. 'As opposed to the dialogue with the patrimonial objects, where I forbade myself any preconceived ideas, here we absolutely need a political bias. As we couldn't exhibit the lies of the regime, we tried to exhibit its ugliness' (Bernea and Nicolau 1998: 225).

Have the curators of this exhibition managed to bypass the paper paradigm that seems to dominate both communist and anti-communist Romanian museum displays? Have they eliminated paper from their exhibition? On the contrary, no, there is an impressive amount of paper (mainly newspapers) and written text in *The Plague*. However, it is displayed as a signifier of the era, not as an exhibit in itself. It is attached to the walls and it surrounds everything, just as it did in communist museums during state socialism. This approach presents a meta-discourse: a critical museum discourse not only on communist reality but also on communist museums.

The prison museum

Among communist museums of the pre-1989 era, there was one type of museum that had some objects on display. The prison museum that existed in most communist countries has the advantage of being itself a museum exhibit. A prison, where communist fighters were imprisoned and tortured, and where they continued their ideological fight in harsh conditions, was a perfect place for a communist museum. After 1989, it became the perfect place for an *anti*-communist museum. The Romanian variant of the prison museum is the Doftana Museum, a building 100 km from Bucharest, in a beautiful natural setting, which functioned as a prison from 1895 to 1940 and was transformed into a museum by the communist regime in 1949. The two main figures of Romanian communism were imprisoned there: Gheorghe Gheorghiu-Dej and Nicolae Ceaușescu. This, supposedly, was where they made each other's acquaintance. The 1940 earthquake affected Doftana and some prisoners were killed in this disaster. They later became martyrs of the communist faith.

The exhibition in Doftana was more vivid than those in other communist historical museums. Some of the cells were re-created; the isolation room and torture objects were on display, and guided tours were given to the visitors explaining the horror of life in Doftana and the struggle of the communist prisoners. There is some debate on whether the objects on display at Doftana – the handcuffs, the torture objects – were authentic prison objects or just brought there when the museum was created. This debate on the authenticity of museum displays also occurs in more recent anti-communist museums, such as Terror Haza in Budapest (Rév 2005: 298) or the Sighet Memorial in Northern Romania. The Sighet Memorial functioned as a museum of the prison where important political figures were exterminated in the 1950s but it has subsequently developed into a museum which aims to display every aspect of the communist regime (Cristea and Radu-Bucurenci 2007: 295–301). The memorial-museum currently has around eighty former cells, displaying subjects as varied as communist kitsch, the Prague Spring, everyday life in the 1980s, and the Bărăgan and Bessarabia deportations.

The visual discourse of these communist and post-communist prison museums is sometimes so strikingly similar. Figure 16.6 shows the re-creations of a cell in three such museums, the Sighet Memorial, the Doftana Museum and National

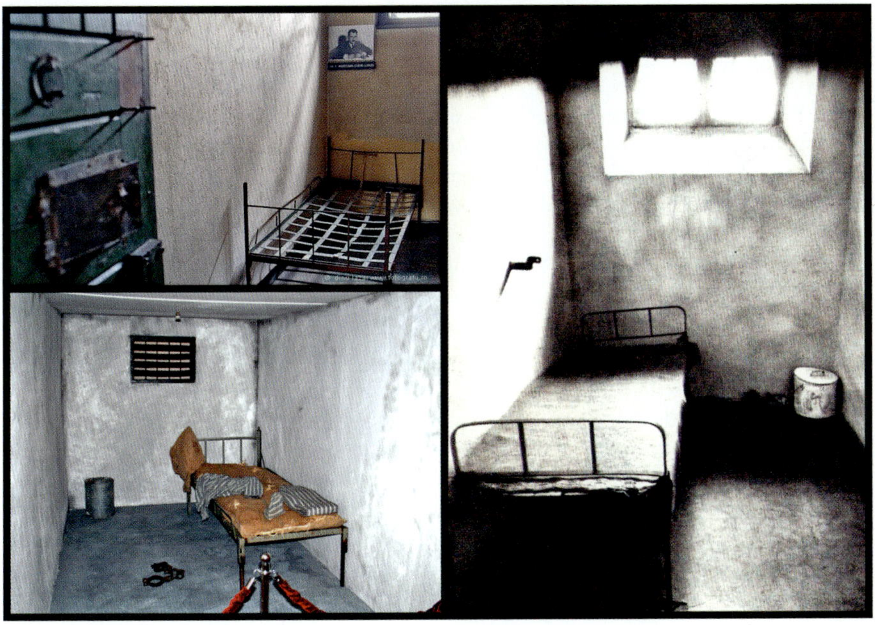

Figure 16.6 Re-created prison cells in three Romanian museums: Upper left: Sighet Memorial. Right: Doftana Museum. Lower left: National History Museum, part of the exhibition *Communism in Romania (1945–1989)*. Photos: Copyright Upper left: Dinu Lazăr; Right: Ardeleanu; Lower left: Muzeul Național de Istorie a României

History Museum. The label is the only obvious answer and solution to this conundrum. It is only the label, the text, the explanation that can lead the visitor's understanding in the 'right' direction. And thus, 'democratic' and anti-communist museums – museums dedicated to revealing the horrors of communism – meet their enemy using the same technologies. Was it not communist museology that advised, 'In order to help the visitor examine the exhibits in a correct manner and to draw the correct conclusions, texts-quotations and texts by the curators are introduced into the exhibition' (Galkina *et al.* 1957: 177)? Exhibitions dealing with communism during the communist regime and in post-communism are indeed – for the most part – 'the same exhibition with different labels'. The narrative and methodology is the same; only good and evil are reversed. The object – as the focus of free thought – is in these museum discourses, lost.

Notes

1 The formulation was suggested to me in 2007 by Vintilă Mihăilescu, Romanian social anthropologist and director (2006–2010) of the Romanian Peasant Museum.
2 'Vulgar sociologism' is a term coined in 1930s USSR in order to condemn the sociological approach to Marxism.
3 Ioana Popescu, unpublished discussions during the colloquium *Museums and Society* at the Romanian Peasant Museum, Bucharest, October 2006.

Bibliography

Apor, P. and Sarkisova, O. (eds) (2007) *Past for the Eyes: East European Representations of Communism in Cinema and Museums after 1989*, Budapest: CEU Press.

Ardeleanu, I. (1978) *Muzeul Doftana*, Bucharest: Meridiane.

Bernea, H. (2003) *Câteva gânduri despre muzeu, cantități, materialitate și încrucișare*, București: Liternet.

Bernea, H. and Nicolau, I. (1998) 'L'installation: Exposer des objets au Musée du Paysan Roumain', *Martor: The Romanian Peasant Museum Anthropology Review* 3.

Boym, S. (2001) *The Future of Nostalgia*, New York: Basic Books.

Cloșcă, C. (1972) 'Cîteva probleme legate de prezentarea muzeistică a istoriei contemporane locale', *Revista Muzeelor* 1: 34–36.

Cristea, G. and Radu-Bucurenci, S. (2007) 'Raising the cross: Exorcising Romania's communist past in museums, memorials and monuments', in Oksana Sarkisova and Peter Apor (eds) *Past for the Eyes: East European Representations of Communism in Cinema and Museums after 1989*, Budapest: CEU Press, pp. 274–303.

Dobrenko, E. (2008) *Stalinist Cinema and the Production of History: Museum of the Revolution*, trans. Sarah Young, Edinburgh: Edinburgh University Press.

Foucault, M. (1977) 'Nietzsche, genealogy, history', in D. F. Bouchard (ed.) *Language, Counter-memory, Practice: Selected Essays and Interviews*, Ithaca, NY: Cornell University Press, pp. 139–64.

Galkina, P. I., Gardanov, V. K., Ivanitkin, I. P., Miteaev, K. G., Novitki, G. A. and Plavilscikov, N. N. (1957) *Bazele Muzeologiei Sovietice* (Bases of Soviet Museology), Bucharest: Romanian Ministry of Culture.

Georgescu, F. (1974) 'Foreword', *Muzeul Naţional* (*The National Museum*) 1:1–2.

Georgescu, F. (ed.) (1975) *Muzeul Naţional de Istorie al Republicii Socialiste România*, Bucharest.

Groys, B. (1994) 'The struggle against the museum; or, the display of art in totalitarian space', in Daniel J. Sherman and Irit Rogoff (eds) *Museum Culture: Histories Discourses Spectacles*, London: Routledge.

History Museum of the Romanian Workers' Party (ed.) (1950s) *History Museum of the Romanian Workers' Party*, Bucharest.

History Museum of the Romanian Workers' Party (ed.) (1960) *Muzeul Doftana*, Bucharest: Editura Politică.

James, B. (1999) 'Fencing in the past: Budapest's Statue Park Museum', *Media Culture Society* 21: 291–311.

Lungu, I. and Leahu, D. (1982) 'Expoziţia "Preşedintele Nicolae Ceauşescu şi pacea lumii"', *Revista Muzeelor şi Monumentelor* 3: 3–16.

National Office for Tourism (n.d.) *Muzeele oraşului Bucureşti* (Museums of Bucharest), Bucharest.

Nicolau, I. (1996) 'Moi et le musees du monde: L'histoire d'une experience museale dans un pays de L'est', in *New Europe College Yearbook 1994*, Bucharest: Humanitas, pp. 15–44.

Nicolau, I. and Huluţă, C. (2003) *Dosar sentimental* (Sentimental dossier), Bucureşti: Liternet.

Nicolescu, C. (1979) *Muzeologie generală*, Bucharest: Editura Didactică si Pedagogică.

Pavel, A. (1972) 'Consideraţii privitoare la organizarea expoziţiilor de istorie contemporană', *Revista Muzeelor* 4: 306–08.

Rév, I. (2005) 'Underground', in *Retroactive Justice: Prehistory of Post-Communism*, Stanford, CA: Stanford University Press, pp. 240–305.

Sherman, D. J. (1994) 'Quatremère/Benjamin/Marx: art museums, aura, and commodity fetishism', in Daniel J. Sherman and Irit Rogoff (eds) *Museum Culture: Histories Discourses Spectacles*, London: Routledge, pp. 123–43.

Ştefănescu, A. (1984) *Ghidul Muzeelor* (Museum Guide), Bucureşti: Editura Sport-Turism.

Tarabukin, N. (1980) *Le dernier tableau*, Paris.

Voukov, N. (2007) 'The unmemorable and the unforgettable: museumizing the socialist past in post-1989 Bulgaria', in Oksana Sarkisova and Peter Apor (eds) *Past for the Eyes: East European Representations of Communism in Cinema and Museums after 1989*, Budapest: CEU Press, pp. 307–34.

Chapter 17

RE-THINKING KOREAN CULTURAL IDENTITIES AT THE NATIONAL MUSEUM OF KOREA

Sunghee Choi

I HAVE OFTEN TURNED TO the National Museum of Korea in Seoul as a kind of shelter to escape from daily stresses and let my spirit breathe (Figure 17.1). Viewing the masterpieces there, both for my studies and for recreation, enabled me to escape the worries of the outside world. Then, in 2001, when I started working in the curatorial department of the museum, my feelings about the museum changed. To my mind, there seemed to be quite fundamental problems with a visitor experience where museum professionals produced and delivered a strongly aesthetic product for public consumption. So much seemed to be unsaid in this museum; so many voices were excluded. Three years later, when I was studying art education at Pennsylvania State University, I came to believe that many of the problems found at the National Museum might be resolved by the adoption of a more open, forum-like mode of delivery, of the kind promoted by Lisa Roberts (1997) and Duncan Cameron (2004). This seemed to offer a more democratic vision of national culture by ultimately challenging the authority of institutions such as museums (Choi 2005, 2006). But then I hesitated. Could such a model really work in South Korea? Would it not undermine the National Museum's vital mission to convey the cultural, social and aesthetic identity of Koreans who now find themselves freed after a long period of brutal colonialism, war and military control?

I realized that there was an inherent tension between building a democratic vision of this kind and building the national identity of Korea. This chapter explores this tension and the contradictory desire that has resided inside of me, of which I was not initially aware. It is a record of my inner journey to national museums as a visitor and researcher and it deploys a methodology which needs a little explanation.

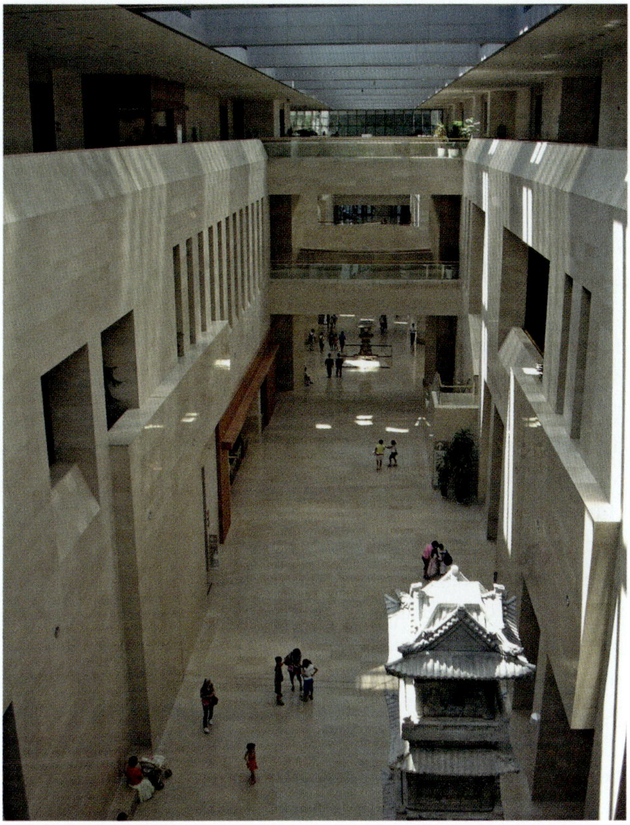

Figure 17.1 National Museum of Korea. Photo: author

Insider's perspectives: autoethnography

Many authors have commented upon and taken a position on the uncritical and politically ideological practices of some national museums, and particularly those that are attempting to escape a colonial past (Ashley 2005; Bennett 2006; Cvoro 2006; Dean and Rider 2005; Jones 1992; Macdonald 2003; Mason 2004; Rassool 2006). This chapter is also concerned with this so-called 'postcolonial' context. My interest is, however, in exploring the memories of the colonized because, while museum displays are constructed by museum professionals, it is only through the embodied experiences of individual visitors that these narratives acquire meaning. In order to locate these embodied experiences, the research here has applied autoethnography or 'insider ethnography' – a qualitative methodology in which the writer's own subjective autobiographical experiences are the focus of the research (Chang 2008; Ellis and Bochner 2003; Ellis 2004, 2009; Jones 2005; Spry 2006). Ellis (2009: 15) argues that autoethnographic reflection helps a researcher interpret how 'micro-events play out in and teach us about macro-structures and processes', thereby connecting the personal to the social, the cultural and the political. According to Spry (2006: 189), the autoethnographic method confers on the researcher the 'narrative authority' of experience, making him or her 'the epistemological and ontological nexus upon which the research process turns'. This approach enables me to deconstruct the voices, and undercurrent, of authority I have unconsciously granted to Korean national museums. It also enables me to discuss personal vulnerabilities without disguising them in a universal and objective language and without pretending that such experiences are neutral and objective. By this means, my personal experiences become 'a subject to look into', and at the same time, they become 'a lens to look through to gain understanding of societal culture' (Chang 2008: 49).

This research methodology is rather different from those which privilege 'personal context' or 'entry narrative' (Falk and Dierking 2000; Doering and Pekarik 2000). Personal contexts and entry narratives represent the sum total of personal and generic history that an individual carries with him or her into a meaning-making situation; it includes a visitor's motivations, expectations, prior knowledge and interests. Such personal histories are based on what visitors can express and recognize consciously. The lived experience of a visitor in my research refers to more than just an individual's consciously recognized experiences and thoughts; it includes ambiguous, uncertain and complex conditions of one's environment and understanding. Personal contexts or entry narratives are concepts from which museum professionals may draw assumptions in order to shape visitor experiences. Lived experiences are concepts felt or perceived by visitors themselves as they try to determine who they are, where they are and why.

This study, then, considers my lived experience as a visitor – my ambiguous and contradictory attitudes regarding the national museum – as the main source of research material. It is focused on my own experiences of the National Museum of Korea.

Korean sorrow: the imposition of identity

I remember I was taught as early as elementary school that the artistic essence of Korean artefacts lies in their embodiment of sorrow – 'Han'. When I came to study Korean literature as an undergraduate, I frequently heard the same discourse. I was taught that the aesthetic essence of sorrow is crystallized in the form of Korean white porcelain known as 'Bakja'. Initially, I believed this idea was common knowledge: everybody knows what 'Han' means to Koreans and how 'Bakja' represents the aesthetic beauty of Korea. I later found that this discourse of sorrow had been developed by a Japanese art critic, Muneyoshi Yanagi, in 1920, during the Japanese colonization of Korea and that it had been supported by many colonialists and the Korean public until quite recently. I then began to question how this sense of sorrow could represent the artistic essence of Korean artefacts when court paintings had been produced to document court ceremonies, rituals, or banquets and to decorate the courts' interiors. How could we Koreans generate those brilliant colours, energetic actions, and ambitious visions if it was sorrow that was being represented? Later still I discovered that Yanagi's logic could be understood as a kind of Orientalism; his definition of Korean identity became the aesthetic of sorrow: the sad, the hopeless and the helpless (Brandt 2000).

When one is asked to explain his or her personal identity, the easiest and most effective way to respond is to differentiate oneself from others through one's uniqueness. If such questions are posed under a power structure in which the self has only limited powers of response, then the questions may play a role in controlling the self. If one cannot respond to these inquiries adequately, then the questioned may – from the questioner's perspective – simply disappear. This can occur because the inquirer preconditions the uniqueness of the one being questioned in terms of how this person or people can be distinguished from the empowered questioner.

One can apply this thinking to Koreans under Japanese colonial rule. At that time, Japan tried to obliterate Korean identity by assimilating it into that of the Japanese, consistently arguing that Korea had no culture to differentiate it from Japan. If Korea had any culture at all, it was inferior to Japan's and, accordingly, of no significance. Even under this contempt, Muneyoshi Yanagi praised Korean art forms, especially Joseon white porcelain, defining, as I have discussed, its beauty as sorrow (Yanagi 1972). At that time, Koreans needed such cultural forms in order to distinguish themselves from the Japanese. The Japanese colonial framework – namely its power to decide inclusion and exclusion – created this logic; thus, Koreans incorporated the logic of others in order to define themselves. They defined themselves with others' words.

Defining the self by borrowing the words of others is hazardous, not because the definition comes from the framework of others but because it is a reaction to the actions of others: the reaction is not created autonomously according to the need of the self but is pre-framed by the action. The self's identity is changed at the moment that one chooses to respond to the pre-framed question, even though one is not aware of the question's capacity to transform.

Pratt (1992) demonstrates well how the colonized cannot avoid incorporating others' words when defining themselves and how this incorporation is developed, thereby transforming their original identity into a cultural mixture. Such amalgamation can be seen in Alger and Welsh's (2000) work with Native American teenagers. Alger and Welsh provided a hybrid space in an art museum in an attempt to empower these teenagers by making their voices heard. In this hybrid space, one of the subjects of the study realized that the only way to voice her opinions was to use Anglo-American communication, and not her own. She finally realized that this was the easiest and most effective way, even though this might end up changing the communication of her people. In this kind of power structure, the identity of the marginalized shifts and adapts to the dominant cultural system. In this way, identity construction of the colonized responds to, and is interwoven with, that of the colonizers because it is formed through a chain of actions and reactions, of shaping and being shaped.

Korea's cultural identity has in part been developed through this 'reactionary system'. Koreans are endlessly discussing the cultural uniqueness of Korea, even though the original identity – or the uniqueness that Koreans are endlessly talking about – has been transformed into a complex hybrid from the moment Korea responded to the colonizer's question. The root of this kind of thinking is a belief that emphasizing the authentic culture of Korea, produced and consumed before the colonial period, would compensate for the lost (and negated) history of Korea under Japanese rule. However, this type of thinking is itself an imposition of the Japanese regime, for it denies Koreans the possibility of possessing, or being aware of the reality of, ever-changing identities. Colonialism entails not just a clear definition of the self and Other but also the negation of the ambiguous definition of the middle ground between the two, which is inscribed into the memories of the colonized both during and after the colonization period.

Speaking the language of a national museum

Due to the limited hours permitted for tours, we could not stay in the archaeology gallery we had first entered. I asked my companions, two American professors visiting Korea, to move up to the third floor to see the fine art gallery, which displays traditional paintings and the internationally famous ceramics collections, because my expertise in art history lies in the traditional paintings of the Joseon Dynasty. I also assumed that my guests might be interested in art rather than archaeology because their expertise also lies in art, history, and education. We rode an escalator up to the fine art gallery, which, in contrast to the crowded atmosphere in the archaeology gallery, seemed to be very calm and dark. We enjoyed the tranquillity. At first we were the only people there, but then a few others arrived.

Amongst the objects we viewed was an eighteenth-century folding screen which documented the ceremonies of the royal court. At first, I explained what the subject matter of the scenes were – what the meaning of a ceremonial ritual for the royal family was, for example – and then I explained how the events described in the folding screen were reliable and verified by historical facts. I explained that the folding screen is not an imaginary painting but a documentary painting. And then, I gave a lengthy explanation of how the Joseon Dynasty, the precedent of Korea, had developed its artistic visions as a means of self-discipline, keeping the ethos of neo-Confucianism, which embodied its moral principles in political praxis and limited the articulation of artistic development beyond the implementation of self-discipline. I explained that both China and Japan were followers of neo-Confucianism, but that both nations had developed their artistic forms significantly beyond the boundary of self-discipline, thus breaking the ethos of neo-Confucianism. Yet Korea did develop the unique tradition of documentary paintings, which maintained the ethos of neo-Confucianism as self-discipline. I added that the number of participants in that celebration – in addition to the number of items, such as dishes, vases, tables, or presents at that celebration – were exactly identical to that recorded in published historical books. Finally, I commented that the documentary paintings and the written documentary records were unique to the Joseon Dynasty, which illustrated the essence of neo-Confucianism well. After I finished my talk, one of the professors told me that I looked very different from when I was in her office in the United States. She said that I looked very confident. I just shrugged my shoulders in response to her unexpected comment, and we went on.

Although this story reflects my fundamental academic interests, it also reflects my belief that the cultural essence for explaining Korean identity lies in fine art, and in paintings and ceramics, rather than in archaeological artefacts. However, on a second level, this story shows that I picked a specific approach to explain the documentary painting we first encountered at the fine art gallery: I was obsessed with explaining *the cultural essence of Korean identity* to Others. This was the first time I had been put in this position and my response was prompt, unexpected, and unplanned; something that resided in me had simply motivated me to act in this way. I could have chosen to take a formalistic approach to the documentary paintings, I could have introduced this particular painting with an interesting anecdote, but I chose to compare three countries in order to emphasize the uniqueness of Korean cultural identity. I did not even notice the subtle change in myself – my increased self-confidence. I probably wanted to give a brief but strong impression of Korean cultural traditions to cultural outsiders during the short amount of time given to me. While the Chinese and Japanese cultures are relatively familiar to most Westerners (both effectively being world cultures), Korean culture is much less known. To raise the importance of Korean cultural achievements and clearly differentiate Korea from other countries, I seemed to have chosen the most effective way to demonstrate our cultural advancement and differences. My discussion of the screen may have met those goals, thus making me more confident than before.

There is another intriguing factor that needs to be noted. The shaping of my choices might have also been reinforced by the narratives of the National Museum itself. The National Museum produces context-free displays, focusing on accentuating the visual value of objects rather than facilitating visitors' understanding of them. The fine art gallery takes this aesthetic approach to the extreme. Each object here is removed from its original context. Descriptive wall texts and object labels provide the title, period, material, and technique without any further visual cues or questions to stimulate a reaction. To highlight their visual characteristics, the museum uses spotlights on specific objects. Curators or exhibition designers seem to want artistic objects to reveal their own power by achieving a visual purity through the removal of any visual disturbances or too much interpretative intervention (Vergo 1994). It is an approach widely used to create spaces for contemplation (Noordegraaf 2004). If I had not pursued my personal interests in learning about traditional Korean art history, it might have been difficult for me to identify what the subject matter of this gallery was at a first glance, even though the wall texts provide a lengthy and detailed description concerning the objects and their contexts.

This focus on the visual quality of objects as art works is also incorporated in the archaeology gallery of the National Museum of Korea (Figure 17.2). Even though I could see secondary materials, such as maps, timelines, and pictures to help

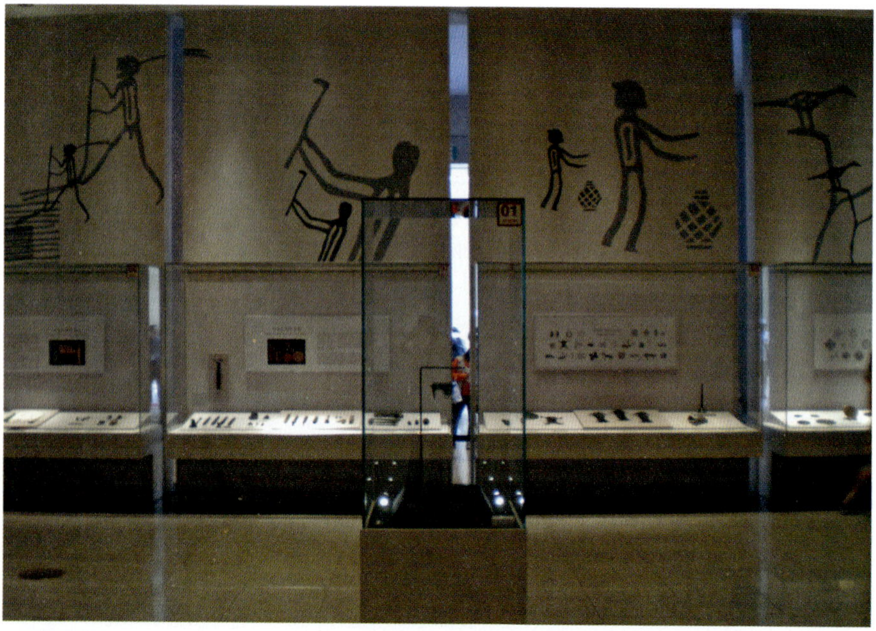

Figure 17.2 Aesthetics in the archaeological gallery of the National Museum of Korea. Photo: author

visitors' contextual understanding in the archaeology gallery, I was still unable to see any step-by-step approaches or interpretative devices to help visitors understand the content. Instead, I saw the same aesthetic approach. These galleries induced visitors to stand as uninstructed 'vacant gazers' under the illusion of a clear and transparent vision for placing things into a single, ordered historical understanding, as Tony Bennett puts it (2004: 161). Here the object can be sanctified through deferential and reverential observation.

So, what did the National Museum of Korea intend to show through their 'narratives'? I found one possible answer in the introduction to the museum's exhibition strategy provided on their website:

> The exhibitions in the National Museum of Korea are designed for both domestic visitors and foreigner visitors to systemically identify the characteristics of the Korean cultural tradition among the world cultures. . . . To make visitors enjoy the exhibition more easily and more interestingly, we incorporate various techniques for exhibition and high technology systems for museum tours.[1]

It would seem that when I introduced this Korean cultural tradition to these foreign visitors by explaining the objects' unique characteristics and comparing Korean traditions with those of China and Japan, I actually internalized intentions written into the displays. In a sense, it may seem like I created my own explanation for the object; however, my choice was actually shaped by the interpretative approach of the museum. My confidence may have resulted from my weaving of the museum's interpretative approach into my lived experiences because the goal of the National Museum is to make Korean visitors feel proud of their culture.

At a micro level, then, it seemed that the National Museum of Korea succeeded in shaping me and my memories by providing me (and other visitors) with only a few available ways to interpret the objects. Because my memories were shaped covertly by the indirect input of the general discourse of Korean cultural identity and directly by the narrative of the National Museum which emphasized Korea's uniqueness, I seemed to voluntarily follow the path the museum makes available to visitors. However, on a macro level, the National Museum also seems to follow the path that Japanese colonialism made available to Korea beginning in 1920, which pushes Koreans to define themselves in terms of qualities which can be differentiated from others.

In the third space

Criticizing the modernistic assumptions that current museums are based on, Hooper-Greenhill (1992) deconstructed the dominant cultural codes then present in museums, using the framework of Michael Foucault's effective history. She aspired to liberate visitors from the legacy of modernity and enable them to rewrite their own meanings into the narratives of museums: 'Effective history teaches us that, because meanings and interpretations are endlessly rewritten, we too can seize the opportunity to make our own meaning, and find our own relevance and significance' (Hooper-Greenhill 1992: 215). However, Spivak (1999) argues that the oppressed may not be inclined to speak because speaking itself summons vulnerable or painful memories. Speaking would not be rewarding if these individuals are not provided with a safe space in which their painful memories can be cured. There are, then, good reasons for them to remain silent: they are smart not to speak. By not speaking they strengthen their self-protection, even if their choice will not free them or improve their condition in the end. However, this is a tacit choice and not necessarily the best one. Ideally, they should break their silence, but for them to do so effectively they would need to deconstruct the hegemonic discourse they embody; without this they would be oppressed again.

For these reasons, the colonized or the marginalized need a safe space where they can tell their stories freely without being hurt; where they can be respected even if their stories are humble or ambiguous. That safe space legitimizes memories of self-humiliation and the experiences of self-negation. Pratt (2000: 17) calls this space the contact zone, saying that 'Where there are legacies of subordination, groups need places for healing and mutual recognition, safe houses in which to construct shared understandings, knowledge, claims on the world that they can then bring into the contact zone.' In this safe place, visitors can empower themselves by defining themselves with their own words and concerns.

In particular, for people whose colonial legacies are hidden in memories, a subversive space is needed to deconstruct such binary oppositions; this is known as 'the third space'. Bhabha first conceptualized the third space as a self-contradictory space where participants can identify the moment that colonizers become the colonized and the colonized also become the colonizers, complicating 'the binary division of past and present, tradition and modernity, at the level of cultural representation and its authoritative address' (Bhabha 1994: 51). In the third space, the clear boundary between the colonizer and the colonized disappears. For example, my defining myself as being culturally mixed is a way for me to empower myself because my contradictory behaviours are embraced as evidence of cultural mixing and are not negated or ignored: my contradictory behaviours become the objects of inquiry. In this third space, I do not need to develop my logic to survive but to understand the present, enliven it, and communicate with it.

This understanding is possible because the third space is sheltered from 'cultural differences' in order to problematize the issue of cultural authority. It runs counter to the recent museum obsession with cultural diversity as a means to mitigate conflicts between the postcolonial powers and the 'post-colonized'. Bennett (2006: 61) criticizes this phenomenon and incorporates the term, 'zoological multiculturalism' from Ghassen Hage (1998). The two concepts of 'cultural diversity', without deconstructing power, and 'democratization', as increasing choices without investigating political agendas, can be used to justify the moral ethos of social dominance (Ybarra-Frausto 1991; Macdonald 1998). A forum-like museum cannot embrace the current, complicated geopolitical realities of global societies, especially in Korea, which was once colonized and is still suffering from the legacy of colonialism. The legacy of suffering has not disappeared but is embodied through individuals' memories. In a forum-like museum, those who have experienced self-negation personally and collectively cannot empower themselves beyond confirming 'cultural diversity' and are stuck within 'liberal notions of multiculturalism', as Bhabha (1994: 50) states.

Through the act of speaking in the third space, the National Museum of Korea could reiterate its reality as a cultural mixture, where our being – our Korean identity – would not be negated by our failure to identify with a pure and authentic culture. Rather, Koreans could rewrite their being as a 'visual narration of cultural negotiation', as Ybarra-Frausto (1991: 147) puts it, going beyond regenerating the chain of colonialism of the shaper and the shaped.

To resist the path that others have shaped for us, we need to choose another path. This new path would not be in the completely opposite direction from that of the Others because the opposite way is merely another name for the same path through which the Others shaped us. This new path would be a third path, deconstructing the boundary between the two. As a practical way to look for this third path, this chapter concentrates on one obscure and unorganized moment, which was contradictory and trivial among my experiences in the National Museum of Korea. After that experience, I was able to investigate how the moment was connected to the public discourse on Korean cultural identity, to the physical and conceptual narratives of the National Museum of Korea, and, finally, to my personal histories.

This new experiment was possible because narrative-making – the writing of this kind of autoethnographic text – has the power to engage writers (or tellers) in reflecting on themselves and making sense of the world around them. Narrative-making is a way for one to admit the raw source behind herself or himself – emotions, ideas, impressions or thoughts – positively, and thereby to nurture his or her body: the raw source is integrated, amplified, and, finally, embodied through narrative-making. As Bruner (1996) suggests, narrative-making enables one to transform complex, ambiguous, and chaotic encounters with contexts into manageable or understandable accounts. He suggests that through the narrative-making processes a person can 'convert the raw Trouble into a manageable Problem that can be handled with procedural muscle' (Bruner 1996: 99). According to Clandinin and Connelly (2000), the narrative-making process is one way to transform what one finds puzzling, ambiguous, uncertain or complex into something more orderly. Narrative-making has therapeutic power (Pennebaker 1997).

However, this narrative-making process is not possible until each individual can actively engage her or his present life with the past and the future. In other words, narrative-making enables people to 'tell remembered stories of [themselves] from earlier times as well as more current stories. All of these stories offer possible plotlines for [their] futures' (Clandinin and Connelly 2000: 60).

This process requires two conditions: revealing oneself and weaving one's individual personal narratives with the social, cultural and political narratives of museums. The relocation of my personal experiences into the National Museum of Korea required me to recognize and navigate the sources of power that already constitute the relationship between me and the museum, thereby enabling me to rewrite the structure of power.

If the National Museum of Korea is interested in visitor empowerment and is actively seeking alternative practices other than those that result from Japanese colonization, a good way to do so would be to encourage visitors to reveal their personal agendas and connect their personal thoughts and ideas with the museum's narrative by providing a third space, both conceptually and physically, within the National Museum of Korea. Providing such a space would open up a new possibility to help visitors build self-reflections on their museum-going experiences, in turn helping them contemplate their lived experiences, which may appear to be obvious but might also include contradictory factors. This empowerment process might be initiated by museum professionals as a top-down approach toward a shared authority between national museums and visitors, but it could also include numerous bottom-up approaches to help visitors envision their future positively in the long run. These numerous bottom-up approaches, however contradictory or obvious they may be, would be seeds to enliven the democratic vision of national museums and finally nurture it.

Note

1 Introduction of Exhibition Strategies (n.d.) The National Museum of Korea, <http://www.museum.go.kr/kor/exh/exh_arc_mai.jsp?orgmenu=2> (accessed 29 September 2008).

Bibliography

Alger, A. B. and Welsh, P. H. (2000) 'Creating hybrid space: The native American high school student guide program at the Heard Museum', *The Journal of Museum Education,* 25(3): 14–19.

Ashley, S. (2005) 'State authority and the public sphere: Ideas on the changing role of the museum as a Canadian social institution', *Museum and Society*, 3(1): 5–17.

Bennett, T. (2004) *Pasts beyond Memory: Evolution, Museums, Colonialism,* London: Routledge.

Bennett, T. (2006) 'Exhibition, difference, and the logic of culture', in I. Karp, C. A. Kratz, L. Szwaja and T. Ybarra-Frausto (eds) *Museum Frictions: Public Cultures/Global Transformations,* Durham, NC: Duke University Press, pp. 46–69.

Bhabha, H. K. (1994) *The Location of Culture,* London: Routledge.

Brandt, K. (2000) 'Objects of desire: Japanese collectors and colonial Korea', *Position,* 8(3): 711–46.

Bruner, J. (1996) *The Culture of Education*, Cambridge, MA: Harvard University Press.

Cameron, F. D. (2004) 'The Museum, a temple or the forum', in G. Anderson (ed.) *Reinventing the Museum: Historical and Contemporary Perspectives on the Paradigm Shift,* Walnut Creek, CA: Alta Mira Press, pp. 61–73.

Chang, H. (2008) *Autoethnography as Method*, Walnut Creek, CA: Left Coast Press.

Choi, S. (2005) 'Visitors or objects: Attitudes of museum educators to museum education in Korea', unpublished master's thesis, Pennsylvania State University.

Choi, S. (2006) 'The visitor-centred museum: Interpretation and pedagogy', *Journal of Research in Art and Education,* 7(2): 121–38.

Clandinin, D. J. and Connelly, F. M. (2000) *Narrative Inquiry*, San Francisco: Jossey-Bass.

Cvoro, U. (2006) 'Monument to anti-monumentality: The space of the National Museum of Australia', *Museum and Society*, 4(3): 116–28.

Dean, D. and Rider, P. E. (2005) 'Museums, nation and political history in the Australian national museums and the Canadian Museum of Civilization', *Museum and Society,* 4(3): 35–50.

Doering, Z. and Pekarik, A. (2000) 'Questioning the entrance narrative', *The Journal of Museum Education,* 21(3): 20–23.

Ellis, C. S. (2004) *The Ethnographic I: A Methodological Novel about Autoethnography*, Walnut Creek, CA: Alta Mira Press.

Ellis, C. S.(2009) *Revision: Autoethnographic Reflections on Life and Work*, Walnut Creek, CA: Left Coast Press.

Ellis, C. S. and Bochner, A. P. (2003) 'Analyzing analytic autoethnography: An autopsy', *Journal of Contemporary Ethnography,* 35(4): 429–49.

Falk, J. H. and Dierking, L. D. (2000) *Learning from Museums: Visitor Experiences and the Making of Meaning*, Walnut Creek, CA: Alta Mira Press.

Hage, G. (1998) *White Nation: Fantasies of White Supremacy in a Multicultural Society,* Annandale, MA: Pluto Press, quoted in Bennett, T. (2006) 'Exhibition, difference, and the logic of culture', in I. Karp, C. A. Kratz, L. Szwaja and T. Ybarra-Frausto (eds) *Museum Frictions: Public Cultures/Global Transformations,* Durham, NC: Duke University, pp. 46–69.

Hooper-Greenhill, E. (1992) *Museums and the Shaping of Knowledge*, London: Routledge.

Jones, J. P. (1992) 'The colonial legacy and the community: The Gallery 33 Project', in I. Karp, C. M. Kreamer and S. D. Lavine (eds) *Museum and Communities: The Politics of Public Culture,* Washington, DC: Smithsonian Institution, pp. 221–41.

Jones, S. H. (2005) 'Autoethnography: Making the personal political', in N. K. Denzin (ed.) *The Sage Handbook of Qualitative Research,* 3rd edn, Thousand Oaks, CA: Sage Publications, pp. 651–79.

Macdonald, S. (1998) 'Supermarket science?: Consumers and the public understanding of science', in S. Macdonald (ed.) *The Politics of Display: Museums, Science, Culture,* London: Routledge, pp. 119–38.

Macdonald, S. (2003) 'Museums, national, postnational, and transcultural identities', *Museum and Society,* 1(1): 1–16.

Mason, R. (2004) 'Nation building at the Museum of Welsh life', *Museum and Society*, 2(1): 12–34.

Memmi, A. (1965) *The Colonizer and the Colonized*, Boston: Beacon Press.

Noordegraaf, J. (2004) *Strategies of Display: Museum Presentation in Nineteenth- and Twentieth-century Visual Culture*, Rotterdam: NAi Publishers.

Pennebaker, J. W. (1997) 'Writing about emotional experiences as a therapeutic progress', *Psychological Science*, 8(3): 162–66.

Pratt, M. L. (1992) *Imperial Eyes: Travel Writing and Transculturation*, London: Routledge.

Pratt, M. L. (2000) 'Arts of the contact zone', in J. M. Wolff (ed.) *Professing in the Contact Zone: Bringing Theory and Practice Together,* Urbana, IL: National Council of Teachers of English, pp. 1–18.

Rassool, C. (2006) 'Community museum, memory politics, and social transformation in South Africa: Histories, possibilities and the limits', in I. Karp, C. A. Kratz, L. Szwaja and T. Ybarra-Frausto (eds) *Museum Frictions: Public Cultures/Global Transformations,* Durham, NC: Duke University, pp. 286–321.

Roberts, L. C. (1997) *From Knowledge to Narrative: Educators and the Changing Museum,* Washington, DC: Smithsonian Institution.

Smith, C. S. (1997) '"In a new light": Issues in the display of portraits at the National Portraitist Gallery, London', *Museum Management and Curatorship,* 16(4): 371–82.

Spivak, G. C. (1999) *A Critique of Postcolonial Reason: Toward a History of the Vanishing Present,* Boston: Harvard University Press.

Spry, T. (2006) 'Performing autoethnography: An embodied methodological praxis', in S. N. Hesse-Biber and P. Leavy (eds) *Emergent Methods in Social Research,* Thousand Oaks, CA: Sage, pp. 183–212.

Vergo, P. (1994) 'The rhetoric of display', in R. S. Miles (ed.) *Towards the Museum of the Future: New European Perspectives,* London: Routledge, pp. 149–59.

Vogel, S. (1991) 'Always true to the object, in our fashion', in I. Karp and S. D. Lavine (eds) *Exhibiting Cultures: The Poetics and Politics of Museum Display,* Washington, DC: Smithsonian Institution, pp. 191–204.

Yanagi, M. (1972) *Unknown Craftsman: A Japanese Insight into Beauty*, Tokyo: Kodansha International.

Ybarra-Frausto, T. (1991) 'The Chicano movement/the movement of Chicane art', in I. Karp and S. D. Lavine (eds) *Exhibiting Cultures: The Poetics and Politics of Museum Display,* Washington, DC: Smithsonian Institution, pp. 128–150.

Chapter 18

THE AESTHETICS AND NARRATIVES OF NATIONAL MUSEUMS IN CHINA

Marzia Varutti

A VISIT TO SHANGHAI MUSEUM is an experience designed to please the senses. Situated on one side of Shanghai's majestic People's Square, its stunning architecture, symbolising the round sky over a square earth of Chinese cosmology, is as distinctive as any of the world's great museums (Figure 18.1). Inside, one enters a marbled hall filled with light and marked by sharp architectural lines. Beyond it, shadowy, intimate and almost secretive, the galleries entice the visitor. In these exhibition rooms the soft light is complemented by velvet furnishings and carefully placed crystal cases, their treasures disclosed in an atmosphere that is still and sumptuous, almost sacred. The objects themselves are placed in pools of light, which sculpts their form and profile. Against muted backdrops, the objects emerge as absolute and discrete, possessing a material potency that seems to owe little to time and space. The objects are silent, eloquent only in their beauty, with nothing to distract the attention away from this encounter with the masterpiece. By these means, the museum seduces the senses and forces the visitor to suspend any remnant of disbelief. By these means, exhibitions at the Shanghai Museum use the wordless language of beauty to communicate the magnificence of Chinese civilisation.

Although the Shanghai Museum is likely to remain unsurpassed among Chinese museums for its architectural grandeur and quality of its collections, it is by no means isolated in its choice of a museological model which emphasises and invites an aesthetic approach to objects. Since its opening in 1996, an increasing number of national museums in China have followed its lead by reorganising their displays in ways that give a renewed importance to aesthetics. While this is a trend seen in other parts of the world, its presence in a country which only a few decades ago framed its collections in the rhetoric of Marxist-Leninist social theory deserves closer investigation. This chapter explores this phenomenon in China's national museums but in doing so also reflects more generally on this museological trend

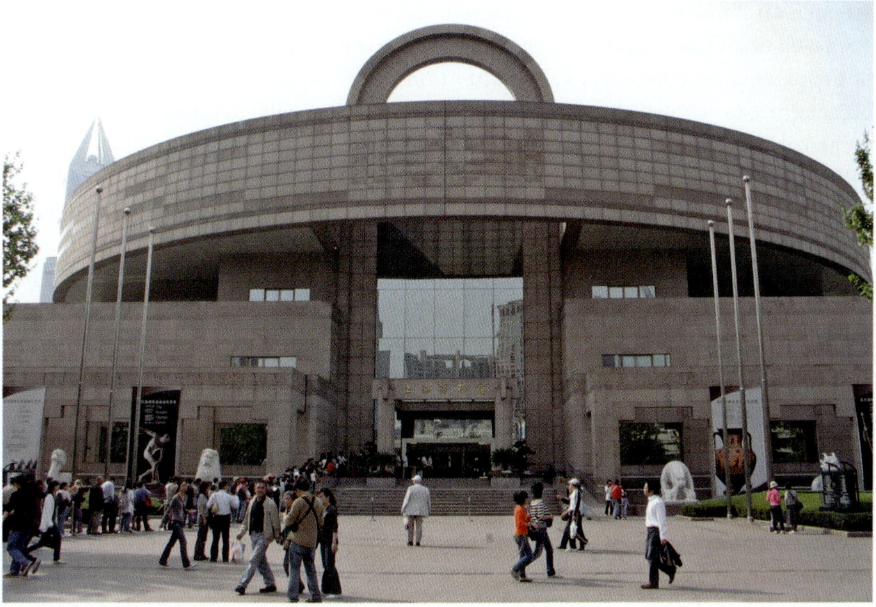

Figure 18.1 The Shanghai Museum. Photo: author

(Clifford 1997, 2007; Errington 1998; Naguib 2004; Price 1989, 2007). Its aim is to contribute to understanding the ways in which national museums throughout the world are reshaping their narratives of the nation in response to contemporary social phenomena, such as transnational social mobility and multiculturalism which seem to challenge the very notion of national identity. Discussion here will focus on the Shanghai Museum, Beijing's National History Museum, Military Museum, Capital Museum and China Millennium Monument, and the Sanxingdui Museum in Sichuan. It is based on direct observation and interviews with museum academics and professionals, as well as the analysis of ephemeral media.

The unifying aesthetic

The aesthetic focus here is not on beautiful things, but on the act of bringing out the aesthetic qualities of objects, using specific display techniques which draw the attention of the viewer to the object's formal characteristics by inviting an appreciative and contemplative approach. If one accepts that the aesthetics of objects are historically and culturally specific (Schaeffer 2004; Descola 2007), then there is good reason to examine systems of representation and viewing which exploit aesthetics (Shanks and Tilley 1987: 68) in order to create meaning and deploy it within narrative. The aesthetics of a museum exhibition can, in these circumstances, be understood not to arise from the intrinsic properties of objects but from institutional discourse (Vickery 2007: 227). The approach seems to adopt what

Stephen Greenblatt (2004: 546) refers to as 'wonder' displays: 'the power of the displayed object to stop the viewer in his or her tracks, to convey an arresting sense of uniqueness, to evoke an exalted attention'. This capacity is not restricted to art objects; it can involve the isolation of any object in order to admire its material, technical and formal characteristics.

As to the intentions of such exhibitionary acts, Donald Preziosi (Chapter 3, this volume) is in no doubt: 'museums exist in the first place to manufacture belief'. What beliefs, then, are being manufactured in the aesthetics of Chinese museums?

One particular intention or implication of the approach has been the ability to transcend historical, cultural and technical specificities. For example, in the temporary exhibition, *Imperial Treasures from China,* held at the China Millennium Monument in Beijing in November 2004 to promote the 2008 Olympic Games, bronzes from the Shang Dynasty (1500–1050 BC) were displayed next to calligraphy from the Qing Dynasty (AD 1644–1911). Aesthetic display creates a sense of coherence and continuity out of the heterogeneous and fragmentary. In the national museum this interlinking and conflation of time produces an abstract sense of antiquity, beauty and craftsmanship which projects a nation defined by longevity, refinement and ingenuity. By this particular means of objectification, an equation is established between museum object, culture and nation.

The aesthetics of tradition and progress

Chinese artistic traditions are known for their refined and elaborate aesthetic canons which draw on such ideas as balance, harmony, liveliness, spontaneity, and propriety. Karl-Heinz Pohl (2008: 90) has remarked that 'these features served, well into the modern period, as fundamental elements of a Chinese cultural identity'. Is it possible, then, that the display and reception of Chinese material culture in modern-day national museums is informed by these aesthetic principles; that these museums are primarily exhibiting a deeply embedded continuity of thinking and feeling?

Chang Wan-Cheng argues convincingly that in the Chinese cultural system, objects tend to be considered important for their cultural, symbolic and ethical values. In other words, the values that can be transmitted through objects and heritage are more important than their material form (Chang 1999: 86). Chang suggests that value is intimately linked to the notion of tradition (*chuantong*), and ideas of transmission (*chuang*) and relation (*tong*). The valuable object contributes to the transmission of fundamental values for the purposes of cultural continuity and reproduction (Chang 1999: 86). In the course of China's long history, valued objects and artistic treasures have been subject to varying 'regimes of visibility': alternately protected from general view as private and personal, and then publicly exposed as emblems of cultural refinement and political authority. In both settings, access to the object had important social and political meanings. For example, during the Shang (1500–1050 BC) and Zhou Dynasties (1050 BC–AD 475), ritual bronzes were exposed to public view in temples to demonstrate an individual's links to ancestors and so legitimise his or her political authority. Conversely, the private art collections that flourished among aristocratic and intellectual elites during the Song Dynasty (AD 960–1279) were mostly kept private, if not secret, reserved for the gaze of the collector himself and few others (Capon 1977: 16; Clunas 1997: 48). In this case, the aesthetic object was meant to enhance the pleasure of life and be a source of personal delight.

The museum entered Chinese culture much later, the first being created in 1905 in Nantong, not far from Shanghai, by Zhang Jiang, a local politician, reformist and entrepreneur. The Nantong Museum fully appropriated the rhetoric and organisational structures of museums founded in foreign settlements in Shanghai at the end of the nineteenth century. It possessed that comprehensive aspect first developed in European museums which systematically arranged collections of artistic, archaeological, historical and natural science material. By these means it possessed new notions of order and hierarchy which incorporated a new progressive system of knowledge. It was, then, not simply a reflection of China's glorious past but a statement of the wealth of the nation, its moral and intellectual stature, its technological potential and scientific achievement. It aimed to educate its public about the Chinese nation through display (Claypool 2005: 569–70). As such, it marked a shift away from symbolism and aesthetics towards science and civic education. It permitted the museum to perform the narrative of the Chinese nation in a language that possessed Western scientific authority.

The opening of the Nantong Museum contributed to cultural reforms that would see China enter the world stage. Following this museum's lead, the first national exhibition, in Nanjing in 1909, combined aesthetic exhibits which should 'please the eye and warm the heart', with accessibility, education and scientific rigour – displayed products would be 'understood at a glance' (Fernsebner 2006: 109).

The creation of the National Museum of Chinese History in 1912, like other museums in the first years of the Republic of China, contributed to the establishment of knowledge as a civic right and a public good. Adoption of the *public* museum also paved the way for the dismantling of the imperial collections. The transformation of the Imperial Palace into a public museum in 1925 was a symbolically charged gesture: the former residence of the Emperor, his personal belongings and the most intimate aspects of his daily life were now exposed to everyone's gaze and judgement. The Palace Museum, through its appropriation of imperial accoutrements, made manifest the power of the state to desacralise the Emperor's persona and to delegitimatise his rule (Hamlish 2000). In this reformist setting the granting of public access to elitist art, and all that it stood for, was perceived as an act of modernisation on a new progressive path for the Chinese nation.

Communist aesthetics

Under communist rule, the values attributed to museums, to their objects and to aesthetic appreciation were subject to significant redefinition. In his famous 1942 speech, *Talks at the Yan'an Forum on Art and Literature*, Mao Zedong promoted the primacy of revolutionary ideals over artistic expression. After the establishment of the People's Republic of China on 1 October 1949, museums were shaped by Maoist principles. There was little room for aesthetic appreciation, now considered a bourgeois activity and an elitist prerogative. In accordance with the principle of historical materialism, museum objects were reconceived as scientific evidence and material proof of social theories (Su 1995). Museums became utilitarian – they served a social function – and objects were required to 'teach lessons'. China's new national museums, such as the National History Museum and the Military Museum, built in the 1950s, were based on these principles. Their collections were (and in many instances still are) centred on revolutionary memorabilia of all kinds, made historically significant by virtue of their association with significant persons or events. It is not so much the artistic, historical or scientific importance of these things that made them museum worthy, but their emotional and evocative power; their qualities as 'witnesses of history'. Rather than wonders, these objects possessed 'resonance' (Greenblatt 2004).

The last decade of Mao's life (1966–76) was marked by the Cultural Revolution, in which Mao sought to destabilise an emergent elite and reorient the country towards the principles of the revolution. Cultural relics previously admired and collected (such as bronze vessels, calligraphy scrolls, jade and porcelain) were now tokens of a 'backward', 'feudal' society; emblems of a bygone, dark era. Museum exhibitions were remade to popularise the revolutionary experience, communist ideology and a shared national mythology based on the spirit of sacrifice and martyrdom. Art was subverted to support the communist cause and adopted the ideological aesthetic of propaganda. In the aftermath, following Mao's death, museum exhibitions changed again and new museums based around science and technology began to appear, extolling the 'modernity' of the Chinese nation.

The aesthetics of cultural nationalism

It was only in the mid-1990s that a new category of national museum – of which the Shanghai Museum is the most iconic example – became established in China. These possessed aestheticising approaches to display which served to uphold a very specific political discourse, which had its roots in the ideological crisis of the early 1990s.

The dissolution of the Maoist economic system and the inception of Deng Xiaoping's economic reforms, the events of 1989 in Tiananmen Square and the collapse of communism in Eastern Europe undermined the political stability of the Chinese government. Abandoning the rhetoric of a communist utopia, a new ideological foundation was created in nationalist discourse, of which cultural nationalism was a major tenet. The old was renewed (Claypool 2005: 589). China's immense cultural and historical richness now became a source of political legitimation and the museum a new political tool (Varutti 2008).

In these new museums, ancient China was no longer presented as a dark era of oppression. On the contrary, it was considered of pivotal importance as supplying Chinese civilisation with its roots. This re-evaluation of Chinese history and heritage provided a platform for the revival of traditional Chinese aesthetics in what became known as the 'culture craze' or 'culture fever' (*wenhua re*) (Lin with Galikowski 1999: 54; Pohl 2008: 93; Wang 1996). A rhetoric of national pride and cultural prestige was translated in national museums into the aestheticisation of the material culture of ancient China (Su 1995). This same aesthetic approach also affected relics of the revolution: revolutionary museums were called on to revitalise Chinese patriotism through 'attractive' displays that enabled visitors to perceive history 'directly through the senses' (Denton 2005: 568).

The transition to aesthetic displays was accompanied by – and expressed through – a shift from chronological to thematic approaches. These thematic displays reshuffled object associations, distancing them from the historical, cultural and socio-economic conditions of their production. Such remodelling may, however, result positively in new and unexpected connections between diverse types of object. In this way visitors could enjoy new pleasurable experiences rather than simply endure an education in ideology. As Guo Qingshen, an educator at Shanghai Museum, remarked, the museum had become customer-oriented and reflected on individual needs, desires and expectations.[1] National museums were well placed to transform themselves in this way as at this time they could attract prominent funding, government endorsement and public visibility.

The aesthetics of the new national museums

The majority of China's recently built and rebuilt national museums state their aesthetic intent in their architecture. The Shanghai Museum, the Sanxingdui Museum in Chengdu and the Capital Museum in Beijing, for example, each possess aesthetically pleasing architecture far removed from the Soviet-style buildings of former days (Figure 18.2). This aesthetic sensibility permeates their interiors. The website of the Shanghai Museum draws attention to this fact, telling potential visitors that the ceramics are 'displayed on a green background, giving a relaxed and harmonious atmosphere'. Sculpture is displayed, 'Using gold, red and black as three basic colors, the gallery is full of a warm and serious atmosphere'; jades are exhibited 'Using specially designed stands and fiber optics for illumination, the exhibits provide the audience with a sparkling and smooth appearance.' All the details of fine design are perfectly exposed.[2]

Explanatory panels, photographs, charts, video screens and other information devices tend to be physically separated from the object, as if not to intrude on its aura. At the Shanghai Museum, for example, panels introducing the content of the exhibition galleries are located in the entrance to the gallery and at the beginning of each gallery section. No other information interrupts the visual flow of objects. The most spectacular exhibits (bronzes, ceramics and sculptures) are displayed in

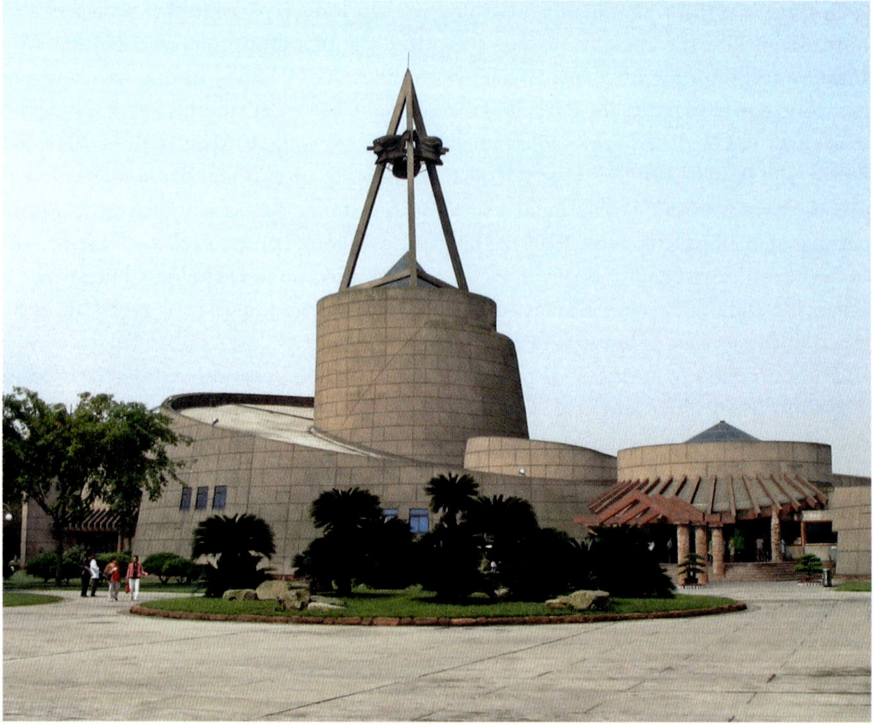

Figure 18.2 Exterior of the Sanxingdui Museum, Guanghan, near Chengdu. Photo: author

individual cases situated at the centre of the exhibition space. The only information available for these objects is provided by the optional audio-guide which is limited to the highlights of the collection.

Intriguingly, in some instances display techniques tend to vary according to the type of material on show. At the Sanxingdui Museum in Sichuan, for example, which displays artefacts of the ancient Shu civilisation, the galleries devoted to ceramics, pottery and gold artefacts are highly contextualised, with panels and charts providing information on their archaeological significance. In contrast, the bronzes tend to be accompanied by minimalist labels, but set dramatically – almost theatrically – against huge panels decorated with large reproductions of bronze masks. The Shu bronzes are given this kind of prominence because they are radically different from those of the Shang Dynasty excavated in Northern China, yet also show advanced casting techniques (Clunas 1997: 19). The Shu remain a relatively mysterious and enigmatic people – a sense enhanced by the museum's displays which portray the civilisation as a 'controversial legend'; 'a rare treasure of ancient humanity' that provides a 'missing link in the evolution of Chinese culture'.[3] By these means the museum amplifies the charismatic aura of the objects, adding to their significance. Here aesthetic impact is in part negotiated to address a deficiency of historical information, but one also senses that this lack of data has been romantically deployed to connect history to legend. There is, then, a literary aesthetic at play.

In common with some major provincial museums in China, objects – such as bronzes – which possess important ritual, mythical, historical, political and social meanings receive only minimal historical and cultural interpretation, as if their relevance and meanings were self-evident. Other, more functional, objects do receive interpretation. The distinction is deeply rooted in realms of connoisseurship. Outside China, displays of Chinese material culture, which mainly focus on the art of the court, tend to be deeply imprinted with aesthetic approaches (see Varutti 2007). Chinese material culture is labelled as 'Asian Art', 'Oriental Arts' or 'Far-East Asian Antiquities'; it is 'High Art', the province of connoisseurs and art experts, rather than ethnographers (Fiskesjö 2007). This aesthetic approach also reflects the market in cultural artefacts which favours aesthetics and intrinsic decorative elements over context and data, and which has fuelled the illicit trade in artefacts. These aesthetic values – of high art and the art economy – have undoubtedly been imported back into China, but clearly they are not the only reason why Chinese national museums have adopted an aesthetic approach.

Behind and beyond the aesthetic display

Aesthetics constitute a mode of representation that cuts through different epistemological registers. The aesthetic gaze is allegedly 'neutral' vis-à-vis different cultural and disciplinary constructions (historical, ethnographic, scientific or artistic). Its application means that museums can dispense with object meanings and instead substitute aesthetics. Deborah Root notes:

> Once an object is named as art either by the relevant specialists or by the market, it tends to be exhibited alone, displayed as an aesthetic form rather than a source of cultural information. . . . Art universalizes the object and places it beyond culture.
>
> (Root 1996: 111)

But this study shows that aesthetics can be used to place objects within a culture, even if mythological and unspecific. In China that cultural setting might be understood as 'the nation', defined by beauty, craftsmanship and longevity. At the Shanghai Museum, the visitor is informed that the bronzes on display 'epitomiz[e] Chinese ancient civilization and rich cultural heritage'.[4] Similarly, at the National History Museum in Beijing, visitors can admire a selection of masterpieces including ceramics, bronzes, sculptures, jewellery and lacquer collections. The introductory panels state that 'people find themselves overwhelmed by pride when they walk around the exhibition hall and feel the aesthetic rhythm from precious relics'.[5] Yet it is paradoxical that this framing of objects as masterpieces, rather than as cultural relics, produces a flattening of history. It contrasts dramatically with the historical materialism that informed museum displays during the Maoist era. Visitors are no longer invited to appraise, for instance, the ritual and political roles of bronze vessels in ancient China, or to appreciate the meaning of the various inscriptions and seals on paintings and scrolls. Instead the objects are dehistoricised (Jewsiewicki 2007).

As Shanks and Tilley (1987: 73) remark, 'the aesthetic artefact is an escape from the nightmare of history'. The past is thus re-actualised in an aseptic, unproblematic form and the national museum can distance itself from the object's historicity and obviate potentially contentious issues that might undermine or threaten authoritative narratives, such as those centred on the unity of the Chinese nation. And because of its aesthetic emphasis, creating objects and notions which are absolute and universal, this exhibitionary mode is particularly suitable to unfold the myth of a similarly absolute and universal entity, the nation. Indeed, the absence of significant hiatuses over 5000 years of history is one of the grand narratives of the present Chinese government. A glorious past is unproblematically actualised, and historical discontinuities and cultural differences transcended.

Conclusion

In the ancient past, China engaged in aesthetic appreciation, both in personal and public spheres. This aesthetic appreciation has continued to play a central role in practices of artefact collection and display, and even in the depths of communism it persisted in Chinese communities and amongst Sinophiles elsewhere in the world. The concern for aesthetics in many contemporary Chinese national museums exhibiting antiquities might then be understood as expression of this historical and cultural tradition, perhaps reintroduced through outside influence. It is, however, possible to locate a continuous if radically changing approach to aesthetics in museums from the moment of the establishment of the first Chinese public museums. Seen in overview, these developments – from the exclusive pleasure of private collections to the establishment of public museums, from the communist propaganda to the emphasis on science and technology, up to the current revival of aesthetic appreciation – might suggest that the situation has gone full circle. Yet this would not be an accurate interpretation. It is true that the legitimacy of aesthetics has been reinstated but, in the process, aesthetics has become imbued with new meanings. The new aesthetics may, for example, be understood as a facet of individualism and consumerism. They may also be perceived as central to the state's new national narrative which has again found a role for the past and for beauty.

The Chinese situation is not disconnected from a broader worldwide trend of aestheticisation in museums. Indeed, there are grounds for arguing that aesthetics has become a mainstream exhibitionary mode in the new millennium, and this is especially the case for the *mise-en-scène* of national identity and of national past. Engagement with the consequences of colonialism and a new social mobility have resulted both in collaborative practices and a retreat into aesthetics, permitting a disengagement from contestations of the past (Clifford 1997, 2007; Errington 1998;

Naguib 2004; Price 1989, 2007). As Donald Preziosi (Chapter 3, this volume) remarks, 'great national museums exist precisely in order to foster and perpetuate the belief in the truth of abstractions such as national identity, character, mentality, or ethnicity'. The popularity of aesthetics as a mode of exhibition of the nation, in China if not elsewhere, can be explained at least in part by the fact that aesthetics work as a 'belief enhancer' by seducing the viewer, by pleasing the senses and by reaching for the individual, the personal and the intimate – even if ultimately a collective goal is desired.

Acknowledgements

I would like to acknowledge the support of the Swiss National Science Foundation (SNSF) and the Swiss Agency for Development and Cooperation (SDC-KFPE), which provided grants to conduct research in China.

Notes

1 Professor Guo Qingshen, Department of Education, Shanghai Museum. Personal communication, March 2004.
2 From the Shanghai Museum website. Bronze Gallery introduction: http://www.shanghaimuseum.net/en/display/display-2.html; sculptures gallery introduction: http://www.shanghaimuseum.net/en/display/display-5.html; jades gallery introduction: http://www.shanghaimuseum.net/en/display/display-6.html (accessed 27 February 2009).
3 From the museum panels of the Sanxingdui Museum, visited in May 2006.
4 Introductory panel to the Bronze Gallery, Shanghai Museum. Last visited April 2008.
5 National History Museum, Beijing. Last visited May 2006.

Bibliography

Capon, E. (1977) *Art and Archaeology in China*, London: Westerham Press.

Chang, W. (1999) 'Esquisse d'une histoire du concept chinois de patrimoine', *Publics et Musées,* 15: 81–117.

Claypool, L. (2005) 'Zhang Jian and China's first museum', *Journal of Asian Studies,* 64(3): 567–604.

Clifford, J. (1997) *Routes: Travel and Translation in the Late Twentieth Century*, Cambridge, MA: Harvard University Press.

Clifford, J. (2007) 'Quai Branly in process', *October Magazine,* 120: 3–23.

Clunas, C. (1997) *Art in China*, Oxford: Oxford University Press.

Denton, K. (2005) 'Museums, memorial sites and exhibitionary culture in the People's Republic of China', in M. Hockx and J. Strauss (eds) *Culture in the Contemporary PRC, The China Quarterly Special Issues,* Cambridge: Cambridge University Press.

Descola, P. (2007) 'Passages de témoins', *Le Debat. Le Moment du Quai Branly,* 147: 136–53.

Errington, S. (1998) *The Death of Authentic Primitive Art and Other Tales of Progress*, Berkeley: University of California Press.

Fernsebner, S. R. (2006) 'Objects, spectacle, and a nation on display', *Late Imperial China,* 27(2): 99–124.

Fiskesjö, M. (2007) 'The trouble with world culture', *Anthropology Today*, 23(5): 6–11.

Greenblatt, S. (2004) 'Resonance and wonder', in B. Messias Carbonell (ed.) *Museum Studies: An Anthology of Contexts*, Oxford: Blackwell.

Hamlish, T. (2000) 'Re-membering the Chinese Imperial Collections', in S. Crane (ed.) *Museums and Memory*, Stanford, CA: Stanford University Press.

Jewsiewicki, B. (2007) 'La mémoire est-elle soluble dans l'esthétique?', *Le Debat. Le Moment du Quai Branly*, 147: 174–77.

Lin, M. with Galikowski, M. (1999) *The Search for Modernity: Chinese Intellectuals and Cultural Discourse in the Post Mao Era*, New York: St Martin's Press.

Naguib, S. (2004) 'The aesthetics of otherness in museums of cultural history', *Tidskrift for Kulturforskning*, 3/4: 5–21.

Pohl, K. H. (2008) 'Identity and hybridity: Chinese culture and aesthetics in the age of globalization', *Intercultural Aesthetics*, 9: 87–103.

Price, S. (1989) *Primitive Art in Civilized Places*, Chicago: University of Chicago Press.

Price, S. (2007) *Paris Primitive: Jacques Chirac's Museum on the Quai Branly*, Chicago: University of Chicago Press.

Root, D. (1996) *Cannibal Culture: Art, Appropriation and the Commodification of Difference*, Oxford: Westview Press.

Schaeffer, J.-M. (2004) 'Objects esthétiques?', *L'Homme,* 170: 25–46.

Shanks, M. and Tilley, C. (1987) *Re-constructing Archaeology*, Cambridge: Cambridge University Press.

Su, D. (1995) 'Museums and museum philosophy in China', *Nordisk Museologi,* 2: 61–80.

Varutti, M. (2007) 'Representing and "consuming" the Chinese Other at the British Museum', in A. Bugge Amundsen and A. Nyblom (eds) *National Museums in a Global World*. NaMu III, Department of Culture Studies and Oriental Languages, University of Oslo, Norway, 19–21 November 2007, Linköping Electronic Conference Proceedings, No. 31, Linköping: Linköping University Electronic Press. Available online: http://www.ep.liu.se/ecp/031/.

Varutti, M. (2008) 'Which museum for what China? Museums, objects and the politics of representation in the post-Maoist transition (1976–2007)', unpublished PhD thesis, Graduate Institute of International and Development Studies, University of Geneva.

Vickery, J. (2007) 'Organising art: Constructing aesthetic value', in S. Knell (ed.) *Museums in the Material World*, London: Taylor & Francis.

Wang, J. (1996) *High Culture Fever: Politics, Aesthetics, and Ideology in Deng's China*, Berkeley: University of California Press.

Chapter 19

THE DUTCH NATIONAL HISTORICAL MUSEUM

A national museum for the twenty-first century

Gwenny van Hasselt

A PROPOSAL FOR A NEW Dutch National History Museum emerged in parliament in the early 1990s. It arose from debates surrounding the rebuilding of the Rijksmuseum and a report on the Dutch educational system by the Council of Education that indicated a lack of knowledge of national history amongst the Dutch public and particularly the Dutch youth. The idea of a new national history museum seemed, to some, anachronistic in an age where society was transgressing national boundaries and stereotypes. It seemed to hark back to an old territorialism, to grand narratives and definitive truths. Museums had moved beyond such narratives (Hooper-Greenhill 1995; Knell 2003; Witcomb 2007). The new museum began a new debate about form, content and location; about the purposes of such museums. This chapter will discuss the issues surrounding the building of this new national museum and give insights into how the Dutch have tried to solve these problems. In order to achieve this, I shall discuss the process of development, the changing cast of actors, the shifting focus of debate and the decisions made. The project has been fraught with controversy but has nevertheless continued to progress towards completion.

A place for a nation's history

In 1993, the Director of the Rijksmuseum – the most visited and prominent museum in the Netherlands – announced that it would undergo major rebuilding. This was to have taken two years, but the scale of the works required continued to increase (Anon. 1993; Bailey 2009), until activity halted in 2003. The museum was then closed to the public and was not expected to reopen before 2013 (Figure 19.1).[1] Some of the museum's most famous paintings were then temporarily displayed in another wing of the museum.

The Rijksmuseum's role is to exhibit Dutch art (de Jong, *et al.* 1985). In this respect it is a national gallery quite unlike the Louvre or the British Museum, both of which exhibit international art rather than art of the home nation. There are, however, many other national museums and galleries which take this more internalised national view of art (see, for example, Bennett 1996; Mason 2007). The focus on authentic Dutch objects was established from the very beginning of the Rijksmuseum in 1800. The first objects, the five hallows of the Dutch nation, were exhibited both for their aesthetical value and their historical significance, and in its early years, the museum included a department of national history (van der Ham 2000). This connection between art and history came under threat when the rebuilding commenced in 1993. The remodelled museum was to no longer possess a history wing and the art and history collections were to be combined (de Lange 2008). Recognised as the nation's museum, the loss of the history wing was met with criticism from MPs, museum professionals and cultural commentators alike (Bailey 2009). The museum's protracted reconstruction now encouraged a debate about the possibility and the necessity of a new, separate national historical museum. Many in the Dutch press thought such a development unnecessary; the Rijksmuseum

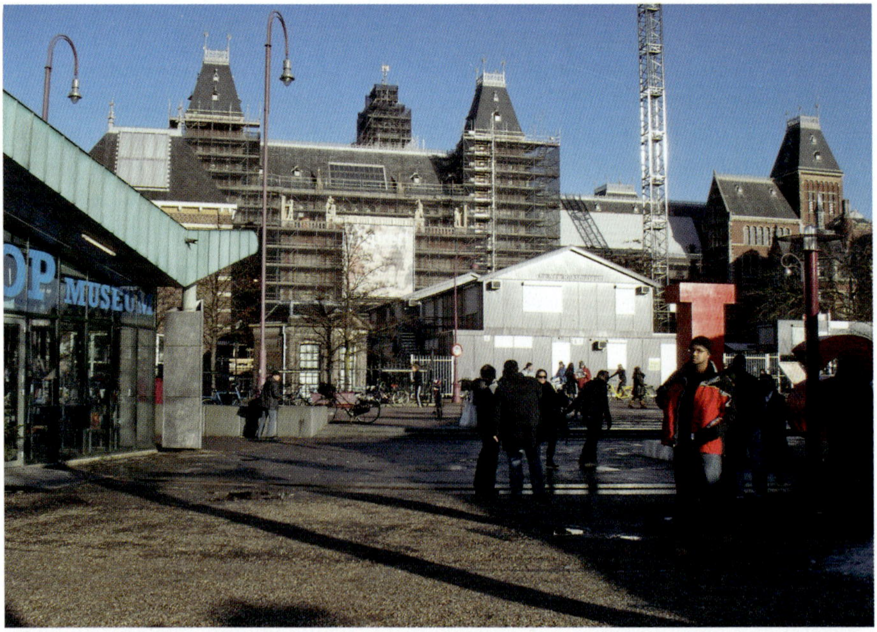

Figure 19.1 The Rijksmuseum in Amsterdam undergoing extensive rebuilding. Photo: M. E. Halbertsma

was already the museum of national history. However, others argued that it was principally an art museum and that Dutch history deserved a museum of its own:

> Another reason for the pleas for a new museum of Dutch history is that the Rijksmuseum removes more history than that it adds. The board has proposed to take away the department of national history and integrate it with art history and applied art. Historical objects and paintings have to tell the story of the Netherlands.
>
> A growing number of people criticise these plans: the Rijksmuseum places national history beneath art history in their eyes. So these plans also play a part in the pleas for a separate national historical museum.
>
> (van Schoonhoven 2001)

The fires of this debate were fanned by the publication of an evaluation of the Dutch educational system by the Council of Education. Revealing a failure amongst the Dutch to know their own history, politicians saw this as a more worrying indication that the Dutch did not really know what it meant to be Dutch anymore. The Dutch parliament considered three solutions: a cultural and historical canon or narrative, educational reforms and/or a national historical museum. In 1994, it was decided that educational reforms were necessary throughout the school system. The new curriculum structure was introduced in 1999 (Onderwijsinspectie 1999). Despite these reforms, debate continued concerning the Dutch historical consciousness. Four years later, the Dutch Ministry of Education, Culture and Science brought into being a commission for the development of a cultural and historical canon in order to 'strengthen the historical, cultural and societal consciousness of (young) citizens' (van der Hoeven 2005; van Oostrom 2006a: 11).

The notion of a national canon was expressed as the need for 'a new story of the Netherlands' which would contribute to a Dutch identity (van der Hoeven 2005) and unite native Dutch people and newcomers. In her declarations to parliament, Maria van der Hoeven, the then Minister of Education, Culture and Science linked this lack of historical consciousness and a national canon to another political failing: Dutch multiculturalism.

The multicultural state

In the 1990s, the Dutch experienced increasing unease in the face of multiculturalism, globalisation, and political and economical instability (Wittenberg 2008). Multiculturalism had spawned vehement debate. Politician Pim Fortuyn had become the spokesman for a group critical of this social model. His assassination in 2001 sent a shockwave through Dutch society but his ideas were not lost and were soon adopted by others, including Rita Verdonk of the Proud of the Netherlands Party, Geert Wilders of the Party of Freedom and Theo van Gogh, a filmmaker and opinion writer.

The Netherlands had barely recuperated from the murder of Fortuyn, when Theo van Gogh was assassinated by a Muslim extremist in broad daylight on the streets of Amsterdam in 2004. The socio-political climate in the Netherlands now became increasingly unstable, with multiculturalism as an issue of national debate. Nationalistic sentiments grew and incentivised the search for a coherent Dutch identity. The idea of a national historical museum developed an increasingly firm foothold in the national conscience and several parties argued that immigrants should learn about Dutch history as part of a process of integration. Bas van der Vlies, of the Reformed Party (SGP), stated in June 2006,

> If we are not sufficiently aware of our history, we are unfamiliar with our present and we stare uncertainly into the future. It is of great importance that newcomers from other cultures are familiar with this [history], so there can be a common experience of citizenship.
>
> (Tweede Kamer 2006)

This connection between historical consciousness, identity and good citizenship would later prove to be problematic for the National Historical Museum. A historical consciousness built on military victories, heroes and highlights reminded many of the fascism that led to the Second World War. In the Netherlands at this time, as in other European liberal democracies, nationalism and fascism were synonymous (see, for example, Schulte 2006). This provided the counter-argument to the museum and to the proposed national canon (Gorak 1991). Nevertheless, the government established the Committee for the Development of the Dutch Canon which presented its final report in July 2007. Amongst its outputs was a wallchart (Figure 19.2) showing fifty canonical moments. These consist of events, persons and objects from different disciplines, including science, literature and economics. Although focused on Dutch history, some items are concerned with providing an international context (van Oostrom 2006b). The Committee wanted a canon that would be welcomed by schools without any special justification (van Oostrom 2006a: 24).

Despite work on the Dutch canon and the educational reforms, parliament remained dissatisfied with the level of historical awareness amongst the Dutch public. It was of the opinion that more action was required and called for an institution that could consolidate and advocate the content of the canon: a national historical museum. On 27 June 2006, Maxime Verhagen of the Christian Democrats, Jan Marijnissen of the Socialist Party and others submitted a motion for the creation of a national historical museum (Verhagen and Marijnissen 2006). They also made a public plea for such a museum in an article entitled 'Save our historical consciousness' in the national newspaper *Trouw*. This united parties across the political spectrum:

> A people without history do not exist. Every people, including the Dutch, have a history. . . . A National Historical Museum that coherently exhibits

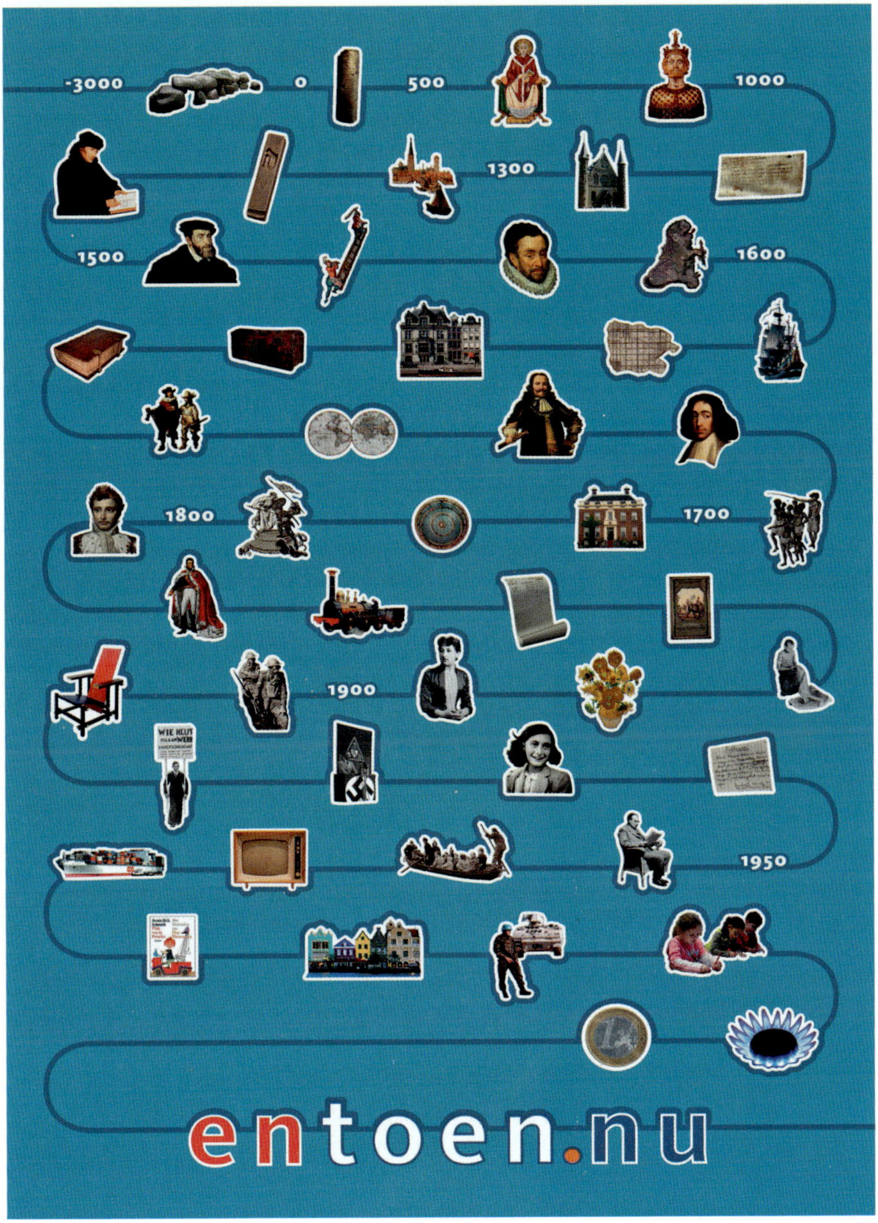

Figure 19.2 Dutch Cultural and Historical Canon Poster. Photo: Copyright foundation entoen.nu

> our history does not even exist. The historical consciousness in our modern society will disappear if we do not change this. . . . Due to the Internet and globalisation, but also through the individualisation of society and the arrival of more people with different cultural backgrounds, it seems we do not share a common identity any longer.
>
> (Marijnissen and Verhagen 2006)

With the new museum approved, a new fear emerged amongst museum professionals and a wider Dutch public that the government might use the museum as a political instrument to create a wave of nationalism. Both the canon and the museum were political initiatives; they did not arise from academic or educational institutions. This is not, however, an unusual public response to the formation of a national museum (Mason 2007; Petrescu 2007). In the Netherlands, the ambiguous relationship between government as funder and ideological instigator would continue to be a point of debate throughout the entire development process. The Dutch remained concerned about what 'national' might mean and what might be the relationship between a national museum and nationalism. Educational psychologist, Johannus Schroots, remarked: 'It is a hazardous affair to specifically look for new binding factors in the national history, because all simplifications and -isms, for example nationalism, are lurking around the corner' (Schroots and Langemijer 2004: 366). Nevertheless, these 'binding factors in the national history' would drive the museum project (Verhagen and Marijnissen 2006).

The Dutch government persevered and soon found that this debate over the politics of history gave way to a new argument concerning the museum's location.

The battle of the cities

Development of the National Historical Museum was overseen by Van der Hoeven's successor at the Ministry of Education, Culture and Science, Ronald Plasterk. After a fierce competition between the cities of Amsterdam, The Hague and Arnhem – referred to as the 'battle of the cities' – on 2 July 2007, Plasterk selected Arnhem as the museum's location (Plasterk 2007). His decision was considered surprising, as many thought such a museum should be in the nation's capital, but the 'battle' had woken up the strong sense of local and regional identity felt by the Dutch.

Amsterdam had been considered the most obvious choice. It offered the new museum profitable connections with the Rijksmuseum and other important cultural institutions, including the Van Gogh Museum and the Palace on the Dam. The Alderwoman of Culture, Caroline Gehrels, argued that '"the best of the Dutch museum collections" is present in Amsterdam' (Hollak 2007). The Hague, as the residence of the government and the Queen, offered the opportunity to associate the museum with democratic citizenship. It is also home to important cultural institutions including the Mauritshuis, the Peace Palace and the City Museum where many renowned Dutch and foreign art works can be viewed. The almost equal importance of these cities for Dutch cultural and political life had produced a natural rivalry.

Arnhem was, by comparison, the underdog. It is the biggest city in the east of the Netherlands and situated in the geographical centre of the country. It can boast the Open Air Museum (Figure 19.3) and the popular Burger's Zoo, and, nearby, the Netherlands' biggest nature park, Palace Soestdijk and the Kröller-Müller Museum. But the Arnhem scheme did not succeed on geographical grounds alone, it surpassed its rivals conceptually. Its proposal had been written by, amongst others, the director of the Open Air Museum, Jan Vaessen, who was respected for his vision and expertise throughout the museum world. His plan gave a tangible picture of a national museum in which the Dutch canon could be exhibited. Vaessen suggested it be built next to the Open Air Museum, which had won the prestigious European Museum of the Year Award in 2005. Plasterk thought that this museum already attracted the target demographic of the new museum; it would attract school-going children and their parents. The local combination of culture and nature produced an ideal leisurescape. Although not everyone applauded his decision, Plasterk stood his ground.

The strength and power of local identities had developed in response to the national cultural and historical canon. Local, provincial and city canons were

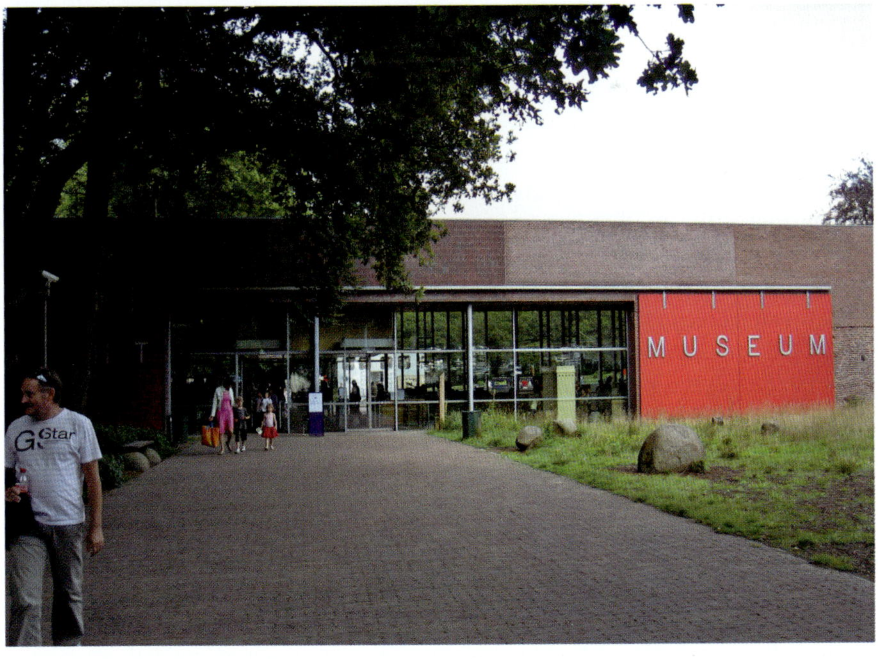

Figure 19.3 The Dutch National Historical Museum will be built next to the Open Air Museum in Arnhem. Photo: H. Verreyke

created. The process made the Dutch even more resistant to central government initiatives. It is a national aversion which has slowed the development of the new museum, but has not rendered the museum incapable of situating itself in the Dutch cultural landscape. As cultural expert, Han Bakker, remarked in the national newspaper, *de Volkskrant*:

> A national museum should be a 'watchtower'; it should bring the visitors into contact with all the 'lieux de mémoire' in the Netherlands. 'Imagine that you are native to the province of Limburg. That province has a very complex and strongly divergent history, for a long time it did not even belong to the geographical part of the land which we now call the Netherlands. . . . The history of the Netherlands is the sum of its local and regional parts.'
>
> (Ramaer 2007)

The placement of the National Historical Museum in Arnhem, in the province of Gelderland, is a clear sign the government wants the museum to be Bakker's 'watchtower'. By choosing Arnhem, Plasterk has tried to give the museum a chance to rise above provincial differences and properly operate on a national level. With the location firmly established, discussions now focused increasingly on the content of the museum.

The exploratory phase: the project sessions

In January 2008, Plasterk appointed Liberal Party MP, Atzo Nicolaï, to manage the development of the museum, until a board could be appointed (Plasterk 2008). Nicolaï's appointment met with suspicion and even outrage from several political parties in parliament. Hans van Leeuwen of the Socialist Party, in particular, objected and motioned for another appointee so as to avoid 'any appearance of political involvement' (van Leeuwen 2008). Newspapers were also critical: 'A dubious appointment . . . He [Nicolaï] will also remain a Member of Parliament. This is an unfortunate choice. These two offices are not to be held by one person' (Schulte 2006). The government remained resolute, however.

Under Nicolaï's supervision, a project team was assembled which would cooperate with an independent project development agency, Tinker Imagineers, to set up so-called 'project sessions'. Professionals from various disciplines were invited to take part in these sessions to develop a concept for the museum. Each session was given a theme: who (is the public?), positioning, mission, manifestation, transference, content, education and style. Different professionals were invited according to the theme of the session. One of the directors of the National Historical Museum later referred to these sessions as the 'inventory' phase.[2] At this stage, the emphasis on the national historical canon became much less pronounced. Indeed, some professionals doubted whether an overview of Dutch national history was enough to stimulate interest in history.[3] The same concerns that had arisen in the debate over the development of the Dutch cultural and historical canon were now repeated during the project sessions by historians, museum professionals and educational experts. They questioned the intentions of parliament and debated the role of local and regional history in the new museum. Even the location came in for critical debate. The sessions also introduced a new element into the controversy: that of global history and international context. This could not be ignored if the nation was to be understood (Nationaal Historisch Museum Stuurgroep 2008). The sessions revealed that the set of issues that arose in the initial critical debate would remain to shape the museum; that these issues had to be faced up to on a practical level. This more practical realisation began with the appointment of a board.

The concept phase: a two-man board and the vision document

The project sessions produced a *Source Document* which was followed by the appointment of a two-man board consisting of Valentijn Byvanck and Erik Schilp, former directors of smaller local historical museums, the Zeeuws Museum in Middelburg and the Zuiderzeemuseum in Enkhuizen respectively. Interestingly, both had come from museums situated on the cultural periphery of the Netherlands (Hannerz 1992). Byvanck and Schilp are considered fresh and inventive but also outspoken. They had a proven track record of success in public engagement through imaginative interpretation which might be imported directly into the new museum.[4]

Byvanck and Schilp now began the production of a *Vision Document* inspired by the project sessions and their own backgrounds. The *Vision Document* aimed to give a preliminary framework for the museum. It speaks of a museum exhibiting Dutch history in five worlds: me and we, land and water, war and peace, rich and poor, and body and mind. These five worlds together form the core of the museum, each with its own exhibition space. Through these five worlds the national history of the Netherlands will be shown thematically. Within each world, history will also be represented chronologically.

Next to these five worlds, an additional room will be constructed, to show the visitor that the Netherlands could have looked very differently if history had taken another path. The museum will provide some scenarios, but the intention is for visitors to share their own ideas with others. That way, the public will play a game with each other, by asking, for instance, 'What if the Netherlands never had any colonies?' (Byvanck and Schilp 2008: 44–53).

The board will also try to market the museum's brand; to make the museum the first thing that comes to the public's mind when they think 'history'. The museum is to function as a historical navigation device; it will point to history in the Dutch landscape and centres of culture. In 2008, this vision was opened to public discussion (Byvanck and Schilp 2008). It resulted in the resurfacing of three problems: disagreements on the use of the national canon; the use of chronology; and the importance of an international context. A group calling itself the 'Young Historians' even released a counter-concept for the museum: *Statement XX*. It promoted a more conservative interpretation centred on the 'national story' of the Netherlands emphasising chronology.[5] This group argued that the vision lacked the kind of international context that was needed to understand Dutch multicultural society, a theme fundamental to the establishment of the museum. Other historians remained worried about the possible nationalistic effect of the museum (Jonge Historici 2008). Historian Jan Derksen commented that the lack of global history in the museum could lead the visitor to think they were enforced with a narrowly defined national identity. An increasing number of Dutch citizens were defining themselves as citizens of the world rather than of a country.[6]

Byvanck and Schilp remained of the view that there was no shame in focusing on Dutch national history in the museum. It is a national museum after all; they reasoned its primary task is to focus on national history. Yet the two had also countered many of the concerns about the politics of the national canon: 'The canon is silently present in the museum . . . but we do not consider it to be the task of the museum to give a visual record of the canon' (Anon. 2008a). This statement was met with surprise and even anger from historians and politicians. Some MPs felt the board had deviated from its original assignment; that the move away from the national canon was too radical. Many historians agreed, though for rather different reasons. They believed the canon gave history a strong coherence through its use of chronology. To negate one was to lose the other. The canon has been created not least to teach children the chronological order of important events in Dutch history. Several educational professionals and the former director of the Rijksmuseum, Henk van Os, shared these views (Anon. 2008a, 2008b). Museum professionals, on the other hand, disagreed, and considered that while chronology was useful in the classroom it was not necessarily an effective educational tool in the museum. The chairman of the Committee which produced the canon, Frits van Oostrom, agreed, stating on multiple occasions that 'the canon is developed for schools and not for a museum' (Hollak 2008).

In order to resolve matters, Byvanck and Schilp participated in several public debates. The matter of chronology was also discussed extensively in Dutch newspapers and magazines. These activities encouraged the board to backtrack and place rather more emphasis on chronology within the five themes and create an additional room in which a brief chronological overview of the history of the Netherlands would be exhibited.

Problems and solutions

The development of the new National Historical Museum produced a number of issues or problems which the museum had to resolve. Amongst these was a Dutch antipathy towards anything labelled 'national'. This caused a resurgence of local historical sentiment and calls for international contextualisation. If these were, in some respects, historiographic arguments about 'national history' then they also included the debate surrounding the place of chronology. The other aspect of these arguments related to nationalism more generally which reappeared in the Dutch conscience as a result of the extraordinary murders of two prominent politicians. The history of the development of the museum, thus far, has seen these issues repeatedly come to the fore but in doing so they have moved from general political arguments to more grounded manifestations in the production of museum spaces and narratives. One senses the role of the museum professional as mediating the political influences of government, professional groups, the media and the wider public. And that this process of mediation is reproduced in a museum which aims to transcend this debate, even if it will inevitably remain criticised in certain quarters. This, it seems, is the manner in which a national museum is produced in a liberal democracy in the twenty-first century.

Notes

1 Rijksmuseum, www.rijksmuseum.nl, retrieved on 1 February 2009.
2 Interview with Schilp, E. and Byvanck, V., 15 April 2009, Arnhem.
3 Observation at a project session, 23 June 2008, Arnhem.
4 Schilp, E. at the Museum Congress, 3 October 2008, Groningen.
5 Observation at public debate between the Young Historians, Byvanck, V. and Schilp, E., 14 January 2009, Amsterdam.
6 Idem.

Bibliography

Anon. (1993) 'Publieksactie voor verbouwing zuidvleugel rijksmuseum', *ANP*, 3 March.
Anon. (2008a) 'Nieuw museum negeert canon', *Dagblad Van Het Noorden*, 11 December, 22.
Anon. (2008b) 'Canon hoort in nieuw museum', *NRC Next*, 23 December, 1.
Bailey, M. (2009) 'Making order out of chaos at the new Rijksmuseum', *The Art Newspaper*, 20 January.
Bennett, T. (1996) *The Birth of the Museum*, London and New York: Routledge.
Byvanck, V. and Schilp, E. (2008) *Visie document: Het nationaal historisch museum stimuleert de historische verbeelding*, Enschede: Drukkerij Roelofs.
Gorak, J. (1991) *The Making of the Modern Canon: Genesis and Crisis of a Literary Idea*, London and Atlantic Highlands, NJ: Athlone Press.
Grever, M. and Stuurman, S. (eds) (2007) *Beyond the Canon: History for the Twenty-first Century*, Basingstoke and New York: Palgrave Macmillan.
Ham, G van der (2000) *200 jaar Rijksmuseum*, Zwolle: Waanders.
Hannerz, U. (1992) *Cultural Complexity: Studies in the Social Organization of Meaning*, New York: Columbia University Press.
Hirsch, E. D. (1987) *Cultural Literacy: What Every American Needs to Know*, New York: Vintage Books.
Hoeven, M. J. A. van der (2005) *Instellingsbesluit commissie ontwikkeling nederlandse canon*, No. BOA/2005/37330, Ministerie van Onderwijs, Cultuur & Wetenschap.
Hollak, R. (2007) 'Vragen over onze identiteit', *NRC Handelsblad*, 28 June, 11.
Hollak, R. (2008) 'Opgeleukte thema's ruïneren opzet NHM', *NRC Handelsblad*, 22 December.

Hooper-Greenhill, E. (ed.) (1995) *Museum, Media, Message*, London and New York: Routledge.

Jong, E. de, Lemmens, G. T. M. and van Thiel, P. J. J. (eds) (1985) *Het rijksmuseum: Opstellen over de geschiedenis van een nationale instelling*, Weesp: Fibula–van Dishoeck.

Jonge Historici (2008) Statement XX: NHM op een kruispunt!, www.geschiedenis.nl.

Knell, S. J. (2003) 'The shape of things to come: Museums in the technological landscape', *Museum and Society*, 1(3): 132–46.

Lange, H. de (2008) 'Het dolfinarium-gevoel', *Trouw*, 1 July, 2.

Leeuwen, H. van (2008) *Motie van het lid van leeuwen*, Ministerie van Onderwijs, Cultuur & Wetenschap.

Marijnissen, J. G. C. A. and Verhagen, M. (2006) 'Red ons historisch besef; Huis van de geschiedenis', *Trouw*, 13 May, 14.

Mason, R. (2007) *Museums, Nations, Identities: Wales and its National Museums*, Cardiff: University of Wales Press.

Nationaal Historisch Museum Stuurgroep (2008) *Bouwstenen*, Badhoevedorp: Koenders & van Steijn.

Onderwijsinspectie (1999) *Analyse studielast leerlingen in de tweede fase*, Inspectie van het Onderwijs.

Oostrom, F. van (2006a) *De canon van nederland: Rapport van de commissie ontwikkeling nederlandse canon, deel A*, Commissie Ontwikkeling Nederlandse Canon.

Oostrom, F. van (2006b) *De canon van nederland: Rapport van de commissie ontwikkeling nederlandse canon, deel B*, Commissie Ontwikkeling Nederlandse Canon.

Petrescu, D. (2007) 'Communist legacies in the "new Europe": History, ethnicity, and the creation of a "socialist" nation in Romania, 1945–1989', in K. H. Jarausch and T. Lindenberger (eds) *Conflicted Memories: Europeanizing Contemporary Histories*, New York and Oxford: Berghahn Books, pp. 37–54.

Plasterk, R. H. A. (2007) *Vaststelling van de begrotingsstaat van het ministerie van onderwijs, cultuur & wetenschap (VIII) voor het jaar 2007*, no. 157, No. 2006-2007, 30 800 VIII, nr. 157 DCE/07/26244, Tweede Kamer der Staten-Generaal.

Plasterk, R. H. A. (2008) *Nationaal historisch museum*, No. DCE/08/3598, Ministerie van Onderwijs, Cultuur & Wetenschap.

Ramaer, J. (2007) 'Stel historisch museum uit', *De Volkskrant*, 14 May, 11.

Schoonhoven, G. van (2001) 'Twee schatkamers', *Elsevier*, 24 February.

Schroots, J. and Langemijer, M. (2004) 'Wie was adolf hitler ook alweer?', *NRC Handelsblad*, 4 May, 9.

Schulte, A. (2006) 'Handen af van geschiedenis: Politiek mag geen visie op verleden voorschrijven', *Het Parool*, 14 October, 99.

Tweede Kamer (2006) *Debat over de oprichting van een museum voor de nederlandse geschiedenis*, No. TK 95 95-5876, Tweede Kamer der Staten-Generaal.

Verhagen, M. and Marijnissen, J. G. C. A. (2006) *Motie 249*, Ministerie van Onderwijs, Cultuur & Wetenschap.

Witcomb, A. (2007) 'The materiality of virtual technologies: A new approach to thinking about the impact of multimedia in museums', in F. Cameron and S. Kenderline (eds) *Theorizing Digital Cultural Heritage: A Critical Discourse*, Cambridge, MA, and London: MIT Press, pp. 34–47.

Wittenberg, E. (2008) 'Het prettige idee dat je ergens bijhoort: In tijden van grote veranderingen wordt het verleden van holland uit de kast gehaald', *NRC Next*, 12 September, 20.

Chapter 20

WHO AUTHORS THE NATION?

The debate surrounding the building of the new Estonian National Museum

Pille Runnel, Taavi Tatsi and
Pille Pruulmann-Vengerfeldt

ESTONIA, THE NORTHERNMOST of the three Baltic states, in 1991 regained its independence as a nation state, which it had lost with the Nazi–Soviet pact in 1939 and the Soviet occupation of Eastern and Central Europe during the Cold War. Its earlier period of independence, from 1918 to 1940, was short-lived, and from the outbreak of hostilities in the Second World War the country suffered the loss of a significant portion of its ethnic population through emigration and deportation, only to have it replaced by Soviet migrants and military units. For the majority of ethnic Estonians much of the twentieth century was an immensely traumatic experience and with the restoration of independence in 1991 came the hope that the nation might pick up from where it left off more than fifty years before. This chapter concerns the Estonian National Museum which was created as a part of the national movement in 1909 and which then established itself as an important symbol of national memory and identity. In the early 1990s, in the 'period of national awakening' when the country underwent major reform, there developed the idea of building a new Estonian National Museum. It arose in that period of hope and ideals, which straddled the moment when independence returned, but it soon found itself locked into a period of pragmatism and economic reality (Runnel *et al.* 2009). Indeed, with large-scale economic turmoil sweeping Europe in 2009, doubts and questions began to emerge concerning the future of the project to build the museum and the value of a national museum to modern Estonian society.

The debate surrounding the erection of the new Estonian National Museum took place within what Michael Kennedy (2002) called the 'transition culture' that enveloped post-communist nations in Central and Eastern Europe in the late 1980s and early 1990s. It involved fundamental changes to social order, technology and infrastructure. Lauristin and Vihalemm (2009) interpret this phenomenon as a field of mediation, where external demands from the powerful international agencies are, through a specific 'learning process', turned into the value preferences and codes of behaviour of the actors within the transforming state; external definitions of 'success' and 'failure' are by these means 'naturalised' in evolving values and practices. The resulting blurring of internal and external agendas in this field of mediation has certainly been apparent in the processes involved in the establishment of new cultural institutions (Tali and Pierantoni 2008). Recent Western interest in major museum-building projects, for example, has resulted in much debate about the spatial conceptualisation and impact of these institutions (Stead 2004) which has spilled over into Estonian discussions of its new national museum. These have fostered expectations which variously see the museum as part of the rapidly developing creative industries, a contributor to the knowledge infrastructure, and a home for civic and ethnic nationalism.

Estonian conceptualisation of the museum and its social purpose has also undergone change. In particular, there has been the necessity to define identities at a time of rapid change by locating and securing old values and repaying history's debts. The latter, in particular, had wide support, as the social disruption caused by the Soviet occupation destroyed lives, artefacts and institutions; now public opinion was in favour of re-establishing those things lost in both the cultural and political arenas. The reinvention of the museum has been closely connected to these questions of collective memory and collective identity, which have in turn also been affected by a return to the European fold.

This study has emerged from the fields of media and communication research and the anthropology of cultural production. It has applied an ethnographic methodology involving participant observation in meetings and in regular working practices, and analysis of different media sources. It draws upon Peterson's (2003: 177) Bourdieuian analysis of production culture: 'a complex network of relations between various institutions and agencies that have various kinds and degrees of power over aspects of media production', which involves 'an ongoing construction of social actors working in it'. Of particular interest to us has been the role of authorship and its ownership amongst various actors, and the manner in which audiences are permitted to participate in decision-making and to what extent this participation is merely a token gesture or pseudo-participation (Carpentier 2007). Participation here can arise from an actor's membership of various communities.[1] Thus, we were also interested in how audience engagement is envisaged within the field of production, the composition of that audience and how it participates in constructing the museum's narratives. We had in mind van Mensch's (2005) assertion that museums need to be laboratories and meeting points for discussion and new initiatives. In an Estonia recently released from the shackles of the Soviet system this was for many a very novel conceptualisation of the museum.

Authoring the nation

The location of the original museum building became historically significant to Estonians for many reasons. Raadi Manor (Figure 20.1) on the outskirts of Tartu, the second largest town in Estonia, prospered under the Baltic-Germans in the nineteenth century and enabled the owner to develop his interest in flying. This resulted in the establishment of a small airfield, which decades later made the area militarily attractive, and which would play an important role in the future conceptualisation of the museum. During the nationalist reforms of the early years of independence in the early twentieth century, the manor was given over to the University of Tartu, which agreed to share part of it with the Estonian National Museum, which was actively seeking a building of its own at that time. The museum situated itself in the less than ideal main building, where a popular permanent ethnographic exhibition focusing on peasant culture opened in the early 1920s. The manor's extensive grounds became a favourite spot for excursions and walks and, with its museum, established Raadi as an important symbolic place for the whole of Estonia. However, the outbreak of the Second World War saw the evacuation of the collections and Soviet bombing of Nazi troops quartered there, which left the building in ruins and the whole area in the hands of the Soviet air force.

For Estonians, this military and destructive influence on the Raadi area, which turned Tartu into a closed city, was a poignant metaphor for the overall situation Estonia had found itself in during the fifty years of Soviet rule. In April 1988, at the beginning of the public movement against the Soviet occupation, a demonstration was held calling for the Estonian National Museum to be returned to Raadi. The following year, the Soviet army made a partial withdrawal from the property and land was appropriated for the 'Estonian National Museum at Raadi'.

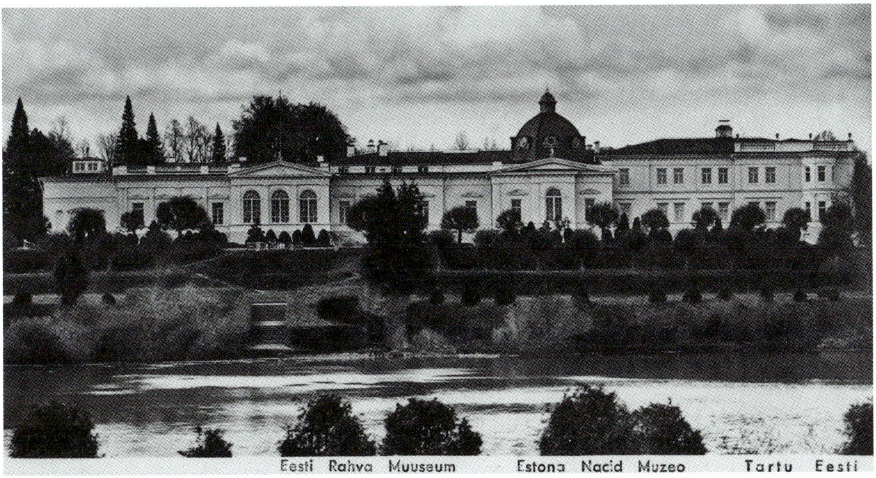

Figure 20.1 Raadi Manor, 1938. Photo: Eduard Selleke, Estonian National Museum

From 1993 to 2005, discussion of the future location of a more permanent museum building revealed no consensus, even within the museum itself. For many, Raadi was in all senses a ruin: polluted both physically and symbolically.

In 1994, early in the transition period, an exhibition on Estonian culture was established in a temporary building at the centre of the town. The exhibition, in keeping with the dominant sentiment of the day, was deliberately ethno-romantic with its focus on the display of nineteenth-century peasant life. The curator responsible for the exhibition reflected a decade later that the role of the Estonian National Museum was to maintain Estonian identity, and its task was to find different ways to accomplish this objective.[2]

The Raadi area was not entirely forgotten though and a competition for future development of the manor was arranged. The resulting ideas were diverse and ranged from a museum of Baltic German culture to a museum of Estonian life histories and a multifunctional cultural centre. In 1998, the central manor area was returned to the museum by order of the Tartu city government but by then there existed plans to erect a new museum building in the centre of the city. An architectural competition was held for the Estonian architects, which resulted in choosing a winning project and preparing the building site. However, delays in the construction work and a pressing need for space[3] led to the museum building its stores at Raadi, beginning the work in 2000.

Building the stores at Raadi certainly facilitated turning opinion in favour of building the new museum at Raadi slightly later, following a number of newspaper articles published between 2001 and 2004, which reawakened a nostalgic reflection on that special place in the minds of an older generation of Estonians: 'From now on, the museum's Raadi-era started, the remembering of which brings a sparkle in the eye of the older generation.'[4] In the national newspapers, the new museum became a symbolic object of national importance, and the Estonian government's decision in 2003 to rebuild the museum in the Raadi area[5] was interpreted as a triumph of justice. The museum now became the material, bodily manifestation of the nation:

> The return of the Estonian National Museum to Raadi is also the return to the home of the nation, which should give us certainty of our national cultural survival. There is no future without the past for a nation, let us take care of the past.[6]

In 2005, an international architecture competition was announced.[7] Estonian historian, Marek Tamm (2005), described the Estonian National Museum as 'one of the most important displays of Estonian culture', but emphasised in the introductory text to the competition that the museum's identity was bound to change:

> It is clear that today the golden age of nationalism is now behind us, and that a museum dedicated to displaying objects from a single nationality (and other Finno-Ugric nations) does not fit in with the terrain of other European

> museums. The ENM's future should be, before all else, to operate as an ethnological museum with an open spirit and a diverse collection, which will gather, store, research, and display different cultural inheritances, not so much on a national as a scientific basis.[8] But also, the ENM should preserve the knowledge of its historical role in the creation of the Estonian nation and through its collection investigate and display this role.[9]

The competition attracted Europe-wide interest and, quite unexpectedly, settled on a winning concept which positioned the museum building as an extension to the runway on the airfield at Raadi. The concept, developed by three young architects from Paris, Dan Dorell, Lina Ghotmeh and Tsuyoshi Tane, entitled 'Memory Field',[10] completely ignored the common public understanding of the museum as the repository of the country's romantic peasant past. Instead, the winning design attached itself to wider contemporary historical debate and something the media had not even considered when discussing the general vision for the museum. It had seemed unthinkable that the Soviet occupation could be part of the discourse on Estonian identity. The judging committee remarked:

> The ideological premise behind this entry is somewhat unexpected and surprising, given Estonia's dramatic recent history – the devastating Soviet occupation lasting more than half a century. This history cannot and must not be banished from the nation's memory by denying the traces still present; rather, these traces should be given a new meaning that inspires hope. This is a design that opens up discussions.
>
> (ENM 2006)

Dan Dorell, one of the architects of the design, said that their inspiration had come from Berlin where the Berlin Wall had been transformed from being a metaphor for Eastern Bloc repression into art that integrated history into the urban space, giving it new meanings.[11] Contemporary Berlin now had numerous objects and spaces that connected the city and nation to its recent history in the lived environment.

Andres Kurg, an Estonian architectural historian, saw in the new museum's design a tension between 'appropriate' and 'inappropriate' elements, as they might be termed in the collective memory of Estonians. He noted that for these Western European architects, the airfield signifies the polarity of the Cold War, rather than the occupation of Estonia: 'For locals it also signifies pollution, confinement, Dzhokhar Dudayev[12] or something else.'[13] Kurg saw the airfield (Figure 20.2) as a scar left from a century of modernisation which was situated in the midst of conflicts and their after-effects but yet which were set to continue to shape the modern nation. He felt that collecting this history together and explaining it was very much a task for the future Estonian National Museum.

The architects understood that recent historical events evoked strong emotions in contemporary Estonians but their desire was to locate a physical and mental space

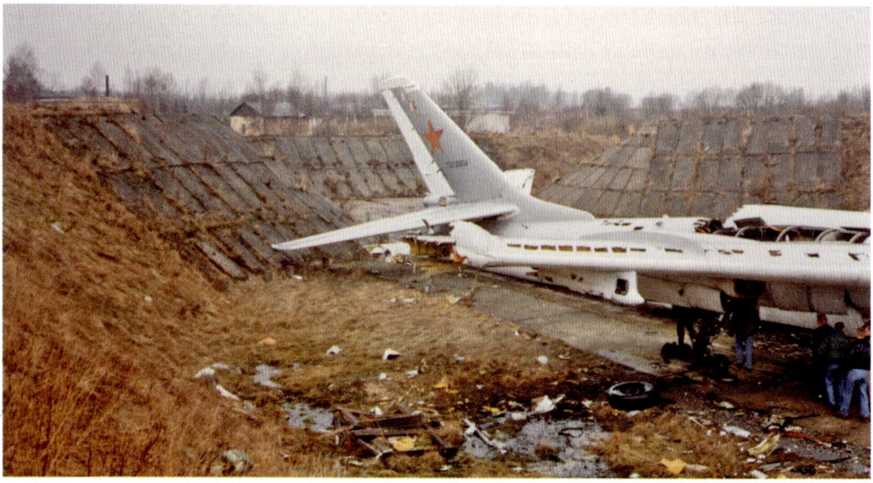

Figure 20.2 Raadi Airfield. Photo: Museum of the Estonian National Defence College

which opened up modes of meaning-making beyond that of explicit condemnation and the eradication of all of residues of the Soviet past. It was seen as an important new role for the Estonian National Museum. The museum also had to recognise that it had entered an increasingly heterogeneous world, that it existed in a state now fully engaged in Europe and with global socio-cultural and intellectual trends. This, however, brought its own challenges, for Estonians also needed to restore and consolidate their own ethnically based sense of nationhood; the world was open to them after fifty years of closure, but they still had work to do at home.

Not everyone saw the plan for the museum positively. Some felt it perpetuated the occupation, and that the imposition of these connections on the very fabric of the national museum only served to humiliate and undermine Estonian identity.[14] The nation had developed strong online communities, and the web soon became the hub for debates concerning the museum's future. Here the tension between the popular image of the ethnographically based museum – that secure monument to the romantic national era which underpinned a long-held national identity – and the winning project, relying as it did on controversial and negative symbols, produced considerable argument. This popular debate indicated that, while the project had its supporters, the critical voices of public opinion dominated. One commentator wrote, outraged at the architects' suggestion that the project offered therapy for Estonians, 'this Soviet airfield runway is symbolic suicide. . . . Frustrating ignorance regarding the common cultural memory!!'[15] The idea seemed an imposition which ignored the public's collective memory altogether. Several groups felt that the museum's representations of the past should be a string of 'beautiful events and secure symbols', which ignored bloodier history. The airfield plan served to glorify the occupation and open up old wounds in a kind of 'psycho-analysis' of the past. Others thought such attitudes were akin to burying one's head in the sand. The museum had already embarked on capturing history in all its forms and circumstances:

> Estonian cultural heritage, for the documentation and preservation of which the Estonian National Museum has been established, as far as I know, is a way more diverse phenomenon than Jakob Hurt[16] and the Finno-Ugric. The museum is for example, actively collecting material from the Soviet period. . . . If the ENM building adopted the calming and comforting mode of the barn swallow,[17] it would mean that we would still be able to identify ourselves only with the secure self-image and the image of history. Which is unfortunately just an illusion.[18]

The architectural competition had effectively disrupted the process of authorship which had begun with popular opinion and the reawakening of a nation seeking to connect to its pre-Second World War past. The successful architects saw the project in terms of prestige – giving Estonia a landmark cultural building that could sit beside those in other European nations. Naturally, these architects thought spatially and sought to highlight the spatial authorship over which they had control. Although they had no expertise as historians, curators or Estonians, they aimed to find a symbol through which to open up the contested issues of the recent history of the nation and give the control back to the people through spatial means. For the general public, the restitution of the nation concerned issues of historical memory. Space was a subordinate matter. Debate in the media reflected the positions of both these parties, but it did not create debate between them or result in public participation. Between these two groups sat the museum's curators who were primarily interested in the internal spaces of the museum and their authorship. For them, restitution was a point of observation rather than participation and their interests focused on the academic understanding of folk culture as an important underpinning of national identity.

Participation

What emerges as critical to the production of the new museum is this distinction between producer (author) and consumer (audience); in what ways was the audience being engaged in the production process and did this constitute participation in the act of authorship? In early discussions, it appears that the public only had a voice in the various media and not in the project itself. This permitted this potential audience the role of commentator on a predetermined end product, using their resources of historical memory and its fixed meanings and leisure consumption as a predesigned experience. The disappointment in the grand narrative of participation was itself a product of Estonia's recent cultural transformation; the approach simply had not found a foothold or become established as a social norm (Runnel *et al.* 2009; Kalmus *et al.* 2009). Discussing such issues with the public would have been unthinkable in the Soviet period and the rebirth of cultural participation has been slow after the crisis in early 1990s, after the mass movements of independence had met the harsh realities of the new state. But even in the West, architects are known for their possession and control of landmark projects. The new Estonian museum had become, first and foremost, a piece of architecture (Figure 20.3).

In discussions between the architects and the jury, the audience remained a vaguely conceptualised entity. It was to be the *target* of meaning-making and strongly guided by the architectural idea rather than by audience study. The jury felt they were delivering what they considered the audience needed, indeed what the nation needed: something that could have been considered progressive and 'trendy'. A member of the architectural jury, Winy Maas, stated:

> As a museum of national heritage, displaying the life and culture of the nation in its temporal, spatial, and social diversity, it complements and at the same time competes with emerging pop culture as expressed through the media, fashion, travel and lifestyle. Therefore, the committee was looking for a design that would give the new museum an active role in the new global pop culture that transcends national boundaries, and that would also transform the rather passive, 'dusty' attitude towards the museum into an active, 'hip' presence, that would attract the younger generation and be functional and competitive in the international context.[19]

As this quote highlights, the audience was kept at a distance – socially and conceptually – and remained as abstractly and taxonomically conceived in this process of conceptualisation as in the architects' and engineers' plans for the museum's public services and spatial logistics, which divided visitors up into a general audience, those with special needs and those classified as children. In this preliminary phase, meetings between engineers, museum employees and architects never involved discussion of the audience or its potential involvement and it never appeared as a topic on meeting agendas.

The Director of the Estonian National Museum, Krista Aru, expected the new museum to convey a story that 'should not be a dull monologue, but a lively dialogue, inspiring participation, research, discourse and continued study. A dialogue that attracts people, sparks ideas and encourages new endeavours' (Aru 2006: 9). These words indicate a desire for the museum to be open, inclusive and engaging but there is no indication of how such a dialogue would be produced. As in museum culture generally, this Estonian director had the power to shape the institutional outlook,[20] but all such directors work within cultures possessing established professional outlooks and long-term careers which can be resistant to change.

Some of the museum's curators and project managers certainly considered audience dialogue a new and challenging development. These curators retained an expert relationship with their audience which puts them in the position of authorship with regard to the museum's various messages. Curatorial interaction with the public follows formats established in the early years of the museum: distributing

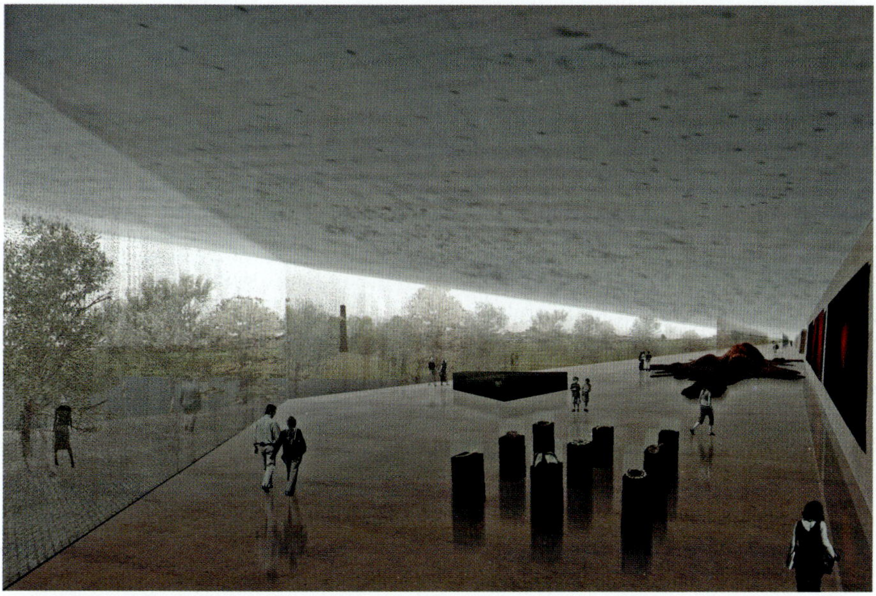

Figure 20.3 Visualisation of the new Estonian National Museum in the winning architectural project. Photo: Dorell.Ghotmeh.Tane/ Architects

questionnaires to the museum's 'network of correspondents'[21] in order to collect ethnographic information; conducting ethnological fieldwork for researching and preparing temporary exhibitions, etc. This kind of audience engagement, despite containing dialogical elements, regards people rather as sources of information and authenticity for the museum, while not leaving much space for active interpretation by the audience or shared authorship.

Perception of the needs of the audience by those controlling the project were mainly based on what they had seen in museums around the world, together with their own particular consumption preferences and cultural tastes. Thus land use, politics and archaeology were deemed 'boring', while volunteer fire-fighters during the first period of independence and the gendered division of everyday life were seen as 'sexy', in terms of audience appeal. When dealing with audience-related matters, staff attention centred on target groups with whom the museum had long had contact. Family members of curatorial staff also appear in conversations and were used to represent the public more generally.

These passing references to the museum's audience, however, did nothing to address the 55 per cent of people living in Estonia who had never heard of the Estonian National Museum.[22] Among this number, the Russian-speaking minority in Estonia is disproportionately significant. The presence of other nationalities in Estonia was discussed with regard to the future museum but they were not integrated into the project. The draft of the profitability and feasibility analyses of the future museum declared that future exhibition spaces would tell the story of non-Estonian groups, particularly Russian speakers, for the first time in the nation's history. The execution of this agenda for minority engagement seems an obvious moment to introduce shared authorship but it would need to overcome the established discourse of ethnic nationalism and the museum's passive engagement with the audience. The ethnographic nature of the Estonian National Museum positioned the audience as a subject and source of information; its marketing efforts saw them only as notional 'target groups'. Perhaps for the museum to realise its potential, change was required not just internally but in the audience itself which had to learn to become active in civic and cultural participation.[23]

Towards innovation and participation

The word 'nation' in the name of the Estonian National Museum indicates that this is an institution serving a living population and a living culture facing the turmoil of today while preparing itself for tomorrow. It is not a word that merely signals that it possesses the residues of earlier or vanished manifestations of the nation. It indicates that the museum has a role in opening up and supporting change, and particularly in addressing an Estonia that is multicultural and situated in a modern global environment. It is in the phase of production that brings into being the new museum that such reconfiguration can take place, but in the case of the Estonian National Museum this phase has produced two quite separate discourses which have failed to communicate. One, popular, vernacular and traditional, is possessed by the public; the other, intellectual, academic and architecturally postmodern, is in the possession of professionals.

In its current form, the Estonian National Museum remains home to the first discourse: an explanatory, fixed view of history based on essentialist views of culture. Although developed by museum curators who researched their interpretations, the exhibition nevertheless communicates a commonsense discourse so well established in Estonian society that one might consider it the national canon. It perceives culture as a bounded unity – something to be secured and protected, using the physical evidence of material objects – which helps express a wider distinction between 'them' and 'us'. In 1994, when the current exhibition was opened, this seemed the correct approach as the country had only recently regained independence and needed to rapidly recover its identity in a shared sense of belonging. As such, it reflects popular conceptions of national culture which, dominated by cultural anxieties, sought security in a timeless romantic peasant past. If the museum was to reinvent itself with this discourse, it would remain centred on a depository for, as one newspaper remarked, 'where the treasury of the nation is located, is also its mentality'.[24] This process of collection-building would result in ritualised confirmation of (ethnic) nationhood and national identity. From the debate surrounding the new museum, it is clear that the Estonian public would consume such an exhibition as authentic.

The professionals involved in shaping the new museum, however, have become attached to a discourse which is intellectual, academic and postmodern (see, for example, Bhabha 1994; Hann 1994; Baumann 1996). It favours negotiation, and the collection becomes rather less central to the museum's purpose. This vision privileges the museum as a particular kind of communication institution, a place that potentially enables changes in what we know and how we think about things, a place that influences attitudes and becomes a laboratory of value systems and identities.

The ethnology curators who are to create the new permanent exhibitions exist between the public and the leaders of the architectural and conceptual project for the museum as a whole. Because of the expectations of the general public, they will, however, be attached to some extent also in the future to a view of culture that is static, fixed, objective, consensual and uniformly shared by the majority of the (ethnic) community. This is seen, for example, in the folk art consultation centre, to be opened in the new building, which would draw its knowledge from the museum collections and give the public advice concerning the 'rights and wrongs' of how to wear ethnic dress, amongst other things.

The missing ingredient at the heart of these issues is the audience itself. Without dialogue during the preparation of the museum, a lot depends on guesswork – or rather, the preferences of those in control on the basis of their professional experience and greater exposure to museums as creative and evolving institutions. Of course, the team developing the museum may have genuine concerns that such consultation will result in compromise and the diminution of concept. The only time the audience was more fully considered was during the writing of the EU application to secure funding for the new museum building. But even then the project-writing company undertaking this work were interested only in market targets and cost-benefit – the audience as numbers rather than people.

A dialogical national museum permitting shared authorship calls not only for the modernisation of the museum's communication and consultation apparatus, it may also require a fundamental shift in the underlying concept of 'Estonianness'. Without these changes the museum cannot be reinvented, but rather would continue to perpetuate itself in its own values. Yet, audience awareness shows that this does not reflect the modern Estonian state, the realities of the past or the nation's modern context. The key issue in the transformation process, or 'reinvention' of the Estonian National Museum, is whether the museum will be able to substitute or enrich ethnic nationalism with civic nationalism. This could be done by supporting and fostering participation; offering audiences an opportunity to engage in rewriting stories about the Estonian past. This would also enable the Estonian National Museum to broaden the default concept of Estonian identity. The role of the museum is, in this setting, to provide *all* audiences with thought-provoking materials rather than just comfort blankets. If the national museum is able to take on this active role in reimagining the nation in this way, then rather than simply being a site of social memory, it can also be a place of cultural innovation and cohesion.

Acknowledgements

The writing of this article was supported by the Estonian Science Fund grant no. 8006 and the Estonian Ministry of Education and Science, target financing project no. SF0180002s07.

Notes

1 Graffman, K. (2004) 'The cruel masses: How producers at a Swedish commercial television production company construct their viewers'. Available at: <http://www.media-anthropology.net/graffman_thecruelmasses.pdf > (accessed 15 May 2008).

2 This point was made in round table discussions in 2006 about the different exhibitions in Estonian museums.

3 The museum needed to remove its collections from temporary stores in different churches in Tartu and return the buildings to the congregations. The buildings were leased/rented until 2005.

4 Mikelsaar, Raik-Hiio (2003) 'Rajame Raadile estoloogiakeskuse ja mini-Eesti?', *Tartu Postimees* 98(1438)/23.05.03, 2.

5 After the thorough discussions in the councils for museums, architecture and heritage protection, the Raadi area was chosen. On 31 July 2003, the Ministry of Culture confirmed the decision.

6 'Ja saagu muuseum!' , *Postimees*. Available at: http://vana.www.postimees.ee/index.html?op=lugu&id=110181&number=870&rubriik=6 (accessed 20 March 2010).

7 The international, public architectural competition started on 22 June 2005, organised jointly by the Ministry of Culture and the Estonian Architects Union.

8 Tamm, M. (2005) 'The status of the ENM within Estonian culture'. Available at: <http://www.museumcompetition.org/en/museum> (accessed 27 October 2008).

9 Native English speakers should note that in continental Europe the term 'scientific' is not used in the same narrow and particular sense in which it is applied in the UK, USA and elsewhere. Here it refers to rigorous academic study and thus embraces such subjects as history without implying a narrowly 'scientific' conceptualisation.

10 The architects, originally from Italy, Lebanon and Japan, established themselves in Paris.

11 Aesma, M. (2006) 'Võitjaarhitekt Dan Dorell: Raadist võib saada põnev paik', *Postimees*. Available at: <http://tartu.postimees.ee/170106/tartu_postimees/uudised/188990.php> (accessed 27 February 2008).

12 Dzhokhar Dudayev (1944–1996) was the first president of Chechnya and before that a Major-General of Raadi air base in the last years of the Soviet Union. His actions during that period are seen to have favoured the Estonian nationalist movement over the Soviet authorities.

13 Kurg, A. (2006) 'Hoolikalt polsterdatud arhitektuurivaidlus', *Eesti Ekspress*. Available at: <http://www.ekspress.ee/viewdoc/EC4C730C771D9B82C22570FF00619456> (accessed 27 October 2008).

14 Hallas-Murula, K. (2006) 'Võidutöö tekitas masendust', *Postimees*. Available at: <http://tartu.postimees.ee/180106/tartu_postimees/arvamus/188991.php> (accessed 27 October 2008).

15 Hallas-Murula, K. (2006) 'Võidutöö tekitas masendust', *Postimees*. Available at: <http:

//tartu.postimees.ee/180106/tartu_postimees/arvamus/188991.php> (accessed 27 October 2008). Reader's comments on the article.

16 Jakob Hurt (1839–1907), one of the central figures in the Estonian national awakening movement, initiated a massive folklore-collecting campaign in 1888, thus paving the way for the collecting activities that led to the founding of the Estonian National Museum in 1909.

17 The barn swallow (*hirundo rustica*) was 'appointed' to the position of Estonian national bird in 1962, at the peak of socialist optimism in the Soviet Union when the authorities launched campaigns to invent national symbols. The barn swallow, like a few other national symbols of the Soviet era, has not lost its meaning, its silhouette today marking locally produced food products, for example.

18 Maas, W. and Alver, A. (2006) 'Muuseum ei pea viitama suitsupääsukesele', *Eesti Päevaleht*, 17 January 2006, http://www.epl.ee/?artikkel=309709. Reader's comments on the article.

19 Paulus, K. (2006) 'Mälu lahinguväljad: Intervjuu Winy Maasiga', *Eesti Ekspress*. Available at: <http://paber.ekspress.ee/viewdoc/82D7129DB71C5B0CC22570F8005ECDFA> (accessed 27 February 2009).

20 Tali, M. and Pierantoni, L. (2008b) 'Ambitions and realities: Developing exhibition programs in new museums'. NaMu conference: European national museums encountering a globalised culture (18 November). Presentation, Oslo.

21 The correspondents' network of the Estonian National Museum was founded in 1931 in order to collect extra data about the objects collected for the museum. Today the correspondents' network comprises about 700 contributors both from Estonia and abroad, answering the questionnaires regarding both modernity and the recent past. The collection of materials on the topics chosen by the correspondents also continues.

22 Turu-Uuringute AS (2008) 'Eesti Rahva Muuseum 2008. Elanikkonnaküsitluse tulemused'.

23 Morrone, A. (2006) *Guidelines for Measuring Cultural Participation*, UNESCO Institute for Statistics. Available at: http://www.uis.unesco.org/template/pdf/cscl/framework/CUL_particip.pdf.

24 Mikelsaar, R.-H. (2003) 'Rajame Raadile estoloogiakeskuse ja mini-Eesti?', *Tartu Postimees* 98(1438)/23.05.03, 2.

Bibliography

Aljas, A. (ed.) (2006) *Estonian National Museum Open International Architecture Competition: Catalogue*, Tartu: Estonian National Museum, p. 25.

Aronsson, P. and Nyblom, A. (eds) (2008) *Comparing: National Museums, Territories, Nation-building and Change, NaMu IV, Conference Proceedings 18–20 February, Linköping University*, Norrköping: Sweden.

Aru, K. (2006) 'The new Estonian National Museum building – an opportunity', in A. Aljas (ed.) *Estonian National Museum Open International Architecture Competition: Catalogue*, Tartu: Estonian National Museum, p. 9.

Baumann, G. (1996) *Contesting Culture: Discourses of Identity in Multi-ethnic London*, Cambridge: Cambridge University Press.

Bhabha, H. (1994) *The Location of Culture*, London: Routledge.

Black, H. (2005) *The Engaging Museum: Developing Museums for Visitor Involvement*, London: Routledge.

Bourdieu, P. (1993) *The Field of Cultural Production: Essays on Art and Literature*, New York: Columbia University Press.

Carpentier, N. (2007) 'Participation and interactivity: Changing perspectives: The construction of an integrated model on access, interaction and participation', in V. Nightingale and T. Dwyer (eds) *New Media Worlds: Challenges for Convergence,* Oxford: Oxford University Press, pp. 214–30.

ENM (2006) 'General committee's remarks concerning the winning entries', in A. Aljas (ed.) *Estonian National Museum Open International Architecture Competition: Catalogue*, Tartu: Estonian National Museum, p. 25.

Hann, C. (1994) *Social Anthropology*, Lincolnwood, IL: NTC Publishing Group.

Kalmus, V., Pruulmann-Vengerfeldt, P., Runnel, P. and Siibak, A. (2009) 'Mapping the terrain of "Generation C": Places and practices of online content creation among Estonian teenagers', *Journal of Computer-mediated Communication*, 14 (4): 1257–82.

Kennedy, M. (2002) *Cultural Formations of Post-Communism: Emancipation, Transition, Nation, and War,* Minneapolis: University of Minnesota Press.

Lauristin, M. and Vihalemm, P. (2009) 'Internal and external factors influencing Estonian public agendas during two decades of post-communist transformation', *Journal of Baltic Studies*, 40(1): 1–28.

Maas, W. (2006) 'The final summary by the general committee –16 January 2006', in A. Aljas (ed.) *Estonian National Museum Open International Architecture Competition: Catalogue*, Tartu: Estonian National Museum, p. 23.

Mensch, P. van (2005) 'Annotating the environment: Heritage and new technologies', *Nordisk Museologi,* 2: 17–27.

Peterson, M. A. (2003) *Anthropology and Mass Communication: Media and Myth in the New Millennium*, New York: Berghahn Books.

Runnel, P., Pruulmann-Vengerfeldt, P. and Reinsalu, K. (2009) 'Estonian Tiger Leap from post-communism to the information society: From policy to practices', *Journal of Baltic Studies*, 40(1): 29–51.

Stead, N. (2004) 'The semblance of populism: National museum of Australia', *The Journal of Architecture,* 9(3): 385–96.

Tali, M. and Pierantoni, L. (2008) 'The processes of contemporary museum constructions: Designing public space and engaging audiences', in P. Aronsson and A. Nyblom (eds) *Comparing: National Museums, Territories, Nation-building and Change, NaMu IV, Conference Proceedings 18–20 February, Linköping University*, Norrköping: Sweden.

Chapter 21

RECOUNTING HISTORY

Constructing a national narrative in the Hong Kong Museum of History

Emily Stokes-Rees

JEROME BRUNER, IN HIS BOOK *Acts of Meaning* (1992) observes that human beings are natural storytellers, that they make sense of the world and themselves through narrative. This chapter is concerned with the narratives in Hong Kong's Museum of History and their use in the production of 'national' identity. As a knowledge-producing institution, the museum offers a location in which 'we can study the making of culture in its concrete reality' (Ames 1992b: 47). In the Museum of History a range of technologies and museological techniques are used to create this concrete reality. In an era when so much is made of personal meaning-making in museums, it is interesting to ask, as Corrine Kratz (2001: 217) did when writing about a travelling photographic exhibition, 'How can exhibitions encourage some ways of understanding rather than others?' As the exhibit moved from venue to venue, she reflected on the ways slight physical modifications to the displays produced a totally different effect at each location. Taking into account the different audiences that viewed the show, she used this information to analyse and demystify the exhibition process. Donna Haraway (1989: 35), to cite another example, in her analysis of the American Museum of Natural History, refers to the methods she observed being used to articulate particular forms of knowledge or power as 'the techniques of effecting meanings'. The aim of this chapter is to apply these ideas to a discussion of some of the 'techniques' at work in the Hong Kong Museum of History, analysing how certain design elements impact on the construction of a particular national narrative for postcolonial Hong Kong.

Material for this study was gathered through detailed analysis of the museum's displays and texts, the observation of museum visitors, and discussions with the curators who were in charge of the construction of the (hi)stories presented. Within museums, display (whether object-based or otherwise) remains the predominant visual form of communication. It is the experience of the displays that, for most visitors, defines the museum, and it is through displays that museums and galleries produce and communicate knowledge (Hooper-Greenhill 1992). For this reason, the most important aspect of this methodological approach is to 'read' the museum exhibits like texts, subjecting their 'grammar' to detailed, critical analysis. The English term 'gaze' refers to 'a particular kind of looking – a steady, intense or absorbed kind of vision . . . that relates seer, seen, the conventions of seeing, and the physical, ritual, and historical contexts of seeing' (Morgan 2005: 4), and this has strongly informed the 'seeing' in this chapter. While it is true that there is no single or correct interpretation or reading of a museum display, of what it 'means' or what it is 'saying', when the exhibits are discussed as producing certain effects, the intention is to speak as much as possible for 'the audience', offering a reading of the museum from the standpoint of a reflective visitor as well as a trained researcher, who observes and absorbs the museum's messages. Approaching the analysis of display in this way offers unique insight into the particular ways museum stories are shaped and narratives framed.

The emergence of Hong Kong identity

In 1997, Hong Kong was returned to China after 156 years of British rule. An event widely reviewed as 'one of the final chapters in the worldwide history of colonialism and postcolonialism', it elicited an abundance of articles speculating on the shaping of Hong Kong's identity under the new regime (Yeoh 2001: 458).[1] Local post-colonial theorist, Akbar Abbas (1997) drew attention at that time to the ways in which the culture of Hong Kong embodies Jackie Chan and John Woo, British colonial architecture, postmodern skyscrapers, Chinese *feng shui* and Western business sense. Hong Kong, he wrote, was 'floating between two worlds' – colonial Britain and communist China – its real identity perhaps disappearing in the process. Through the dualities of East and West, and modern and traditional, Hong Kong's real transition has been complex and profound, and concerned rather more with identity than sovereignty (Abbas 1997: 5–6).

Ironically, it was not until the people of Hong Kong were faced with the imposition of mainland power, which came with the signing of the 'Sino-British Joint Declaration' in 1984, that they began to discuss the idea of there being a 'Hong Kong identity' (Abbas 1997: 3–4). Writing in *China Quarterly* at that time, Hugh Baker described 'The Emergence of Hong Kong Man' as follows:

> He is go-getting and highly competitive, tough for survival, quick thinking and flexible. He wears Western clothes, speaks English or expects his children to do so, drinks Western alcohol, has sophisticated taste in cars and household gadgetry, and expects life to provide a constant stream of excitement and new openings. But he is not British or Western (merely

> Westernised). At the same time he is not Chinese in the same way that citizens of the People's Republic of China are Chinese. . . . Hong Kong man is *sui generis* and the problems of the territory's future are more difficult to resolve because of it.
>
> (Baker 1983: 478–79)

Through this transition period, the belief that 'the Hong Kong experience has been a unique one in the history of modern China, and perhaps in the history of the world, and it deserves, and needs to be, recorded' became central to Hong Kong identity (Young, quoted in Evans and Tam 1997a: 14). Indeed, five years after the handover, a news article by Hong Kong journalist, Klaudia Lee, observed: 'It appears that Hong Kong people are becoming more patriotic and aware of their history . . . and increasingly identifying themselves as Hong Kongers' (Lee 2003). This identity reflects that of a territory which even under British rule had established itself as a world city while still retaining aspects of its indigenous culture, and it is this complex meeting of East and West, and of tradition and modernity, that has been uniquely packaged in the new Hong Kong Museum of History.

Building the Museum of History

In 1962, a temporary museum and gallery occupied the top two floors of City Hall in the Central district of Hong Kong. The museum consisted of a seemingly random display of cluttered objects which ranged from local paintings to various types of archaeological material with little interpretation. The museum guidebook read: 'Historical objects collected range from invaluable philatelic items and coins to old account books and commercial letters rescued from dustbins, from an arrowhead to a peak tram' (Hong Kong Museum of History 1991: 4). In 1975, the Urban Council decided to split the museum into the Hong Kong Museum of History and the Hong Kong Art Museum and relocate the former to a rented site near 'Star House' in Tsim Sha Tsui (Watson 1998: 172). The museum moved again to a converted army barracks in Kowloon Park in 1983, and in 1988–89, the exhibition space in this museum was extended to almost double its previous size, to 1000 square metres.

With growing public interest in the city's origins and identity, and with strong government backing, the Hong Kong Museum of History installed in 1989 a temporary exhibition outlining 6000 years of the region's history. *The Story of Hong Kong*, as it was called, began in traditional fashion by laying the foundations of that history in nature. Archaeological finds provided evidence of the earliest known human activities in the territory, the first time such material had been exhibited in Hong Kong.[2] The exhibition also introduced Hong Kong's different ethnic groups, their means of livelihood and traditional customs. The main thematic focus of the exhibition, however, was the transformation of Hong Kong from a fishing village to a modern metropolis (see Cheung 2001: 565), including the cession of Hong Kong Island and the building of the City of Victoria, industrial development, the Japanese occupation, and the growth of the modern city. The displays featured historical photographs, documents and objects, supplemented by audio-visual devices. In view of the limited space available, however, this grand narrative could not be covered in detail and it was acknowledged that a new and larger building was required to do justice to 'Hong Kong's Story' (Watson 1998: 173).

The construction of a permanent national history museum commenced in 1996, the year before the handover, under the direction of Dr Joseph Ting, the Chief Curator, and using André and Associates, a Canadian design company, to develop the interpretation plan. The new Hong Kong Museum of History (Figure 21.1) was to house one exhibition, *The Hong Kong Story*, which aimed to explain the history of this tiny enclave from the Devonian period, 400 million years ago, to the return of sovereignty to China in 1997. The completed exhibition was hugely ambitious and occupied almost 7000 square metres of floor space. Chief Curator Ting said at the time of its opening in August 2001:

> We believe that *The Hong Kong Story* has succeeded in recreating Hong Kong's past, and visitors can gain an understanding of Hong Kong's historical development and its old ways of life. . . . The museum is a cultural resource centre – a place for holding exhibitions and activities, complementing classroom learning as well as providing lifelong learning opportunities for the public.[3]

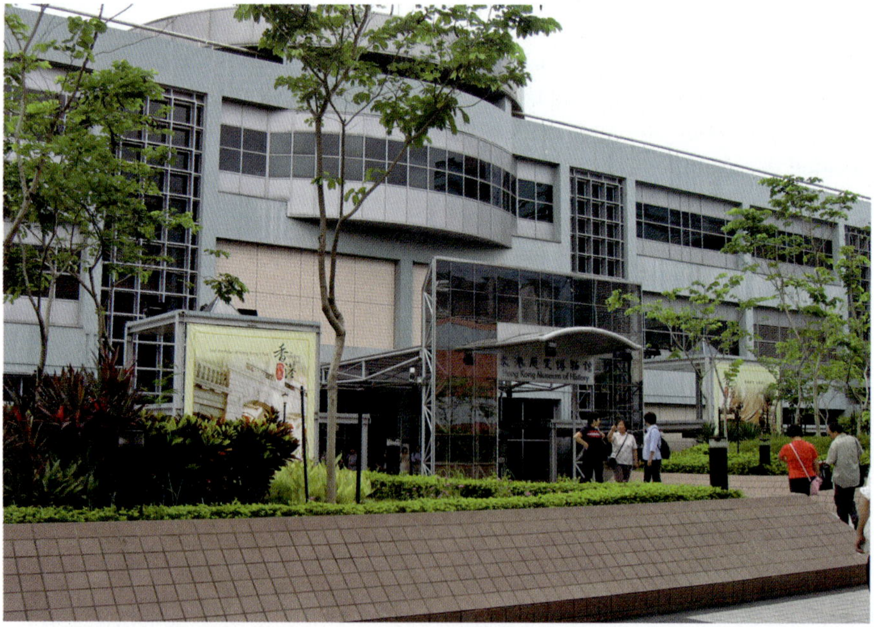

Figure 21.1 The Hong Kong Museum of History. Photo: author

The Hong Kong Story emerged during a period of huge political and ideological change. It sought to locate an identity for the people of Hong Kong which was independent from those of both the departing British and incoming Chinese of the Peoples' Republic. More than a container to house the nation's relics, the exhibition sought to reinforce a distinctly national story.

Designing the national story

In the Hong Kong Museum of History, national history is told like a story divided into chapters, with a carefully controlled pathway guiding the visitor through the narrative.[4] This narrative is broadly that of the experimental temporary exhibition but now extends over eight galleries and two floors, and tells the story of the miraculous transformation of 'a few insignificant fishing villages and farming hamlets to an international metropolis'.[5] The exhibition narrative is progressive and positive: 'We will take you back to Hong Kong's beginnings and then *lead you forward* through the *dramatic events* that have marked her long history.' The direction of travel is carefully signposted with little scope for wandering freely. From the outset one senses a journey towards a point of culmination which calls the visitor's attention to acts of moving and viewing, to ordering one's experience of the nation, to embracing a continuous and unified story of the nation's progress, to a celebration of Hong Kong's modern uniqueness.

Hong Kong's geological past forms the bedrock of this story and is incorporated into the history of the territory to provide drama, a basis in nature and narrative length. It permits Hong Kong to exist before an era of human intervention. In a chapter called 'The Natural Environment', the gallery, comprising two large areas, focuses on the length of Hong Kong's history and, equally, the dramatic nature of its past. The storyline aspect is pushed even at this early point in the exhibition, and a text panel here reads: 'The rocks at Ping Chau are *pages* in Hong Kong's history' (Figure 21.2). It allows Hong Kong to place itself within the landscape, this being a key defining feature of national identity, 'for it is the possession of a territory, a homeland, that most forcefully expresses the identity of those that live within its borders' (Palmer 1998: 183). This sense of being rooted in the territory itself – 'securely fixed to something unquestionably non-artefactual – nature' (Preziosi, Chapter 3, this volume) – is reinforced in the section dealing with prehistory. Here the myth of the nation is being founded in 'a natural image of reality' (Barthes 1989: 155); the myth transmutes what 'is essentially cultural (historical, constructed and motivated) into something which it materialises as natural (transhistorical, innocent and factual)' (Lidchi 1997: 181–82).

If prehistory and nature are unified to establish Hong Kong in the natural landscape, the qualities of Hong Kong people begin to emerge when the exhibition reaches Dynastic history and folk culture. Here tribal clothing, artefacts of jade and porcelain, indigenous baskets and tools predominate and, in an adjoining space, a life-size diorama permits visitors to visualise Hoklo people making salt 'in the old

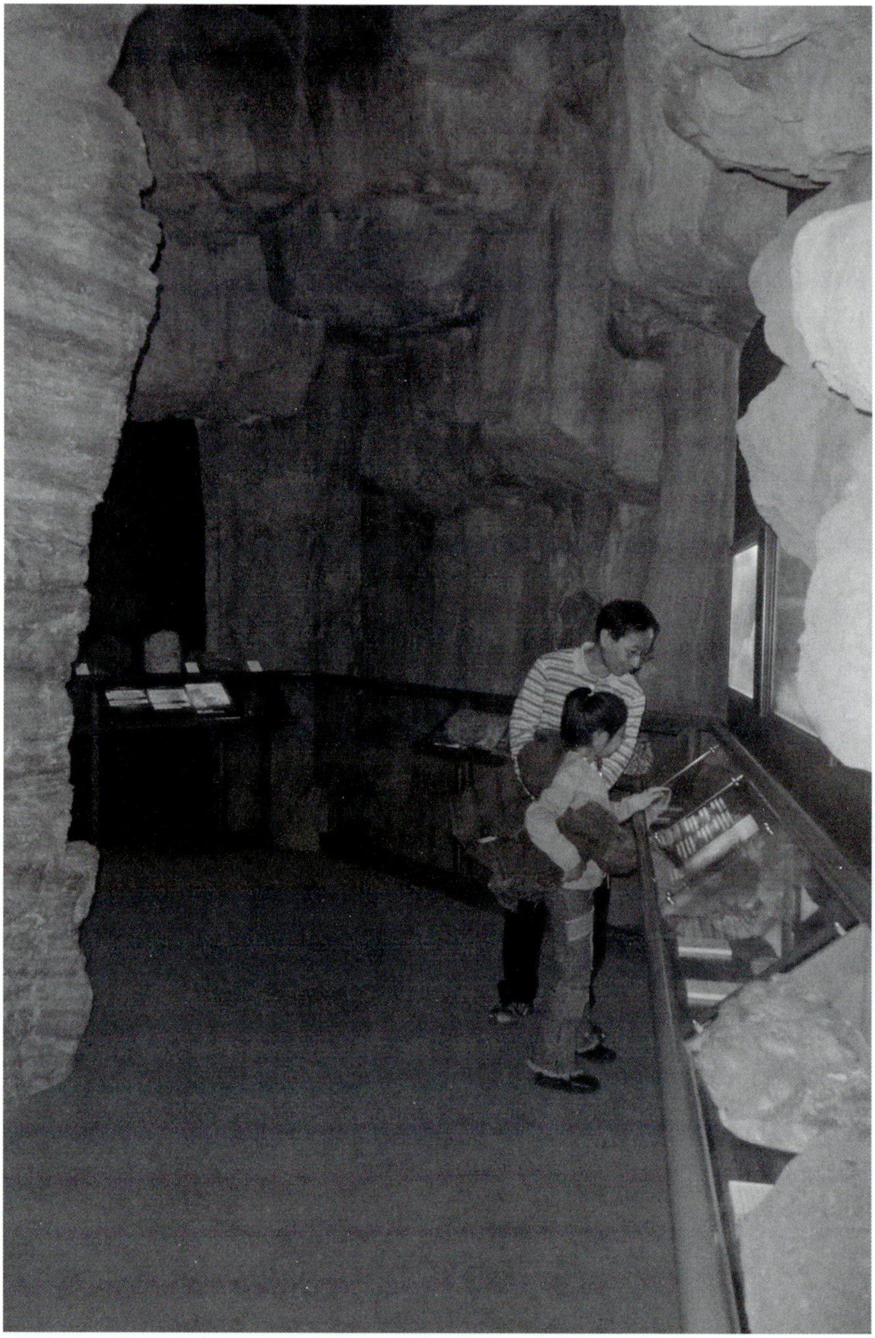

Figure 21.2 The 'rocks at Ping Chau' ground history in an ancient and natural past. Photo: author

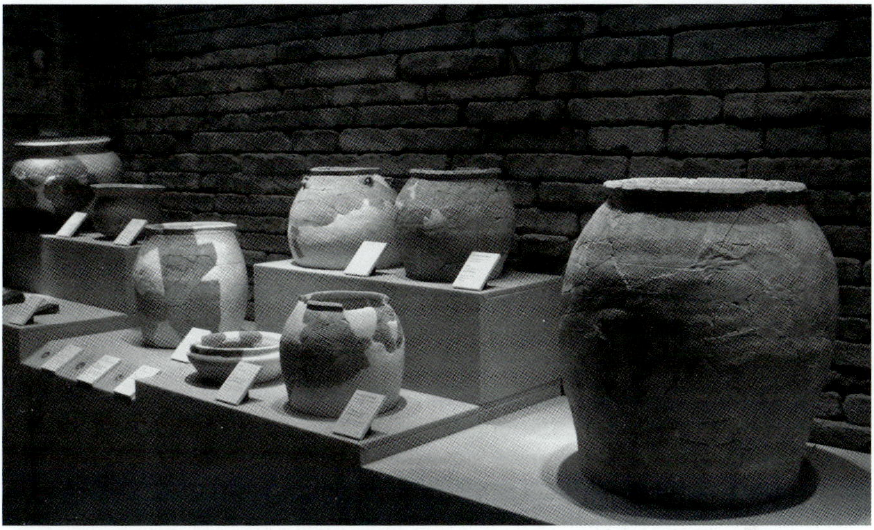

Figure 21.3 Local pottery on display, representing 'tribal traditions'. Photo: author

days'. Their lifestyle is also presented in a model of a sparsely furnished peasant family's dwelling. These exhibits emphasise 'tribal traditions' (Figure 21.3) and 'Chinese resourcefulness',[6] presenting Hong Kong as a nation with a rich cultural history, while also setting the stage for a contrast to be made with its modern image on the upper level of the museum.

On the second floor, the exhibition continues the narrative to the modern day. One of its most arresting exhibits concerns 'High Rise Life Today'. Here, a model of a typical modern high-rise flat, with a floor area of just 11 square metres, gives a vivid impression of the effects of Hong Kong's post-war economic and social development. The contrast between the 'traditional' displayed on the ground floor versus the 'modern' exhibited the level above play off each other in such as way as to reinforce the national narrative of development and success.

In places the architecture and displayed objects are virtually inseparable. Large reconstructions of street scenes, for example, form a physical part of the building; they are immovable and unchanging. In other places, display cases are set into the walls, and so unify the building and its contents into one entity, one idea. At the end of the journey, amongst displays of contemporary Hong Kong, visitors can peer down to the floor below, and reflect on those early beginnings. Here one sees an attempt to 'spatialise material knowledge . . . by the way in which words and texts [and objects] are linked together and arranged in space' (Hooper-Greenhill 1992: 90).

As one moves along the route of *The Hong Kong Story*, there are few opportunities to skip ahead a few pages; the visitor navigates a set path, picking up elements of the story's vignettes along the way. The extent to which this kind of predetermined route is enforced is explained by Tony Bennett's 'exhibitionary complex' – an efficient and disciplinary way of getting visitors to internalise meaning: 'to comply with a programme of organised walking which transform[s] any tendency to gaze into a highly directed and sequentialised practice of looking' (Bennett 1995: 186–87). Eilean Hooper-Greenhill once described the museum's exhibition practices as follows:

> A division is created between knowing subjects, between producers and consumers of knowledge, expert and layman [*sic*]. In the public museum the producing subject is located in the hidden spaces of the museum, while the consuming subject is located in the public spaces. . . . The seriated public spaces, surveyed and controlled, where knowledge is offered for passive consumption, are emblematic of the museum as one of the apparatuses that create 'docile bodies' in the disciplinary society.
>
> (Hooper-Greenhill 1989: 71)

A conversation with a Hong Kongese woman suggested that such a reading of the museum did not belong to the past:

> While there are some charming shop fronts and promising-looking displays of living conditions, this is certainly an example of dumbing down. Visitors are expected to be passive viewers of someone else's cautious and not very challenging view of the past. There are no substantial extra rooms where, for example, a curator in 2004 might be able to explore the long-term effects of the economic decline of 1997–1998, the changing use of the waterfront, the 1998 elections or the impact of returning to China. . . . Well, we wanted 50 years of no change and that's what the history museum is showing us.[7]

Visitors to the museum appeared to move through the galleries quite quickly, rarely stopping to look at something closely, rather, moving along steadily with the flow of people, pausing at larger exhibits such as a full-scale model of a Tanka boat or streetscape reconstruction, but otherwise taking in the whole environment rather than following the carefully produced narrative. After about thirty minutes or so, most visitors stop lingering and simply start walking along, following others and taking note of certain displays that catch their eye, but rarely stopping for more than a moment or two. One senses that despite the museum's efforts to write the story, most people go home with a general sense of the big picture of Hong Kong's 'miraculous transformation' and an appreciation of the nation's success, without having taken in much detail.

The exhibition's historical timeline is reinforced as each display refers to the next in the sequence and relies on the ones before. So, 'Prehistoric Hong Kong' flows logically out of 'The Natural Environment' and transitions smoothly into 'The Dynasties'. Later on in the displays, 'The Japanese Occupation' merges seamlessly into 'The Modern Metropolis', implying that perhaps Hong Kong would not have developed in the same way without the influence of the war. Even if a visitor were to attempt to deviate from this predetermined route, display cases, stand-alone objects, and benches are poised to obstruct an attempted escape. As Donald Preziosi observes: 'a rather extraordinary fusion takes place in institutional practice between historical time rendered in space as *sequence and succession* and a certain sense of time as aspect: time as a syntactical relation between events connected in a causal relation' (Preziosi, Chapter 3, this volume). In this way, the national narrative of transformation from fishing village to metropolis is constructed and maintained, the physical structure of the narrative adding to the effect of creating a continuous, authentic story of national development.

Even in the museum shop one senses the centrality of the traditional/modern tension that runs through the museum's narrative – postcards of Hong Kong's spectacular skyline, empirical proof of its miraculous transformation, versus sepia-toned pictures of boat people drying fish and 'natives' emphasising 'traditional roots' (Figure 21.4). By drawing attention to the conflicting experiences of local cultures and global capitalism, the museum simultaneously exploits the existing social desire for a sense of communality, while promoting modernism and globalisation in a time of extraordinary change.

The museum thus represents a commitment to the production of a unique national image, fashioned so as to situate the nation in a story of its own rather than one rooted in wider Chinese or European narratives. Here the identity of the people of Hong Kong is legitimised in narrative space; Hong Kong and its people are undeniably unique and identifiable. However, compressed into a single exhibition space, carefully narrated and arranged, the visitor is given limited possibilities for

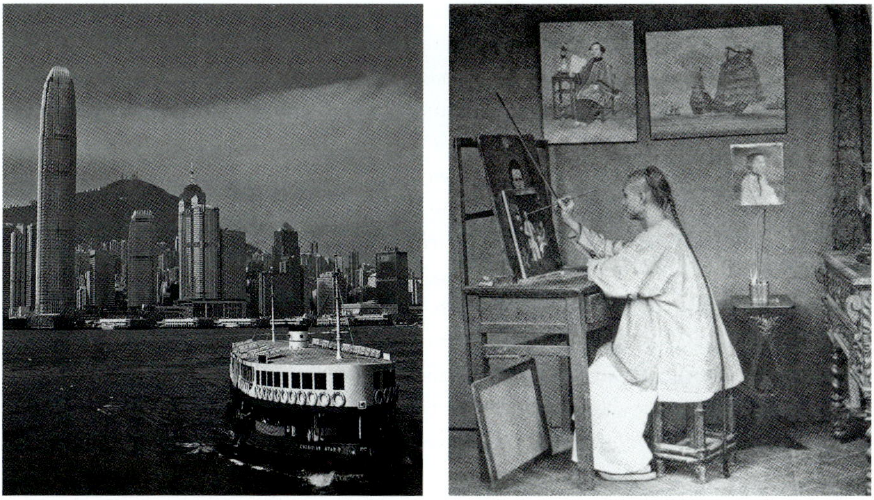

Figure 21.4 Postcards of Hong Kong's spectacular skyline contrast with sepia-toned pictures of former times. Photo: Copyright Hong Kong Museum of History, 2002

movement – physically or mentally – beyond the intended experience. This installation of a single story was an intention of the museum, as Dr Yim Shui-yen, Director of the Hong Kong Heritage Museum in *Sha Tin*, makes clear:

> Why did they call it 'The Hong Kong Story'? Why not '*A*' Hong Kong Story or 'Hong Kong Stories'? I guess there is a need for an 'official history' these days. With so much change and uncertainty people want something they can be sure of.[8]

Massey (1995: 182) notes, 'identity of place is very much bound up with the *histories* which are told . . . *how* these stories are told, and which history turns out to be dominant'. It is important to note, however, recalling Sharon Macdonald's observations at *Aros* on Skye, that 'in using the term "story", [a museum] signals [its] awareness of the fact that the exhibition . . . is a particular *version* of history, told from a particular "point of view". It is also only a very small part of what might potentially be told' (Macdonald 1997: 161). National history thus becomes more than a simple sequence of events; it becomes an *authored* story with direction and purpose.

Authenticating and enhancing the national story

In addition to the physical layout of the narrative, the museum also employs a number of 'technologies' which aid in reinforcing the progression and, importantly, the impact of the national story being told. Contributing to the effect of movement through history, for example, the museum uses the concept of 'time travelling' by displaying its prehistory in a tunnel-like passage, a technique now widely used in museums.[9] As a text panel in the section on 'Prehistoric Hong Kong' attests, 'Archaeological evidence reveals that Hong Kong has a long history, like the rest of China'.[10] The length of Hong Kong's history established in the sections dealing with nature is thus reinforced in the prehistoric section: 'The painstaking excavations of dedicated archaeologists have yielded a wealth of priceless cultural relics confirming human habitation in Hong Kong going back thousands of years.'[11] The effect of the text here, using words like 'relic', which suggests 'old' objects, and phrases such as 'thousands of years' reinforces the passage of time – the modern nation has a *long* history – one that has steadily and linearly progressed through to the present.

The exhibition also shifts focus from traditional *in-context* displays, which depend on the inclusion of artefacts, to *in-situ* displays which, in privileging experience, rely more on display enhancements and effects rather than objects (Kirshenblatt-Gimblett 1998: 3). Writing in the *South China Morning Post* in 1998, journalist Victoria Finlay remarked:

> On plans appear a forest of plastic trees, a fleet of model boats, 6,000 square metres of space, and a bottom line of $580 million. . . . elsewhere, happy

> to be of service, our designers have given us concrete (or man-made) rocks. . . . The rest of the first floor is taken up by the Yue, Sui and Tang dynasties, a full-size model of a Tanka fishing junk, 10 metres long, reproductions of ancestral halls, and a six-room Hakka house.
>
> (Finlay 1998)

There is no shortage of dioramas and reconstructions: a replica teashop, a traditional Chinese wedding parade, a colourful Cantonese Opera and a 42-metre-long beach diorama on which several activities, such as fire-making, house-building and fashioning ornaments with stone are represented (Figure 21.5). Although they rely on the presence of *some* artefacts, it is not important whether they are real or reproduced; it is their presentation in an accurate context and authoritative setting that creates the effect of realism in such a powerful way.

Dioramas have long been criticised for presenting simplistic and linear history; restricting the number of interpretations that can be applied, of telling just *one* story (Coombes 1994; Haraway 1989). And indeed, this is surely the purpose at the Hong Kong Museum of History, where the advancement of its national narrative of tradition versus modernity is presented through a singular storyline, the dioramas acting as illustrations for those disinclined to read. Although the exhibition also highlights the authenticity and significance of many objects using lighting, high-quality cases and labels, the museum perhaps follows Lowenthal (1992: 25) in believing that 'artefacts on their own fail to convey the pulse of life . . . the relics selected for museum display need further assistance to become persuasively lifelike'.

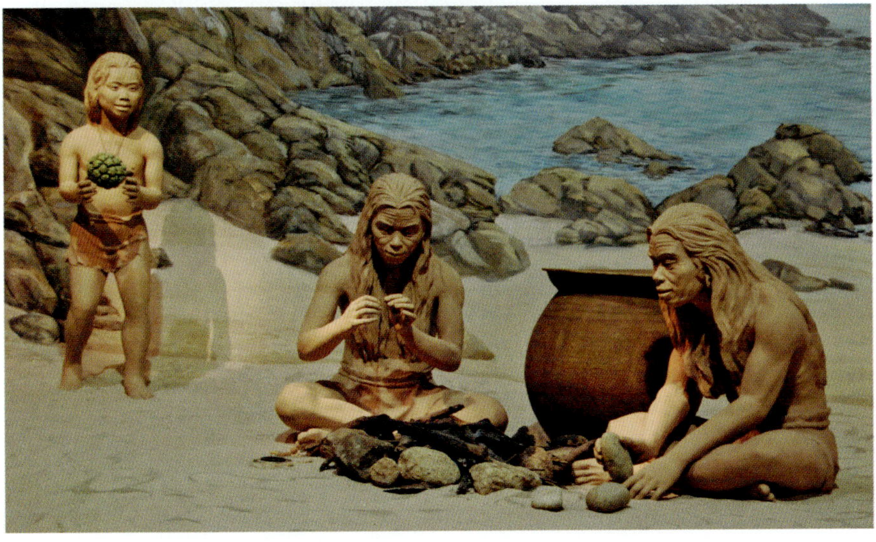

Figure 21.5 A large diorama depicting fire-making, house-building and other activities produces a sense of realism through accurate historical contextualisation. Photo: author

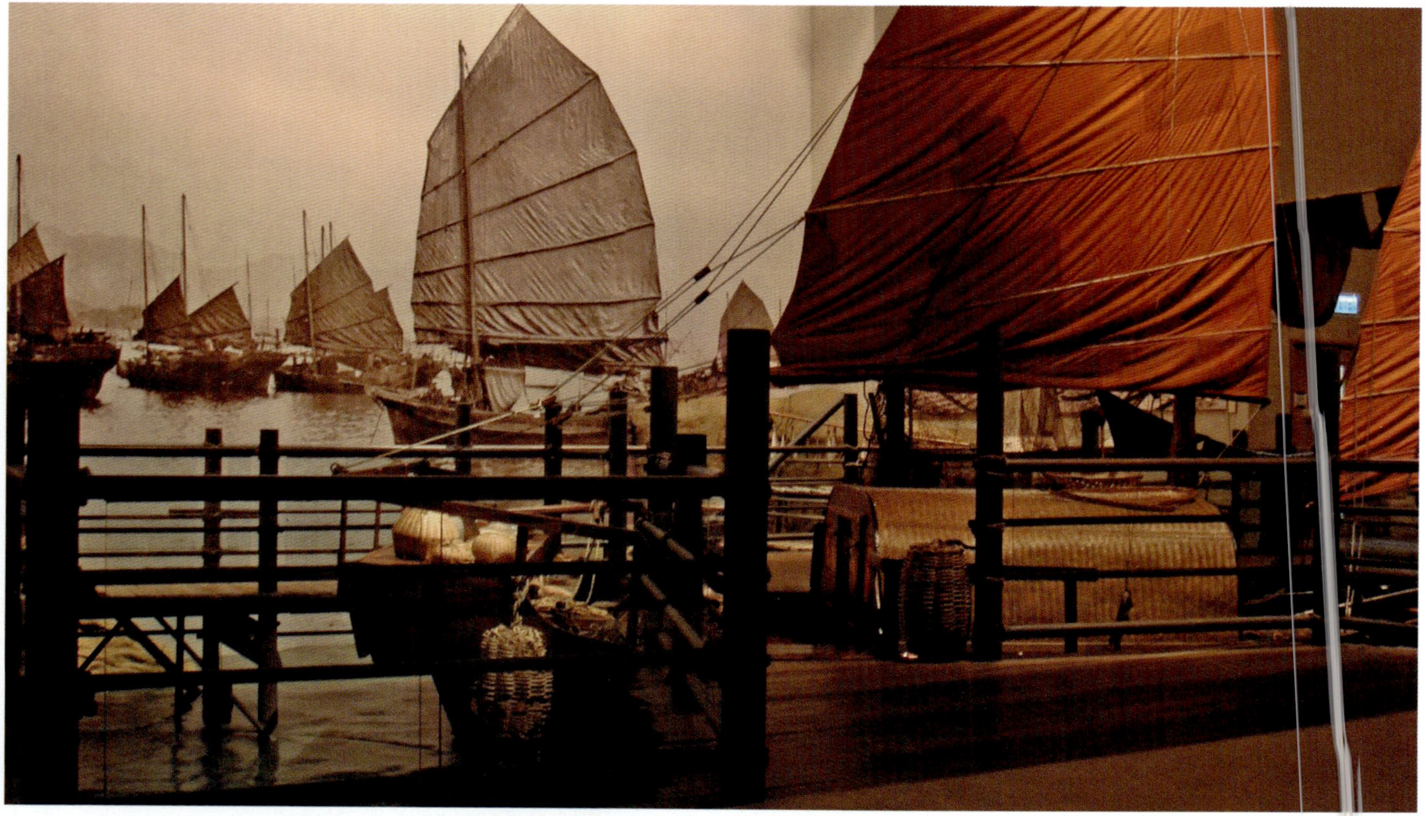

Figure 21.6 Tanka fishing junk. Photo: author

One of the most popular dioramas in the museum depicts Tanka fishermen and their livelihoods, representing them as 'traditional' in opposition to modern industrial Hong Kong (Figure 21.6). In the display, life-size fishermen throw out their nets from a small junk. What can be observed here, I would argue, is a construction of modern Hong Kongese identity as counterposed to a traditional, 'exotic Other' (cf. Karp and Kratz 2000: 196). Ironically, the development of urban, industrial Hong Kong was made possible by exploiting these fishermen, placing strict restrictions on fishing, for example, forcing them to have to work as ferry drivers to make a living.[12] 'Their communities have steadily been dissolving and the Tankas have been paying the price for the rise of modern Hong Kong' (Hung 1998: 11). The museum remembers them differently, however, as making modernisation possible through the positive influence of their traditional entrepreneurship.

The story also draws upon more readymade (primarily male) heroes in the form of Bruce Lee, Chow Yun Fat and Jackie Chan. These two narrative elements of self-sufficient, hardworking fishermen and upwardly mobile, 'heroic' celebrities, seem to connect Hong Kong's narrative of success linking the past to the future, but they also reveal a tension between a local 'indigenous' identity and the global future of a world city.

The construction of readymade images is also achieved by the use of film throughout the museum. It has become a much-used technique to explain the arcane science of geology and is used in Hong Kong to similar effect. Film gives substance to the unimaginable: 'Once we have imagined the past and translated this into imagery, it effectively becomes what we pictured it to be' (Moser 1998: 11). Chris Sia, one of the exhibition's multimedia designers, said:

> Multimedia was a design element from the very start . . . [we] believed that sight and sound was needed to appreciate Hong Kong's story. From its wild, tropical natural history displayed through large dioramas, to its incredibly colourful and ceremonious tribal human beginnings, which we hear in context and we see theatrically, to its modern development; movement, colour and sound have always described Hong Kong.[13]

Sia's words indicate that the intention was not simply to narrate the story but to perform it in the museum as theatrical space. Frey and Kirshenblatt-Gimblett (2002: 59) describe this practice as '"performing museology", reveal[ing] the very nature of the museum as an active agent in constituting knowledge and experience' of the nation. This 'performance', in other words, involves the use of 'time travelling', dioramas and re-creations, and the visual medium of film to communicate meaning, pointing to a shift from the explanation of objects, to the interpretation of narratives or stories. Moreover, by objectifying the idea of the nation in visual form, national values and ideologies are made relevant to modern-day visitors, offering an experiential reality centred on the museum's overarching narrative of Hong Kong's metamorphosis from a tiny fishing village to a modern metropolis, and enhancing a sense of collective nationhood.

Conclusions

The Hong Kong Museum of History has used spatial design, storytelling, mixed media and strategic sequencing to perform its narrative of the 'nation' and in the process produces a singular identity for the territory and its people. By dictating a particular path through the nation's history, and controlling the visitor's direction of movement, ways of seeing, and the objects and displays around which this vision is directed, *The Hong Kong Story* imposes its authority. Indeed, as they move through the spaces of the museum in such a way, visitors, whether consciously or subconsciously, participate in the establishment of Hong Kong's rags-to-riches story as national canon. The inclusion of visual technologies to enhance the telling of its story draws attention to the authority of images, which 'are inextricably interwoven with our personal identities, narratives, lifestyles, cultures and societies as well as with definitions of history, space and truth' (Pink 2001: 17).

Defining the history of Hong Kong in this way cuts to the very heart of the city's ongoing dilemma: how to maintain the distinctive identity of this dynamic outpost of Western capitalism, while at the same time embracing a 'motherland' that many view as threatening to absorb that uniqueness. A national narrative 'is a path carved, a journey charted and projected through the world' (Hearn 2002: 749), and 'a thread that unites otherwise disparate happenings into the significance of a development, a directionality or a destiny' (Kerby 1991, quoted in Ku 2002: 343). One senses that the museum forms a memory of all that Hong Kong was before it became China, and as such it is intended to preserve in this essentialised narrative the very essence of all that is unique.

Notes

1 For examples of articles published in the wake of the handover, see Macpherson (1997), Mathews (1997) and Man and Lo (1998).
2 Chief Curator, Joseph Ting, Hong Kong Museum of History, personal communication, August 2002.
3 Ting quoted in 'Stories of Hong Kong . . .', Hong Kong Museum of History, museum clippings, 2000.
4 Many museums have adopted a storytelling approach. See Macdonald (1997) on *Aros: The Skye Story,* as well as Mupira, P. (1998) 'Why do all museum displays look alike?' (unpublished paper), Zimbabwe: Mutare Museum. Available online at: http://www.natmus.cul.na/projects/samp/mwwtheme/th1mupira.html (accessed 24 November 2004) for his related observations about museum displays in Zimbabwe. Also: *The Story of the Filipino People* at the National Museum of the Philippines (1998), *The Story of Mann* (1992), *Homelands: The Story of the Czech & Slovak People* at the National Czech and Slovak Museum (1998), and *The Singapore Story* (1997).
5 Text from 'The Introductory Gallery', Hong Kong Museum of History, 2002.
6 Gallery text, 'Folk Culture in Hong Kong', 2002.
7 Conversation with a Chinese visitor, female, aged 45–50, Hong Kong Museum of History, August 2002.
8 Interview with Yim Shui-yen, Director of the Hong Kong Heritage Museum, August 2002.

9 This idea of 'time travelling', and the use of a tunnel-like effect can be seen, for example, in the History Hall of the Canadian Museum of Civilization in Ottawa, as well as the new Australian National Museum.

10 Panel text, 'Prehistoric Hong Kong', Hong Kong Museum of History, 2002.

11 Ibid.

12 Hayes, J. (nd.) 'Hong Kong's own boat people' (unpublished paper), Hong Kong Heritage Museum clippings file.

13 Sia, quoted in Goodyer, T. (2003) 'Making history'. Available online at: www.proaudioasia.com/paa/article (accessed 15 March 2005).

Bibliography

Abbas, A. (1997) *Hong Kong: Culture and the Politics of Disappearance*, Minneapolis: University of Minnesota Press.

Ames, M. (1992a) 'The development of museums in the Western world: Tensions between democratization and professionalization', in M. Ames, *Cannibal Tours and Glass Boxes: The Anthropology of Museums*, Vancouver: University of British Columbia Press, pp. 15–24.

Ames, M. (1992b) *Cannibal Tours and Glass Boxes: The Anthropology of Museums*, Vancouver: University of British Columbia Press.

Baker, H. (1983) 'Life in the cities: The emergence of Hong Kong Man', *China Quarterly*, 95: 467–79.

Barthes, R. (1989) *Mythologies*, London: Paladin.

Bennett, T. (1995) *The Birth of the Museum: History, Theory, Politics*, London: Routledge.

Bruner, J.S. (1992) *Acts of Meaning*, Cambridge, MA: Harvard University Press.

Cheung, E. M. K. (2001) 'The Hi/stories of Hong Kong', *Cultural Studies*, 15(3–4): 564–90.

Coombes, A. (1994) *Reinventing Africa: Museums, Material Culture and Popular Imagination in Late Victorian and Edwardian England,* New Haven, CT: Yale University Press.

Evans, G. and Tam, M. (1997a) 'Introduction: The anthropology of contemporary Hong Kong', in G. Evans and M. Tam (eds) *Hong Kong: Anthropology of a Chinese Metropolis*, Richmond: Curzon, pp. 1–24.

Evans, G. and Tam, M. (eds) (1997b) *Hong Kong: Anthropology of a Chinese Metropolis*, Richmond: Curzon.

Finlay, V. (1998) 'A poor reflection of history', *South China Morning Post*, Hong Kong, 13 March.

Frey, B. S. and Kirshenblatt-Gimblett, B. (2002) 'The dematerialization of culture and the deaccessioning of museum collections', *Museum International*, 54(4): 58–63.

Hall, S. (ed.) (1997) *Representation: Cultural Representations and Signifying Practices*, London: Sage.

Hallam, E. and Street, B. (eds) (2000) *Cultural Encounters: Representing Otherness*, London: Routledge.

Haraway, D. (1989) *Primate Visions: Gender, Race and Nature in the World of Modern Science*, London: Routledge.

Hearn, J. (2002) 'Narrative, agency, and mood: On the social construction of national history in Scotland', *Journal of the Comparative Study of Society and History*, 44(2): 745–66.

Hong Kong Museum of History (1991) *A Brief Guide to the Hong Kong Museum of History*, Hong Kong: Urban Council.

Hooper-Greenhill, E. (1989) 'The museum in a disciplinary society', in S. Pearce (ed.) *Museum Studies in Material Culture*, Leicester: Leicester University Press, pp. 61–72.

Hooper-Greenhill, E. (1992) *Museums and the Shaping of Knowledge*, London and New York: Routledge.

Hung, H. 1998. 'Rediscovering the rural in Hong Kong's history: Tankas, Hakkas, Puntis and immigrant farmers under colonialism', *Hong Kong Cultural Studies Bulletin*, 8–9: 2–16.

Karp, I. and Kratz, C. (2000) 'Reflections on the fate of Tipoo's Tiger: Defining cultures through public display', in E. Hallam and B. Street (eds) *Cultural Encounters: Representing Otherness*, London: Routledge, pp. 194–229.

Kerby, A. P. (1991) *Narrative and the Self*, Bloomington: Indiana University Press.

Kirshenblatt-Gimblett, B. (1998) *Destination Culture: Tourism, Museums and Heritage*, Berkeley: University of California Press.

Kratz, C. (2001) *The Ones that are Wanted: Communication and the Politics of Representation in a Photographic Exhibition,* London: University of California Press.

Ku, A. (2002) 'Postcolonial cultural trends in Hong Kong', in A. So and M. Chan (eds) *Crisis and Transformation in China's Hong Kong*, London: Eastgate, pp. 343–62.

Lee, K. (2003) 'Identifying more with Sun Yat-sen', *South China Morning Post*, Hong Kong, 2 April.

Lidchi, H. (1997) 'The poetics and politics of exhibiting other cultures', in S. Hall (ed.) *Representation: Cultural Representations and Signifying Practices*, London: Sage, pp. 151–222.

Lowenthal, D. (1992) 'From patronage to populism', *Museums Journal*, 92(3): 24–27.

Macdonald, S. (1997) 'A people's story: Heritage, identity and authenticity', in J. Urry and C. Rojek (eds) *Touring Cultures: Transformations of Travel and Theory*, London: Routledge, pp. 155–75.

Macpherson, K. L. (1997) 'The city and the state: Historical reflections on Hong Kong's identity in transition, 1997 and beyond', *Cities*, 14(5): 279–86.

Man, S. W. and Lo, S. P. (1998) 'Introduction' to the special issue on 'Cultural identities and cultural politics; colonial and postcolonial imaginations in Hong Kong', *Chinese Sociology and Anthropology*, 30(3): 3–12.

Massey, D. (1995) 'Places and their past', *History Workshop Journal,* 39 (Spring): 182–92.

Mathews, G. (1997) 'Heunggongyahn: On the past, present and future of Hong Kong identity', *Bulletin of Concerned Asian Scholars*, 29(3): 3–13.

Morgan, D. (2005). *The Sacred Gaze: Religious Visual Culture in Theory and Practice*, London: University of California Press.

Moser, S. (1998) 'Seeing the past in standard images', *British Archaeology*, 39: 10–11.

Palmer, C. (1998) 'From theory to practice: Experiencing the nation in everyday life', *Journal of Material Culture*, 3(2): 175–99.

Pearce, S. (ed.) (1989) *Museum Studies in Material Culture*, Leicester: Leicester University Press.

Pink, S. (2001) *Doing Visual Ethnography: Images, Media, and Representation in Research*, London: Sage.

'Stories of Hong Kong exhibition in the Hong Kong Museum of Historical Relics' (2002) Newspaper Article (Hong Kong Heritage Museum, museum clippings) (translated by John Lai), Hong Kong, 14 August.

Urry, J. and Rojek, C. (eds) (1997) *Touring Cultures: Transformations of Travel and Theory*, London: Routledge.

Watson, R. (1998) 'Tales of two "Chinese" history museums', *Curator*, 41(3): 167–77.

Yeoh, B. (2001) 'Postcolonial cities', *Progress in Human Geography*, 25(3): 456–68.

PART FOUR OURSELVES AND OTHERS

Chapter 22

KAESŎNG KORYŎ MUSEUM

The place of one Korean nation?

Ruth Scheidhauer

THIS CHAPTER IS CONCERNED with the role of memory – personal memory and national memory – and the negotiation of multiple national identities, each of which might be simultaneously played out in the museum and the individual. As Confino (1997: 1386) reminds us, memory 'has come to denote the representation of the past and the making of it into a shared cultural knowledge by successive generations in vehicles of memory such as . . . museums'. The subject of this chapter is the Koryŏ Museum situated in Kaesŏng, North Korea (Democratic People's Republic of Korea, DPRK), which, it will be argued, demonstrates how museums can simultaneously mediate overlapping and divergent group memories and identities. For nearly five hundred years, Kaesŏng served as the capital of the Koryŏ dynasty (918–1392), a kingdom that encompassed all of present-day South Korea (Republic of Korea, ROK) and most of North Korea. These same territorial lines remained broadly in place during the succeeding Chosŏn period (1392–1910) and the period of Japanese occupation (1910–1945). For a thousand years, the people of this geographically constant Korean state were able to shape a shared sense of identity. With the end of the Japanese occupation in September 1945, the Korean peninsula was divided under the trusteeship of the Soviet Union and the USA. Kaesŏng is located just a few kilometres north of the Demilitarised Zone along the 38th parallel, the ceasefire line established at the end of the Korean War in 1953. Since then, the Korean peninsula has been divided by two different state ideologies, a socialist society in the North, and a capitalist society in the South.

Drawing on some similarities to the German discourse of one cultural nation (*Kulturnation*) as a – contested – unifying force in two statehoods, this chapter aims to emphasise the very sensitive relationship between the narratives of state and nation and their possible deliberation in a museum. When talking about a national museum in either of the two Korean states, does the nation encompass the whole Korean nation or does it exclusively refer to the respective political nation-state?

The unifying rhetoric of the *Kulturnation*

Although Korean cultural history is actively employed as historical evidence for the ideological claims of each political state, it overwhelmingly stresses a national identity that encompasses the people of the whole Korean peninsula throughout Korean history. The Korean nation is seen by most Korean scholars, regardless of their ideological views, as a continuation of longstanding patterns of ethnicity, built on pre-existing geographic or cultural foundations (Connor 1994; Smith 1986, 1991, 2008). This Korean nation seems to be conceived of as a community with a deeply shared cultural and historical identity regardless of, and overpowering, the actual ideological differences across the 38th parallel. In 1947, the prominent South Korean historian, Son Chint'ae, wrote, 'Since the beginning of history, we [Koreans] have been a single race that has had a common historic life, living in a single territory . . . sharing a common destiny' (cited in Duncan 1998: 198). Similarly, the well-known Marxist historian, Paek Namun, noted in 1946, 'The Korean nation is a unitary nation with a common blood, territory, language, culture and historical destiny for thousands of years' (cited in Pang 1992: 124). Like many archaeologists, these historians argue that the Korean nation has been united since the dawn of historical time, if not always as a state then at very least by culture, blood and language. Though they may stand in direct opposition over claims of national legitimacy and representation, political leaders in South and North Korea share the same view on the ethnic homogeneity of the Korean nation (Shin 2006: 5). Accordingly, the former South Korean president Park Chung Hee proudly proclaimed in 1972 that 'we have never given up our pride nor our dignity in being a homogeneous people'. He even called North Koreans 'our brethren in the north of this great Han race' and argued that 'although we are now separated into south and north, we are one entity with a common destiny, bound by one language, and by one history and by the same racial origin. Ideology changes, but the nation stays and lasts' (Park 1975: 179, 22). The theoretical basis for this concept of a 'shared', or 'common', history and culture has, of course, been contested by Western and Korean scholars alike for being false or invented. Grinker even went as far as to argue that the South Korean nation exhibited a collective desire to maintain the dream of unification rather than achieve it (Grinker 1998; Pai 2000). Nonetheless, the Korean *Kulturnation* has become the underlying principle of shared nationhood based upon the unification discourse present within the existing states. Two North Korean authors, for instance, connect the rhetoric of historical homogeneity to the desire for state unification to ensure the unified development of the Korean national culture.

> Our homogeneous nation that lived on the same territory for thousands of years finds itself cut in two; this obstructs a unified development of the national culture and, if it continues, must stifle the common national characteristics formed during a long history. . . . All men of culture in the Democratic People's Republic of Korea are resolved to devote their efforts

> and their talents to reunification, to overcoming the division of their country and ensuring a unified development of the national culture.
>
> (Chai and Hyon 1980: 41)

The unbroken and homogenous cultural community has been conceptualised as the ultimate justification for the Korean unification ever since the division of Korea. With the aim of rapprochement in mind, cultural exchange and joint Korean cultural heritage activities were introduced, as addressed in white papers, inter-Korean joint agreements and the implementation of the *Socio-Cultural Exchanges Promotion Act* during inter-Korean ministerial talks in 1990. Paradoxically, the state-employed images of unity, homogeneity and a cultural nation based on a common history, language and ethnicity have had to ignore the heterogeneous ideological developments of recent history created by the partition (Cho 2002; Ko 2002). In the third part of the French miscellany 'Lieux de mémoire', *Les France: Conflits et partages* edited by Pierre Nora in 1992, the problematic issue of conflicting memories and identities in the French nation-state is addressed. It made necessary the use of the plural *Les France*. In contrast to the dialectic of divided past memories that shape the shared contemporary French national identity, the contemporary divided Korean nation-state identities hold on to the memories of a shared past.

The Koryŏ Museum: from national awakening to national division

The history of Korean museums has always evolved around nation-building politics. The foundation history of the Koryŏ Museum and the concurrent cultural-political developments in Korea reveal this close relationship and contextualise the interest and historical connection of both Korean states to Kaesŏng and the Koryŏ period. At the very end of the Chosŏn period and during the Japanese occupation, a political and cultural awakening of the nation with hopes for sovereignty emerged on the Korean peninsula. The so-called 'cultural movement', a mass popular cultural identification, regarded national cultural consciousness as the unifying force for nation building and for national liberation on the basis of shared cultural characteristics, such as language and common moral aspirations. Set amidst this Zeitgeist came the opening of the Imperial Household Museum in 1909, when Korea was already a Japanese protectorate.[1] Granting public access under state stewardship, it was the first Korean museum that could be defined as a national museum. On 1 December 1915 the Japanese colonial government inaugurated the Chosŏn General Government Museum, and the Imperial Household Museum was demoted and renamed the Yi Royal Household Museum, Yi being the dynastic name of the Chosŏn period. The renaming of the museum demonstrates the political significance in the choice of a museum's name, as it was purposefully chosen to mark the political power of the Japanese colonial government.

The name given to the Koryŏ Museum is important in regard to nation and state identifications. The official historiographies of both Koreas culturally identify with the golden period of the first Korean unified dynasty, the Koryŏ period. In addition to shared cultural identifications with the Koryŏ period, the history of the Koryŏ Museum collection also reveals how it has been a pawn in a changing political scene. The Koryŏ Museum's predecessor, the Kaesŏng Provincial Museum, was founded on 1 October 1930 (Ko 2007: 355). Prior to the opening of the Kaesŏng Provincial Museum, a collection of local Koryŏ period relics were placed in a small gallery in front of Kaesŏng's royal palace site Manwŏltae (Chin 1970: 167). This gallery had long been inaccessible, but in 1912 the Kaesŏng provincial office took an initiative to protect its old monuments and relics and founded the Kaesŏng Safeguarding Committee. Its efforts to provide a space for these artefacts resulted in the establishment of the Kaesŏng Provincial Museum. It is a source of present-day pride that its foundation and administration were Korean. The museum was under the leadership of Korean directors, while during the Japanese occupation most employees at the Chosŏn General Government Museum were Japanese. Furthermore, arguing that the collections were of local significance, the director Ko Yu-sŏp of Kaesŏng Provincial Museum managed to decline requests to become a branch of the Chosŏn General Government Museum.[2] Nevertheless, many of the Kaesŏng artefacts were 'loaned' to the Chosŏn General Government Museum. After the Korean liberation in 1945, Kaesŏng became part of the South Korean provisional division and the Kaesŏng Provincial Museum became a branch of the National Museum in Seoul in March 1946 (Chin 1970: 169).

One year before the official outbreak of the Korean War on 25 June 1950, Kaesŏng, located close to the dividing 38th parallel, was caught between the North and South Korean armies as they continuously attacked each other. Koryŏ ceramics and other important cultural relics were hastily sent to Seoul and the Kaesŏng Provincial Museum closed its doors. The museum director, Chin Hong-sŏb, remained to guard the empty museum. After the Chinese Army entered the Korean War, the return of the battle frontier made the stay in Kaesŏng life-threatening. Director Chin left for Pusan,[3] the southern port city to which the National Museum was evacuated in December 1951, though not before burying a part of the remaining museum collection underground ('Hanguk pakmulgwan 100 nyŏn-ŭi saram-dŭl' 2009). Immediately following this, and as the war continued, in January 1952 the Kaesŏng Koryŏ Museum was reopened by North Korean authorities. At the end of the Korean War, the former Kaesŏng Provincial Museum's collection was returned from its Pusan exile to the museum inside the old Kyŏngbukgung palace in Seoul in August 1953. Thus, many Koryŏ artefacts preserved in South Korea today have a Kaesŏng provenance, while a good many artefacts in the Koryŏ Museum are local post-war finds. In 1988 the Koryŏ Museum was moved into the buildings of the ancient academy Sŏnggyun'gwan and inaugurated under its current name, but apparently without big changes to the exhibition content or staff.[4]

Kaesŏng Koryŏ Museum: a national museum in many ways

The unifying ideology of Koryŏ has been co-opted in the North as in the South for its evident parallels with North–South unification rhetoric. In general, both Korean states look favourably to the Kŏryo dynasty as a period when true unification was achieved (Vermeersch 2008: 1516). The Koryŏ Museum's thematic exhibition acts as an interface where the two Koreas reveal a collective memory of their nation's past. In his book *La mémoire collective*, Maurice Halbwachs already showed that collective memory may diverge sharply between different groups of people, just as a single individual may have contradictory memories of the same event:

> Thus, there are individual memories and collective memories. In other words, the individual has access to two types of memory, but depending on which of these he chooses, he could adopt two very different or even contradictory attitudes.
>
> (Halbwachs 1950: 35)

These two types of memory of an individual can be translated to the sometimes conflicting Korean collective memory of cultural nation and state. Although museum practices and the historical interpretation in the Koryŏ Museum are not the product of a joint Korean effort, the Koryŏ Museum is of particular interest for its role as a contemporary mediator of Korean national and state identities. Just a two-hour bus drive from central Seoul, recent one-day tours of Kaesŏng by South Korean citizens have provided a rare opportunity to reflect on self-identification within the Korean nation. The South Korean company, Hyundai Asan, started the one-day tours as a pilot project in September 2005 but soon afterwards the project came to an abrupt end. In December 2007, the tourist visits were again sanctioned. These were interrupted in December 2008, as a result of escalating tensions as South Korea hardened its policy towards the North under President Lee Myung Bak, and North Korea attempted to subdue his administration and press for a change in South Korean policy.[5]

Along with other scenic and historical spots, the itinerary of a tour of Kaesŏng included a visit to the Koryŏ Museum. For most visitors, the trip was necessarily charged with hopes of national unification and the excitement to experience the 'fraternal north' (World Research 2009: 23).[6] Based on common history and pride in Korean nationhood, these hopes led to common interpretations as well. Indoctrinated with South Korean historical interpretations and political ideology, how did these tourists confront the ideological interpretations of this North Korean museum?

> A museum is more than a place of memory, it is the place where, by *narrative reconstruction*, individual and collective identity can be shaped by presenting the multiple traditions with which it is crossed. The museum contributes to the construction of a community rooted in history.
>
> (Pierron 1996, italics added)

For the purposes of this chapter, I will introduce single examples of the collection to illustrate the articulation of multiple national narrative constructions that shape collective identity, in terms of the cultural, ideological and the emotional. As we have seen, cultural and ideological national identities can overlap in one individual. Although the emotional narrative is contained within the other two, it is important to address it separately, as it binds an individual to another and obliges the individual to communal memories through what Avishai Margalit (2003) calls 'thick relations'.

In order to observe the experience of South Korean tourists to Kaesŏng, I joined the one-day tour starting from Seoul in September 2008. This allowed me to observe the behaviour of visitors and tour guides at first hand, and speak directly to some of the travellers about their experiences. The informality of this research methodology was imposed by the South Korean organisers who prevented more formal methods. However, the organisation did kindly allow me to use their customer satisfaction questionnaire of the tour.[7]

The narrative of the cultural nation

Before entering the main compound of the Koryŏ Museum, the visitor is already confronted with the historical Koryŏ theme of the museum. As the name of the museum suggests, the theme of the exhibitions is the Koryŏ period. The museum is located in the capital city of that period, in a building that like its exhibition content stems from that same period. In both Koreas, Koryŏ is recognised as the first unified Korean period[8] and thus carries significant symbolic weight in contemporary interpretations of the nation.

It was only in 1988 that the museum moved into the buildings of a famed Koryŏ period academy and, accordingly, in both Koreas it is still often referred to as the Koryŏ Sŏnggyun'gwan (Institute of North Korea Studies 2006: 113). The museum display is organised in five exhibition areas, divided by the single buildings of the historical Sŏnggyun'gwan compound and its outdoor space (Figure 22.1). As the architecture is not that of a purpose-built museum, the exhibition space follows a spatial layout shaped by the buildings. The museum tour, which was conducted by a museum guide and detailed in the scant literature on the museum, also followed the architectural structure. However, the ancient building complex is more than an exhibition shell, it is an exhibit in its own right and, as indicated on the signpost at its entrance, designated 'historical remain no. 50' and 'national treasure no. 127'. Interestingly, part of the narrative of the common cultural nation concerns the memory of foreign aggression to which the Korean people were subjected throughout their history. The signpost not only informs visitors about the building's foundation as an educational institution in 992 CE, it also explains its destruction during the Japanese Invasion in 1592 and its subsequent repairs in the Chosŏn period between 1602 and 1610.

Perhaps due to this constant struggle against foreign domination, another narrative of the common nation is the proud comparison of Korea with the international

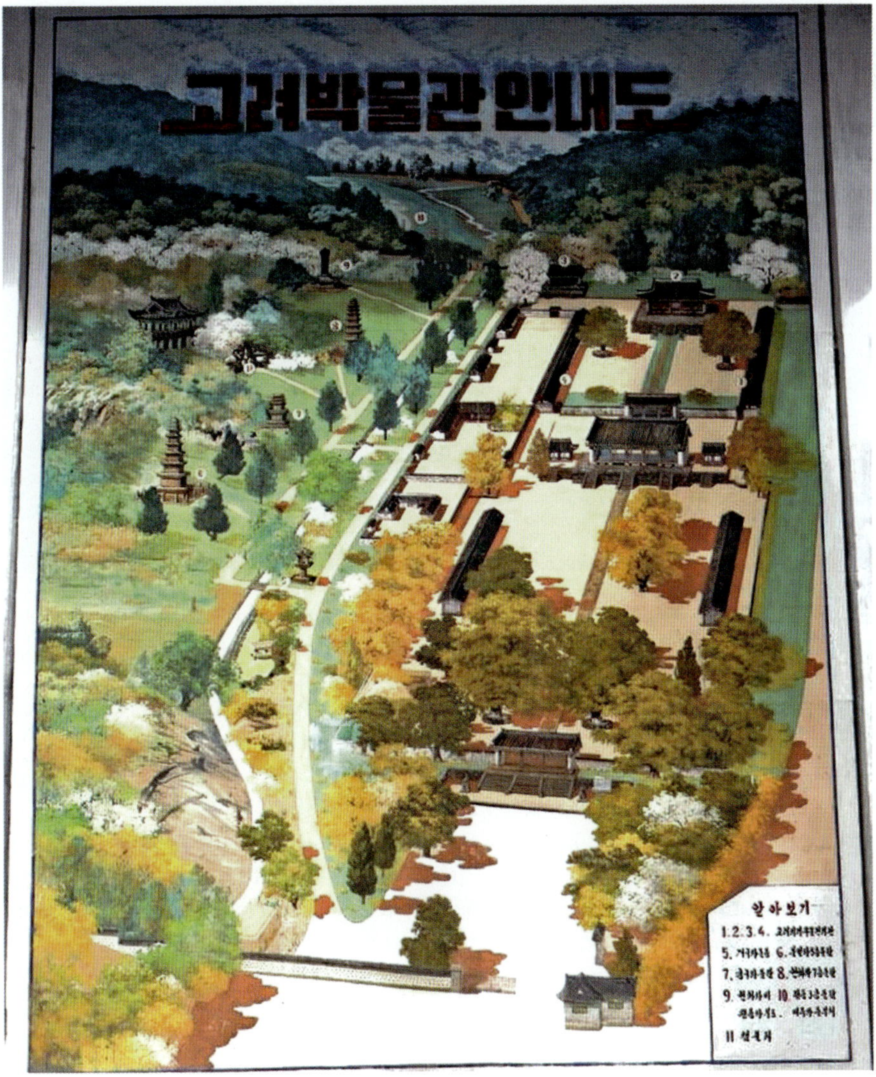

Figure 22.1 A map showing the layout of the Koryŏ Museum. Photo: author

community. The North Korean art historian, Chŏn Ryŏng-ch'ŏl, for example, stresses the importance of the Sŏnggyun'gwan complex in Korean and *world* history. According to him, Koryŏ scientists and historians working in the Sŏnggyun'gwan have been responsible for some of the great achievements and discoveries of their time (Chŏn 2002: 176–78). The Sŏnggyun'gwan compound has powerful historical value, encompassing legacies from a royal palace, guesthouse, higher education institute and a Confucian place of worship with both high cultural and historical value attached to it.

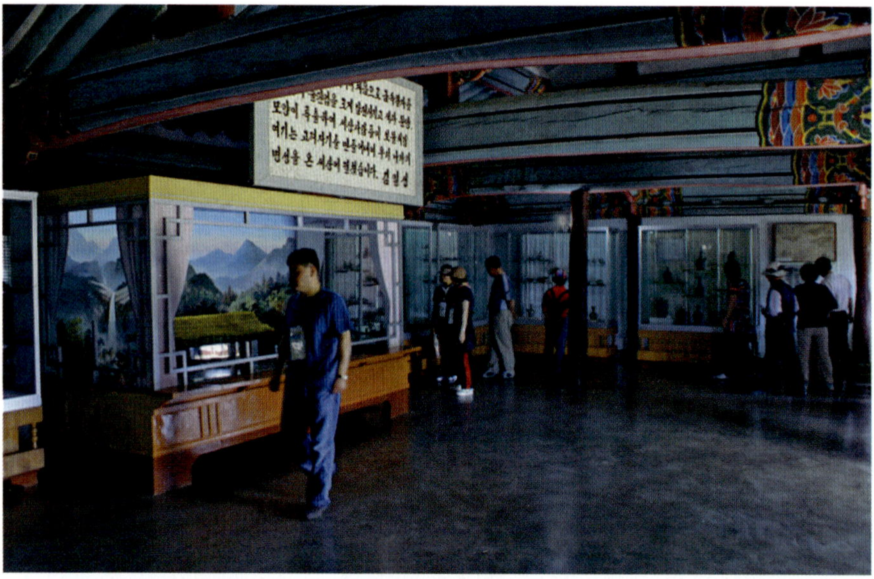

Figure 22.2 The display of Koryŏ celadon ware and movable print letters. Photo: Chosok

On display at the Koryŏ Museum are the world-renowned Koryŏ celadon wares, ceramics recognised for their fine jade-green glaze and elegant shape (Figure 22.2). To this day craftsmen try to evoke their beauty in what is named Kaesŏng porcelain, examples of which can be purchased in the museum shop.

The ceramics are exhibited together with some iron and bronze artefacts, the most prominent being metal print lettering. As in South Korean museums, the Kaesŏng Museum proudly highlights the use of movable print in Korea dating from the early thirteenth century. This fact is fiercely promoted through the display of material evidence. It demonstrates that Koreans used this printing technique two hundred years prior to Gutenberg. On a panel (literally translated: Kim Il-sung teaching) overlooking the exhibition room, Kim Il-sung's citation affirms the glorious achievements of the Korean people and the important role that metal-type printing played for Koreans and the world.

The narrative of the ideological nation

The state-defined display of cultural relics therefore acts as a form of people's educational engagement with the past and their acts of remembering. It is supposed to set an example by stressing the importance of *glorious* past achievements. Visitors are taught and encouraged to engage with this particular understanding of their cultural history. Indeed, North Korea's leading authority, Kim Jong-il, openly prescribed this form of historiography:

> There is no need to rehabilitate all the historical sites and relics indiscriminately. Only representative ones which have educational significance and are capable of showing the excellence of our national culture should be rehabilitated.[9]

This instruction continues to be observed in modern-day North Korea and in the Koryŏ Museum, where the state-controlled national museum acts as a cultural tool for public education and remembrance. Based on the logic of the cultural nation, Koryŏ, as the first unified nation, is also employed to legitimise a Korean unification under its territorial heir the North Korean state. In commemoration of the fiftieth anniversary of the museum, the North Korean News Agency described the importance of the timing of its foundation and the resulting political symbolism:

> 50 years have passed since the Koryŏ Museum was established in January Chuch'e 41[10] (1952). Koryŏ is the first unified state in Korea. The museum is situated in Kaesŏng, which was the capital city of Koryŏ. It was founded in the period of war decisive of the destiny of the country and the nation in Korea.[11]

Compatible with the political ideology, Koryŏ's historical past is explained explicitly to serve the state. This creates a paradox. The history of Koryŏ serves as a symbol of Korean unification sentiments, yet is concurrently used to consolidate the ideological narratives that separate North and South. One such narrative in North Korea is the Soviet-style Marxism that tends to treat culture as being merely a secondary epiphenomenon of economic relations, of class, and of modes of production. Stalin's '*Short Course*' *History of the Soviet Communist Party* in 1938 introduced five successive social formations that were alleged to have defined human history: the primitive stage, slavery, feudalism, capitalism, and socialism (Markwick 2006: 284). In conformity with this historical theory, one poster in the exhibition of the Koryŏ Museum illustrates the shortcomings of the Koryŏ feudal society. It compares the low market price of a human slave to the significantly higher one of an ox.

However, Petrov (2006) observed in North Korean historiography after the Korean War an increasing nationalistic trend that emphasised national heroism and unique cultural traditions. A new generation of North Korean scholars was encouraged to stress patriotism and nationalism in national history, something that clearly moved away from the traditional internationalist Marxist dictum that 'Die Arbeiter haben kein Vaterland' (the workers have no fatherland). Kim Il-sung, leader of the ruling Korean Workers Party (KWP) started the new era of *Chuch'e*, the national ideology of self-reliance (Petrov 2006: 14).

> We are not engaged in any other country's revolution, but solely in the Korean revolution. This, the Korean revolution, determines the essence of

> Chuch'e in the ideological work of our Party. Therefore, all ideological work must be subordinated to the interests of the Korean revolution.
>
> (Kim 1982: 396)

Understanding the capacity of history and culture as nation-building tools, Kim Il-sung criticised the lack of books on the *precious* cultural heritage and history of the national liberation struggle. As a result, from the end of the 1950s, Kim Il-sung's *Chuch'e* theory took over historical scholarship, eliminating the voices of old school historians, who were removed from central research and educational bodies. The word *Chuch'e* refers to the empowerment of the working masses of the people, the subject of the revolution. The term is used to roughly imply 'self-reliance', but has many socio-political undertones, with the ultimate being the centrality of the Korean people in the fight against foreign aggression and oppression (Kim *et al.* 2006: 1033).

After the declaration of the new constitution in 1972, *Chuch'e* became the official ideology for all fields of study. In 1977, it officially replaced Marxist–Leninism in the constitution as the guiding philosophy underpinning the North's exclusive form of communism (Kim *et al.* 2006: 1034). Based on this new *Chuch'e* model of national history, *Chosŏn Chonsa* (*The Complete History of Korea*) was published in thirty-three volumes between 1979 and 1982. In the same period, two North Korean authors of the 1980 UNESCO publication on cultural policy in the DPRK emphasised the link between state and culture:

> Today, the Democratic People's Republic of Korea is enthusiastically continuing and developing the splendid cultural traditions created thousands of years ago. Literature and the arts are flourishing, sustained by the government's *chuche* principles. . . . Each province has its own historical museum. The museums are centres of propaganda in favour of the national cultural heritage. . . . [Artists] have striven to present a true picture of reality, throwing light on the corruption and the reactionary essence of the exploiting classes and demonstrating the superiority of our country's socialist regime.
>
> (Chai and Hyon 1980: 12, 17 and 21)

Unsurprisingly, then, the themes of revolution and national independence are highlighted in the Koryŏ Museum. A poster of peasant rebellions between the end of the twelfth and beginning of the thirteenth centuries refers to manifold local uprisings. The historical backdrop for these rebellions is a military *coup d'état* in 1170 that shook the aristocratic society of Koryŏ. The new ruling power deposed and exiled King Ŭijong (who ruled from 1146 to 1170), and killed many civilians in order to set up its government. Corruption and tyranny ran rampant and caused great suffering among the people, especially the peasants and slaves, who rose up in rebellion and fought for freedom. 'On the proper evaluation and treatment of cultural heritage of our nation with correct viewpoint and attitude', the ideological narrative of the museum 'deal[s] with the national cultural heritage on a class basis and appraise[s] it according to the interests of our revolution. . . . We must revive those parts that are beneficial to our revolution and discard those that are detrimental'.[12] The exhibition also illuminated the historical struggle and advancement of the Korean people to defend their independence. The visitor can, for instance, look at drawings of Koryŏ engineering inventions for warfare or an iron helmet of a Koryŏ general that ironically was discovered while ploughing a Japanese-owned field in Kaesŏng in 1933.[13]

Stepping out of the exhibition room into the courtyard, one can admire in its foreground two famous stone dragonheads, the so-called female and the male. The dragonheads come from different sites in Kaesŏng, and were once distinguished by an association with their respective sites. But following a visit to the Koryŏ Museum by Kim Il-sung in commemoration of the 1000th anniversary of the Koryŏ Sŏnggyun'gwan on 5 May 1992, they were incorporated into state ideology for the 'Great Leader' who introduced the now much-used female/male distinction by saying that male animals are better looking than their female counterparts. He then designated the head from the Such'anggun palace as the male one (Chŏn 2002: 194).

During this same visit, Kim Il-sung designated five major Kaesŏng Koryŏ cultural treasures (Sŏnggyun'gwan, Sŏnjuk-gyo, Manwŏltae Palace, T'aejo royal tomb, and Kongmin royal tomb). On a follow-up visit on 1 September 1992, the symbolism of the number five became interwoven into a fanciful story formed around the installation of a granite plate with Kim Il-sung's handwriting at the Koryŏ Museum's entrance. The five letters read 'Koryŏ Sŏnggyun'gwan'. It is said that these five letters also stand for the five periods through which the Sŏnggyun'gwan building complex existed (Koryŏ, Chosŏn, Japanese occupation, South Korean and North Korean) and the five grand Kaesŏng cultural treasures (Lee 2005). Measured by the quantity of physical and oral markers commemorating the sayings, acts or visits of Kim Il-sung surrounding the Koryŏ Museum and its collection, the 'big five' symbolism is only outdone by the greatest of all Kaesŏng treasures, Kim Il-sung himself (Figure 22.3).

The narrative of the emotional nation

This authoritative North Korean ideological narrative is further embodied in the North Korean museum guide who, equipped with a megaphone, leads the visitors through the museum. The traditionally dressed guide entertains the visitor not only with historical facts, but also with historical anecdotes and ideologically charged explanations. Interestingly, the South Korean visitor seems to take this ideological confrontation lightly. Comments in press articles and visitor blogs tend not to refer to ideological problems with the exhibition, but to the 'shameful poverty' and backwardness of the museum display that has apparently not changed since the museum's opening in 1988.

While supposed backwardness is sometimes romanticised as quaint, particularly

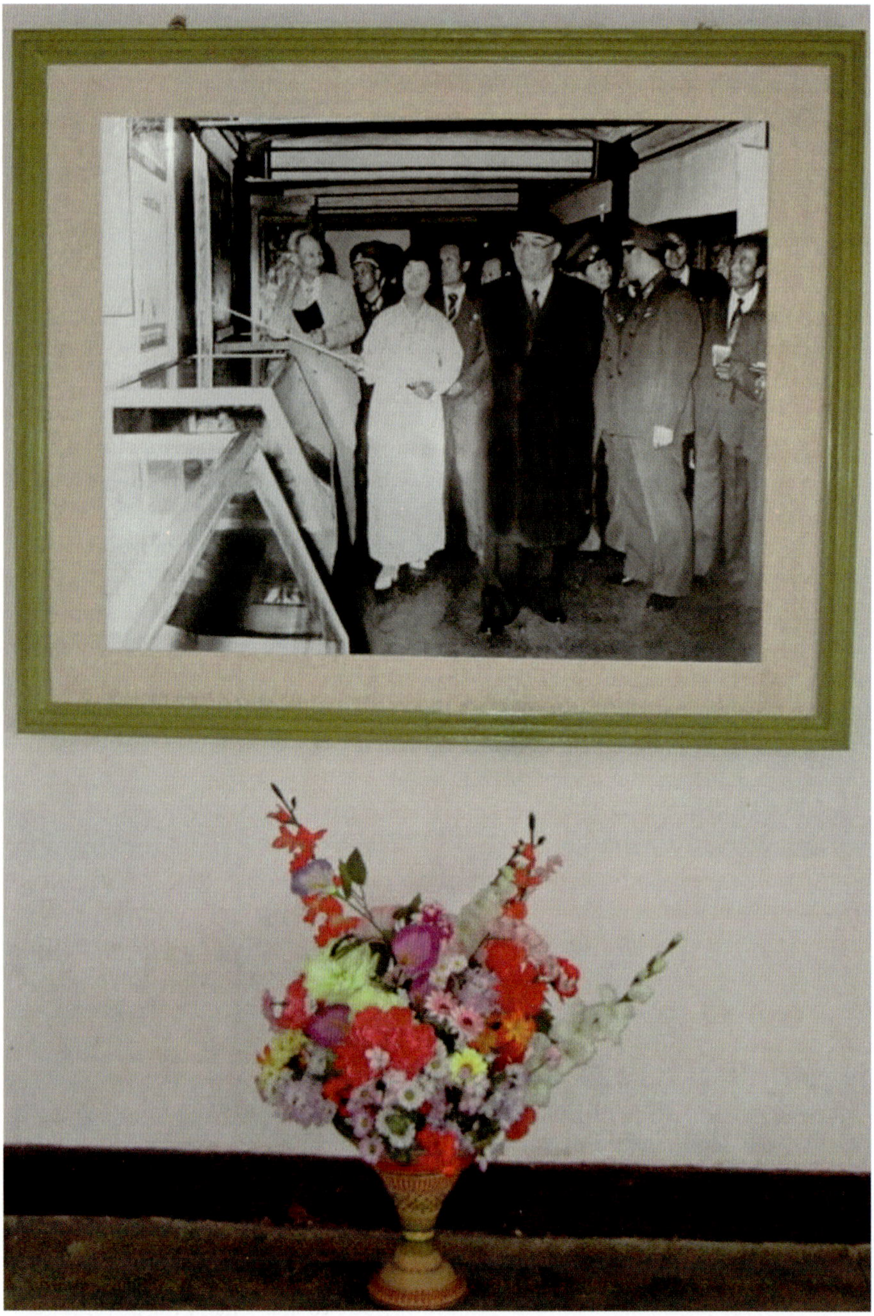

Figure 22.3 Flowers for the picture of Kim Il-sung's visit to the Koryŏ Museum. Photo: author

with such things as food and attitudes,[14] backwardness and poor facilities in the museum were interpreted as a failure to exhibit pride in Korean history. It permitted South Koreans to feel a sense of superiority:

> Compared to our [South Korean] museums, the state of the exhibition is shamefully appalling. As electricity was being saved it was dark, and one light was not enough to illuminate the Koryŏ celadons sufficiently. The stone sarcophagus and bronze censer are unprotected from the hands of passing tourists. Despite the exhibition's lack of refinement, the over one thousand artefacts are a stupendous treasure of great value.
>
> (Lee 2008, my translation)

As was the case with the focus on Koryŏ's past, there predominates among South Korean visitors a curiosity to engage with their North Korean guides and to find commonalities with them. It appears that in the end the memories of these symbolic interactions had a significant impact on visitors. Indeed, most visitors (52 per cent) who took part in the tourist survey stated that their most salient memory of their trip was the chance to see the life of people in North Korea. Next most popular were the scenic spots (25 per cent). The cultural properties of Kaesŏng had a lasting impression on 7.5 per cent of the respondents (World Research 2009: 23).

Unfortunately, there is no way, even superficially, to ascertain the effect of the museum on North Korean visitors. Instigated by the ideological narrative of the nation, it appears that the emotional experience is strongly linked to local popular events. Emotional connection to local events in the Koryŏ Museum is interwoven with ideological symbolism surrounding the figure of Kim Il-sung whose memory permeates the museum. Following Kim Il-sung's visit, a few books have been published in North Korea about Kaesŏng, perhaps to make a little money from the tourists. Among these are books that combine legendary accounts and new stories concerning Kaesŏng.[15] While the historic Koryŏ culture and its relics are the setting for these stories, the living memories and emotions of the popular contemporary culture are most skilfully related back to Kim Il-sung's exemplary actions and his visits to the site. This creation of an emotional link between local memories and official state ideological narrative is a cunning development in government propaganda.

Conclusion

The collection of the Koryŏ Museum, not unlike the Korean people, is intrinsically linked to complex understandings of a Korea of one cultural nation and two political states. During the Korean War, both the Korean people and the Koryŏ Museum collection were torn apart. For the Korean people this division created a bipolar appreciation of state and nation, the logic of which is reflected in the interpretation of the museum and in both governments' aims to unify the Korean peninsula. Both

Korean states have promoted cultural interpretations which aim to support both the state and the nation. The state focuses on the people's identity and affiliation to particular ideologies; the nation focuses on a common and unifying Korean identity which can be reached through cultural interpretation and appreciation. It is for this reason that both states are in consensus over the cultural and historical significance of the Koryŏ period. The Koryŏ theme is an idealised continuation of the cultural nation, and access to the cultural heritage in Kaesŏng had soothed the longing for the right to walk the ground of the country (Korea), expressed in the patriotic idea of *uri nara,* our land.

The Koryŏ Museum's interpretation provides a platform for the negotiation of multiple national narratives; the Koryŏ Museum is a 'vehicle of memory' which seems to reach beyond the politics of state back to a point of shared origin. It offered the opportunity to come and realise, and possibly deliberate national historical memory, state ideology and individual identity. In this way and for a limited time, the Kaesŏng tours permitted tourism to play an important role in nation-building processes and, despite ideological difference, helped ease inter-Korean political tensions.

Notes

1 An official overview of the history of the Museum of Korea is available at http://www.heritage.go.kr/eng/mus/nat_01.jsp.

2 Guk, S. H. (2008) 'Kaesŏng pulip pakmulgwan- gwa pakmulgwan saram-dŭl', *Minsok sosik,* 158: 16–17. Available online: www.nfm.go.kr/downfile/book_mag/434/158.pdf.

3 Also transcribed as Busan.

4 In early 1988, Wang Sŏng-su published a report about Old-Chosŏn relics in the journal *chosŏn kogoyŏngu* with his affiliation stated as Kaesŏng History Museum. A year later in 1989 he published, affiliated to the Koryŏ Museum, an introduction to the newly opened Kaesŏng Koryŏ Museum in the journal *ryŏksa kwahak*.

5 In response, President Lee and his core advisers wrote this off as an 'inevitable period of adjustment in the process of normalising the misguided inter-Korean relations of South Korea being dragged around by North Korea for the 10 years of the Kim Dae-jung and Roh Moo-hyun administrations' ('The five factors' 2008).

6 At the North Korean border control, North Korean-style music with unification lyrics can be heard. South Korean unification organisations came in groups to visit Kaesŏng, posing for pictures in front of unification-slogan boards. Visitors with family ties in the Kaesŏng region represented a significant proportion of the tourists.

7 I am very grateful for the support I received from Mr Kim Sunghun, deputy director in the KTO (Korea Tourism Organisation) Frankfurt office, who not only forwarded and coordinated my requests to the KTO head office, but also provided me with a free Kaesŏng tour ticket and the KTO questionnaire results.

8 This has occurred only recently in South Korea, as previous governments have tried to emphasise the unification of three kingdoms under the Silla in 668 CE centred on the modern-day territory of South Korea. The two Koreas have also established the idea of a legendary prehistoric foundation of the Korean nation.

9 Kim, J. I. (1964) 'On improving party guidance relating to the preservation of historical sites and relics', 16 September 1964. Available online (outside ROK): www.korea-dpr.com/lib/Kim%20Jong%20Il%20-%201/ON%20IMPROVING%20PARTY%20GUIDANCE%20RELATING%20TO%20THE%20PRESERVATION%20OF%20HISTORICAL%20SITES%20AND%20RELICS.pdf.

10 Since 1997, the North Korean government has officially used the *Chuch'e* calendar, under which years are numbered from 1912, the date of Kim Il-Sung's birth.

11 Koryŏ Museum (2002) *Korean Central News Agency*, 17 January. Available online (outside ROK): www.kcna.co.jp/item/2002/200201/news01/17.htm.

12 Kim, J. I. (1970) 'On proper evaluation and treatment of cultural heritage of our nation with correct viewpoint and attitude', 4 March 1970. Available online (outside ROK): www.korea-dpr.com/lib/Kim%20Jong%20Il%20-%202/On%20Proper%20Evaluation%20and%20Treatment%20of%20the%20Cultural%20Heritage%20of.pdf.

13 Koryŏ Komul palgul T'ugu-wa chagidŏng (1933) *Donga Ilbo*, 2/4, 4 November.

14 Some South Korean tourists told me that North Korean dress, food, liquor, cigarettes and sweets reminded them of a Korea in the past that evokes nostalgic and romantic feelings.

15 See, for instance, (1999) *Songaksan-ŭi ŭiroun saram-dŭl*, Pyongyang: Pyongyang Publisher; Li S.D. (2002) Songaksan, Kaesŏng: Munhakyesul Publisher; Li U.H. (2002) Songdo chŏnsŏl, Kaesŏng: Munhakyesul Publisher.

Bibliography

Breuker, R. (2004) 'History with a capital H: Kaesŏng's forgotten claim to capital history', *Acta Koreana,* 7(2): 65–102.

Chai, S. S. and Hyon, J. H. (1980) *Cultural Policy in the Democratic People's Republic of Korea*, Studies and documents on cultural policies, UNESCO Paris. Also available online: http://unesdoc.unesco.org/images/0003/000370/037091eo.pdf.

Chin, H.S. (1970) 'Kaesŏng Pakmulgwan-ui', in Woo M.-H. (ed.) *Kaesŏng*, Seoul: Yaesulch'unch'usa, pp. 167–71.

Chin, P. S. (2002) 'Sŏnggyun'gwan-ŭi ŭnhaeng namu', *Songdo chŏnsŏl,* Kaesŏng, 13–20.

Cho, H. B. (2002) 'Chŏngsang hoedam ihu sahoe munhwa kyoryu-ga Pukhan sahoe-e mich'in inghyang', in T'ongil yŏn'guwŏn (ed.) *Nambuk kwangye palchŏn-gwa hanbando p'yŏnghwa kŏngch'ak*, Seoul.

Chŏn, R. C. (2002) *Kaesŏng-ŭi yet chach'ui-rŭl dŏdumŏ*, Pyŏngyang.

Confino, A. (1997) 'Collective memory and cultural history: Problems of method', *The American Historical Review,* 102: 1388–89.

Connor, W. (1994) *Ethnonationalism: The Quest for Understanding*, Princeton, NJ: Princeton University Press.

Duncan, J. (1998) 'Proto-nationalism in Premodern Korea', in S. O. Lee and D. S. Park (eds) *Perspectives on Korea,* Sydney: Wild Peony Press.

Grinker, R. R. (1998) *Korea and its Futures: Unification and the Unfinished War*, New York: St Martin's Press.

Halbwachs, M. (1950) *La mémoire collective,* Paris: Presses Universitaires de France.

Hanguk pakmulgwan 100 nyŏn-ŭi saram-dŭl (2009) *Donga Ilbo*, March 12. Available online: www.donga.com/fbin/outputŏn=200903120031.

Institute of North Korea Studies (2006) 'Koryŏ Pakmulgwan', *Pukhan*, May.

Kim, I. S. (1982) *Works*, 9, July 1954 to December 1955, Pyongyang: Foreign Languages Publishing House.

Kim, S. S., Timothy, D. J. and Han, H. C. (2006) 'Tourism and political ideologies: A case of tourism in North Korea', *Tourism Management*, 28(4): 1031–43.

Ko, Y. G. (2002) 'Nambuk hoe munhwa kyoryu hyŏmnyŏg-ŭi hyŏnhwang-gwa kwaje', in Kunsan Taehakkyo and T'ongilmunje yŏn'gu hyŏbŭihoe (eds) *Nambukhan kyoryu hyŏmnyŏg- ŭi alsŏnghwa pangan*, Seoul.

Ko, Y. S. (2007) *Songdo-ŭi kojŏk – Uhyŏn Ko Yu-sŏp chŏnjip* 7, Paju: Youlhwadang Publisher.

Lee, P. T. (2005) 'Koryo Songgyungwan Iyagi', Tongilnews, 6 September. Available online: http://www.tongilnews.com/news/articleView.html?idxno=59368.

Lee, S. W. (2008) 'Koryŏ-ga sara itta 'toshi pakmulgwan' Kaesŏng-sok-ŭro', *Hanguk Ilbo*, 4 January. Available online: http://news.hankooki.com/lpage/health/200801/h2008010417453584510.htm.

Margalit, A. (2003) *The Ethics of Memory*, Cambridge, MA: Harvard University Press.

Markwick, R. D. (2006) 'Cultural history under Khrushchev and Brezhnev: From social psychology to mentalités', *The Russian Review*, 65(2): 283–301.

Nora, P. (1992) *Les France: Conflits et partages,* Paris: Gallimard.

Pai, H. I. (2000) *Constructing 'Korean' Origins: A Critical Review of Archaeology, Historiography, and Racial Myth in Korean State-formation Theories*, Cambridge, MA, and London: Harvard University Press.

Pang, K. J. (1992) *Han'guk kŭn-hyŏndae sasangsa yŏn'gu: 1930, 40-nyŏndae Paek Nam-un-ŭi hangmun-kwa chŏngch'i kyŏngje sasang*, Seoul: Yŏksa p'ip'yŏngsa.

Park, C. H. (1975) 'Let us turn the Joy of Liberation into Joy of Unification, 1972 August 15' and 'Let us consolidate legitimate National Identity on the 4304th anniversary of the founding of the Nation, 1972 October 3', in *Major Speeches*, Seoul: Samhwa Publishing Company, pp. 19–24, 178–81.

Petrov, L. A. (2006) 'Turning historians into Party scholar-bureaucrats: North Korean historiography in 1955–1958', *East Asian History*, 31(6): 101–24.

Pierron, J. P. (1996) 'Le musée: lieu de mémoire, horizon d'avenir?', *Hermes*, 20: 125–31.

Shin, G. W. (2006) *Ethnic Nationalism in Korea: Genealogy, Politics, and Legacy*, Stanford, CA: Stanford University Press.

Smith, A. D. (1986) *The Ethnic Origins of Nations,* Oxford: Blackwell.

Smith, A. D. (1991) *National Identity (Ethnonationalism in Comparative Perspective)*, University of Nevada Press.

Smith, A. D. (2008) *The Cultural Foundations of Nations: Hierarchy, Covenant and Republic*, Oxford: Wiley-Blackwell.

The five factors (2008) *The Hankyoreh,* 26 November. Available online: http://english.hani.co.kr/arti/english_edition/e_northkorea/324056.html.

Vermeersch, S. (2008) 'Yŏngt'ong-sa and its reconstruction: A medieval Buddhist site as a space for North–South Cooperation', *Proceedings of the 4th World Congress of Korean Studies 2008*, Seoul, 22–24 September.

World Research (2009) *Kaesŏng Kwangwang chosa*, unpublished survey report commissioned by the Korean National Tourism Organisation in 2008.

Chapter 23

THE NATIONAL MUSEUM AS PALIMPSEST

Postcolonial politics and the National Museum of Korea

Jung Joon Lee

OVER THE LAST quarter century, some the world's great museums have reinvented themselves in new architecture: the Louvre, the Guggenheim in Bilbao, Tate Modern in London and the Museum of Modern Art in New York, to name some prominent examples. Amongst the most extraordinary of these reinventions, however, has been the National Museum of Korea which, in 2005, was transformed into a vast monolithic building with an open plaza, and six annexes. One of the most remarkable museum developments of the new millennium, it publicised itself as 'the twenty-first century's foremost cultural complex and museum of international stature'.[1] Built on land in Seoul's Yongsan district once used as a US airbase, the museum's relocation was intended to signal a new beginning.[2]

The museum's construction, together with the demolition of the old museum building, has generally been viewed as a manifestation of the country's cultural and technological advancement. However, this raze-and-build approach arose from the museum's complex past. It is a past that survives and, today, hovers over the new museum building and its exhibition spaces; the museum has become a palimpsest of the colonial and postcolonial history which has brought it into being. That past has literally been written into the museum's buildings and is deeply rooted in attempts to build a nation that can perform on the global stage.

Much has been written about the National Museum of Korea's approaches to marketing, education and exhibition but there has been little discussion of the postcolonial politics that have contributed to its remaking. The political, cultural, social and financial specificities of a national museum shaped by the nation-building process should not be undermined by the political naivety and essentialising nationalism that produce the museum solution. In this chapter, the 'national' character of the National Museum of Korea is examined across the period of its making and remaking, from its foundation as a colonial museum through the various changes it underwent as part of postcolonial nation-building. A particular focus will be placed on the complex nationalist discourse which culminated in the decision to dismantle the old national museum building in 1995 – an event that produced a bipolar public reaction. Government has played a leading role in the reinvention of South Korea as a world player; this chapter will end by asking when, in the development of a liberal democracy, a government needs to step back from making the nation in the arts.

The museum as palimpsest

The origins of the National Museum of Korea date back to 1908 when King Soonjong of the Joseon Dynasty (1392–1897) founded the Museum of the Yi Royal Family within the Changgyeong Palace complex. That museum opened its doors to the public the following year. In 1910, Korea signed the Japan–Korea Annexation Treaty and came under Japanese colonial rule. The Joseon Governor-General's Museum[3] was established in 1915 as a cultural auxiliary to the Governor-General's Office. The museum existed for thirty years under the Japanese occupation, and in September 1945, a month after the nation regained its independence, it was renamed the National Museum, which was merged with the Museum of the Yi Royal Family in 1969.[4] In June 1950, having never escaped the legacy of its colonial past, the country entered the turmoil of the Korean War. The museum was closed and its collections moved to Busan, Korea's second largest city. Situated on the southern tip of country, the museum was now as far as it was possible from the conflict zone. After the war, the museum moved four times within Seoul, mostly around the Royal Palace complex, before being returned to the former Governor-General's building[5] in August 1986.

The Governor-General's building was one of the most ambitious of the Japan's colonial architectural projects (Figure 23.1). Desiring a central administrative building in Seoul that spoke of the new status of the nation under Japanese colonial rule, the colonial government commissioned the German architect, Georg de Lalande, to design the building in 1912 (Song 1996: 72). Two years later de Lalande was dead and the project was passed to the Japanese architect, Ichiro Nomura (who also designed the Governor-General's building in Taiwan) and the Korean architect, Gil-yong Park (Song 1996). Under Nomura's supervision, construction of the building began in 1916 and was completed in October 1926 (Joo 1996: 74).

The site chosen was located between Gwanghwamun, the front gate of the Gyeongbok Royal Palace complex – the largest palace complex built during the Joseon Dynasty – and the main Palace building of the complex, Geunjeongjeon (the King's Chamber). To provide ample space for the Governor-General's building,

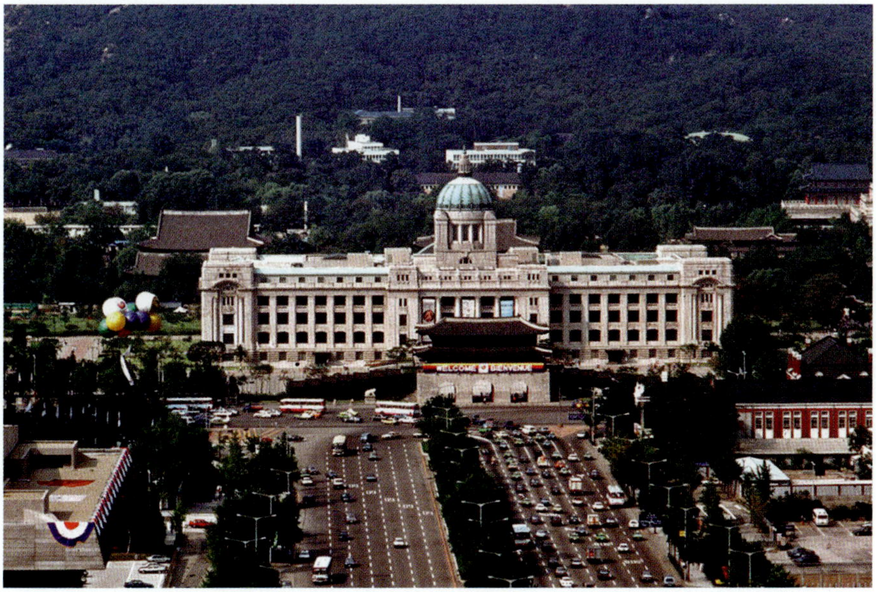

Figure 23.1 The National Museum of Korea, Seoul, *c.*1991. Gwanghwamun, the main gate of the Gyeongbok Palace complex, is seen in front of the building. Photo: Courtesy of National Museum of Korea

the main gate that separated the Palace complex and Gwanghwamun was demolished and reconstructed in a new location.[6] Several other buildings in the complex were demolished to secure the approximately 97,000 square metres needed for the building (S. C. Kim 1995a). Construction of the vast new building required the mobilisation of 50,000 Korean labourers (Y. Chung 1995). In Korea at this time, the scale of this construction was unprecedented.

Rather than constructing a building that would harmonise with the traditional Korean architecture of the Palace complex, the Japanese chose to use a neoclassical style for the new building borrowed from Europe. It included such neoclassical stylistic elements as a central copper dome and Doric and Corinthian columns, but its rigid exterior design was not entirely composed of the elaborate styles of French Beaux-Arts architecture. A number of Korean historians have argued that the building's neoclassicism was a Japanese reinterpretation of nineteenth-century German military architecture (Y. Chung 1995).[7] The building's symmetrical plan included a central dome and two parallel wings to each side, each attached to a courtyard in the French Beaux-Arts style (Zanten 1997: 114–15).[8] It has been suggested by historians in Korea and Japan that the floor plan for the building was in fact based on the character, 日. In Kanji – the Chinese script adopted by the Japanese – Japan, or Nippon, is written 日本 (Y. Chung 1995; Joo 1996; C. Choi 1995). In other words, far from being a simple imposition of Western architecture on a site dominated by the traditional designs of the Korean palace complex, this building had its imperial agenda written into it – quite literally.

The Governor-General's building existed for seventy years until its demolition in 1995. It served as a colonial administrative office until the Japanese surrender in 1945. Indeed, the Japanese surrender was announced from this building. In 1948, the building was the venue for some of the most important events in the modern history of Korea, including the declaration of the first constituent assembly, the inauguration of the first President, Syngman Rhee, and the declaration of the first government of Korea. Rather ironically, this symbol of colonial imposition became the home of the independent government of Korea and the building was renamed Joongangcheong (the Central Hall).

On 29 September, in the year of the outbreak of the Korean War, the building was occupied by the North Korean Communist Army. It was recaptured by allied forces under the command of General Douglas McArthur and then served as a military office until the end of the war in 1953. It was then used as the administrative office of central government until 1961 when it was briefly turned into the Korean War Memorial. From 1962, however, it was reused as a subsidiary office of government, and many rooms were turned into storage spaces for historically important documents and artefacts.

The building underwent major restoration during the military regime of Doo Hwan Chun and then became the National Museum of Korea, the largest museum in South Korea, at the peak of democratisation movements against the military regime in the 1980s. The most prominent cultural institution during the military regime, the building at this time acquired yet another layer of socio-political complexity. This complexity was not assessed nor understood by the military government. The museum building now signified not only the colonial past but also the cultural hegemony of the military regime.

When the President, Young Sam Kim, was freely elected in 1993, after thirty-two years of military rule, his administration reviewed the use of the building as a matter of some urgency. It soon became set on a course to demolish the building, a plan for which it gained overwhelming congressional support. It was argued that the building's destruction was long overdue; it was time for a brand new, state-of-the-art, and truly 'Korean', building. The proposal to demolish the building was accepted in November 1993. Less than two years later, on the fiftieth anniversary of Independence Day, an elaborate ceremony was held to mark the dismantling of the building in front of Gwanghwamun.[9] Thousands gathered to bear witness. The Korean media gave the event wide coverage and almost unanimously described it as an extremely important and meaningful event which united the Korean people. The public and political activists saw in President Kim's actions the symbolic overthrow of past colonialism and military dictatorship. Demolition was completed in late 1996. President Kim's administration realised the political opportunity the demolition presented and launched a powerful political campaign that would boost his popularity and gain the trust of the public.

Erasure or restoration?

Not everyone was in tune with Kim's plan. A group of intellectuals argued that the building was – regardless of its politics – part of Korean culture and history, and could continue to teach a powerful lesson about colonialism for generations to come (Y. S. Choi 1995). When the honeymoon with Kim's administration ended in corruption and the catastrophic economic crisis of 1997, others began to re-evaluate the demolition of the building. Cultural critics wrote about the importance of preserving colonial architecture and artefacts in order to remember the past, however uncomfortable it might be. They argued that it was important to understand them as part of the nation's history rather than simply negating them. The economist, Jeongho Choi, also argued that these sites possessed economic value as tourist attractions (J. H. Choi 2006). A nation's history simply could not be rewritten through the demolition of a building, Choi wrote. Choi's analysis, however, tends to overlook the specificity of Korean history, not to mention the government's mobilisation of 'nationalism as a force of modernity' in order to justify its political agenda (G.-W. Shin 2006: 11). For over a millennium, Korea had maintained 'a coherent political community within a stable territorial boundary with a well-established agrarian bureaucracy' (G.-W. Shin 2006: 18). This contrasts with the history of Western Europe, where, according to Gi-Wook Shin (2006: 18), 'a more territorial notion of nation took place and ethnicity was attenuated or even suppressed by modernizing forces or nation-building efforts'. Korea was a colony of Japan during the formative years of nation-building and built its national imaginings around the idea of ethnic homogeneity that distinguished the native population from its colonial rulers.

The National Museum was not the only large colonial building in downtown Seoul. The old Seoul Station and City Hall, both completed in 1925 to 1926, were built in the same Japanese colonial style. City Hall served its function from the nation's independence in 1945 until 2008, when it was closed for renovation and additional construction work. The old Seoul Station was transformed into a transport museum. Both buildings no longer serve their original purposes, yet their historic and educational value was considered sufficiently significant to keep them intact. Their survival further underlines the exceptional politics of the old building of the National Museum of Korea. As a central administrative building of the old colony, it occupied the most conspicuous lot in the Gyeongbok Royal Palace complex. It was situated close to where Empress Myeongseong of the Joseon Dynasty was assassinated by Japanese colonists in 1895. The Palace was originally planned and built, in 1394, in accordance to the principles of *pungsu* (*feng shui*, geomancy) which gave the Palace clean and strong *gi* (in the West, *chi*, the flow of spirit or aura and its power) from the Bukhan Mountain to the north. This *gi* would then spread out into the rest of Seoul through Gwanghwamun, the front gate. The colonial building was not only out of scale with the Palace buildings but also blocked the flow of *gi*. *Pungsu* experts claim that the Japanese colonists were very aware of this, and they aimed at 'shutting up the mouth of the Palace and pressuring the neck of the Joseon Dynasty' (C. Chung

1995; C. Choi 1995). The building became a symbol of the foreign power that destroyed the sovereignty of Korea. The building acted like a curtain obscuring the view to the north of Gwanghwamun and the legacy of the Joseon Dynasty.

The Japanese colonists used the building as a political centre from which they expanded their powerbase not only in Korea but across East Asia. Erected in the name of 'colonial identity' and 'colonial unity', the building became closely tied to imperialists set on reinforcing 'patterns of intercultural dominance and submission' (Vale 1992: 315). The building's function as a colonial machine made it symbolic of the violence inflicted both on the people and the land. This symbolism drove support for the demolition of the building; most Koreans felt that the conversion of this building into the National Museum damaged Korean cultural heritage – it placed Korean artefacts in a building that represented the painful colonial past.

Renationalising the National Museum

The three elements – *pungsu*, its colonial identity, and the complex history that it symbolised – played major roles in reaching a consensus to demolish the building. President Kim saw this as more than the demolition of a colonial building; it was an initiation of a great restoration project for the Palace complex and the museum. The plan had been announced as an inaugural project for the re-establishment of a strong sense of nationality, a notion generally overlooked by a large proportion of the general public who had spent the past three decades focusing on industrialisation and 'catching up' with the rest of the world. At the same time many students and activist groups had been struggling to bring in a democratic government. In order to assuage a public which had lost faith in the previous military government, the new government used globalisation as a motto for foreign affairs while, at home, it assessed and castigated the former regime. A series of government hearings were held to bring that regime to justice; these hearings acted as a consolation to Koreans who had for so long endured oppression.

It was in this political climate of reinvention and erasure that the decision had been taken to demolish the old building. Prominent historians and writers, such as In Ho Choi (1995), argued that the event should be used as a starting point to study and locate remedies to atrocities carried out by the military regimes. The event was to be no mere political postcolonial gesture but the re-strengthening of a nation that was entering the world stage of global competition. This 'renationalisation' was orchestrated in three major cultural events in 1995: the fiftieth anniversary of Independence Day, the inauguration of the Gwangju Biennale and the demolition of the National Museum building.

A number of revisionists have argued that this was merely the collective fervour of a 1990s 'bubble'. Their arguments were supported by the ways in which many Koreans view the colonial period. Although the industrialisation and modernisation of Korea was jump-started with Japan's colonisation, most Koreans focused either on colonial oppression or the independence movement (Cumings 2002: 69–94).

They were generally reluctant to associate the colonial period with progress or positive influence, including in the field of civil engineering and architecture. Nevertheless, this tendency did not mean that people were unwilling to deal with the past, or that 'nationalism and its fascist potential' prevailed in the cultural politics of postcolonial Korea (G.-W. Shin 2006: 15). Others have argued that modernisation began before Japanese annexation, and that most colonial infrastructures were destroyed along with much of the country during the Korean War (Cumings 2002: 71–72).[10] The notorious US-backed military dictatorship also enforced a rapid and gruelling process of modernisation from the 1960s; the scars left by that regime ran deep in Korean society. The people's persistent and collective effort to gain democracy was strengthened by the anti-imperialistic aspect of nationalism in Korea. It could be argued that the moment of demolition was no mere bubble but represented a deep-rooted legacy of oppression and atrocity.

The rebirth and diversification

Since 1995, three different political administrations have overseen the rebirth of the National Museum which resulted in its grand architectural remaking, completed in 2005 (Figures 23.2 and 23.3). The new National Museum of Korea then faced a different set of issues and challenges to those that led to the old building's demise. With its rapid economic recovery, South Korea was – like many European nations – experiencing large-scale, labour-related immigration, mainly from other parts of Asia. It was no longer the ethnically homogeneous country it once considered itself to be. Korea also possesses a global diaspora of expatriates, who left the country for various reasons but who are still socially, economically and culturally active in Korea.

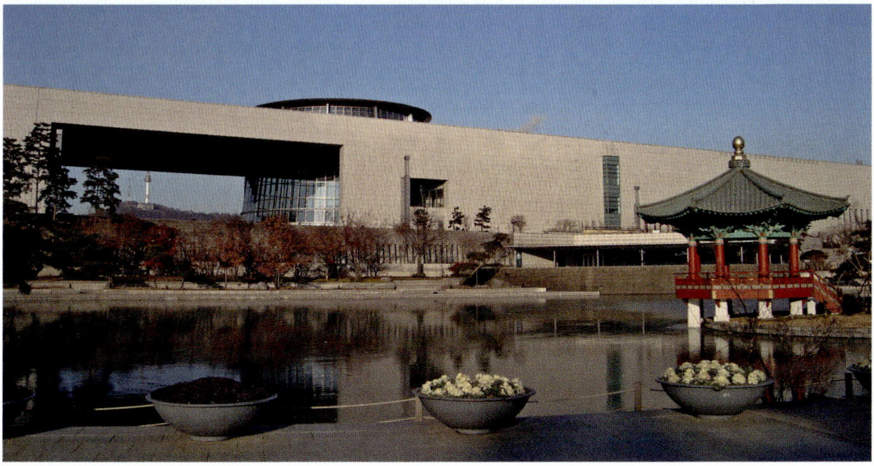

Figure 23.2 The National Museum of Korea, Seoul, Junglim Architecture, 2009. Photo: author

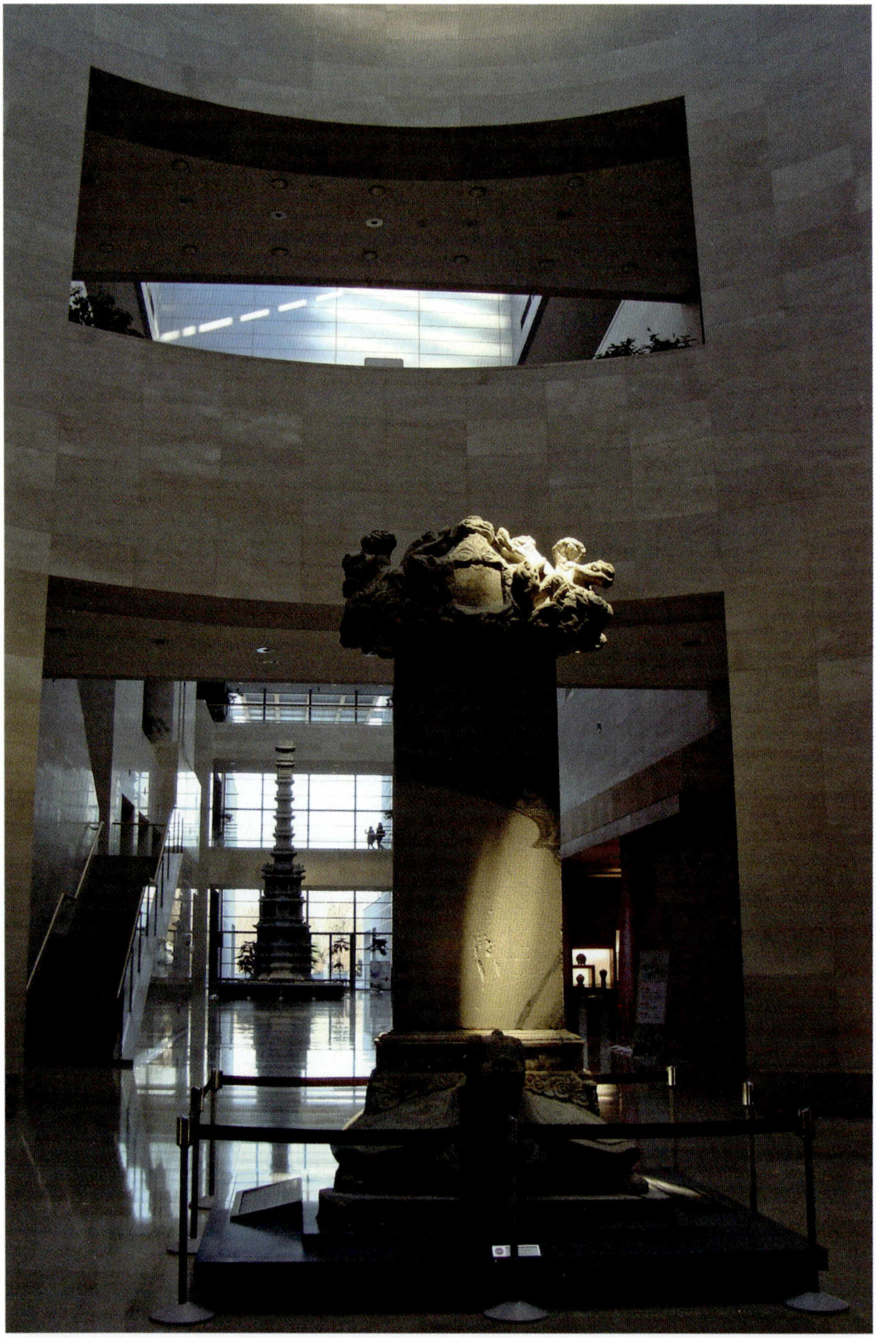

Figure 23.3 Interior of the National Museum of Korea, Seoul, Junglim Architecture, 2009. Photo: author

The concept of the 'nation' has by these means acquired yet another level of complexity. At the same time, the South Korean government has begun a process of diversifying cultural provision; of producing subsidiaries to the national museum and inventing other cultural forms or performances which function, in terms of cultural purpose and impact, in a fashion similar to those of the national museum.

Amongst these acts of diversification were those of the National Museum of Contemporary Art. There has also been a move towards smaller, more specialised, institutions, such as Insa Art Space, and the inauguration of events including the Gwangju Biennale, the Busan Biennale and Media City Seoul. Although these events uniquely exist outside the institutional umbrella of national museums, they perform similar functions. They are ways in which the Korean government has adapted to the globalised industry in culture; they are also manifestations of the government's approach to regional economic development.

These developments have written into them attempts to overcome a historical legacy, which persists in the political and educational system, and seek renewal through cultural development. Perhaps unsurprisingly, the inauguration of the Gwangju Biennale coincided with the fiftieth anniversary of Korea's Independence Day. It marked the beginning of a cultural engagement with 'the developed world'. The government wished the Biennale to be the largest cultural event in Asia and attract professionals and visitors around the world who would contribute to the economic rejuvenation of the city of Gwangju. However, a group of leading fine art and art history intellectuals, including Wan Kyung Sung, Professor of Art and Art Education at Inha University (and ironically the artistic director of the 2002 Gwangju Biennale), was opposed to the government's plan. They argued that such state-controlled and economically driven events would not contribute to the enrichment of culture or the lives of the public.[11] The group argued that it was mere gesture politics which sought to appease the people of Gwangju, a city deeply scarred by a military government-led massacre in May 1980. The Biennale, however, was inaugurated with such success that other cities, including Seoul, Busan and Jeonju, began to organise their own biennials and international film festivals, each establishing a public foundation to receive a large part of their budgets from the government. State-sponsored culture had again become political.

Insa Art Space began life as a subsidiary of the National Museum of Contemporary Art until its sudden change of management and organisation in May 2009, when it was merged into Arts Council Korea. This small multi-storey exhibition space, situated close to the centre of Seoul's most popular tourist area, was run as an independent organisation with a diverse and international outlook. In 2006, Insa Art Space became part of an international collaboration initiated by the New Museum of Contemporary Art, New York, under a programme called 'Museum as Hub'. The programme's conceptual direction was introduced on the New Museum's website:

> A partnership of five international arts organizations, Museum as Hub is a new model for curatorial practice and institutional collaboration established

> to enhance our understanding of contemporary art. Both a network of relationships and an actual physical site located in the fifth-floor New Museum Education Center, Museum as Hub is conceived as a flexible, social space designed to engage audiences through multimedia workstations, exhibition areas, screenings, symposia, and events. Initiated by the New Museum in 2006, this partnership includes Insa Art Space, (Seoul, South Korea); Museo Tamayo Arte Contemporáneo (Mexico City, Mexico); Townhouse Gallery of Contemporary Art (Cairo, Egypt); and Van Abbemuseum (Eindhoven, The Netherlands).
>
> (New Museum of Contemporary Art)[12]

Rather different is Korea's National Museum of Contemporary Art, which is located in Gwacheon, a satellite city outside Seoul, off the tourist trail and beyond the reach of most Koreans. Unlike Insa Art Space, which promotes individual artists and collaborates with artists from other countries, the National Museum of Contemporary Art organises 'national' exhibitions, such as on Russian Abstract painting and Thai contemporary art, which the museum argues enhance international relations through cultural exchange; these are often called 'Amity' exhibitions, and included, for example, *The Exhibition for the Fiftieth Anniversary of Korea–Thailand Amity Agreement: Daily Life in Korea*.[13] In this way, the museum directly supports the government agenda for nationalistic understandings of culture and cultural exchange. It is therefore unsurprising to find that audience figures for such exhibitions depend heavily on school or community excursions and field trips. These exhibitions seem to have little appeal for the informed contemporary art and culture audience. Many recognise fundamental problems with an approach which owes so much to state policy and supervision, and prefer the contemporary independence of Insa Art Space. It was with some surprise and disappointment, then, that this audience heard the news that this innovative and lively institution was to be merged directly into Arts Council Korea. It was to become a rentable space; its re-designation a factor in widespread consolidation activity in government.

Conclusion

The relocation and rehousing of the National Museum of Korea in 2005 marked a new era for an institution which had accrued layer upon layer of historical, cultural and political complexity. The new museum was built in a South Korea engaging in increasingly liberal politics which wished to see the nation as a major player on the stage of global cultural politics. The new National Museum of Korea has yet to demonstrate how it will continue to contribute to the nation-building process that took a new turn in 1995. Its monolithic grandeur emerged from a nation which feels it has a great deal to prove, but it is now a player in a cultural world which brings its own complexities, its own specialist events and institutions. Of these, it was the small Insa Art Space that successfully incorporated and revealed such complexities through its collaborative work with other global institutions. In contrast, the National Museum of Contemporary Art tends to show the legacy of a top-down approach to cultural politics and production. This approach sees the Museum, as well as other numerous government-funded art events, as a space for displaying and institutionalising a cultural remaking of the nation. In a country that strives to be an international centre of the arts and technology, yet whose contemporaneity is still marked by a palimpsest of colonialism and dictatorship, the evolving definition and implication of 'national' in the National Museum begs a reassessment from within the institution. Insa Art Space's attempts to perform within an arts democracy seem at first sight to be far removed from that moment in 1995 when the government's crane decapitated the old museum, but they each reflect the way in which nations seek to place themselves through culture.

Notes

1 Kwang-shik Choe, 'Director's message', National Museum of Korea, Online. <http://www.museum.go.kr/eng/about/greeting.jsp> (accessed 8 May 2009).

2 Throughout this paper, I follow the Revised Romanisation of Korean System (in effect since 2002) rather than the McCune-Reischaur system commonly used in North America.

3 Recently referred to as Joseon Government-General Museum in English.

4 Mun-seong Cha, 'Establishment of Museums under Japanese Occupation', *Electronic Journal Gaeksa*, Online. <http://uriul.or.kr/zeroboard/zboard.php?id=co1&page=2&sn1=&divpage=1&sn=off&ss=on&sc=on&selectarrange =headnum&desc=asc&no=16> (accessed 2 March 2007).

5 Also recently referred to as the Japanese Government-General Building.

6 Although the name Joongangcheong was designated to the former Governor-General's building in the 1970s, using different names for the building may cause confusion; thus I address the building as the Governor-General's building or the former Governor-General's building depending on the period (colonial or postcolonial) I refer to.

7 'Japanese Colonial Architecture', *Things Asian*, Online. <http://thingsasian.com/stories-photos/3439> (accessed 26 March 2007).

8 Jean-Nicolas Huyot's restoration site plan for the Temple of Fortune at Praeneste in 1811, for example, shows two grand courtyards on the East and West wings of the building on the lower level.

9 For planning of the new Gwanghwamun Square, see Sung Jin Park, 'Detailed plans and prospects for the new Gwanghwamun Square', *Maeil Business News* (27 December 2006), Online. <http://news.mk.co.kr/newsRead.php?year =2006&no=562674> (accessed 23 March 2007), and Jae-ki Lee, 'The new Gwanghwamun Square'; *CBS Nocut News*, Online. <http://www.cbs.co.kr/Nocut/Show.asp?IDX=393892> (accessed 24 March 2007).

10 Cumings provides 'evidence supporting the Korean claim that their route to modernity was not facilitated but was instead derailed and hijacked by Japan' and that Seoul's technological development such as electricity, trolley cars, telephones and telegraph systems already existed prior to Japanese colonisation.

11 Interview with the author, 13 February 2001.

12 'Museum as Hub', *The New Museum of Contemporary Art*, Online. <http://www.newmuseum.org/learn/museum_as_hub> (accessed 29 July 2008).

13 The exhibition was held at Bangkok Queen's Museum from 2 September to 30 September 2008. The exhibition included works by sixteen Korean artists in their mid-career.

Bibliography

Alsayyad, N. (ed.) (1992) *Forms of Dominance: On the Architecture and Urbanism of the Colonial Enterprise*, Aldershot: Avebury.

Anon. (1995a) 'Disappearing Japanese war symbols', *Chosun Ilbo* (14 August), final edition, sec. 4.

Anon. (1995b) 'Regaining the nation's pulse', *Dong-A Ilbo* (16 August), final edition.

Balibar, E. and Wallerstein, I. (1991) *Race, Nation, Class: Ambiguous Identities*, London and New York: Verso.

Bennett, T. (1997) *Korea: Caught in Time*, Reading: Garnet Publishing.

Cho, S. G. (1995) 'Demolition of the Governor-General's Building', *Kukmin Ilbo* (5 June 1995), final edition.

Choi, C. (1995) 'Reopening the mouth of our nation', *Joongang Daily News* (14 August), final edition, sec. 9.

Choi, I. H. (1995) 'A new Gyeongbok Palace song', *Hankook Ilbo* (16 August), final edition.

Choi, I.-J. (1992) *Hankuk Shinmun Sajinsa (The History of Photojournalism in Korea)*, Seoul: Youl Hwa Dang Publishers.

Choi, I.-J. (1997) *Hankuk Sajinsa (The History of Korean Photography)*, Seoul: Nunbit.

Choi, J. H. (2006) *CEOs, Read Cultural Codes: Decoding Cultural Trends*, Seoul: Life and Dream Publishing.

Choi, J. Y. (1995) 'Dismantling of the steeple by Jeong Pyung Lee', *Kyunghyang News* (16 August), final edition.

Choi, J. Y. (1997) 'Iron posts in the basement of the former Governor-General's building', *Joongang Daily News* (8 May), final edition, sec. 6.

Choi, Y. S. (1995) 'Is the Kim Administration a Feng Shui Administration?', *Monthly Chosun*, 16: 156–79.

Chung, C. (1995) 'Re-establishing identity through restoration of the Gyeongbok Palace', *Joongang Daily News* (5 August), final edition.

Chung, Y. (1995) 'Pressed by the Governor-General's building for 50 Years', *Chosun Ilbo*, (17 August), final edition, sec. 9.

Cumings, B. (1981) *The Origins of the Korean War: Liberation and the Emergence of the Separate Regimes*, Princeton, NJ: Princeton University Press.

Cumings, B. (1997) *Korea's Place in the Sun: a Modern History*, New York: W. W. Norton.

Cumings, B. (2002) *Parallax Visions: Making Sense of American–East Asian Relations*, Durham, NC: Duke University Press.

Halliday, J. and Cumings, B. (1988) *Korea: The Unknown War*, New York: Pantheon Books.

Harvey, D. (2005) *The New Imperialism*, Oxford: Oxford University Press.

Hitchcock, P. (2003) 'The genre of postcoloniality', *New Literary History*, 34: 299–330.

Hitchcock, P. (2007) 'Postcolonial failure and the politics of nation', *South Atlantic Quarterly*, 106: 727–52.

Hyun, K. D. (1995) '50,000 gathered with tears and cheers', *Kyunghyang News* (15 August), final edition, sec. 4.

Joo, N. (1996) 'The old Governor-General's building must be demolished', *Architecture*, 35: 74–77.

Kang, K. W. (1995) 'Against the demolition', *Joongang Daily News* (5 August), final edition.

Kim, M. S. (1995) 'Japanese Embassy blocking the Ki of Bukhan Moutain', *Kyunghyang News* (1 March), final edition, sec. 1.

Kim, S. C. (1995a) 'Demolishing the 69 years of Japanese Governor-General', *Chosun Ilbo* (16 August), final edition, sec. 4.

Kim, S. (1995b) 'Origin of the cruelty and exploitation', *Chosun Ilbo* (16 August), final edition, sec. 4.

Kusno, A. (1998) 'Beyond the postcolonial: Architecture and political cultures in Indonesia', *Public Culture*, 10: 549–75.

Mongia, P. (ed.) (1996) *Contemporary Postcolonial Theory: A Reader*, London: Arnold; New York: St Martin's Press.

Nalbantoğlu, G. B. and Wong, C. T. (eds) (1997) *Postcolonial Space(s)*, New York: Princeton Architectural Press.

Pelley, P. M. (2005) 'Review essay: Asian cities in a comparative light', *Journal of Urban History*, 31: 736–45.

Shin, G.-W. (2006) *Ethnic Nationalism in Korea: Genealogy, Politics, and Legacy*, Stanford, CA: Stanford University Press.

Shin, G.-W. and Robinson M. (eds) (1999) *Colonial Modernity in Korea*, Cambridge, MA: Harvard University Press.

Shin, Y. (1995) 'After the removal is more important', *Joong Ang Daily News* (14 August), final edition, sec. 8.

Shin, Y. H. (1994) 'Removal of colonial residues', *Dong-A Ilbo* (15 August), final edition, sec. 20.

Song, M. (1996) 'A proposal for relocation of the old Governor-General's Building', *Architecture*, 35: 72–73.

Spivak, G. C. (1999) *A Critique of Postcolonial Reason: Toward a History of the Vanishing Present*, Cambridge, MA: Harvard University Press.

Vale, L. (1992) 'Designing national identity: Post-colonial capitols as intercultural dilemmas', in N. Alsyyad (ed.) *Forms of Dominance: On the Architecture and Urbanism of the Colonial Enterprise*, Aldershot: Avebury.

Williams, P. and Chrisman, L. (eds) (1994) *Colonial Discourse and Post-colonial Theory: A Reader*, New York: Columbia University Press.

Yoo, H.-T. (2006) 'Why is the new City Hall not getting rid of 本shaped floor plan?', *Joong Ang Daily News* (16 September).

Zanten, D. van (1997) 'Architectural composition at the Ecole des Beaux-Arts from Charles Percier to Charles Garnier', in A. Drexler (ed.) *The Architecture of the Ecole des Beaux-Arts*, Cambridge, MA: MIT Press.

Chapter 24

EXHIBITING CHINA IN LONDON

Amy Jane Barnes

FOR CENTURIES THE WEST has been fascinated with China or, at least, an image of China mediated through exoticised cultural imaginings and fuelled by fantastical semi-mythical accounts. Until the mid-nineteenth century and the advent of the popular press, the material products of China – silk and later, porcelain and tea – or those manufactured in Europe in the 'Chinese style', remained the main means by which the vast majority of Europeans could conceive of China. Indeed, porcelain became so synonymous with its country of origin that in the West it came to be known simply as 'china'; a term now applied as a generic descriptor for all types of ceramic, from within and without China, but which nevertheless ascribes an aura of rarity, value and status. Exotic and evocative, and produced by unknown and thus seemingly magical technologies, these products, enthusiastically consumed by fashionable Europeans, came to symbolise an imagined China and sparked successive phases of imagineering, alternating between fascination, ambivalence and distrust.

In the wake of the Opium Wars (1839–43 and 1856–60), and coinciding with colonial expansion and heightened popular patriotism and confidence in the British Empire, ideas about China began to shift from the admiration of the previous centuries. In this chapter I shall contend that this new, negative attitude was consolidated by the international exhibitions and newly established public museums that became such a defining feature of Victorian cultural life. The popular, sometimes ghoulish interest in the perceived barbarity and cruelty of the Chinese propagated by the Victorian popular press, coupled with paradigm shifts in the intellectual field and a growing awareness of museums and exhibitions as effective vessels for the dissemination of state propaganda, infiltrated interpretations of the material culture of the Chinese Other in exhibitionary contexts. For the most part, the British populace derived their vision of China from the illustrated news and popular press. But increasingly, the populist exhibition took a role in the process of image creation and consolidation, serving up to the British public a packaged sampler of Chinese culture, society, faith and artistic endeavour, as interpreted against a backdrop of racism and conflict. Exhibiting institutions were taste-makers for the nation. Royal patronage lent weight to the perceived representation of truth embodied in their interpretations and juxtapositions of objects selected for display, establishing fixed ideas about Chinese visual culture: what it was and what it looked like. In turn, exhibitions reflected back a particular self-image and conception of British nationhood. It is this contention which forms the crux of my argument in this chapter; that the British nation was 'exhibited' to itself through its representations of China. Taking examples of nationally appealing displays of Chinese material (and ideas about China) in the capital, I will locate representations of China, and the ideas of Self and Britishness, implicitly promoted by each incidence of display. The chapter predominantly focuses upon the now little-known *Celebrated Chinese Collection*, but also considers the display of Chinese material at the Great Exhibition of 1851; the South Kensington (later the Victoria and Albert) Museum; and the symbolic representation of China at the Albert Memorial. Each case study will be contextually fixed within the wider, historical Sino-British political and cultural relationship.

The *Celebrated Chinese Collection*

> At no period in the history of the world, has the attention of civilised nations been so fully directed towards China, its early history, and modern position, as at the present moment.
>
> (Langdon and Dunn 1842: XIX)

During the nineteenth century, commercial exhibitions appealed across the social classes as populist entertainment (Pearce 1995: 124–29). Capable of attracting a large audience, and in direct contrast to the 'scientific' sterility of more respectable establishments, their success was rooted in the romanticised view of the world they presented (Pearce 1995: 129–31). These exhibitions interpreted objects with showmanship and sensationalism; they were the antithesis to modernist interpretations of the world. The *Celebrated Chinese Collection* – frequently referred to at the time as 'Ten Thousand Chinese Things', was one such exhibition. Housed in a purpose-built hall on Hyde Park Corner, it had the added, popular appeal of representing the culture and society of a recent adversary.

The Opium Wars had been forced by Britain (casting itself as moral arbiter) in retaliation for the Chinese court's refusal to allow the legal importation of opium from the Indian poppy fields. The 'opening' of China was sold to the British public, not as a conquest, but almost as a 'gift' to the Chinese people; Britain was a paternal, liberating saviour removing China from despotism and bringing beneficial technological advancement, thereby inducting it into the modern world. But these military campaigns were the culmination of decades of growing mutual disenchantment. In contrast, eighteenth-century ideas about China were largely positive, with the Chinese political system and Confucian tenets held up by European philosophers as representative of the acme of human civilisation (and a nation from which the West could learn a great deal). China was the stuff of romantic cultural imaginings; the luxurious, exotic and mysterious Chinese goods which reached Europe were principal image-makers. *Chinoiserie* – decorative motifs developed in Europe to emulate and evoke the China of the imagination – supported the myth. But the 1800s were, conversely, 'the years of estrangement' (Sullivan 1989: 117).

The Opium Wars were enthusiastically reported at home by the popular press. Publications like the *Illustrated London News* – 'the publishing success story of the Victorian era'[1] – reached a 'remarkable' proportion of the population. Coverage was xenophobic at best and overtly racist at worst: the Chinese were presented as stupid, lawless, lazy, cruel and barbaric savages. By resisting the indomitable march of progress (the moral justification for British incursions into Chinese territory) they displayed, it was asserted, conceited notions of their own self-importance and power. But, despite illustrations like 'A Chinese Cage' (1842) which depicts a European woman interred in a bamboo cage while a Chinese family look on,[2] the *Illustrated London News* was, in comparison with other publications fairly liberal, and more balanced in its coverage and representation of foreign conflicts than its contemporary rivals.[3] One can only assume that other quarters of the media were even more vociferously anti-Chinese. Indeed the visual representation of the Chinese family in 'A Chinese Cage' is fairly sympathetic when compared with the Punch cartoon, 'A Lesson to John Chinaman' (9 May 1857), which depicts Lord Palmerston – egged on by Mr Punch – wielding a cat-o'-nine-tails at a hideous, wizened 'Chinese' man with claw-like hands and an exaggeratedly long queue (plait). The man wears a placard around his neck that reads 'The Destroyer of Women and Children'.

Although the image of China had been tarnished by the Opium Wars, Chinese products were, paradoxically, regarded rather more positively. War created a demand by raising the profile of China in the popular consciousness and limiting the passage of trade goods (Pagani 1998: 28) like porcelain, tea and silk. A new wave of Sinomania showed that 'China was regarded as a marketable commodity just as were her products'; people could buy, consume or wear a little bit of China, and by doing so were participating in 'feelings of cultural and economic superiority which came about through victory in the Opium War' (Pagani 1998: 29). As a case in point, the *Celebrated Chinese Collection* exploited the British public's passion for all things Chinese in the 1840s and sold 300,000 catalogues (Pagani 1998: 35). The exhibition operated as a 'tangible point of intersection' for the two strands of the paradoxical image of China in the mid-nineteenth century: a nation of 'unenlightened savages with whom Britain was recently at war and one which produced objects worthy of admiration' (Pagani 1998: 36).

The exhibition comprised the collection of an American merchant, Nathan Dunn (1782–1844). Originally from Philadelphia – where his collection was first displayed before its transfer to London at the request of an unknown, but apparently 'eminent Englishman' (Saxbee 1990: 26) – Dunn had lived and worked in China for twelve years, during which time he had amassed an enormous quantity of diverse material, aware that he was well placed in his endeavours to furnish the Western world with an encyclopaedic array of Chinese things: a 'Chinese World in Miniature' (Langdon and Dunn 1842: vii). There is some confusion with regards to the date of opening; Pagani (1998: 36) gives 23 June 1842, whereas my research has placed the date as early as 1840 (if not the full collection, the exhibition existing in proto-form), but what is not in dispute is that by 1842,[4] a building – 'in the style of Chinese Architecture' (Langdon and Dunn 1842: 11) – had been erected in St George's Place, at Hyde Park Corner, to house the collection.[5]

Although Dunn's collection comprised the largest assemblage of Chinese objects to have ever been exhibited in London, many were familiar to the audience (Figure 24.1). A glance at the catalogue indicates that therein displayed were heavily decorated lacquer furniture, polychrome porcelains, intricately carved ivory models, embroidered textiles, religious figures, and opium pipes. Staged dioramas featuring appropriately dressed mannequins according to rank and trades – mandarins, a Buddhist priest, a 'literary gentleman', a Chinese lady, actors, a barber at work, and a boatwoman, amongst others – presented many of these objects *in situ* for everyday use (see Langdon and Dunn 1842). The exhibition reflected and consolidated the skewed images of China already present in the British popular

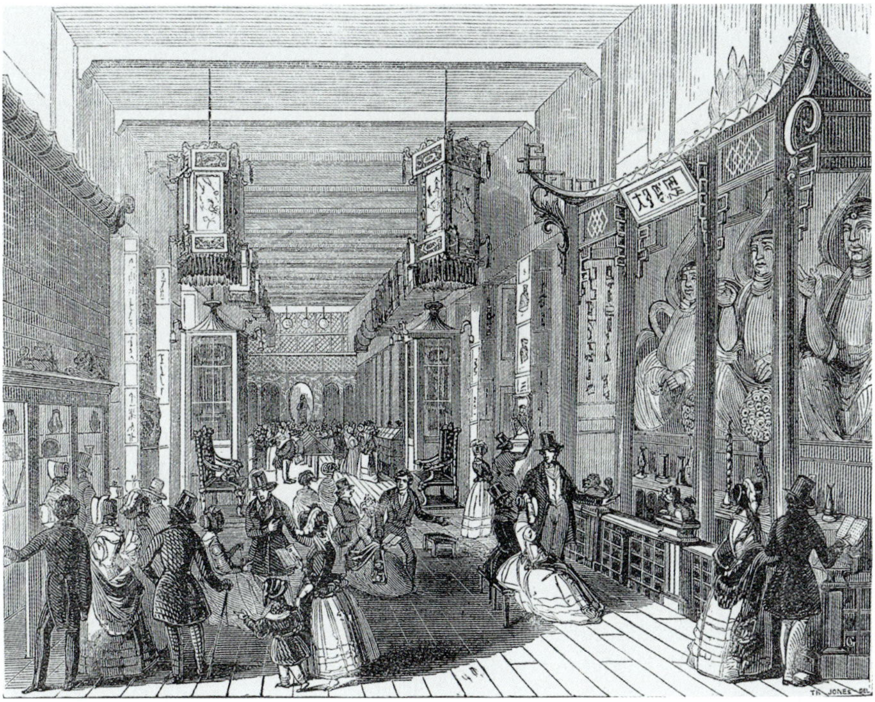

Figure 24.1 A contemporary view of the *Celebrated Chinese Collection* from the *Illustrated London News*, 1842. Photo: Copyright *Illustrated London News*/Mary Evans

consciousness, which were based upon export ware (made for the European market and thus pandering to European tastes and ideas of Chineseness) and *chinoiserie*.

It would be erroneous to paint an image of an entirely uncritical and unaware audience. Indeed, by virtue of Dunn's mercantile career, much of the collection – albeit presented as representative of the wider Chinese nation – had its origins in the principal trade and manufacturing port of Canton (modern-day Guangzhou) and this was commented upon by observers at the time (Saxbee 1990). Nevertheless, the catalogue declared that visitors might experience all of China through the exhibition: an impression emphasised by the comments of a journalist:

> A few hours spent in studying the contents of this collection, with the aid of the descriptive catalogue, which is full of information, much of it original, will possess the visitor with an idea of the Chinese almost as complete and vivid as could be formed by a voyage to China.
>
> (*Spectator*, cited by Pagani 1998: 38)

Indeed, after a private viewing of the exhibition, Queen Victoria is said to have remarked that 'one could have almost fancied oneself in China'.[6]

It had long been accepted that it was possible to know a culture through its objects. Indeed, from the Renaissance cabinet of curiosities, the museum has been viewed as a microcosm: 'an ordered representation of the world' (Barringer and Flynn 1998: 11). In much the same way as knowledge was collected, collated and disseminated under the modernist paradigm, 'the world itself [was] . . . ordered up as an endless exhibition . . . to generate an unprecedented effect of order and *certainty*' (Mitchell 1992: 290 (italics added)), thus rendering the observer's (usually the Westerner's) gaze (of the 'Other') 'objective'. This was a panoptic way of seeing the world; of viewing the unfolding scene unobserved. The phenomenon of international and colonial exhibitions in the century from 1851 is, perhaps, its ultimate expression and reflects the self-confidence of British nationhood. One can imagine visitors to the Great Exhibition, for example, gazing upon the assembled material culture of the Empire and its colonial possessions, from a viewing platform constructed for that very purpose, and gleaning from that experience a sense of ownership, power and destiny.

The popularity and notoriety (and likely perceptual influence) of the exhibition in the cultural landscape of mid-nineteenth century London is evidenced by the references to it, found in eclectic and diverse contemporary records. For example, Charles Dickens writes of its 'Feast of Lanthrons [lanterns]' in a letter to Daniel Maclise, dated 31 October 1840 (Dickens *et al.* 1969: 143). The *London Saturday Journal* published a short, highly complimentary description of a 'gas-light' visit to the collection in 1841, advising readers that: 'the exhibition will be very attractive to the gentler sex; and as a consequence, very gratifying to visitors of a stronger growth' (Timbs 1842: 50).[7] A report on palaeontological specimens dating to 1846 remarks upon the existence of particular fish fossils in Dunn's collection (see

Richardson 1846), thus demonstrating the scope of objects presented for public consumption (and the importance of the natural sciences).

While the exhibition sought initially to educate the British visiting public, to 'unveil' China (Langdon and Dunn 1842: vi), after Dunn's death in 1844 – and, perhaps, as attitudes hardened and negative images of China became entrenched in the British collective psyche – aspects of melodrama and sensationalism crept into the presentation. In 1843 a display on opium smoking was added. In 1851, by which time the exhibition was housed in a new location, following an unsuccessful attempt to sell the collection to the British Museum and a subsequent British tour and brief stop-off in Philadelphia,[8] human exhibits were on display. The 'Celestial Belle', Pwan-Ye-Koo, became a particular draw for the crowds, with her two-and-a-half-inch 'golden water-lily' bound feet (see Pagani 1998: 38). That this 'by-line' evokes the impression of a circus freak show is not entirely erroneous; by this time, the *Celebrated Chinese Collection*'s counterpart institution in Philadelphia had been leased by P. T. Barnum.

The thirst for the exotic which Nathan Dunn's exhibition sought to slake not only reflected the idea that it was possible to know another country through its material products, but also reinforced the inextricable link between images of China and its material culture. By virtue of Dunn's position, first as merchant and later as showman, the exhibition once again asserted the link between China and its goods: 'Commodity and China became inextricably linked: Dunn *read* Chinese culture *from* these commodities, and in his presentation of them, he *made* Chinese culture *into* a commodity' (Pagani 1998: 37). It must be recognised, however, that while after his death (under the direction of the curator William Langdon) the exhibition took an increasingly money-spinning, sensationalist direction, in his lifetime, Dunn – a Quaker – only wished to 'cover the current expenses' and that the Queen believed that any profits would be provided for charitable purposes.[9] Nevertheless, it was Dunn who set the initial entrance fee at half-a-crown (much higher than most comparable exhibits in London).[10]

That China and its material culture had been thoroughly conflated by the second half of the nineteenth century is further advanced by its symbolic representation, in sculptural form, at the Albert Memorial in Hyde Park (1876) (Figure 24.2) located not far from the original location of the *Celebrated Chinese Collection* at Hyde Park Corner. Placed at the principal compass points around its boundary, four sculptural groups represent the continents; Europe, the Americas, Africa and Asia. The Asia group is dominated by the figure of an Indian princess seated on a recumbent elephant, reflecting the pre-eminence of the Indian subcontinent among Britain's imperial possessions in the region. Extra-colonial China (that is, partially subjugated but not colonised by Britain, see King 1999: 21) is represented by a young Chinese male wearing a long queue and 'typical' costume. He holds a porcelain vase. Another is placed at his feet. This representation of China – below both Indian and Arabia – is a visual manifestation of China's submission to the might of the British Navy. Together these references and the image of China disseminated by Dunn's exhibition confirm that by the second half of the nineteenth century and

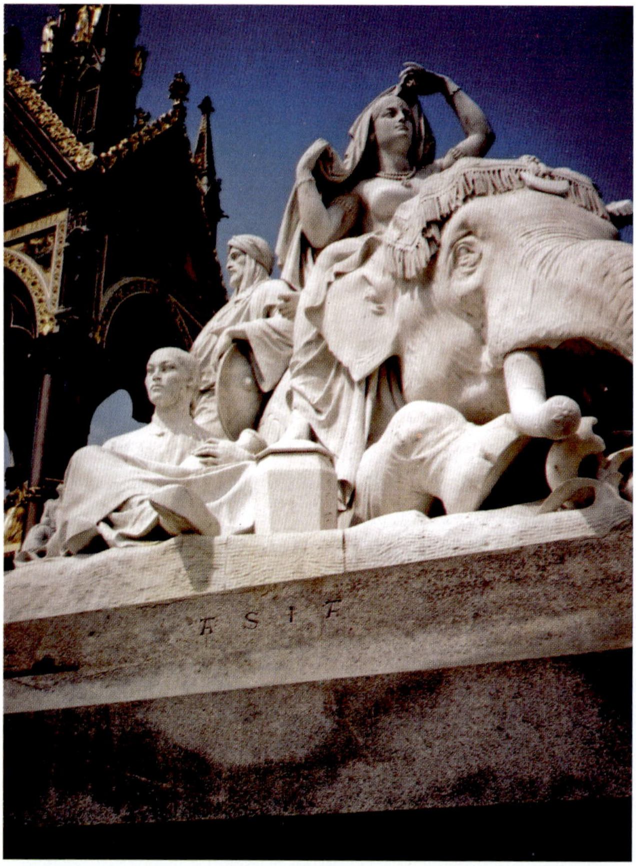

Figure 24.2 Asia, the Albert Memorial, Hyde Park, London. Photo: author

after the forcible opening of China to trade, the symbolic identification of luxury trade goods (specifically porcelain) with (an imagined) China was complete: exotic, different and commodified for Western consumption.

The Great Exhibition of 1851

The aim of the Great Exhibition – the brainchild of Albert, the Prince Consort – was to present the best examples of design and manufactured products from around the world, with an emphasis on the British Empire. The explicit objective of the organising committee was to inspire British artists and designers and to effect an improvement in quality (see Hobhouse 2004: xx). This goal reflected the contemporary belief that technological progress would bring social progress, and that exposure to great art would improve the masses. On another level, the Great Exhibition was a triumphal expression of the British Empire, upholding the perceived benevolence of British imperialism that brought social and technological benefits to all peoples of the world (Maxwell 1999: 7). As was its earlier counterpart – Dunn's *Celebrated Chinese Collection* – the Great Exhibition was a resounding success, visited by 6,039,195 people (Gibbs-Smith 1981: 24). It had a profound and lasting impact on the nation's self-image and that of its empire.

Much scholarly time has been devoted to the study of the development and execution of the exhibition in recent years (see, for example, Auerbach 1999; Purbrick 2001; Leapman 2002; Hobhouse 2004; Auerbach and Hoffenberg 2008). This chapter does not intend to duplicate this research. Instead, it focuses upon the small display of Chinese material at the exhibition; an analysis of which may help to elucidate contemporary attitudes towards China in Britain, and consolidate the hypothesis posited at the beginning of this chapter, that exhibitionary representations of China may tell us a great deal more about Britain.

China was accorded a small display at the Great Exhibition, which, in contrast to those representing other countries, was not sponsored by the subject nation's government. Given the animosity between Britain and China at the time, this should come as no surprise. Instead, the selection of Chinese art was made from extant collections in Britain, from private individuals, the diplomatic service and importers (see Pagani 1998: 29; Yapp 1851: 214–16) although, as this quote from a contemporary appraiser of the Chinese section suggests, visitors were not aware of this fact:

> Allusion has been already made to the strange collection sent to the Great Exhibition from the Celestial Empire. The curious workmanship of many articles bears witness to the sort of instinctive learning which the Chinese have for the most difficult and delicate manual labour; but we need envy nothing that they have, unless it be the abundance of some natural productions, especially silk. Their porcelain has been known since time immemorial and in everything else the Chinese are so stationary that they may be considered as the most ancient workmen on earth. Among the articles which they displayed were some which were produced at a period nearly as remote as that of the deluge, and which, in truth, did not appear to be very dissimilar to those which they manufacture at the present day.
>
> (Anon., cited by Gibbs-Smith 1981: 56)

As a result, the art objects on show (there were also specimens of plants, crops and other natural products) were unrepresentative of native Chinese taste, and comprised *chinoiserie* and export ware made for the European market. The official catalogue (Yapp 1851: 214–16) reveals these to have included: lacquered furniture inlaid with mother-of-pearl; bronze and silver censers; tableware and porcelain vessels; embroidered upholstery, costumes and shoes; and intricately carved ivory balls (Figure 24.3). As described earlier in this chapter, these kinds of objects were the luxury commodities often made and decorated in response to European misconceptions of Chinese art. Thus, their display consolidated pre-existing images of Chinese art, which were legitimised here by their close association with the monarchy. Comments, such as those expressed above by a visitor to the exhibition, suggest a passing interest in the intricacy of the workmanship (Pagani 1998: 29), but also a weary sense of disengagement; it had all been seen before. Such criticisms of Chinese makers reveal more about British tastes for the generically 'oriental' and prevailing misconceptions of Chinese aesthetics than any reality of Chinese art (an argument which will be developed in the next section of this chapter). However, making accurate representations of non-Western art and design was not really the point of the Great Exhibition. Above all, it accorded an opportunity for the British to display their national pride and, in so doing (and with particular reference to the Chinese section), patronise a display comprised of objects from a country defeated (Pagani 1998: 37).

Whereas Nathan Dunn's commercial exhibition, a decade earlier, had met with enthusiasm, by the 1850s, as British reliance on Chinese goods fell,[11] so did the overall interest in Chinese art and manufactured products: 'The excitement concerning the Chinese and their products generated at war's end had again changed. The British were bored: the euphoria of the post-war boom was over' (Pagani 1998: 39). As was so often the case in preceding centuries, the Sinomania of the mid-nineteenth century was, thus, just a passing fad, lending weight to Hung's (2003) assertion that Western ideas about and attitudes towards China can be directly correlated with what was happening at home, in the intellectual, economic and political fields.

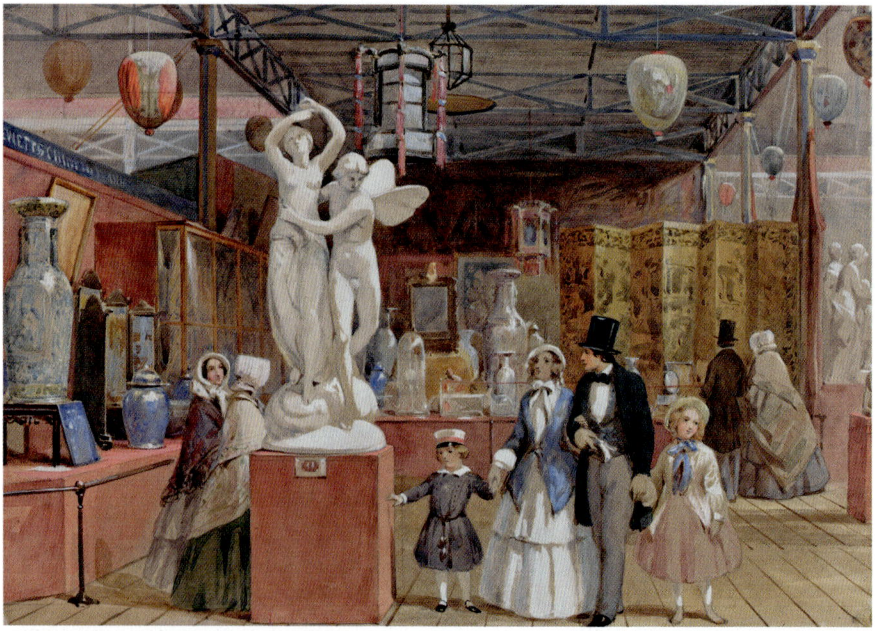

Figure 24.3 *Part of the China Court,* John Absolon, 1851 (watercolour). Photo: Copyright Victoria and Albert Museum

The South Kensington Museum

The ideals of the Great Exhibition found their permanent expression at the South Kensington Museum (renamed the Victoria and Albert Museum in 1899), the influence of which, upon Western scholarship and the representation of Chinese material culture, cannot be overestimated. As a national museum it was simultaneously patronised and supported by the British state. The profits of the Great Exhibition paid for the purchase of the land for the new museum (which opened to the public in 1857) and the early collections were partially derived from those objects displayed in 1851 at the Crystal Palace (Barringer and Flynn 1998: 13).

The primary purpose of the museum was didactic. Indeed, its founder Henry Cole wrote that it should be 'an impressive schoolroom for everyone' (Cole, cited by Barringer and Flynn 1998: 15). Its collections were to act as inspirational resources for students, artists and designers, as well as offering the public the opportunity for, in today's parlance, self-directed learning. Within three years of its opening, the museum was attracting over 600,000 visitors per annum (Darby *et al.* 1983: 5). To encourage working people to attend, the galleries were opened in the evenings and, in 1858, it was reported that visitors to these sessions exceeded daytime visits by five-to-one (Waterfield 1991: 10). The potential influence on the visiting public of the museum's interpretive strategies and representations of Other cultures, therefore, is indisputable.

Contemporary perceptions of Chinese art – which suffered from the continued influence of distorted images derived from *chinoiserie* and export ware – exerted, perhaps, the strongest influence on the South Kensington Museum. Additionally, the ways in which Chinese art was misread meant that it was never assessed on its own terms, using native aesthetic values as the basis for criticism. Traditional Chinese aesthetics are bound up in ideas which are difficult to translate or express in English; a philosophy at odds with Western conventions. It was, in particular, the privilege of expression over representation, that as Clunas (1999: 126) states, was the concept that the European academy found most difficult (or unwilling) to grasp, prior to the twentieth century. For a start, contemporary Western art history demoted calligraphy from its pre-eminent status in China, privileging painting instead, which better reflected that genre's position in the Western canon (Clunas 1999: 122). It is unsurprising, then, that this clash of cultures would result in confusion and misunderstanding. However, as King and others (1999: 13) remind us: 'Being a European colonizer meant being powerful enough to maintain the fiction of speaking objectively for all humanity and possessing knowledge that was not bound by one's cultural and geographical location.' The presumption that the West was entitled to speak for the rest of the world; that Western social and ethical mores, that Western aesthetics were the only right and proper way of representing the world, reveals the contemporary arrogance and self-importance of the West.

Chinese art was thus problematic for nineteenth-century art historians. It belonged, if not to an entirely subjugated nation, to a culture which was, according to apparently scientific hierarchies, inferior to the West. The usual way of dealing with the art of Other, supposedly inferior cultures was to classify it as 'artefact' or 'material culture', whereupon it became the subject of science – anthropology, ethnography, archaeology – rather than aesthetics (King 1999: 8); fields which were heavily influenced by Social Darwinism. But China had a written art history and theory that pre-empted the Western canon by centuries, and could not, therefore, be so easily dismissed. The solution was to essentialise (that is, ignore diversity and artistic innovation) and classify Chinese art as different from 'progressive' Western art, as 'a way of reducing the threat it [posed] to *the* single, linear history of art as a story that takes place above all in Europe' (Clunas 1999: 135). And yet 'The theories of separate cultural development were circulated by people who were borrowing from every culture on the globe' (King 1999: 13): an inbuilt hypocrisy which meant that China was never, and could and would never be considered on a par with the West, which would always maintain economic, social and cultural superiority.

Paradoxically, Chinese art was criticised too for its apparent stagnation: while China was forcibly encouraged to accept 'progress', the West mourned the very customs and traditions it supplanted. To paraphrase Stewart (1993, cited by Pierson 2007: 139) objects became the means of authenticating a non-existent (imagined) past, desirable because it was absent. By design, the West preserved, appropriated and controlled orientalist imaginings in perpetuity, to be recalled and evoked in nostalgic remembrance.

Yet, at the South Kensington Museum, art history had less impact than the nineteenth-century framework of taxonomy, which 'exercised a powerful hegemony over the ordering of man-made products as well as those of nature' and influenced the implicit aims of the South Kensington Museum to 'make the historical and geographical series of all decorative art complete' (Clunas 1998: 44). In reality, this meant the exclusion of the material cultures of those classified 'primitive' – Africans, Native Americans, Pacific Islanders and aboriginal peoples – who, with neither art nor history (according to contemporary Western understanding), fell under the auspices of the human sciences of ethnography and archaeology. But China, by virtue of its long tradition of art history and criticism, and its visual culture which to some degree correlated with Western notions of 'art' was, along with that of other Asian cultures, permitted *limited* entry into the realm of the decorative arts. 'A Chinese picture could be bad art, failed art, but it could not cease to be art at this point' (Clunas 1998: 45). Perceived as a three-dimensional colonial archive – 'a fantasy of knowledge made into power' (Barringer and Flynn 1998: 11) – the museum was to be interpreted as a microcosm of the world, reflecting and consolidating the contemporary belief that it was possible to know Other cultures through their visual culture.

In 1862, the Asian collections were designated their own wing. Interiorly decorated by the designer Owen Jones, the galleries were said to be evocative of the 'Orient': the fabric of the building took on an interpretive role, enhancing the Otherness of the objects on display (Barringer and Flynn 1998: 15). Hierarchies of civilisation were mapped out in the physical geography of the museum. The marginalisation of Chinese, Japanese, Indian, Korean and Persian art within the accepted canon of art was emphasised by separation from the rest of the collection. Displayed together they were perceived as one, homogeneous artistic tradition. Said has commented that this project asserted the absolute difference between the Orient and the West, at the same time as collapsing differences within the term 'oriental art' (Said 1978, cited by Barringer and Flynn 1998: 16). Indeed, John Ayers, a former curator of the Victoria and Albert Museum's collection, echoes Saidian discourse by defining the widely used term 'oriental', as not representing any real cultural identity but existing as a defining mirror image for the Occident (Ayers, cited by Darby *et al.* 1983: 289).

Indeed, a fair proportion of the Chinese material that found its way into British collections of the mid-nineteenth century was looted in the aftermath of military campaigns. Emperor Qianlong's lacquered throne, which remains a highlight of the Victoria and Albert Museum's collection, was itself 'taken' from the *Nan Yuan* hunting park in 1901 (Clunas 1998: 42). In the wake of the Opium Wars and the 'opening' of China, this once 'vaunted repository of an exquisite aesthetic, [was] downgraded to a mere flea market' and 'handled and picked over' by Westerners (Thurin 1999: 43). Displayed at the colonial centre (Baker and Richardson 1997: 221), these trophies of war become symbolic of the defeat and humiliation of their country of origin. By possessing them, manipulating them and subverting their meaning, their new masters (re)gained the control and power of their old symbolic or ideological meanings. Drawing upon Bourdieuian discourse, this process may be explicated as a transference of symbolic capital; that is, the honour, prestige and legitimacy held by an authoritative individual, or institution (see Forest and Johnson 2002), here serving to enhance the association of Chinese things with a victorious Britain.

Conclusion

This chapter has shown that by analysing the representation of an Other culture in national museums and nationally significant exhibitions, one can reveal something of the host country's self-identity. But what it has also demonstrated is the extent to which, in Britain, during the latter half of the nineteenth century, institutions with an exhibitionary function reflected and consolidated extant images – held in the wider intellectual and popular consciousness – of China, its art and its material products. The increased contact with Chinese things, brought by war and the clamour for information about the enemy, and the selection – consciously or otherwise – of particular styles and types of object for display, contributed to the perpetuation of an image of China as moribund, unchanging and preoccupied with infantile trifles, in direct contrast to the perceived virility and morality of a strident Britain and its empire.

Notes

1 Bennett, T. (2003) 'Japan and the *Illustrated London News*: 1851–1905', *Japan Society*, 1. http://www.japansociety.org.uk/lectures/030916terrybennettlecture.html (accessed 22 November 2006).

2 This illustration accompanied a sensationalised report of the capture and internment in the aforementioned bamboo cage of Mrs Noble, wife of the captain of the armed brig, *The Kite*, which was wrecked in 1840. While Mrs Noble was indeed caged, it was for approximately thirty-six hours and not the six weeks reported by the *ILN,* two years after the event (see Pagani 1998: 30).

3 Bennett, T. (2003) 'Japan and the *Illustrated London News*: 1851–1905', *Japan Society*, 1. http://www.japansociety.org.uk/lectures/030916terrybennettlecture.html (accessed 22 November 2006).

4 British History Online with English Heritage (2000) 'Hyde Park to Wilton Place'. http://www.british-history.ac.uk/report.aspx?compid=45896&strquery=hyde%20park%20corner (accessed 11 February 2010).

5 The 'pagoda' entrance hall of the building was sold in 1847 and relocated to Hackney to begin a new life as a garden ornament. It was demolished in 1956. The main building continued to be used as an exhibition hall (known as St George's Gallery) until the mid-1850s, after which parts were slowly demolished for redevelopment. The last trace disappeared in the late 1960s/early 1970s (see ibid.).

6 Cited by ibid.

7 It is this pair of references, from credible sources, which suggest that the exhibition existed in some form at the same location earlier than 1842.

8 British History Online with English Heritage (2000) 'Hyde Park to Wilton Place'. http://www.british-history.ac.uk/report.aspx?compid=45896&strquery=hyde%20park%20corner (accessed 11 February 2010).

9 *Spectator*, cited by ibid.

10 Ibid.

11 In particular the successful cultivation of tea in India resulted in a dramatic fall in exports of tea from China.

Bibliography

Anon. (1851) 'Chinese Exhibition', *The Times*, 30 April.

Auerbach, J. A. (1999) *The Great Exhibition of 1851: A Nation on Display*, New Haven, CT: Yale University Press.

Auerbach, J. A. and Hoffenberg, P. H. (2008) *Britain, the Empire, and the World at the Great Exhibition of 1851*, Aldershot: Ashgate Publishing.

Baker, M. and Richardson, B. (eds) (1997) *A Grand Design: The Art of the Victoria and Albert Museum*, New York: Harry N. Abrams Inc.

Barringer, T. and Flynn, T. (eds) (1998) *Colonialism and the Object: Empire, Material Culture and the Museum*, London and New York: Routledge.

Ching, J. and Oxtoby, W. G. (eds) (1992) *Discovering China: European Interpretations in the Enlightenment*, Rochester, NY: University of Rochester Press.

Clunas, C. (1998) 'China in Britain: The imperial collections'. In Barringer, T. and Flynn, T. (eds) *Colonialism and the Object: Empire, Material Culture and the Museum*, London and New York: Routledge, pp. 41–51.

Clunas, C. (1999) 'What about Chinese art?' In King, C. (ed.) *Views of Difference: Different Views of Art*, New Haven, CT, and London: Yale University Press pp. 121–41.

Darby, M., Burton, A., Haskins, S. and Ayers, J. (1983) *The Victoria and Albert Museum*, New York: Viking Press.

Dickens, C., House, M. and Storey, G. (eds) (1969) *The Letters of Charles Dickens, 1840–1841*, Oxford: Oxford University Press.

Dirks, N. B. (ed.) (1992) *Colonialism and Culture*, Ann Arbor: The University of Michigan Press.

Forrest, B. and Johnson, J. (2002) 'Unraveling the threads of history: Soviet-era monuments and post-Soviet national identity', *Annals of the Association of American Geographers*, 92 (3): 524–47.

Gibbs-Smith, C. H. (1981) *The Great Exhibition of 1851*, London: HMSO.

Hobhouse, H. (2004) *The Crystal Palace and the Great Exhibition*, London: Continuum International.

Hung, H. (2003) 'Orientalist knowledge and social theories: China and the European conceptions of East–West differences from 1600–1900', *Sociological Theory*, 21(3): 254–80.

King, C. (1999) 'Approaching art made in China'. In King, C. (ed.) *Views of Difference: Different Views of Art*, New Haven, CT, and London: Yale University Press, pp. 7–22.

Langdon, W. B. and Dunn, N. (1842) *'Ten Thousand Chinese Things': A Descriptive Catalogue of the Chinese Collection, Now Exhibiting at St George's Place, Hyde Park Corner, London.*

Leapman, M. (2002) *The World for a Shilling: How the Great Exhibition of 1851 Shaped a Nation,* London: Headline.

Mackenzie, J. M. (1995) *Orientalism: History, Theory and the Arts*, Manchester and New York: Manchester University Press.

Mackerras, C. (1999) *Western Images of China* (second edition), Oxford and New York: Oxford University Press.

Maxwell, A. (1999) *Colonial Photography and Exhibitions: Representations of the 'Native' and the Making of European Identities*, London and New York: Leicester University Press.

Mitchell, T. (1992) 'Orientalism and the exhibitionary order'. In Dirks, N. B. (ed.) *Colonialism and Culture*, Ann Arbor: University of Michigan Press, pp. 289–318.

Pagani, C. (1998) 'Chinese material culture and British perceptions of China in the mid-nineteenth century'. In Barringer, T. and Flynn, T (eds) *Colonialism and the Object: Empire, Material Culture and the Museum*, London and New York: Routledge, pp. 28–40.

Pearce, S. (1995) *On Collecting*, London and New York: Routledge.

Pierson, S. (2007) *Collectors, Collections and Museums: The Field of Chinese Ceramics in Britain, 1560–1960*, Bern: Peter Lang.

Purbrick, L. (ed.) (2001) *The Great Exhibition of 1851: New Interdisciplinary Essays*, Manchester: Manchester University Press.

Richardson, Sir J. (1846) *Report on the Ichthyology of the Seas of China and Japan*, London: John Murray.

Saxbee, H. (1990) 'An Orient exhibited: The exhibition of the Chinese Collection in England in the 1840s' (unpublished doctoral thesis, Royal College of Art).

Sullivan, M. (1989) *The Meeting of Eastern and Western Art*, Berkeley, Los Angeles and London: University of California Press.

Thurin, S. S. (1999) *Victorian Travelers and the Opening of China, 1842–1907*, Athens, OH: Ohio University Press.

Timbs, J. (ed.) (1842) 'The Chinese Collection: Hyde Park Corner', *The London Saturday Journal*, 23 July, 50.

Waterfield, G. (ed.) (1991) *Palaces of Art: Art Galleries in Britain, 1790–1990.* London: Dulwich Picture Gallery.

Yapp, G. W. (1851) *Official Catalogue of the Great Exhibition*, London: Royal Commission.

Chapter 25

EXHIBITING THE CONGO IN STOCKHOLM

Lotten Gustafsson Reinius

One Sunday at church, as the sermon had been finished by me and Lars Jönsson, an aged idol priest suddenly rose and proceeded to the pulpit, carrying with him his various gods. We wondered about his intentions. The assembly squatted in the benches for fear.

When the elder finally reached the pulpit, he turned towards the parish and gestured vigorously as he screamed: Sa pi keti I meno [Calm down, it is just me!]. Is it true that I am a priest of idols? Have I been telling you that I am capable of curing the deceased and of uncovering secrets?

THIS FORCEFUL ENTRANCE, as recalled in the autobiography of the Swedish missionary, Svante August Flodén (1933: 112–15), turned out to be the preface to a rite of passage that took place in the Congolese village Nganda sometime in the late 1890s. What was about to be performed publicly was a radical transformation of religious and social identities, effected verbally and materially. Having secured the attention of his audience, the *nganga* displayed his objects, holding each up and stating its function. Afterwards he went on to more palpable acts of deconstruction. Flodén describes in excitement how small clouds of chalk, dried herbs and clay obscured the raging movements by which the converting idol priest was tearing apart compositions once brought together with skill and care. As a final gesture, he handed over to the missionaries a couple of especially significant ritual tools and signs of expert knowledge: a drum and a staff.

The church in question was located in lower Congo, at the heart of a region well known for the cult of different species of *minkisi*: elaborate material compositions created and used in order to make ancestral and other spiritual powers present and communicable (MacGaffey 1991, 1994). The abandonment or destruction of this category of things had played a role in conversions to Christianity since the time of the first Portuguese encounter with the former Congo Kingdom in the late fifteenth century. About four hundred years later, at the time of European colonialism, the production and ritual handling of such objects increased as the local population sought protection and truth in a time of trauma and turmoil (Axelsson 1970; Ekholm-Friedman 1991). Many missionaries concentrated their ambitions and aggressions on transforming these spiritual and material fields of activity.

Flodén is one among many Protestant missionaries who provide evidence of the church's attempts to take 'sinful and dangerous objects' out of circulation and use. Some of these objects were concealed in little storage houses at the missionary stations. Others were sunk in the Congo River, reflecting perhaps both Christian and Congolese beliefs in the purifying and transformative powers of water (Andersson 1947: 135 ff.). Flodén (1933: 165–7) himself arranged public bonfires (Figure 25.1). One event in particular, introduced to the local Congolese as a trial of strength between traditional spiritual powers and 'the Almighty', became known in missionary circles as 'the fire of a thousand burning idols'.

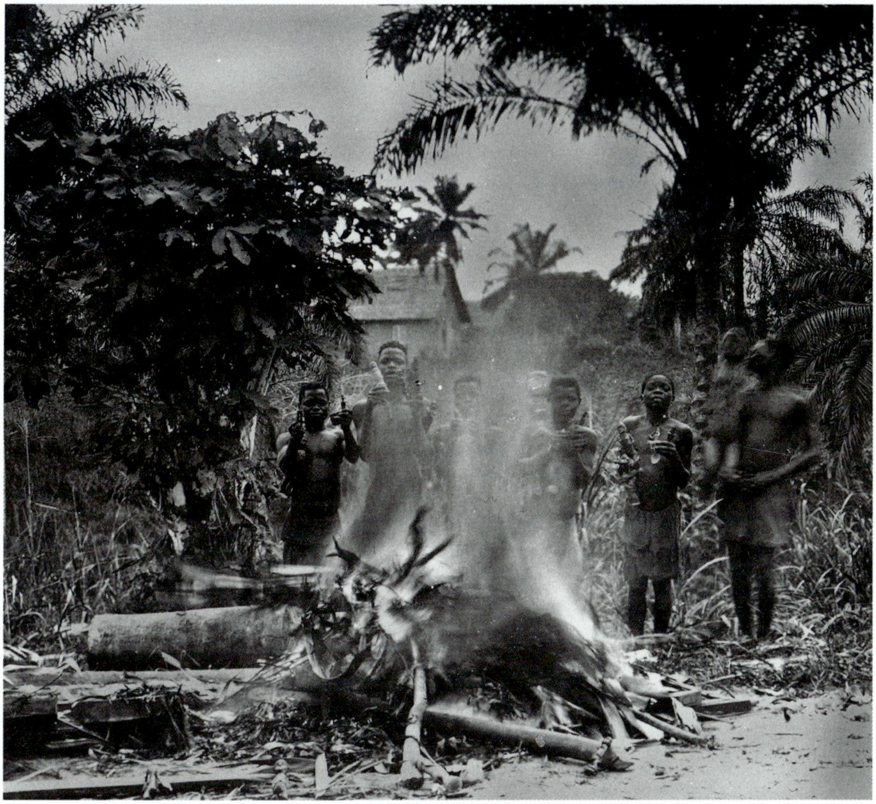

Figure 25.1 Burning of 'idols' taking place at Kingoyi missionary station, 1912 or 1913. Photo: Edvard Karlman. Copy from the archive of the Mission Covenant Church of Sweden. Photographer unknown. Digitally improved by Tony Sandin/Museum of Ethnography

Flodén's discussion of the conversion of the *nganga* reveals that in return for his public sacrifice, the convert received a collection of mass-produced objects of a fundamentally different kind. Calling at the missionary's house that night, the *nganga* collected a table of European type, a full matchbox and a parcel of industrially produced candles. The events form sequences of an almost paradigmatic narrative: a story about promises, demands and roles in modernity. In Sweden, at this time, there were similar performances.

The drum and staff were never consigned to the fire. Instead, they entered ethnographic collections in museums in Gothenburg and Stockholm.[1] Today, they form part of collections of approximately 18,000 objects from areas surrounding the Congo River.[2] This chapter will focus on the ways in which these pieces of Congolese material culture were used and symbolically transformed when unpacked in Sweden. The main actors in this account are material objects which travelled between people and places, functions, genres and meanings, and ended up, for a while, at the *Ethnographic and Missionary Exhibition* that took place in Stockholm in 1907.

A new national narrative

Museum visiting in Europe and the United States has undergone an intense process of development since the late nineteenth century. World fairs and museums transformed into new kinds of public space that played an important role for the development of the modern state (Bennett 1995: 60). Sites formerly reserved for the privileged few were opened to a more democratic public inspection, and visitors of all classes were invited to share the grand narratives of their time – literally as well as symbolically. The *Ethnographic and Missionary Exhibition* was one such site where themes of a national narrative were negotiated. Here, the bringing together of people and Congolese and other artefacts established embodied relations that mediated sentiments, imaginations and ideas more effectively than words; they became 'powerful engines of meaning that mediate messages' (Kirshenblatt-Gimblett 1998: 157). This relationship echoed, but in rather different ways, how people had experienced these objects in Africa.

This account of Swedish experiences draws upon a range of contemporary written and photographic sources, and has involved examination of the material objects themselves (Gustafsson Reinius 2009). A recent inventory discovered that the majority of the approximately 10,000 items from lower Congo acquired by missionaries was donated in the year preceding the exhibition. It is noteworthy that Sweden possessed no colonies of her own at this time, yet these collections found their way into discussions of the Swedish nation. Indeed, during the hundred years preceding the events discussed here, the relatively old kingdom of Sweden had been subjected to repeated and painful territorial loss. Finland, for centuries considered a crucial and natural part of the country, in spite of ethnic and linguistic differences, had been ceded to Russia in 1809. And in 1905 the union with Norway was reluctantly but peacefully dissolved. This drastic shrinkage of territory, historians have often suggested, led to a remarkable strengthening of the Swedish nation, emotionally and symbolically, and Swedish museums and exhibitions played a role in the negotiation and dissemination of new national narratives.

Nordiska Museet and the related open-air museum, Skansen, then abandoned their intentions to cover the whole of Scandinavia, and instead confirmed the new national borders (Grandien 1991; Hillström 2006). Skansen, an open-air museum in the heart of Stockholm, turned agrarian and rural lifestyles into evocative embodiments of Swedishness (Jönses 2005; Kirshenblatt-Gimblett 1998). However, in 1897, this romantic and nostalgic construction was offered a modern progressive counterpoint in Stockholm's first world fair (Ekström 1994).

The *Ethnographic and Missionary Exhibition* was a rather different museological engagement. In England and France, displays of artefacts and people (living or dead) promoted the nation's power and importance (Bennett 1995; Coombes 1994). The exhibition of ethnographic material in Sweden at this time is, however, less easily explained. Why was this exhibition produced and by whom? Why did the Congo play such an important role? And, more importantly, how did it impact upon the contemporary production and negotiation of this newly remodelled nation?

The exhibition

The four-month *Ethnographic and Missionary Exhibition* opened on 15 May 1907 with the Swedish royal family, members of the government, representatives of several scientific and missionary societies, and the city's clergy in attendance. The exhibition was located in the Academy of Sciences, close to the office of the Swedish Missionary Society and, charging a modest entry fee, it soon attracted a socially diverse audience. It had been the brainchild of Erland Nordenskiöld, then Head of the Ethnographic Department at the National Museum of Natural Science. That museum had, however, lacked the space to house it.

A year earlier, and perhaps inspired by similar ventures by the London Missionary Society and the Church Missionary Society (Coombes 1994: 161), Nordenskiöld had approached the Swedish missionary movement and actively encouraged missionaries working abroad to create ethnographic collections. In this task, however, the missionaries were already active, having used them in their schools. It was a task they understood in biblical metaphors, which saw the harvesting of the remains of an ancient culture, while seeding the new order through the Word of God (*Sanningsvittnet*, 23 May 1907). Nordenskiöld's (1907a) own view was little different; he praised the missionaries' 'scientific objectivity' and their desire to 'shed light upon the imprisonment of heathen people in witchcraft, ancestral cult and idolatry'. The exhibition attracted a sizeable audience and received enthusiastic coverage in all the major newspapers (Figure 25.2).[3]

Some of Nordenskiöld's curatorial colleagues were critical and saw the exhibition as an intrusion of 'non-professionals' into the scientific field (Frödin 1913: 53; Lindberg 1995: 132). However, later a number of these missionary collectors could

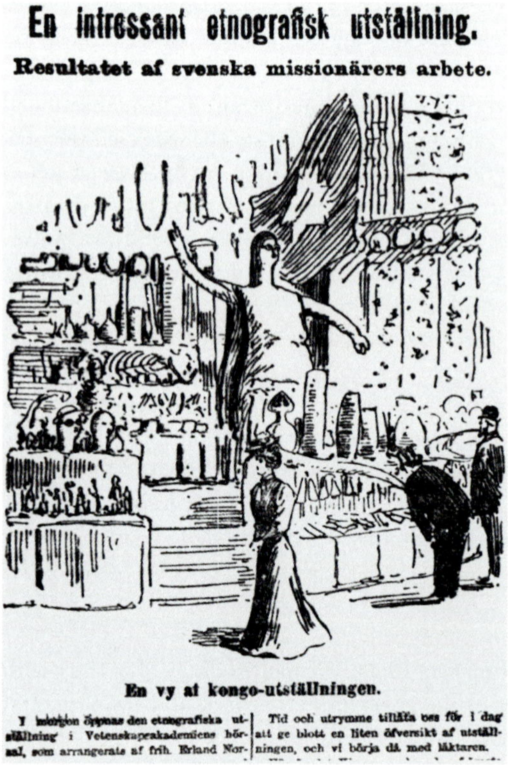

Figure 25.2 Drawing entitled *A view of the Congo exhibit* illustrated in the Swedish daily newspaper *Aftonbladet,* 14 May 1907, the day before the Ethnographic and Missionary Exhibition opened. Illustrator unknown

be seen as forerunners of the Malinowskian turn in anthropology. Living close to the people they studied (in order to convert and civilize), they practised a kind of participant observation and fieldwork. Johannes Fabian (2000: xii) refers to them as proto-ethnographers. This blurring of professional fields is also revealed in several detailed ethnographic descriptions of the lower Congo published in Swedish in connection with the exhibition (Laman 1907). Nordenskiöld himself edited *Ethnographical Contributions by Swedish Missionaries in Africa*, which blended descriptions by missionary ethnographers, such as Karl Edvard Laman and Johan Hammar, with recordings of local verbal art and oral knowledge made by Congolese evangelists, including Titus Makundu and Sila Nimba (Nordenskiöld 1907b; Janzen and MacGaffey 1974).

While the complaints of Nordenskiöld's colleagues may conceal a degree of elitist protectionism, it was nevertheless true that the quality and principles that informed the documentation and selection of ethnographic collections, performed by so many individuals with differing backgrounds, varied significantly. As for the quantitative result, however, the alliance was undeniably a success. Approximately 10,000 objects had been collected in a single year. Unlike the aesthetic and didactic principles that govern the sparse displays in present-day museums, the Stockholm exhibition revelled in the aesthetics of mass effect. Each 'Swedish missionary field' was thus represented by an impressive number of tools, textiles, baskets, weapons, saddles, and other – to the Stockholm audience – exotic objects. The abundance of objects conveyed the industriousness and geographical reach of missionary activities. It also made the case for an independent ethnographic museum where the materials of a rapidly changing world could be preserved.

Swedes in the Congo Free State

The collections on display in the exhibition, where individual objects were grouped into coherent units, embodied an idea, shared by the science of ethnography and the missionary movement, that the world's population consisted of a number of clearly distinguishable peoples. The written guide to the exhibition informed visitors that the collections illustrated, for instance, the Zulu, the Bakongo, the Gonds, the Galla people and the 'likable little Kunama' (Nordenskiöld 1907c). The spatial organization suggested a hierarchical relationship between these peoples, informed by contemporary Western notions of evolution and civilization. Collections from Asia, for example, were displayed on the second floor, whereas those from Africa occupied the ground level. Nordenskiöld had told his collecting missionaries that the future museum would specialize in what he described as 'people poor of culture' (Nordenskiöld 1906a).

Collections from the lower Congo, brought together by the Swedish Missionary Society, occupied almost half of the exhibition space: an entire hall on the ground floor. It was the largest of the missionary fields and its prominence drew comment across the media. Nordenskiöld (1907c: 1) said the display provided insights into the first steps of human culture on the road to development. Geographical distance took on a temporal quality (Fabian 1983) and the Congo held a special place in this imagining.

At this time, the area surrounding the Congo River was plundered for valuable goods, including ebony, rubber and collections of artefacts. It also fed the European imagination with the exotic, strange and dangerous. Through the bestselling travelogues of Stanley and Livingstone and their followers, the Congo developed into an imaginary and exotic entity, symbolically occupying, as Nicholas Mirzoeff (1999: 133) puts it, 'the position zero in the timespace of modernity'. It possessed epic symbolism and particularly for Swedish missionaries. This image remains in the European imagination, and has found expression in such works as Swedish artist Jan Håfström's painting, *The Inner Station* (Figure 25.3). Here, the River Congo, one of the world's great watercourses and on whose banks are spoken hundreds of languages, is portrayed as a thin ribbon of light threading its way into homogeneous darkness.

The first Swedish missionary arrived in Congo in 1881. He worked as a volunteer for the British Livingstone Inland Mission. Five years later, the Swedish Missionary Society became an autonomous organization, and its members became concentrated in an area of the lower Congo, between the river mouth and Brazzaville. Only in

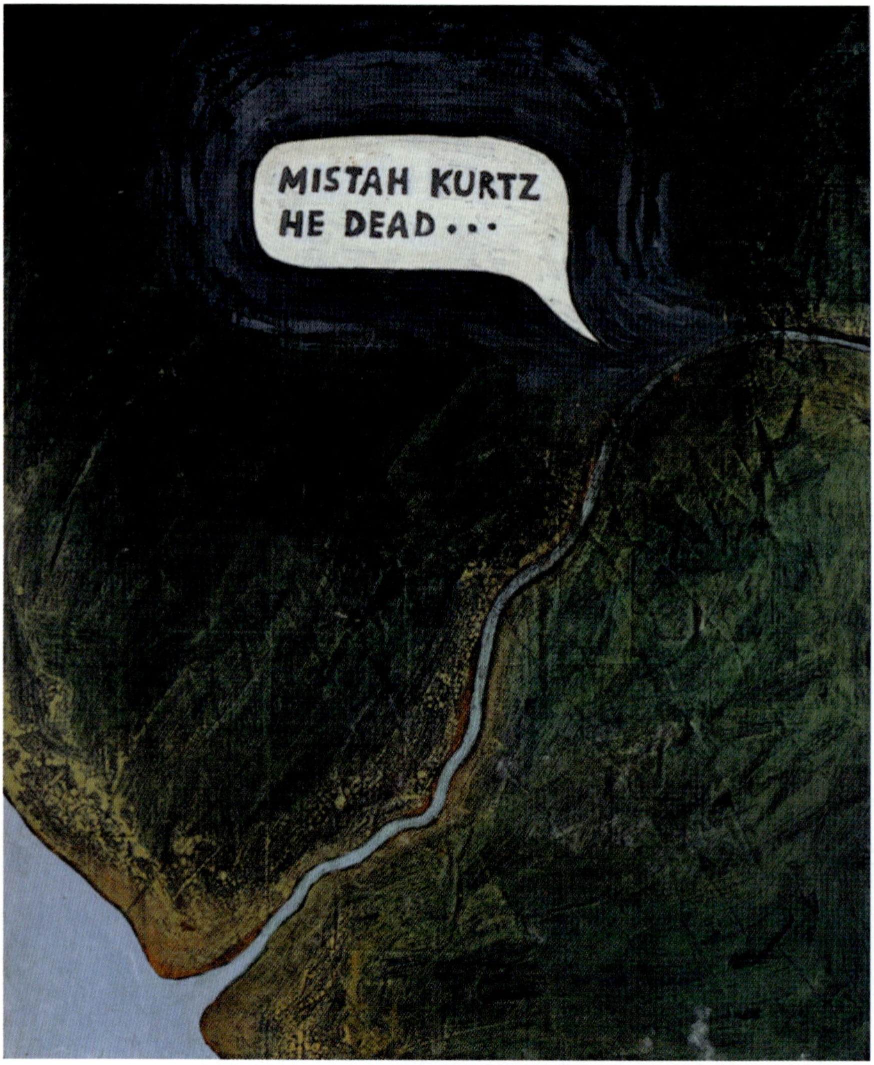

Figure 25.3 *The Inner Station,* Jan Håfström, 2000 (oil on canvas). Photo: Mattias Johansson. Published with permission of the artist

1908, when the Free State became a Belgian colony, did the Swedish missionaries expand their activities north of the river. At the time of the exhibition more than a hundred Swedish missionaries had been sent to the region, and those who had been educated in missionary schools aspired to follow them, to 'spread evangelical light' into the symbolic darkness of supposed isolation and spiritual need.

Congo Free State (1885–1908) already occupied a place in the minds of the Swedish audience which visited the exhibition, and it did so not simply as a result of missionary zeal (Axelsson 1970: 217; Jenssen-Tusch 1902–1905: 705; Tygesen 2005). The 'State' was a colonial project and private venture of Belgian King Leopold II, which later gained international notoriety. Leopold's proposal had been accepted by the Berlin Conference on condition that it would be an international undertaking. Consequently recruits came from various countries to work in the huge territory that had been mortgaged for the king with the help of Henry Morton Stanley. It lay open, therefore, at least theoretically, to Christian missions of all persuasions. For the small Protestant nation of Sweden, without its own colonies and weakened by poverty, emigration and territorial shrinkage, the Congo Free State became a land of economic opportunity and national reimagining.

Contemporary rhetoric called for 'all civilized people' to join the philanthropic battle against 'Arabic slavery, heathendom and primitivism'. What transpired in Congo was an episode of extremely violent exploitation of people and goods. By the time of the exhibition, the atrocities in Congo Free State had already been greeted with public condemnation in Europe (Hochschild 1998). The postcolonial philosopher Valentin Mudimbe (1988) has described colonialism as three interlinked processes: the first, where the control of the foreign territory is secured by military force, the second, where local infrastructure is transformed, and a third, where the entire worldview of pre-colonial life becomes the main target of colonial efforts. Although this is admittedly a simplified model, it still provides a schematic image of the major fields of activities for the hundreds of Swedes participating in this colonial process. Apart from the many young missionaries, there were, for instance, merchants and engineers. The colonial army, *La Force Publique*, recruited Swedish military officers actively, to safeguard its international character and, according to testimonies, because Leopold II had been greatly impressed by the disciplined aesthetics of Swedish gymnastics. For a period the Swedes made up the third largest group in that army; they were outnumbered only by the English and Belgians. An even larger category of imported labourers was that of sailors – more than three hundred Swedish sailors were engaged to maintain the crucial steamboat traffic on the Congo River. The guide to the exhibition (Nordenskiöld 1907c: 9 ff.) mentioned that fifty-two Swedish missionaries were at work in the seven main stations at the time. It also notes, regretfully, that as many as forty-three had passed away. The high death rates among natives were left without remark.

Only a year after the exhibition opened, in 1908, after the death of King Leopold, the Belgian nation was to take over and transform Congo Free State into a more regular national colony, not least in response to growing international criticism. Some Swedes had been ruthless participants in violence against the local people, while others, and in particular missionaries, had been outspoken critics of the atrocities. The curators, however, celebrated a political arrangement that enabled Swedish international activity and brought them such material benefits. Ironically, the source of fear and disgust that seems to have provoked most comment during the exhibition arose not from the much-written-about brutality and hardships of the colony but from the objects on display.

The expressive agency of objects

Nordenskiöld (1906b) had issued missionaries with instructions on how to collect scientifically. He was probably concerned about the tendency of missionary collections to focus mainly upon the religious sphere of life. This kind of biased collecting was well known from the military tradition of bringing trophies home from victories in battle; indeed, both groups probably saw the objects similarly (Cannizzo 1998; Östberg 2002). Nordenskiöld, on the other hand, asked the missionaries to capture the *totality* of the local culture, and on this occasion it seems that many of them did. Religious objects remained the most collected items, but the missionaries also gathered tools and other everyday items used in construction and other practical tasks (Östberg 2002:18). The ethnographically ambitious collections made by Johan Hammar, for example, represented the ways in which people dressed and ate, travelled, fought, hunted, fished, took care of their sick, and so on. This thoughtful engagement with the local population, however, appears not to have made its way into the arrangement of Congolese artefacts at the *Ethnographic and Missionary Exhibition*, where a predominant aim seemed to be to illustrate and dramatize the primitive and heathen (see Figure 25.4).

Important local products such as raffia cloth were reduced to mere decorative backdrops for more striking and evocative weapons. The many wooden bowls were laid out in a decorative pattern that hid their original function as cooking utensils.

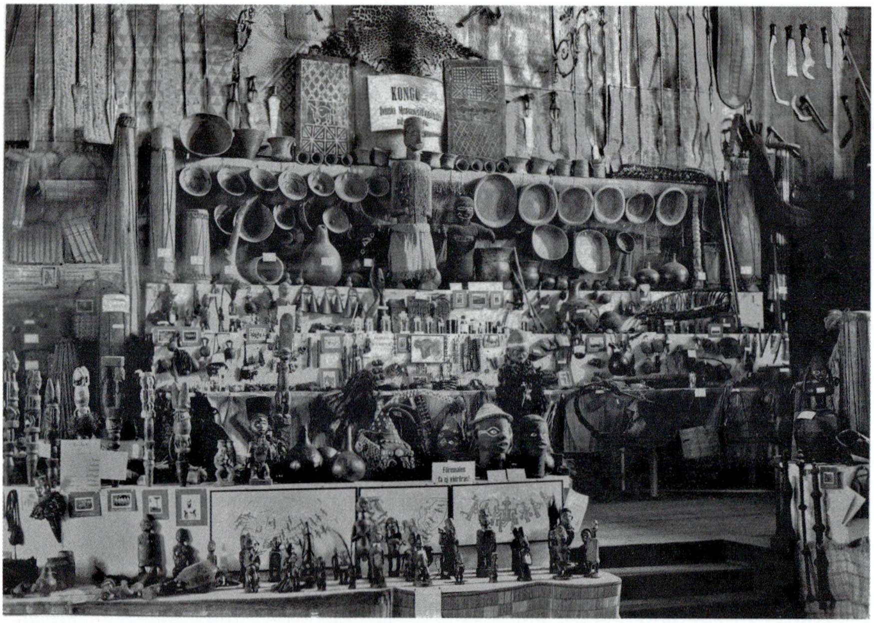

Figure 25.4 The section devoted to the Congo at the Ethnographic and Missionary Exhibition in Stockholm, 1907. Source: Archive of the Mission Covenant Church of Sweden

The exhibition, in this way, served to confirm a stereotypical image of the Congo as a place of warfare, music and magic. To the Swedish audience looking at the large collection of ritualistic objects it was difficult not to imagine the colonized Other as frightening, archaic and inferior. Here they could view the carefully created *minkisi* which were ritualistically destroyed by the missionaries and the local people they converted. Composed of diverse natural materials which are combined to form spiritually forceful entities, often with the most important elements tucked away in vessels, the *minkisi* took on many forms, such as bags and shells. The local press, however, failed to see even the art in these objects, *Svenska Morgonbladet* scornfully remarking, 'a negro as we all know can have just anything as his God'. It is worth noting that the exhibition took place in the same year that the Swedish Missionary Society in Congo sanctioned a new and stricter decree against 'all kinds of community and service with the idols'.[4]

On anthropomorphic sculptures signs of ethnic identification, such as decorative scarification and hairstyles, had been marked with the utmost care. The genitals, in particular, were eloquently fashioned. Eyes, nostrils and ears had been marked and opened: *minkisi* were obviously intended to hear and see. Rows of dark wooden faces met exhibition visitors with a many-eyed gaze. Their facial expressions were serious, authoritative and attentive.

At the centre of this display was a *nkisi nkondi*, or 'nail fetish' as it was referred to in Europe at the time. The hands of this human sculpture were held close to the body. The entire torso was cracked and pierced by numerous nails and projectiles. His posture, his wounded naked body and far-seeing eyes, may have evoked images of Christ. It is uncertain whether the raffia skirt on the lower part of the body is to be interpreted as another such allusion or simply an attempt to avoid embarrassment. To the left of the *nkondi* a smaller funerary sculpture had been placed. Its hands were open, as in a frozen embrace, and the eyes were carefully encircled with white lines traditionally indicating an ability to see in both the world of the living and the world of the dead.

The display was labelled 'The Congo. Collection by the Swedish Missionary Society'. Above this sign was a giant leopard skin and the flags of Sweden and of Congo Free State, both yellow and blue, with daggers and axes surrounding them. It created a sense that here the wild and dangerous had been overpowered, disarmed and tamed. The ordering of the many individual objects into symmetries and other aesthetic patterns may also, of course, be understood as a demonstration of the ability to control and transform into a new aesthetic order what is unordered and wild. In similar fashion, a little row of glass bottles, on a shelf just below this centrepiece of daggers and flags, contained indigenous medical ingredients – or magical potions, depending on who described them – collected by missionaries, captured, ordered and seemingly hermetically sealed. Natural science and hygiene were also to celebrate their victory over superstition, danger and sin – an analogue to the triumph of civilization over the primitive, magical mind. Indeed, in spite of the objects being entirely detached from the social and cultural context where they were originally made and used, they remained – at the exhibition – sacred objects of another kind. The written notice: 'Do not touch the objects!' expressed their new status as ethnographic data. These objects were now highly valued, isolated, classified and preserved. The warning also probably added to the objects' air of frightening potency.

The *minkisi* collected by Karl Edvard Laman and on display at this exhibition may have looked rather different from the others as he had coated them in a shiny varnish (see Figure 25.5). In a letter to Nordenskiöld, the collector himself explains – rather apologetically – that he had initially planned to use these items as decorations in his home.[5] In the varnishing process, utilitarian things made from materials that would decompose relatively fast in a tropical environment were preserved and made expressively static. Under this thin coating the sculptures wore the traces of former and more intimate human handling: they were darkened and stained through repeated application of liquids such as chewed nuts, oil and wine. As such a traditional and regular way of 'feeding the ancestors' had come to a definite end (as naked

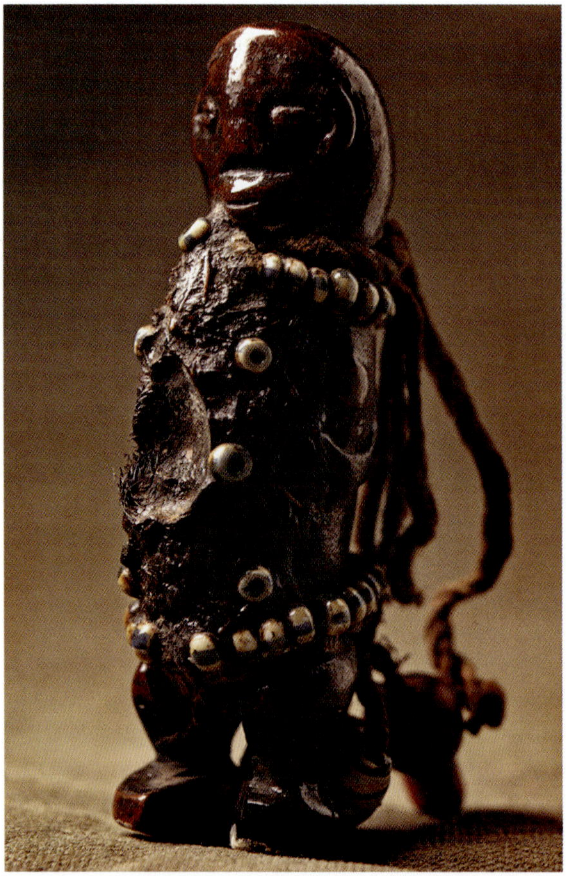

Figure 25.5 Some of the anthropomorphic *minkisi* in the collection of Karl Edvard Laman were covered with an unusual varnish. Object number 1906.46.01 at the Museum of Ethnography in Stockholm. Photo: Tony Sandin/Museum of Ethnography

surfaces and symbolic passages were physically closed and unreachable), and a new kind of handling seems to have been invited (Gustafsson Reinius 2009). The new surface provided an invisible but distinct shelter that granted protection to the object as well as to its holder. In much the same way as the see-through and reinforced glass of a museum showcase, the varnish highlighted and coated the objects in a manner that made them more evocative, yet beyond direct physical contact.

Nordenskiöld's (1906b) fascination with these sculptures was infused with the evolutionary paradigm. Expressive conventions for the creation of postures or sizes were interpreted as signs of the incapacities of 'the primitive mind'. Using such words as 'incomplete', 'raw', 'wrong' and 'exaggerated', his descriptions, however, extended beyond the rational to a subjective assessment provoked by the material. Ironically, the aesthetic and embodied power of the Congolese ritualistic artefacts added to their usefulness in a narrative of their own disarmament. In the exhibition they clearly no longer served their original use. Once mediators between the living and the dead, the *minkisi* at the exhibition were now as abandoned and unused as a broken telephone. Having lost their primary communicative functions, which admittedly depended on context and human agency, the expressive agency of Congolese traditional artefacts was nevertheless still active and skilfully put to new use. The imaginative potential of African art, which strongly influenced the exhibition event, had been recycled.

Exhibiting the Congolese

An exhibition convention quite widespread in nineteenth-century ethnographic displays at world fairs, museums and amusement parks was lacking entirely at the event in Stockholm. This was the display of living people. The absent men, women and children were, however, symbolically evoked in miniature by a diorama at the centre of the exhibition hall (Figure 25.6). As the popular newspaper, *Aftonbladet*, reported on 14 May 1907, visitors were confronted with the quaint and unexpected sight of 'a Congolese village placed directly on a typical Swedish pinewood floor'. The small models illustrated different house types used for living, storage, and keeping poultry and pigs.

The mission used the language of love and Christian duty. In the missionary press, *Sanningsvittnet*, on 23 May 1907, the exhibition was interpreted as a touching appeal by the less fortunate to the Swedish Christian nation, which said: 'Come on over here and help us!' The display of the quaint little village surrounded by a protective rope served well to evoke an image of the Other as needy and childlike. Located in dramatically close proximity to the tiny village was a *niombo* – a funerary shroud made of several layers of red blankets and therefore many times bigger than any ordinary human body – its face and open arms towards it, as if struggling to grab it. It created a sense of threat that might be mitigated by missionary action. A striking transformation of symbols took place once more and for a moment the visitor could behold the world from above – a gaze vaguely reminiscent of the paternalistic but pitying perspective of a distant God.

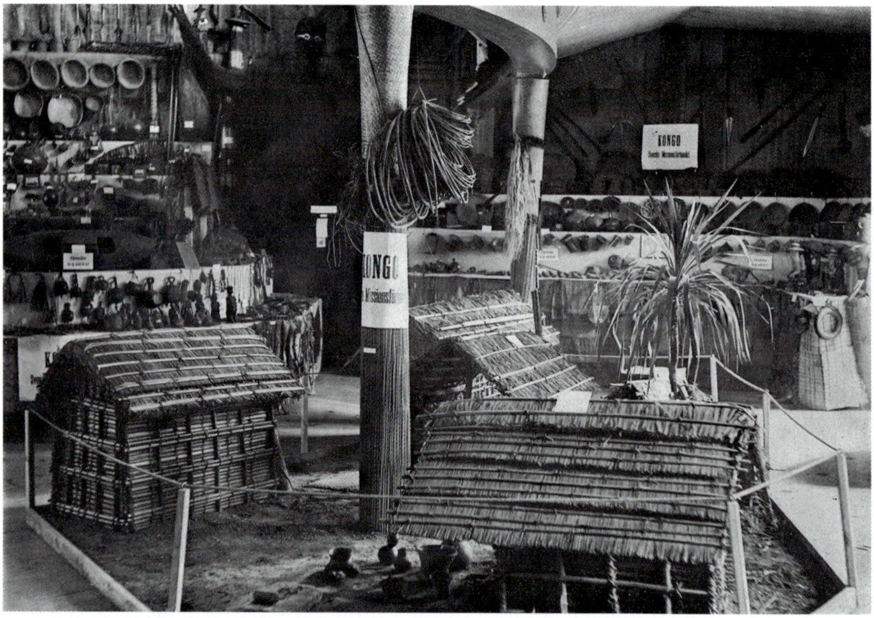

Figure 25.6 The Congolese missionary field in miniature at the Ethnographic and Missionary Exhibition in Stockholm, 1907. Source: Archive of the Mission Covenant Church of Sweden

The miniature is also of a size that epitomizes nostalgia (Stewart 1993: xii). The small buildings, with little pots in the sand reminiscent of absent human residents, may be interpreted as an image of something shrinking and going away. The miniature village allowed a view of a vanishing world. Although there were traces of human presence there – such as the tiny pots – the people themselves were gone. The powerful objects were not the only things abandoned: the display as a whole portrayed a way of life that people had left behind. The heathen and frightening aspects of the Other were materialized as concrete collectable objects, and therefore were also presented as something capable of being handled and taken away. The rhetoric of the mission – differing on this point from contemporary racist theories and echoing instead the utopian visions of the budding working-class movement – taught that the people had the capacity to move on to a higher state. The huge collection, and the little village, illustrated that the Congolese had already abandoned their old ways. A frightful heap of dangerous objects had been left behind.

In the room next to the hall dedicated to collections from the Congo, visitors could see the geographical impact of the Swedish missionary movement in maps and photographs (Figure 25.7). In each location, they could imagine the world undergoing the same dramatic and wonderful change. Recurring images of former 'heathens' in European dress, and recently erected churches and other Swedish small-town buildings, made this act of transformation comprehensible. Pieces of

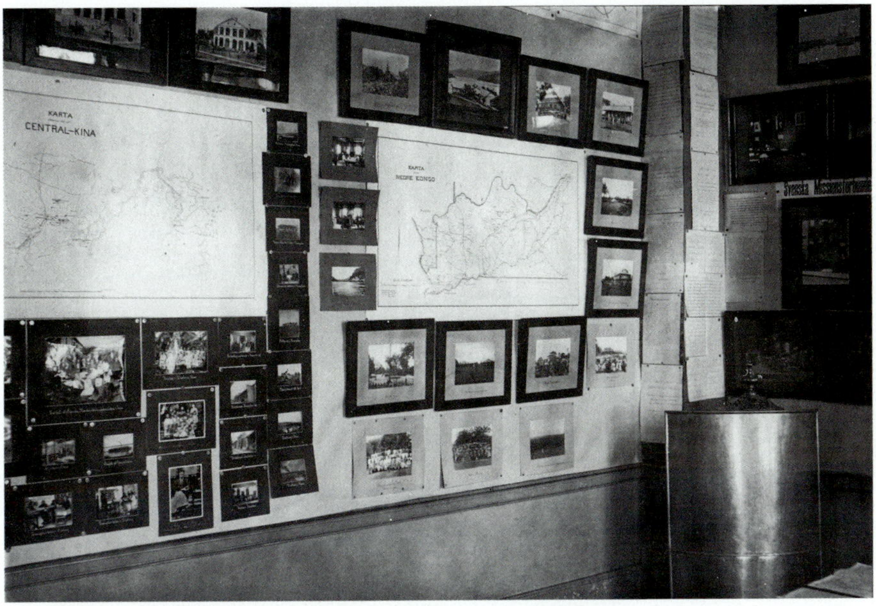

Figure 25.7 The last exhibition hall at the Ethnographic and Missionary exhibition. Source: Archive of the Mission Covenant Church of Sweden

paper filled the corner of the room. Each held the writing of former 'illiterates' converted to Christianity. One of the messages conveyed by this section of the exhibition was that, no matter how different places like China, Persia, Turkistan and South Africa were *before* the arrival of the Christians, they all proved to be equally flexible raw material for the construction of a new and more homogeneous world. In this way, the visitors had been taken on an uplifting journey from the primitive to civilization as a Christian utopia. The mythical temporalities of evolutionism and Protestant awakening fused smoothly in a vision of the final and unavoidable stage of benevolent progress.

Exhibiting the Swedes

The *Ethnographic and Missionary Exhibition*, as its name suggests, merged two fields – one scientific, the other religious – which developed rapidly in the wake of European colonialism. In Sweden, as elsewhere, nationalistic themes were also attached to these fields in order to achieve public recognition and support. At the exhibition, what might seem a paradoxical alliance fused into a unified narrative, conveying temporal sequence, moralities, sentiments and roles. These narrative themes were present in catalogues and contemporary commentaries, and, in the exhibition, through the material rhetoric of objects, bodies and spaces. Indeed, this union of science and religion was significantly aided by the exhibition format.

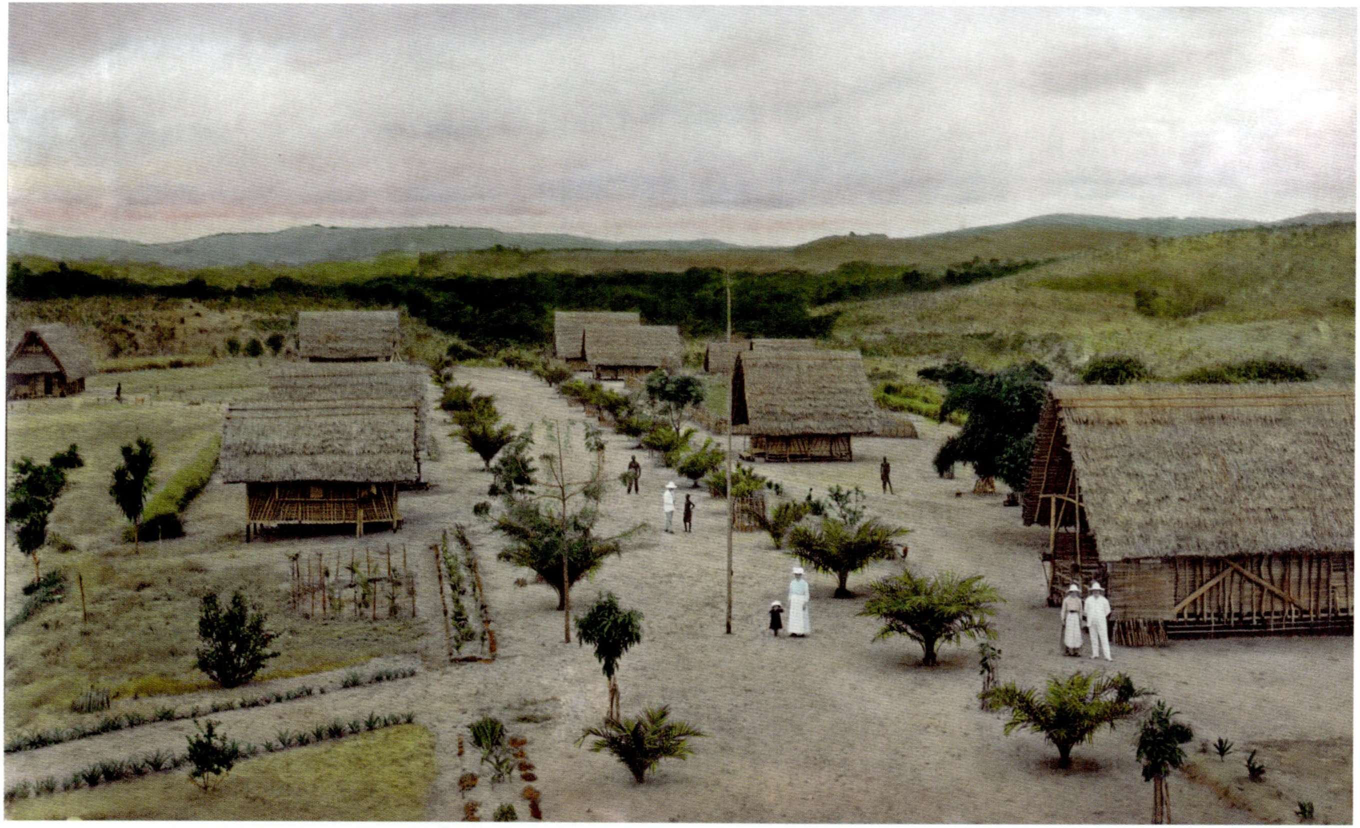

Figure 25.8 Hand-coloured photograph of the Swedish missionary station Madzia in the lower Congo. Source: Museum of Ethnography

Objects are equipped with a generous capacity to harbour ambiguities. On this occasion, ironically, highly expressive Congolese art served as illustrations of the needs of people, perceived as pagan and primitive – deficiencies which both science and religion were compelled to address. There were, then, implicitly written into the narrative, the photographs, maps and possessed objects, portraits of the Swedish nation as voluntaristic, benevolent, sophisticated, charitable, advanced and wise.

Most of the Swedish missionaries had actually come from poor rural backgrounds but as proponents of a new Christian way of life, and equipped with modern medicines, cameras, printing devices, light bulbs and sewing machines, they appeared urbane and modern (see Figure 25.8). Some of these tools must have been as new and fantastic to them as they were to the natives. Indeed, not even the magic and heathen, so evocatively exhibited in Stockholm, were particularly distant in space or time. In Skansen and at the Nordic Museum, curators were busy collecting Sweden's own black magic books and powerful roots from the so-called 'wise' people who had been practitioners of folk magic. Although these traditional ritual experts were declining in number and influence, they nevertheless remained active in the Swedish countryside, performing their healing, divinatory and – at times – intentionally harmful arts (Tillhagen 1958). Their tools, however, were not likely to be seen in any missionary display. Instead, collections representing a declining agrarian Sweden were carefully separated from colonial objects. The visual and functional similarities between, for instance, ritualistic bundles from southern Sweden and the lower Congo were never commented upon.

It could also be argued that the necessity of belonging on the right side of that line between the primitive and civilized – a line marked out in bold at this time – was of particular importance to a people situated on the periphery of Europe. The general interest and pride exhibited in Stockholm at this modest Swedish participation in European colonialism might be understood as symbolically compensating for the loss of power and territory closer to home. This colonial past has rested rather peacefully in collective national memory.[6] The portrait of compassionate Swedes partaking in the deed of making the whole world more homogeneous, moral and modern remains part of this nation's grand narrative.

Notes

1 The staff remains at the National Museum of World Culture in Gothenburg (no. 4072 in the Flodén collection, 1658). The drum (formerly no. 4071/1658), however, was involved in an exchange with an Argentinian museum.

2 Silow, C.-A. (2005) 'Survey of Congo culture in Sweden' (Online), National Museum of World Culture (accessed 30 August 2007).

3 *Aftonbladet, Dagens Nyheter, Svenska Morgonbladet, Stockholmstidningen* and *Stockholms Dagblad,* all 14 May 1907.

4 Archive of the Swedish Mission Covenant Church, Stockholm, Folder: Protocols from the Congo conferences, § 22, protocol from the Congo conference 1907.

5 K. E. Laman to E. Nordenskiöld, 10 September 1906, Archive of the Museum of

Ethnography, Stockholm, Box labelled: Erland Nordenskiölds tjänstearkiv (professional correspondence of Erland Nordenskiöld).

6 Recently, however, the issue has been given more public attention, for example, through the travelling exhibition *Traces of Congo*, on tour in the Nordic countries 2005 to 2007.

Bibliography

Andersson, E. (1947) 'Primitiva religiösa väckelser i Afrika', in *Uppdrag i Afrika: Svenska missionsinsatser i de svartas världsdel*, Stockholm: Svenska missionsförbundet.

Axelsson, S. (1970) *Culture Confrontation in the Lower Congo: From the Old Congo Kingdom to the Congo Independent State with Special Reference to the Swedish Missionaries in the 1880s and 1890s,* Stockholm: Gummessons.

Bennett, T. (1995) *The Birth of the Museum: History, Theory, Politics,* London: Routledge.

Cannizzo, J. (1998) 'Gathering Souls and Objects: Missionary Collections', in T. Barringer and T. Flynn (eds) *Colonialism and the Object: Empire, Material Culture and the Museum*, London and New York: Routledge.

Coombes, A. (1994) *Reinventing Africa: Museums, Material Culture and Popular Imagination in Late Victorian and Edwardian England,* New Haven, CT: Yale University Press.

Ekholm-Friedman, K. (1991) *Catastrophe and Creation: The Transformation of an African Culture*, Chur, Switzerland: Harwood Academic Publishers.

Ekström, A. (1994) *Den utställda världen: Stockholmsutställningen 1897 och 1800-talets världsutställningar*, Stockholm: Nordiska museets förlag.

Fabian, J. (1983) *Time and the Other: How Anthropology Makes its Object,* New York: Columbia University Press.

Fabian, J. (2000) *Out of Our Minds: Reason and Madness in the Exploration of Central Africa*, Berkeley: University of California Press.

Flodén, S. A. (1933) *'Tata' Flodén berättar Kongominnen* I, Stockholm: Svenska Missionsförbundets förlag.

Frödin, O. (1913) *En 'sakkunnig': Prof. C. V. Hartmans sakkunnigeutlåtande vid tillsättandet af intendenturen i etnografi vid Göteborgs museum: En granskning*, Stockholm: Bonnier.

Gell, A. (1998) *Art and Agency: An Anthropological Theory,* Oxford: Clarendon Press.

Grandien, B. (1991) 'Tiden före 1891', in A. Björnstad (ed.) *Skansen under hundra år,* Wiken: Bra böckers förlag.

Gustafsson Reinius, L. (2007) 'Materializing Modernity: Swedish Imaginations and Congolese Objects at the Ethnographic and Missionary Exhibition in Stockholm 1907', in C. Folke Ax *et al.* (eds) *Encountering Foreign Worlds: Experiences at Home and Abroad. Proceedings from the 26th Nordic Congress of Historians, Reykjavik 8–12 August 2007*, Reykjavík: Háskólaútgáfan.

Gustafsson Reinius, L. (2009) 'Sensing through White Gloves: On Congolese Objects in Swedish Collections', *The Senses & Society* 3. 4: 75–97.

Henare, A. Holbraad, M. and Wastell, S. (eds) (2006) *Thinking through Things: Theorising Artefacts Ethnographically*, Milton Park, Abingdon, Oxon: Routledge.

Hillström, M. (2006) *Ansvaret för kulturarvet: Studier i det kulturhistoriska museiväsendets formering med särskild inriktning på Nordiska museets etablering 1872–1919,* Linköping: Linköping University (Diss.).

Hochschild, A. (1998) *King Leopold's Ghost: A Story of Greed, Terror, and Heroism in Colonial Africa*, Boston, MA: Houghton Mifflin.

Janzen, J. M. and MacGaffey, W. (eds) (1974) *An Anthology of Kongo Religion: Primary Texts from Lower Zaïre,* Lawrence, Kansas: University of Kansas Publications in Anthropology.

Jenssen-Tusch, H. (1902–1905) *Skandinaver i Congo: Svenske, norske og danske Mænds og Kvinders Virksomhed i den uafhængige Congostat*, Copenhagen: Gyldendalske Boghandel/Nordisk forlag.

Jönses, L. (2005) 'Den patriotiska promenaden: Skansen och Svenska turistföreningen i konstruktionen av det svenska territoriet', in P. Aronsson and M. Hillström (eds) *Kulturarvens dynamik: det institutionaliserade kulturarvets förändringar,* Linköpings universitet/Tema Kultur och Samhälle.

Kirshenblatt-Gimblett, B. (1998) 'Objects of Ethnography', in *Destination Culture: Tourism, Museums, and Heritage*, Berkeley: University of California Press.

Laman, K. E. (1907) *Några drag ur Kongofolkets lif från Svenska missionsförbundets arbetsfält i Kongo*, Stockholm: Illustrerade Missionsskifter 1.

Lindberg, C. (1995) *Erland Nordenskiöld – en antropologisk biografi*, Lund: Socialantropologiska institutionen.

Lundahl, J. E. (1907) 'Ett besök på Etnografiska Missionsutställningen', *Ansgarius: Svenska missionsförbundets årsbok*, Stockholm: Svenska Missionsförbundets förlag.

MacGaffey, W. (1991) *Art and Healing of the Bakongo*, Stockholm: Etnografiska museet/ Folkens museum.

MacGaffey, W. (1994) 'The Eyes of Understanding Kongo Minkisi', in *Astonishment and Power,* Washington, DC: Smithsonian (exhibition catalogue).

Miller, D. (ed.) (2005) *Materiality*, Durham, NC: Duke University Press.

Mirzoeff, N. (1999), 'Transculture: From Kongo to the Congo', in N. Mirzoeff (ed.) *An Introduction to Visual Culture*, London: Routledge.

Mudimbe, V. (1988) *The Invention of Africa: Gnosis, Philosophy, and the Order of Knowledge*, Bloomington and London: Indiana University Press.

Nordenskiöld, E. (1906a) 'Om insamlandet av etnografiska föremål', appendix to *Ymer* 26, folder no. 4. Stockholm: Sällskapet för antropologi och geografi.

Nordenskiöld, E. (1906b) 'Om Kongofolkens återgifvande af människokroppen i skulptur', *Ymer* 26, Stockholm: Sällskapet för antropologi och geografi.

Nordenskiöld, E. (1907a) 'Den etnografiska missionsutställningen till förmån för Riksmuseum', *Ymer* 27, Stockholm: Sällskapet för antropologi och geografi.

Nordenskiöld, E. (ed.) (1907b) *Etnografiska bidrag af svenska missionärer i Afrika,* Stockholm: Palmquists förlag.

Nordenskiöld, E. (1907c) *Kortfattad handledning för besökande af den etnografiska missionsutställningen till förmån för Riksmuseum i vetenskapsakademiens hörsal Drottninggatan 96 15 maj–15 september 1907,* Stockholm: Naturhistoriska Riksmuseet.

Östberg, W. (2002) *När Afrika kom oss nära: Missionen och den svenska Afrikabilden*, Stockholm: Etnografiska museet.

Stewart, S. (1993) *On Longing: Narratives of the Miniature, the Gigantic, the Souvenir, the Collection,* Durham, NC: Duke University Press.

Tillhagen, C-H. (1958) *Folklig läkekonst*, Stockholm: LTs förlag.

Tygesen, P. (2005) *Kongospår,* Stockholm: Etnografiska museet (exhibition catalogue).

Chapter 26

AFTER THE FALL OF THE BERLIN WALL

Nationalism and multiculturalism at the Bulgarian National Ethnographic Museum

Radostina Sharenkova

FOR FORMER COMMUNIST STATES, the fall of the Berlin Wall marks a divide between ideologically different approaches to political, social and economic change which greatly affected the role and functioning of national museums. This chapter focuses on the Bulgarian National Ethnographic Museum in Sofia after 1989, and pays particular attention to the two conflicting discourses, those of nationalism and multiculturalism, which at the time seemed to offer the museum a new future. Lacking guidance from the state's political institutions, the museum had to locate its own solution. In the new political climate, communism was universally condemned and democracy was celebrated, but neither position pointed to an appropriate discourse for a museum that, as much as the nation, needed to remake itself after 1989. Museum practitioners, themselves, did not see an inherent opposition in the two paths that seemed to be open to them, but behind the scenes the museum had inherited, and indeed had materialised in its collections, a complex relationship between diversity and nation which produced a particular challenge to multiculturalism.

Before the Second World War and the formation of the communist state in 1944, the National Ethnographic Museum was one of three national museums in Bulgaria. The other two focused on the nation's archaeology and natural history. Together, these museums contributed to the historical, cultural and geographical conceptualisation of a nation which according to national ideology comprised the mythical territorial triangle of Moesia, Thrace and Macedonia. From its birth, in 1892, the National Ethnographic Museum had collected objects from beyond the nation's contemporary political borders: from parts of Macedonia, so-called Eastern and Western Thrace (now in Turkey and Greece), the 'Western borderlands' (now in

Serbia) and Northern Dobrudzha (now in Romania). Much of this collecting took advantage of periods of destabilisation and mass population movement during the Balkan wars and after the First World War. The objects it collected provided arguments to support the idea of a historically deep-rooted and geographically large Bulgarian community – the so-called 'Bulgarian ethnic lands'. Consequently, objects gathered from these diverse locations were mixed together and labelled 'Bulgarian'. The purpose was to demonstrate unity, even if represented in a diversity of material styles. The museum was organised into departments (repositories) on the basis of the types of material it possessed and not on the provenance of those objects. In effect, then, the museum's collecting activities involved the appropriation of cultures which were then written into the story of a single people. It is important to note that this imagined nation was not distinguished on the basis of ethnic groups; it was understood that this nation possessed – as a result of history and geography – a diversity of interrelated populations. Having been part of the Ottoman Empire for centuries, Bulgaria and its neighbouring countries (Greece, Serbia and Macedonia) shared (and share) both a common heritage and contested territories. Situated at a crossroads linking the East to the West, the Balkan peoples possessed a feeling of belonging simultaneously to Europe and the Orient. This sensibility has only been challenged with the country's fairly recent admission to the European Union.

The Bulgarian National Ethnographic Museum offers a particularly interesting case study of the changing relationship between the national museum and the nation, and the negotiation of that cultural space that lies between nationalism and multiculturalism. This chapter is based upon unstructured interviews with extant and retired curators, documentation in the museum's administrative archive, and articles, reviews and reports written by the museum's directors and curators.[1] Before considering the post-communist development of the museum, however, it is necessary to briefly outline how the museum was reshaped during the communist era.

Communism, social engineering and the museum

When the communists took control of the country in 1944, they put the three national museums under the control of the institutes of the Bulgarian Academy of Sciences. The National Ethnographic Museum found itself attached to, and subservient to, the Ethnographic Institute. The curators were restricted to practical activities and were not permitted to devote time to academic work.[2] Increasingly, their role was to support the Institute's social and cultural engineering projects. While day-to-day control remained in the hands of the curators and chief curator, curatorial standards fell as new levels of bureaucracy took hold; the preservation and documentation of 'materials', as they were known (rather than 'heritage'), became neglected. At a senior level everything was managed by numbers, and the museum was forced to collect all that was redundant without any thought to its present or future use. At the same time the National Ethnographic Museum became instrumentalised as a propaganda tool of the state. It was perfectly designed for this purpose because its new political masters saw the old social and cultural system as degraded and obstructing progress towards a communist ideal. What better way to demonstrate this than through the collection and exhibition of artefacts of that old order? Such objects served to expose inherited deficiencies and clear the way for social reform.

The curators themselves, however, never saw their mission in these terms, but inevitably some of this thinking could be found in the proposals and reports of the Director of the Ethnographic Institute and Museum.[3] Most of the museum's curators retained an awareness of their role in countering loss, and in this period of mass urbanisation, they began to accumulate traditional cultural materials from villages threatened by imposed social change. They also took to exhibiting objects with the idea of contradicting both cultures – those of the past and present.

During this period, the idea of the nation had a negative connotation and it was no longer visible in the museum's public activities. Instead, the museum followed the policies of state and kept its collecting activities within Bulgarian territory. An alternative version of the community was constructed, examined and put on show: that of the meta-national, communist, 'internationalist', 'Homo Sovieticus'. Guides working in the museum's galleries told visitors of an imagined kinship that united the Slavic peoples.

In the late 1970s, the government began to take an increasing interest in the nation as a concept fundamental to the construction of a socialist country (Kalinova and Baeva 2002). Materials once hidden away in the National Ethnographic Museum were now reassessed and reinterpreted for the public; the 'socialist nation' was reinvented and given integrity. No mention was made of ethnic difference or minority problems, and large parts of the collection, originating from Macedonia, Thrace and elsewhere, which were considered politically inappropriate, remained locked away in the stores. A new and separate department was invented to house them. It was referred to as 'Materials from outside the Bulgarian lands'. Thus, material culture once considered as belonging to the national story was now redefined as foreign, distant and, therefore, unfit for display in Bulgaria's national museums. By these actions, the government intended that the museum present the nation correctly in terms of its contemporary political geography. This 'correctness', however, relied upon a new kind of exclusion that had not been there in earlier imaginings of the nation.

Nationalism or multiculturalism?

With the fall of the communist regime in November 1989, the museum's mission was reformulated in order to escape its discredited communist past. Indeed, everything about the National Ethnographic Museum changed: its role, its collecting and its public activities. Initially, however, the political response was reactionary; against a recent past rather than shaped by a vision of the future. This reaction produced two, partly conflicting, trends. The first was a nationalistic trend designed to rediscover and reinvent the nation. For the museum, as much as for Bulgarians more generally, this called for the rediscovery of 'traditional' objects in an attempt to locate symbols from the past. The second trend to be embraced by the museum's curators was that of multiculturalism (Teneva 2001, 2002; Decheva 2005).[4] In 1995, curators started designing exhibitions focused on the different ethnic communities living in Bulgaria, but in doing so they came across the problem of diversity being concealed by the old policy of labelling all objects as Bulgarian, regardless of their exact provenance. Curators now needed to collect anew and particularly in ethnic and religious communities previously excluded from representation in the museum.

Simply in terms of resources, then, the museum was naturally disposed to follow the nationalist course as it already possessed the collections and expertise to achieve that end (Sharenkova 2007). The museum had, of course, been engaged in nationalistic ethnographic research for more than a century (Kovacheva-Kostadinova 1997) and the development of academic ethnography reinforced the museum's decision to focus on proving the origins of the nation. In the early 1990s, the Institute's staff started to explore different cultures and initiated a research series concerned with ethnic communities. In time this would infect the museum.

The museum remained silent on the subject of multiculturalism until 1995; it may have long been discussed in international ethnology, and cultural and social anthropology, but it was a relatively new topic for Bulgarian researchers emerging from communist isolation. And while it soon gained popularity with Bulgarian academics, it was slow to penetrate the thinking of museum professionals. Exhibitions production tended to focus on collections and archives long before the community was considered. The brief turn to multiculturalism involved rather different forces: the sponsorship of non-governmental organisations, the personal inclinations of the particular museum professionals (rather than institutional mission, debate or resources), and the interests of staff from the Ethnographic Institute.

Nationalism had long shaped the museum's narratives and resulted in an emphasis on national symbols. Exhibitions produced by the National Ethnographic Museum up to the 1990s used to tell one story: that of a single, clearly defined, Bulgarian culture. It ignored other cultures. This nationalist line in museum presentation defined the nation using selected cultural elements and declaring them 'unique'. This manipulation of culture was hardened under communism; the policy with regard to minorities was nothing less than complete exclusion. The museum was then not permitted to make comparisons with neighbouring cultures or locate similarities. Difference and uniqueness were meant to assist the visitor in his or her ethnic self-identification. Whether promoting the expansive nation envisaged prior to the Second World War or the politically defined nation of the communists, the national museum operated close to government policy. There is virtually no documentary evidence, or evidence from the curators themselves, to support the idea that a formal agreement existed between the government and museum.[5] The coincidence could be explained by self-censorship or naturalisation within the contemporary political regime; certainly staff needed to be wary of taking a wrong step and in most nations it would be foolish for a government-funded museum to oppose state policy.

Such nationalistic exhibitions promoted competitiveness with narratives in neighbouring Balkan countries. They were developed in isolation and rarely left the limits of their own secure territory. But such narratives were not simply about asserting territorial claims: the notion of a larger Bulgarian nation, for example, was not meant to result in a contest of land or to provoke neighbours; it was simply something to be believed (Smith 1991; Gelner 1999).

The museum's history reveals an institution which has clearly been attached to an ideology that is oppositional to the basic tenets of multiculturalism: diversity, connections, relationships and exchange as grounds for producing or elaborating cultural differences and/or similarities (Taylor 1999; Ashworth *et al.* 2007). Multiculturalism visibly affects the hierarchy of ethnic communities in a nation-state (Mesa-Bains 2004). It replaces the nationalist's vision of the Balkans as a region of independent nation-states possessing contested border territories and populations with a view that sees the Balkans as a spatial crossroads inhabited by a mix of cultures. The latter permits intercultural communication, and the detection of cultural similarity and interdependence. The problem for the National Ethnographic Museum was that it had long been engaged in confirming established ideas about identity for the Bulgarian majority. The museum maintained a minimum set of national symbols, recognised by the authorities and the educational system, which were transmitted to the public. During public holidays and anniversaries the national museums are the places of national pilgrimage. Under the communist regime there was no concern for neglected communities, but with the fall of the Berlin Wall and consequent freedom of speech and movement, these institutions found themselves burdened with new, morally determined expectations, borne out by the new democratic and tolerant state (Brown 2004).

The admission of multiculturalism into the museum risked the loss of the majority audience. By stating as fact that there were different ethnic groups living in the same country, each of which possessed equal rights, the museum risked contradicting its own strong claims for the uniqueness of Bulgarian culture. It risked undermining public trust in the museum's authority. And lacking the resources to really present the multicultural nation, the museum would need to turn to community involvement and narration – practices far removed from those already established in the museum. Some museum staff felt they had to change and they did so first in the exhibition galleries and then in the collections.

Multiculturalism enters the museum

In 1995, the National Ethnographic Museum staged its first temporary exhibition dedicated to a different culture. *Gypsies in Times Past* explored the life and beliefs of the Gypsies in Bulgaria from the late nineteenth to the early twentieth century.[6] Archival photographs and documents, authentic objects, and the texts of Gypsy songs, tales, legends and myths were used to reveal the way of life in this community. Some years later, the exhibition travelled to a number of Bulgarian cities and to Budapest in Hungary. The exhibit had been initiated by two researchers in the Institute who had written dissertations, books and some articles on Gypsies. Supported by a non-governmental organisation, these researchers produced the exhibition in collaboration with a team from the museum. It was the first exhibition initiated by the Institute and produced cooperatively. Although official public discourse did not tolerate the ethnic name used to describe this community, the museum curator in charge asserted that the designation 'Gypsy' included more subgroups than 'Roma' which is the ethnic name of a small part of the community in Bulgaria.[7] The exhibition gave a holistic overview of community lifestyle but also distinguished the characteristics of different subgroups, using craft objects (Decheva 2005). Aside from these considerations, however, Gypsy culture was displayed as homogeneous. The exhibition dealt only with the past, and ignored modern-day problems faced by the community. The emphasis was on positive images of a romantic past. The museum's decision to follow its natural course of presenting this kind of history deprived visitors of the chance to reconsider the link between past and present or their own present-day encounters with the community.

The similar holistic approach was applied to a range of other temporary displays on different cultures: *The Holy Path: Bulgarian Jews through the Centuries* (1998); *Armenians in Bulgaria* (1999); *The Karakachans in Bulgaria* (2001); and *The Aromanians* (2005). In this way, the museum simply divided multicultural Bulgaria into a number of discrete ethnic communities, focusing on their unique features rather than any relationship to Bulgarians. These exhibitions were based on exhibition formats used in the museum to praise Bulgarian culture; they were not innovative in this sense.

The Aromanians marked the end of this cultural experiment in the museum's galleries. The reasons for cessation were complex but the museum's Director summarised them thus: 'It is time to show more Bulgarian materials, the high potential of the collection, as we have also a Bulgarian public.'[8] She claimed that this was a response to public demand. The decision-making power in the museum became particularly centred on the Director in the early 1990s, especially as the hierarchy above her exerted little control. She was in a position to impose, and certainly took a role in approving, exhibition ideas. Movements in policy – both their initiation and closure – were at the behest of the Director.

Not every ethnic or religious community had been given a place on this public stage, and two of the largest – the Turkish and Bulgarian-speaking Muslim communities – were noticeably absent.[9] However, on the grounds of Bulgarian nationalism, both groups posed problematic subjects; curators at the museum appear to have been influenced by this nationalist discourse and some were certainly attached to right-wing political parties opposed to these groups.

The narratives of so-called Bulgarian exhibits incorporated 'the diversity inside the Bulgarian nation, even considered a variation of the multicultural trend in museum policy'.[10] Most of the exhibitions showed masterpieces from the collections, the principal exhibitions of this type being: *From the Treasures of the National Ethnographic Museum* (1994); *Chiprovtsi Carpets* (1997); *Eugenia Lepavtsova* (1995); *Folk Culture and Christianity* (2000–01); and *Magic of Colours* (1999 and 2002). The last two also mark two sub-narratives within these 'Bulgarian' presentations. The first focused upon Christianity as a subject of discrimination in the communist era, and the second juxtaposed the colours of traditional culture with the greyness of the uniforms and popular culture associated with the period between 1944 and 1989.

Nationalist discourse in the 1990s arose from a desire to contradict museum policy initiatives of the communist era. Historians of twentieth-century Bulgarian politics argue that, in times of crisis, Bulgarian policymakers deny earlier decisions as unsustainable (Kalinova and Baeva 2002; Elenkov 2008; Znepolski 2008). Seeking 'correct' narratives, the National Ethnographic Museum's curators turned to religion, clothing and, lastly, to intangible heritage. Using these topics they tried to build a bridge between the 1990s and the museum's pre-communist past. The nation was presented as a horizontal notion, described in geographical terms, no longer possessing hierarchical or chronological dimensions. The museum's curators were once again inspired by the ideas of a Greater Bulgaria that had been forgotten during communist rule.

Ideologists of this Greater Bulgaria supported their assumptions with evidence derived from the largest of the museum's collections, that focused on traditional clothing from inside and outside the political boundaries of the state (Kovacheva-Kostadinova 1996). They reasoned that since similar objects had been collected from Macedonia, Thrace and other contested lands, these territories should be considered Bulgarian. Following this rationale, some of the exhibits engaged in storytelling while the others placed masterpieces on show. One of the narratives presented the Bulgarian diaspora of people who had emigrated from the Ottoman Empire in the eighteenth and nineteenth centuries.[11] These were reaffirmed as Bulgarians. Another narrative sought to demonstrate the Bulgarian identity of a group of people spanning the Serbo-Bulgarian border.[12] Discussion of the Orthodox Church and its rituals supported similar claims and took them far back in time.

Official state policy on the subject of the nation during the 1990s and 2000s adopted multicultural language and demonstrated a will to fight discrimination. The same is true of the exhibits at the National Ethnographic Museum. Although the 'ethnic' displays were only a short-lived trend, they marked a period of transition and had a clear impact upon other ethnographic museums in the country. The museum continued to follow governmental inclinations regarding cultural policy but also continued to lag behind the multicultural research being conducted by Bulgarian academic ethnologists.

Museum and community

During the 1990s, no visitor studies were carried out in the National Ethnographic Museum and the public was not consulted on the exhibitions that appeared in the museum's galleries. Those staff who appeared in the museum's galleries – security staff and guides – were rarely asked to participate in monthly museum meetings. There was a disconnection between public consumption and the act of authoring the exhibitions. The museum's staff – in common with museum staff around the world – maintained their authority to decide matters of public display. The Director's desire to return to the Bulgarian narrative was possibly prompted by informal discussions with security staff, as interviews have revealed that this group felt the public wanted more 'Bulgarian stuff'. If true, this reaction could be interpreted as a vote against the museum's multicultural experiment. Certainly, the national museum's attempts to represent ethnic and religious diversity provoked a nationalistic response. But on the other hand, these exhibitions attracted segments of Bulgarian society that had never regarded the museum as their own space. The multicultural exhibits demonstrated the will of members of these communities to participate actively in the museum's initiatives, even to the extent of voluntarily donating their objects and stories. Many curators felt great personal satisfaction having prepared the exhibits about Jewish and Armenian culture in association with these groups. These exhibits demonstrated that the museum could actively work with its audiences and become a lively meeting space through the events these exhibitions supported. Although these activities were only temporary, they diversified the museum's audience and extended its outlook.

However, our reading of the museum's actions in the post-communist era needs to take into account the legacy it inherited – as a public institution – from the communist era that could not be so easily shaken off. In the four decades before 1989, the public had acquired a changed attitude towards museum visits because members of the public used to be brought to the museum collectively in organised visits. The initiators were local party organisations or schools. Groups were not allowed unguided tours; the visit was simply a propaganda event of the Communist Party. Without adequate funding, the post-1989 museum was poorly advertised and lacked a PR specialist, and with a residue of bad feeling associated with the institution, visitor numbers were lower than in the communist era. This, too, explains the museum's desire to keep and grow its core audiences; small, ethnically defined audiences offered no such potential.

One further explanation for this change of policy is found in the ease with which the museum could construct these different kinds of exhibit. The 'ethnic' exhibitions were the result of long-term projects to collect narratives and objects; they required resources for planning and arranging the exhibits and for supporting activities. In contrast, the 'Bulgarian' exhibits were taken from the shelf as ready-made ideas for which the collections were perfectly suited. They were, thus, simply less costly and time-consuming exhibitions to produce. With the Ministry of Culture neither forbidding nor offering any support, financial or otherwise for the ethnic exhibitions, the museum was left to make its own decisions about their viability and worth. Financial support for these exhibits came both from the communities themselves and by the museum's budget.

In the 1990s and early 2000s, the idea of the 'national museum' in Bulgaria, as a representative institution serving its audiences, was slightly different from other parts of Europe. Lacking visitor data, besides the positive comments left in visitor books, the National Ethnographic Museum was deprived of the information it needed to modernise and realise more fully those democratic ideals celebrated with the fall of the Wall. Comments received in the museum's visitor books reflected satisfaction with the status quo, with a museum focused on the Bulgarian core, which was conservative in its outlook and its displays. Such data could not drive change and the government itself did not provide policy leadership of a kind which forced change upon the institution – despite its desires for democracy and tolerance. Declarations to fight discrimination remained on the paper they were written on; they were not fully instrumentalised in the museum. This case study shows how a museum attempts to change, but also shows the forces that operate to return the museum to a more risk-averse course. It shows that multiculturalism – recognised by Bulgarian academics and politicians as an important national discourse – simply could enter public debate in the museum without external influence, and particularly the direct support of government.

Notes

1 Interviews were conducted with A. Komitska, NEM, curator of Bulgarian Folk Clothing Department (Macedonia and Thrace Section), 18 October 2005; E. Ivanova, NEM, security, 10 November 2005; I. Philipova, NEM, ex-guide (1995–2003), 1 November 2005; M. Decheva, NEM, curator of Fabrics and Embroidery Department, 1 November 2005; N. Dikanska, NEM, security, 10 November 2005; N. Teneva, NEM, Deputy Director, curator of Woodcarving and Home Crafts and Furniture Departments, 12 October 2005; P. Puntev, NEM, director (1967–87), 16 October 2005; V. Nestorova, NEM, curator of Jewels Department, 12 October 2005. Archival sources used: the National Ethnographic Museum Archive; Scholarly Archive of the Bulgarian Academy of Sciences; National Statistical Institute, http://www.nsi.bg/Census/Census.htm (accessed 3 April 2009); Ethnographic Institute and Museum, http://hs41.iccs.bas.bg/etb/et1.htm (accessed 3 April 2009).

2 Interview with P. Puntev, Director (1967–87), 16 October 2005.

3 Report written by H. Gandev for the Second National Meeting of Historians in Bulgaria, 1964. Scholarly Archive of the Bulgarian Academy of Sciences, fund 88, list 2, file 74, page 118.

4 The NEM's ethnic exhibits influenced the public activities and rearrangement of permanent exhibitions in many regional museums throughout the country.

5 Interview with A. Komitska, NEM, curator of Bulgarian Folk Clothing Department (Macedonia and Thrace Section), 18 October 2005.

6 Ethnographic Institute and Museum , http://hs41.iccs.bas.bg (accessed 3 April 2009).

7 Interview with V. Nestorova, NEM, curator of Jewels Department, 12 October 2005.

8 Interview with N. Teneva, NEM, director (1994–2009), curator of Woodcarving and Home Crafts and Furniture.

9 According to the 2001 official census, the Turks number 746,644 and the Bulgarian Muslims 131,531 (National Statistical Institute).

10 Interview with M. Decheva, NEM, curator of Fabrics and Embroidery Department, 1 November 2005.

11 'From the life of the Banat Bulgarians' (1998).

12 'The cut up yard: Weekdays and holidays of the Bulgarians in the Western border regions' (1996).

Bibliography

Ames, M. (1992) *Cannibal Tours and Glass Boxes: The Anthropology of Museums*, Vancouver: University of British Columbia Press.

Anderson, B. (1983) *Imagined Communities: Reflections on the Origins and Spread of Nationalism*, London: Verso.

Ashworth, G., Graham, B. and Tunbridge, J. (2007) *Pluralising Pasts: Heritage, Identity and Place in Multicultural Societies*, London: Pluto Press.

Bennett, T. (1998) 'Speaking to the eyes: Museums, legibility and the social order', in Macdonald, S. (ed.) *The Politics of Display: Museums, Science, Culture*, London and New York: Routledge, pp. 25–36.

Bicknell, S. (1995) 'Here to help: Evaluation and effectiveness', in Hooper-Greenhill, E. (ed.) *Museum, Media, Message*, London and New York: Routledge, pp. 281–93.

Bicknell, S. and Farmello, G. (eds) (1993) *Museum Visitor Studies in the 90s*, London: Science Museum.

Bouquet, M. (2001) 'The art of exhibition-making as a problem of translation', in Bouquet, M. (ed.) *Academic Anthropology and the Museum: Back to the Future*, New York and Oxford: Berghahn Books.

Brown, C. (2004) 'The museum's role in a multicultural society', in Anderson, G. (ed.) *Reinventing the Museum: Historical and Contemporary Perspectives on the Paradigm Shift*, Walnut Creek, CA: Altamira Press, pp. 143–49.

Cameron, D. F. (1971) 'The museum, a temple or the forum?', *Curator*, 1: 11–24.

Dean, D. and Rider, P. E. (2005) 'Museums, nation and political history in the Australian National Museum and the Canadian Museum of Civilisation', *Museum and Society*, 3: 35–50.

Decheva, M. (2001) 'Ethnographic exhibition as a way of intercultural communication: The experience of the Bulgarian National Ethnographic Museum', in *Néprajzi Értesítő. Annales musei ethnographiae*, 83, Néprajzi museum, Budapest, pp. 117–22.

Decheva, M. (2005) 'The experience of the Bulgarian Ethnography Museum in the process of Roma cultural changes', in *Zborník Slovenského Národného Múzea v Martine*, XCIX, Slovenské Národné Múzeum, Martine, pp. 58–66.

Elenkov, I. (2008) *Cultural Front: Bulgarian Culture during the Communist Era – Political Management, Ideological Basis, Institutional Regimes*, Sofia: Ciela (in Bulgarian).

Gelner, E. (1999) *Nations and Nationalism*, Sofia: St Kliment Ohridski Publishing House (in Bulgarian).

Ginsburgh, V. and Mairesse, F. (1997), 'Defining a museum: Suggestions for an alternative approach', *Museum, Management and Curatorship*, 16: 15–33.

Hooper-Greenhill, E. (1999) *Museums and the Shaping of Knowledge*, London and New York: Routledge.

Hudson, K. (1991) 'How misleading does an Ethnographical Museum have to be?', in Karp, I. and Lavine, S. D. (eds) *Exhibiting Cultures: The Poetics and Politics of Museum Display*, Washington, DC: Smithsonian Institution Press, pp. 457–64.

Kalinova, E. and Baeva, I. (2002) *Bulgarian Transitions (1939–2002)*, Sofia: Paradigma Publishing House (in Bulgarian).

Kovacheva-Kostadinova, V. (1996) Hristo Vakarelski and the National Ethnographic Museum', *Bulgarian Ethnology*, 4: 18–29 (in Bulgarian).

Kovacheva-Kostadinova, V. (1997) 'A century of the National Ethnographic Museum (past and present)', *Rodina*, I–II: 267–91 (in Bulgarian).

Kreamer, C. M. (1992) 'Defining communities through exhibiting and collecting', in Karp, I., Mullen Kreamer, C. and Lavine, S. (eds) *Museums and Communities: The Politics of Public Culture*, Washington, DC: Smithsonian Institution Press, pp. 367–81.

Macdonald, S. J. (2003) 'Museums, national, postnational and transcultural identities', *Museum and Society*, 1(1): 1–16.

Marushiakova, E. and Popov, V. (1995) *The Gypsies of Bulgaria: Problems of the Multicultural Museum Exhibition*, Sofia: Club '90 (in Bulgarian).

Mesa-Bains, A. (2004) 'The real multiculturalism: A struggle for authority and power', in Anderson, G. (ed.) *Reinventing the Museum: Historical and Contemporary Perspectives on the Paradigm Shift*, Walnut Creek, CA: Altamira Press, pp. 99–109.

Prösler, M. (1996) 'Museums and globalization', in Macdonald, S. and Fyfe, G. (eds) *Theorizing Museums*, Oxford: Basil Blackwell, pp. 21–44.

Radonov, Z. (1971) 'Establishment, initial activities and international cultural relations of the National Ethnographic Museum in Sofia (1906–1944)', *Proceedings of the Ethnographic Institute and Museum*, XIII: 245–68 (in Bulgarian).

Sharenkova, R. (2007) 'Cultural policies of the ethnographic museums in Bulgaria in the second half of the 20th century', Sofia (unpublished PhD thesis).

Smith, A. (1991) *National Identity*, London: Penguin.

Taylor, C. (1999) *Multiculturalism*, Sofia: Publishing House Kritika y Humanizam (in Bulgarian).

Teneva, N. (2001) 'National Ethnographic Museum, Sofia: Exhibitions, audience and problems', in *Néprajzi Értesítő. Annales musei ethnographiae*, 83, Néprajzi museum, Budapest, pp. 53–62.

Teneva, N. (2002) 'The National Ethnographic Museum at the border of the millennium', *Bulgarian Ethnology*, 4: 79–90 (in Bulgarian).

Znepolski, I. (2008) *Bulgarian Communism: Socio-cultural Aspects and Power Trajectory*, Sofia: Ciela (in Bulgarian).

Chapter 27

THE IJZERTOREN MEMORIAL MUSEUM

A Flemish national museum?

Karen D. Shelby

IN RECENT YEARS, scholars of the Western Front in the First World War have addressed the implications of the war on modern conditions of nationalism (see, for example, Whalen 1984; Winter 1995; Sherman 1999; Gillis 2000). However, the majority of these studies frame the war within a British, French or German context. Little attention has been given to Belgium, the country that, as a direct result of the war, experienced the most profound ramifications in a nationalist rhetoric that has, in contemporary Belgium, threatened to split the country into two separate states: Flanders and Wallonia. This chapter addresses the IJzertoren (IJzer Tower) Memorial Museum in Diksmuide as an expression of a specifically Flemish, rather than Belgian, nationalism. The IJzertoren first became an important memorial site in 1919 in order to honour the Flemish men who had, in response to King Albert I's appeal to Flemish patriotism, volunteered and died in defence of Flanders. The tower was erected in 1930 to commemorate the sacrifices of these men, and in the course of the ensuing sixty years became the controversial symbol of the separatist Flemish Movement. In the 1990s, in an attempt to defuse what had become a potentially explosive political tension, the tower was transformed into a museum narrating the history of the Flemish nation. Its message became one of peace. Nevertheless, the physical structure of the memorial, within which the museum is housed, remains a powerful symbol of Flemish nationalism (Figure 27.1). National days of commemoration, such as the annual pilgrimage to the IJzer, create common spaces around which national identity is formed and performed (Young 2003). The entirety of the museum – the symbolism of the architectural structure for Flemish partisan politics, the content of the museum's exhibitions and the annual pilgrimage – all serve to construct what is effectively a national museum, though it is not designated as such. Indeed, the IJzertoren Memorial Museum is a private,

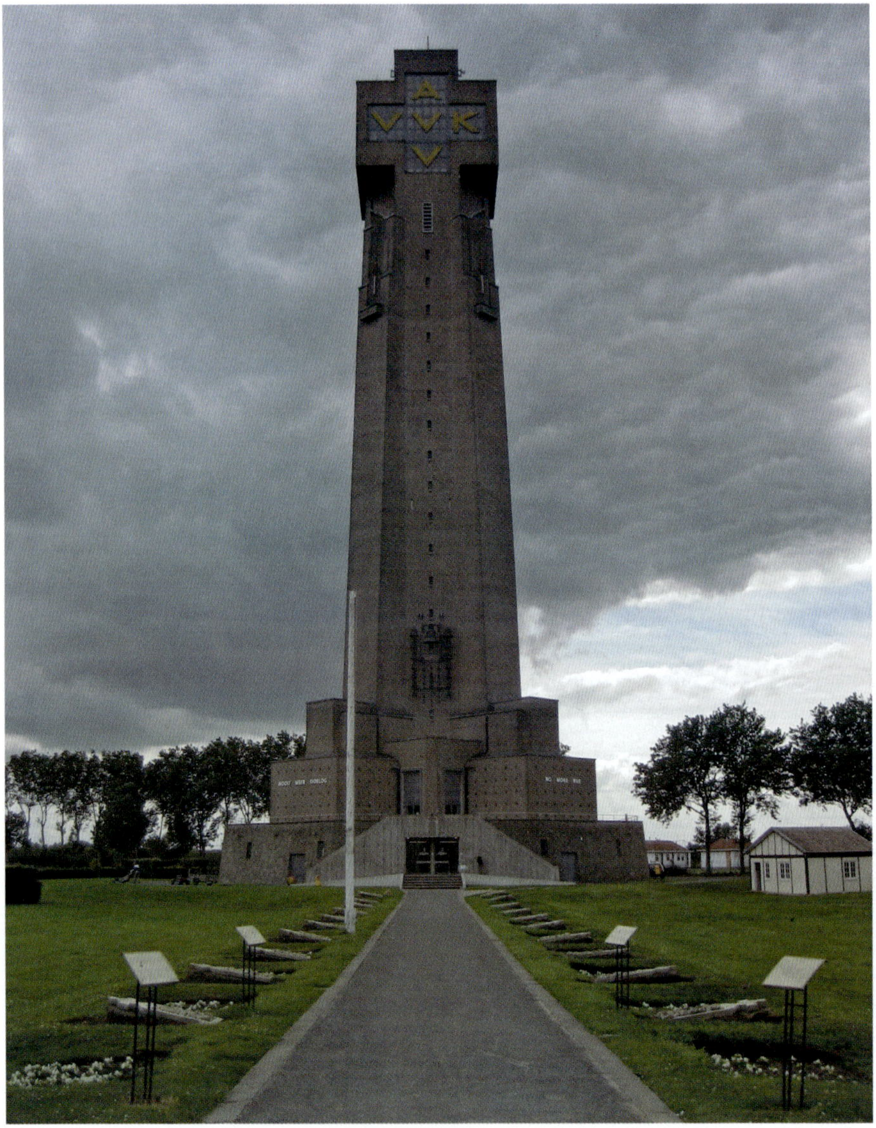

Figure 27.1 The IJzertoren, 1965. Photo: author

not-for-profit institution run largely on visitor entrance fees. Since 1998, the Flemish government has contributed some funding on the basis of annual requests made by the IJzer Pilgrimage Committee but a shortage of resources has generally resulted in a low national profile and constrained activity.

The political significance of this memorial and museum can be understood only in the context of Belgium's difficult internal political history. The Belgian general election on 10 June 2007, for example, resulted in a hung parliament as a direct result of the schism between the two language groups. A crisis developed which ran for 196 days, during which time Belgium possessed a government with no popular mandate. Flemish Interest,[1] a right-wing, anti-immigration political party, claimed as recently as 2008 that Belgium was an artificial state created in 1830 merely as a buffer between France and the other European powers (Frazier 2007). This political and linguistic division in Belgium is thus deep-rooted. Indeed, the country of Flanders originated in the ninth century – almost a millennium before the union of the French-speaking Walloons in the South and the Dutch-speaking Flemish in the North West. Over the intervening centuries Flanders had been ruled by the French, Spanish and Dutch. The Belgian Revolution in 1830, which culminated in revolt against the Dutch government, resulted in the formation of Belgium as an independent, Catholic and neutral state. These arrangements, however, served to favour the smaller region of French speakers and the minority language of French became the official language. Dutch was soon banned in schools, and Flemish, a regional Dutch dialect, was reduced to the status of local patois. A cultural and political movement in Flanders began to address these inequities, which led first to the option, and then the requirement, in 1930, to use Dutch in public life. The language conflict played a considerable role in the rise of the Flemish Movement and the founding of the Front Movement during the First World War (Beheydt 1995; Barbour and Carmichael 2002).

The memories and symbolism ascribed to the IJzertoren are expressions of this long and complex history of the oppression and liberation of Flemish identity and nationalism. Flemish culture is unified in language, Flemish intellectualism and pious Catholicism, and in these separated from French-speaking society quite fundamentally. The building of the IJzertoren Memorial Museum offers insights into this complex cultural and political context where a people feel their identities and autonomy are denied by other national narratives. In a political sense of possession, ideology and voice, this museum has the inescapable qualities of what might be considered a national museum.

The Flemish 'nation' in the First World War

In the summer of 1914, King Albert I appealed to the 'imagined nation' (Anderson 1983) of Flanders and asked its people to fight for its protection, promising Flemish autonomy after the war. The Belgian Army was deployed for a distance of some fifteen miles along the IJzer River from the port city of Nieuwpoort to just south of Diksmuide, the main site of the battles on the Belgian Front in Western Flanders. In October and November of that year, the Battle of the IJzer raged. However, strategic flooding of the IJzer Plain effectively blocked the Germans from crossing the river into French territory in Diksmuide. The Belgian Army became entrenched to the west of Diksmuide for the remainder of the war in what is known as 'The Guard on the IJzer'.[2] This prevented it from engaging in the near constant warfare that characterized the Ieper Salient to the south. In consequence, an ennui gripped the Belgian Front which led to the formation of the Front Movement[3] in 1915, a group initially organized to address the ill-treatment of the Flemish soldiers at the hands of French-speaking officers. Led by a group of university intellectuals and Catholic priests, it quickly became an overt political movement dedicated to the recognition of Flemish autonomy and the primacy of the Dutch language in Flanders.

During the war, the Belgian army had acquired an increasingly disproportionate number of Flemish speakers (72 per cent in 1918), since it could only recruit from refugee populations in France and the UK, and from the unoccupied zone. At the outset of hostilities, more French speakers (making up 44 per cent of the army) had been involved (and lost) in the summer sieges of Liège and Namur in 1914.[4] The conscripted, and partly uneducated, Flemish men were instructed in French during their initial training. Their deaths then played into the hands of Flemish nationalists whose principal concern was language. This linguistic battle now had its Flemish martyrs who would live on in Belgian politics and calls for Flemish independence (Becuwe and De Lentdecker 1993).[5] The Front Movement instituted the unique Flemish heroes tombstone, the heldenhuldezerkjes, so that Flemish-speaking soldiers could be buried beneath an inscription in their native tongue (Figure 27.2)

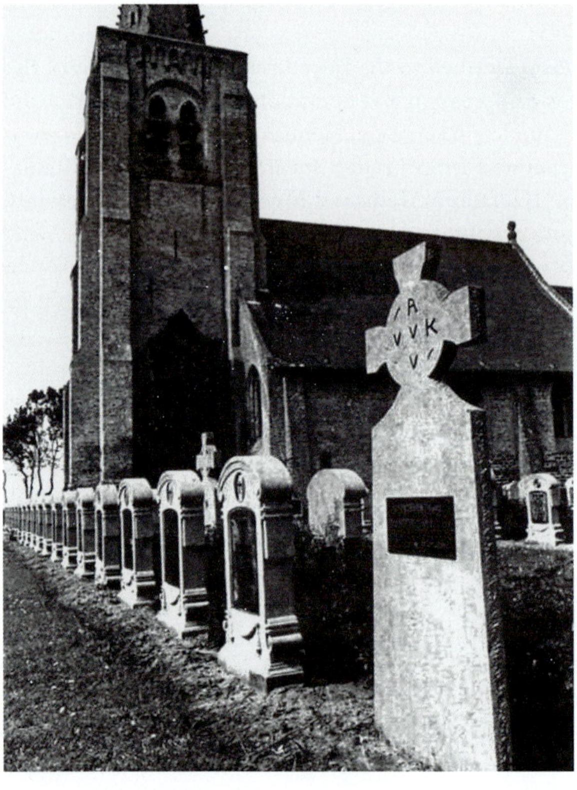

Figure 27.2 Flemish heroes' gravestones. Photo: J. Pingree

(Vanacker 2000; Seberechts 2003). The cross reflected Flanders' Catholic foundation and was carried over into the design of the IJzertoren which has the form of a huge heldenhulde or hero memorial. On it is inscribed 'AVV – VVK', short for 'Alles Voor Vlaanderen – Vlaanderen voor Kristus' (All for Flanders – Flanders for Christ). This overt manifestation of Flemish nationalism would challenge the museum's future hope of being a neutral symbol of peace.[6]

Building the tower

In 1919, a site on the IJzer River in Diksmuide was established as the destination of a pilgrimage, which was to take place annually, in August each year, in honour of the Flemish men who had perished in the war. In 1925 it was suggested that the site should acquire a permanent marker, and by the following year a clear vision for the tower had been produced:

> Planted on the shores of the IJzer, this monument will rise high into a commemoration of the fallen: rejoicing for the triumphant soul of Flanders; confirming their ideal and the witness of their sacrifice: loudly accusing its detractors, . . . [an] unbreakable chain between the dead and the living in the battle of Flanders' liberation![7]

By the end of the year, the tower had risen to 20 metres, and fragments of 120 heldenhulde had been incorporated into the walls of the inner rooms. The Minister of Defence, Albert Hellebaut, had ordered the systematic destruction of these tombstones that year in order to use them in the construction of a road in the town of Adinkerke (Seberechts 2003: 128). Building the heldenhuldes into the fabric of the memorial served both as a symbolic and preservational act. Symbolically, these were signifiers of heroically martyred Flemish men; now they were to support the IJzer monument and represent the strength of Flanders.

On 21 August 1930, on the occasion of the eleventh pilgrimage, the memorial was inaugurated (Figure 27.3). A palpable symbol of Flemish nationalism representing Catholic martyrdom, patriotism and pride, its builders had chosen the inauguration to coincide with the Belgian state's centennial celebrations.

The tower consisted of seven empty floors above the crypt in which several of the 'martyred' men of Flanders were buried. The top floor was an observation platform with panoramic views of the reconstructed Flemish countryside. A map indicated the exact locations of the major battles in the area. On the exterior walls the slogan 'no more war' was written in the four dominant languages of the Western Front: Dutch, French, German and English; four Christian countries now at peace joined beneath the cross (Daels 1933). A brochure, published in 1925, however, leaves the reader in no doubt that the IJzer cross was raised as an enduring memorial to Flemish soldiers who fought in the war and for an autonomous Flanders within Belgium.[8]

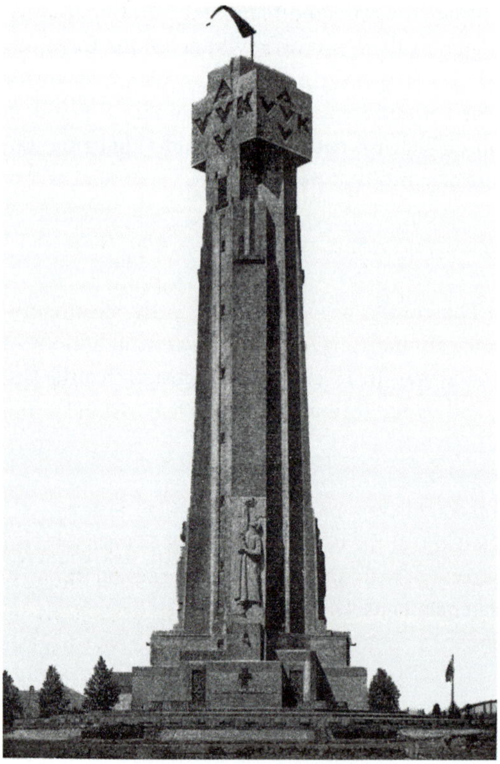

Figure 27.3 Richard and Frans Averbeke's *The IJzertoren*, 1930. Photo: Uitgave van de Vereniging, Bedevaart naar de graven van den IJzer, Diksmuide, 1952

Towering over the IJzer Plain, its verticality seemed to speak of defiance. But it would not remain standing for long. In the late 1930s, the IJzertoren became associated with fascism when a number of Flemish nationalists collaborated with the Nazis in an attempt to further their separatist agenda. In 1946, the tower was destroyed in order to disrupt these fascist connotations. The destruction affected both moderate and extreme nationalists and in time a new tower was built. This one was dedicated on 22 August 1965, during the 38th pilgrimage.

As that century came to a close, the annual pilgrimages sought to locate more positive representations of Flemish nationalism. In 1990, the tower was re-inaugurated as a museum which then set about emphasizing the historic battles fought in the region and stressing the original tenets of the Front Movement (freedom, peace and tolerance). This embodied a vision of a federalist autonomy rather than the separatist agenda espoused by the right-wing faction. It was an attempt to locate the memorial in moderate politics.

The physical form of the museum, its complex military and political history, and its sacred aspect, raise questions about what this museum is or could be. In the context of this book, it also begs the question: In what sense can this be considered a *national* museum? The museum does not and never has dealt with Belgian history in its entirety but this fact seems unimportant. In 1917, Lodewijk Dosfel, a prominent member of the Flemish Movement, argued in his *Katholieke Activistendefensie* (Catholic Activist Defence):

> Belgium is a *State,* not a *nationality.* In the ethnic meaning of the word there are no Belgians. Belgium is a juridical and historic State. . . . There is a Flemish *nationality,* whose most characteristic expression is the *Dutch* language, the language of the overwhelming majority of the inhabitants of the Flemish land, and which in Flanders is truly the national language.
>
> (Dosfel 1922)

It is a rhetoric still embraced by those who espouse an independent Flanders (Kimmelman 2008): Flanders is a nation – Belgium is not.

The narrative of a nation

In the desire to divest the tower of its negative histories, the IJzer Pilgrimage Committee created exhibitions which served to disrupt the old reactive Flemish nationalist rhetoric. It deliberately rewrote the master narrative of the First World War while unfolding a secondary narrative concerned with the history of Flanders in which the Flemish Movement appears out of a rational and morally based humanism. This narrative begins on the 22nd floor, at the very top of the tower. Here, in a repeat of the first tower, visitors are treated to panoramic views of the Flemish countryside still scarred by the First World War. From here, the visitor descends through an exhaustive chronological narrative, and inevitable museum fatigue, to the ground floor. The floors themselves give a structure to this narrative: floors 22 to 14, for example, cover the First World War, which is then followed by the interwar years and the Second World War. Floors 8 and 7 are dedicated to a more general history of Flanders, while floors 6 to 1 return again to the First World War. Within this grand narrative is the sub-narrative of Flanders and its nationalist aspirations. It is this sub-narrative that will be explored in the remainder of this chapter in order to address the contention that this museum is indeed a national museum of a nation.

The narrative really begins on the 20th floor, with images of prominent Flemish figures including, amongst others, the writer and 'Father of the Flemish Movement', Jan Frans Willems (1793–1846); his pupil Jean-Baptist David (1801–66); Hendrik Conscience (1812–83), writer of the influential nationalist text *De Leeuw van Vlaanderen* (*The Lion of Flanders*) (1838); and the poet-priest, Guido Gezelle (1830–99), who was instrumental in the Front Movement and the creation of the first IJzertoren. The presence of these men in the exhibition establishes a long and vital history for the nation of Flanders, but it also places them in a historicized academic, rather than political, context. These activists were instrumental in achieving equal recognition for the Dutch and French languages, at least theoretically (in practice,

French remained dominant in legal and intellectual spheres across the whole of Belgium). The centrality of language to the Flemish national discourse is here apparent but addressed in humanitarian terms, such as in the following wall text:

> By speaking French, by displaying erudition and displaying a bourgeoisie mentality, we distance ourselves from the common people, our inquisitive fellow citizens, workers, farmers and women. Dutch, the language of our ancestors, the language of Freedom. Can you despise this language?

The conditions of service and experience of Flemish soldiers in the First World War are addressed on the 18th floor. Here there is discussion of Joe English's designs for the heldenhuldezerkjes together with his propaganda posters designed to support the war effort: 'houdt u fier, houdt u rein' ('Be proud, be pure'). These are rather muted images, when compared with the nationalist rhetoric of English's heldenhulde, but they encapsulated the moralistic Catholic teachings of the Front Movement that were so integral to the group's strategies for Flemish autonomy. Near a red wall proclaiming: 'Sire, We hebben Geen vertrouwen meer in inze oversten' ('Sire, we have no more faith in our superiors . . . no confidence'), a case holds an enlarged copy of the public letter, known as the *Open Brief*,[9] sent to King Albert by Adiel Debeuckelaere on behalf of the Front Movement. Written only in Dutch, it was sent on the anniversary of the Battle of the Golden Spurs, 1302, a famous victory of the Flemish over the French, commemorated each year by the Flemish in a national holiday and celebrated in the novel, *The Lion of Flanders*. It was by reference to this battle that Albert had called for Flemish recruits to the First World War and promised Flemish autonomy. Now members of the Front Movement were in internment camps.

On this floor, visitors are, however, reminded that the Front Movement sought moral ideals of no more war, peace and federalist autonomy. The exhibition addresses the failure of Albert to keep his promise of 'equality in right and fact' after the war which led to the establishment of the Flemish National Party but then shows how this revolutionary trend was slowed by the introduction of the single vote system in 1918, which opened the way for the larger, more moderate and more democratic Flemish Movement. But the museum cannot always steer clear of controversy, particularly during the Second World War when the Flemish cause was contaminated with right-wing separatists and racists, like Staf de Clercq (1884–1942), leader of the Flemish National League (Vanacker 2000: 167),[10] and other Flemish collaborators. Steven Maes, the Educational Director, however, remarked ironically that there would be no tower without those right-wing nationalists who funded its rebuilding, though most modern-day Flemish people would wish to forget this association. The effect of 300 named collaborators adds to this sense of the narrative becoming mired in ambiguous controversy. The list contains little other information and one wonders if these individuals are being historicized as martyrs to the cause (just as one records the names of the fallen on a war memorial) or listed in shame or to demonstrate distanced academic objectivity. Members of the 55-person committee which oversees the museum appear to possess no consensus view on the matter.

If the museum struggles with the representation of difficult history on the tenth floor, it is on the first floor that the museum's attempt to negotiate a dispassionate history culminates in its most controversial object. Here, one might believe that the Flemish cause is indeed composed of martyred collaborators and de Clercq's extreme nationalism, for on display is one of the greatest and most notorious of Flemish icons. And to critics it was not simply the fact that this object had returned to public display but that it had done so in a building which had Flemish nationalism written into its design and in giant lettering, AVV – VVK.[11] The object in question is Hendrik Luyten's (1859–1945) *The Golden Painting of Flanders* (*Gulden Doek van Vlaanderen*) depicting 115 figures from Flemish history, including de Clercq and his co-collaborators (Figure 27.4). An impressive 3.3m × 4.7m in size, the painting dominates the space, and in doing so seems to sabotage the carefully constructed, politically neutralized and positive narrative of much of the remainder of the museum. Taken directly from Ruben's triptych commissioned for the Antwerp Cathedral, *The Descent from the Cross* (1611–12), it also imitates Raphael's *School of Athens* (1509–11) and Ingres's *Apotheosis of Homer* (1827). It conflates time in order to illustrate a dialogue between generations of Flemish nationals, including some who were subsequently executed for treason. With these stand Rubens, Rembrandt, van Dyck and Desiderius Erasmus van Rotterdam, signifying the linguistic and cultural connection to the Dutch. The artist himself is positioned next to de Clercq; he, too, is remembered as a controversial political figure. On the central table below Christ the documents of Flemish nationalism are being drafted. A garland of red poppies, the dominant symbol of the First World War, follows the bottom edge of the painting. The martyred soldiers of Flanders and their righteous victimization are thus contextualized within the sacrificial tenets of their faith.

When asked if he considered the inclusion of *The Golden Painting* too controversial for the museum's present goals, Maes preferred to see it as an important historical artefact, but added that it would better serve the visitor if accompanying text provided it with an intellectual and political context. By placing *The Golden Painting* in a location of prominence, the museum appears to speak with Luyten's highly charged political symbols and ideals. The painting seems to reinterpret aspects of the displays on the tenth floor. Any visitor thinking this might find their view confirmed in the chapel on the ground floor. Here, above the altar, hangs a crucifix designed by Flemish Nazi collaborator Cyriel Verschaeve; it was he who also laid the first stone of the original IJzertoren in the 1920s. His presence in the museum seemed to add to the sense that this museum had not escaped its past; he was not simply a subject of historical interpretation but seemed to be one of the architects of the modern installation.

The chapel also contains the last of the powerful images that the visitor will see before departing: two sets of stained-glass windows. These windows once again

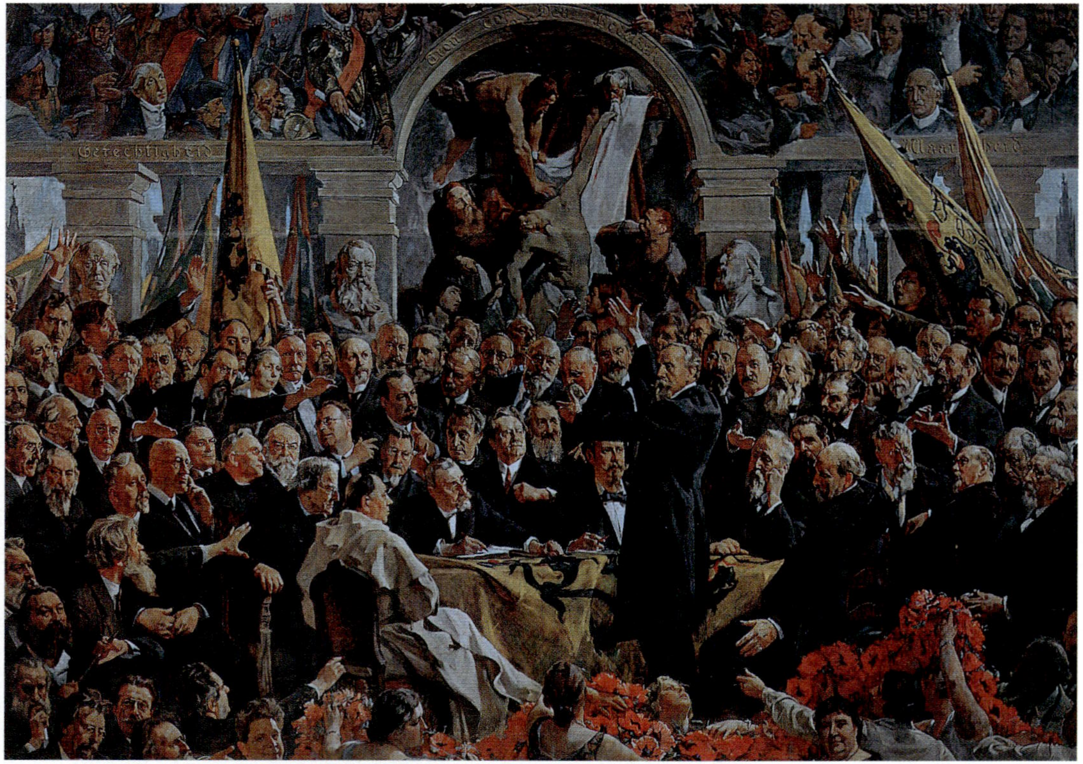

Figure 27.4 Hendrik Luyten's (1859–1945) *The Golden Painting of Flanders (Gulden Doek van Vlaanderen) (1931–44)* depicting 115 figures from Flemish history, including de Clercq and his co-collaborators (oil on canvas), IJzertoren Memorial Museum. Photo: J. Pingree. Copyright The IJzertoren Museum

underscore the foundations of Flemish nationalism in the liturgy of the Catholic faith. The first set of windows is based on sketches by Joe English. Eugene Yoors (1879–1975) transformed these into stained-glass cartoons in 1959–60, following the reconstruction of the second IJzertoren. Jan Wauters, who also designed and executed a second set of windows, set Yoors's cartoons into the stained glass. The English windows emulate the Stations of the Cross, but the figure of Christ, typically represented moving towards his destiny as the first martyr, is replaced by the First World War soldier crawling in the mud of the Front, guided by the AVV–VVK, towards his own submission to the cause of the Flemish Movement (Figure 27.5). The soldier/martyr finds solace in Mary and Christ. Yoors chose to execute the windows in the colours of the Front, shades of brown and grey with a dull golden light, perhaps echoing the ubiquitous mud at the Front.

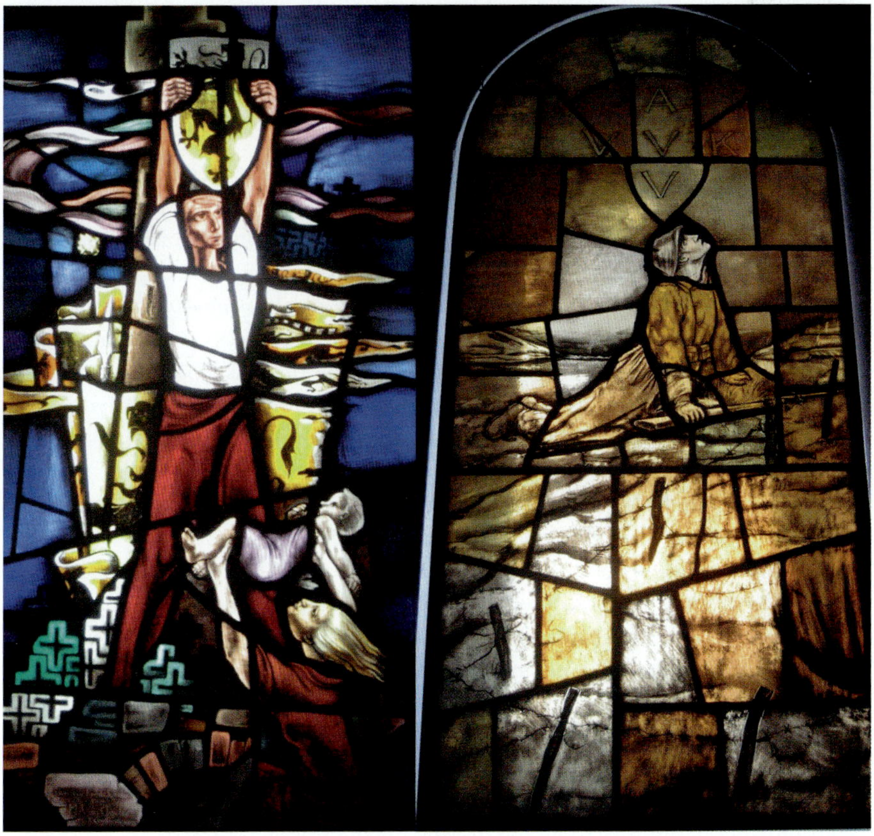

Figure 27.5 Stained-glass windows in IJzertoren chapel. Right, those by Eugeen Yoors, 1959–60, based on the First World War sketches by Joe English. Left, the windows by Jan Wauter, installed in the 1960s. Photo: J. Pingree

Wauters's windows are, in contrast, composed of the bright colours of medieval stained glass. They depict the First World War soldiers as medieval warriors, strong and decisive, perhaps in reference to Robrecht van Bethune, the Lion of Flanders, and the Battle of the Golden Spurs. These windows conflate the medieval and romantic aspirations of a sovereign Flanders with the labour of the Flemish soldiers in the war working for a vital and autonomous Flanders. The noble and inspiring figure of the Flemish medieval soldier is often juxtaposed with the Flemish flag with the black lion against the golden background. One window depicts the triumphant reconstruction or resurrection of the IJzer cross. The Flemish soldier stands tall atop the rubble. He is synonymous with the IJzer cross behind him as he holds aloft the shield with the Lion of Flanders. A multitude of small crosses and the pile of rubble from the first IJzertoren at his feet suggest the many deaths that make his triumph possible. A woman holds her baby up as an offering to the Flemish ideal. In another of Wauter's windows, a soldier looks to Mary as she holds a broken heldenhulde adjacent to the scales of justice. The entire building of the IJzertoren, from the heldenhulde crown to the chapel foundation, is thus integrated into its initial programme as a redemptive and political holy site.

Conclusion

All 12- to 13-year-old Flemish school children are required to visit the museum and memorial in order to locate themselves within a specifically Flemish, rather than Belgian, history. Flemish visitors predominate and there is limited interpretation beyond that in Dutch. Although the museum has bravely attempted to locate and promote a moral and cultural basis for Flemish national identity, it constantly faces the challenge of presenting this history in a building that has a political resonance which has not diminished. Objects like *The Golden Painting* feel the catalytic effect of being in such a powerful building; would any amount of interpretation really neutralize such objects? While the visitor might feel a sense of objective engagement in the upper floors, he or she is likely to leave having been exposed to the unmediated nationalism and the mythological effect of the soldier martyrs.

In order to move beyond the negative connotations of the Second World War years which produced the most extreme manifestations of Flemish nationalism, the current curatorial staff have aligned the morally based aims of the Flemish Movement with the challenges faced by Ireland and Tibet. There is, then, a vision that reaches beyond nationalism to social conscience, but one senses Belgium's own unsettled political divisions and the residues of past nationalism and its present-day currency as potentially subverting more humanistic connections. The memorial inevitably shapes the museum; it is a political actor and not a representation of neutral interpretation and this appears to be the destiny for the museum too.

Notes

1 Vlaams Belang.
2 The Wacht aan de IJzer, November 1914–November 1918.
3 Frontbeweging.
4 Percentages of participation, P. Chielens, personal communication, 9 August 2007.
5 See, for example, 'Preparing for Flemish Independence', http://members.lycos.nl/vlaamssiteje/history.htm (accessed 27 March 2008).
6 S. Maes, personal communication, 15 July 2008; P. Chielens, personal communication, 7 August 2007.
7 Archief Documentatie en Onderzoekscentrum (ADVN) D, Clemens De Landtsheer. Brochure, *Oproep tot het oprichten van Vlaanderen's IJzergedenkteken*, Temsche, 1926. AMVC, L, 196.
8 Ibid.
9 Open Brief van de Frontbeweging aan de koning der Belgen, Albert I (Open Letter from the Front Movement to the King of the Belgians, Albert I), 11 July 1917.
10 The Vlaamsch National Verbond (VNV).
11 Rob Gollin, 'Oorlogssmet kleeft aan opgeknapt doek', *Die Volkskrant*, 29 September 1998 (updated 16 January 2009). http://www.volkskrant.nl/archief_gratis/article775941.ece/Oorlogssmet_kleeft_aan_opgeknapt_doek (accessed 3 March 2009).

Bibliography

Anderson, B. (1983) *Imagined Communities: Reflections on the Origin and Spread of Nationalism*, New York: Verso Press.

Barbour, S. and C. Carmichael (2002) *Language and Nationalism in Europe*, Oxford: Oxford University Press.

Becuwe, F. and L. De Lentdecker (1993) *Van IJzerfront tot zelfbestuur*, Diksmuide: vzw Bedevaart naar de Graven aan de IJzer.

Beheydt, L. (1995) 'The linguistic situation in the new Belgium', in S. Wright and H. Kelly (eds) *Languages in Contact and Conflict: Contrasting Experiences in the Netherlands and Belgium,* Exeter: Short Run Press, pp. 48–64.

Bennett, T. (1995) *The Birth of the Museum: History, Theory, Politics*, London: Routledge.

Daels, F. (1933) *Voor mijn Volk in Nood (For My People in Distress)*, Tielt: Lannoo.

Debeuckelaere, A. (1917) 'Open brieven van de Vlaamsche Frontpartij in het jaar 1917 en vertoogschriften van het Vlaamsche Frontverbond' ('Open letters of the Flemish Front Party in the year 1917 and representations of the Flemish Front League') (The Hague, 1919), trans. Jane Fenoulhet, in T. Hermans (ed.) (1992) *The Flemish Movement: A Documentary History*, London: Athlone Press, pp. 227–69.

Dosfel, L. (1922) 'Katholicisme and Nationalisme. Antwoord op de vragenlijst rondegestuurd dor het Fransch tijdscrift *Les Lettres* te Parijs (October 1922)', in *Verzameld Werk (Collected Works)*, Vol. 3 (Tielt: Lannoo, 1928), trans. George Robinson, in T. Hermans (ed.) (1992) *The Flemish Movement,* London: Athlone Press, pp. 264–9.

Duncan, C. (1995) *Civilizing Rituals: Inside Public Art Museums.* New York: Routledge.

Farrar, L. L. (1995) 'Nationalism in wartime: Critiquing the conventional wisdom', in F. Coetzee and M. Coetzee (eds) *Authority, Identity and the Social History of the Great War,* London: Berghahan Books, pp. 133–52.

Frazier, I. (2007) 'Belgium? Something that does not exist: Political fault lines divide a nation', *Guardian*, 17 September, http://www.guardian.co.uk/world/2007/sep/17/iantraynor.international> (accessed 17 September 2007).

Gellner, E. (1983) *Nations and Nationalism: New Perspectives on the Past*, New York: Cornell University Press.

Gillis, J. R. (ed.) (2000) *Commemorations: The Politics of National Identity,* Cambridge: Cambridge University Press.

Halbwachs, M. (1992) *On Collective Memory*, trans. Lewis Coser, Chicago, IL: Chicago University Press.

International Council of Museums (2009) Conseil international des musées, Paris, France, http://www.icom.org (accessed 3 August 2009).

Kimmelman, M. (2008) 'With Flemish nationalism on the rise, Belgium teeters on the edge', *New York Times,* 4 August, nytimes.com (accessed 5 August 2008).

Seberechts, F. (2003) 'Slechts de graven maken een land tot vaderland', in F. Seberechts (ed.) *Duurzamer dan Graniet: Ongeveer Monumenten en de Vlaamse Beweging*, Antwerp: Lannoo, pp. 123–54.

Sherman, D. (1999) *The Construction of Memory in Interwar France*, Chicago, IL: University of Chicago Press.

Vanacker, D. (2000) *De Frontbeweging: De Vlaamse strijd aan de IJzer*, Koksijde: De Klaproos.

Whalen, R. (1984) *Bitter Wounds: German Victims of the Great War, 1914–1939*, Ithaca, NY: Cornell University Press.

Winter, J. (1995) 'Communities in mourning', in F. Coetzee and M. Coetzee (eds) *Authority, Identity and the Social History of the Great War*, New York: Berghahn Books, pp. 325–56.

Young, J. (2003) 'Monuments and memory', in R. Nelson and R. Schiff (eds) *Critical Terms for Art History*, Chicago, IL: University of Chicago Press, pp. 234–50.

Chapter 28

POSTCOLONIALISM, ETHNICITY AND THE NATIONAL MUSEUM OF IRELAND

Alan Kirwan

> The distinctiveness of Ireland's history of colonial domination precludes any so direct an application of generalizing or 'transferable theories'. As at once one of the earliest colonies caught up in Europe's westward expansion and a society intimately bound to the cultural and, especially, the religious context of Western Europe, Ireland has always been both a template and an anomaly.
>
> (Lloyd 1993)

THIS CHAPTER ASSESSES the extent to which postcolonial theory has influenced representations of ethnicity within national museums of so-called settler and non-settler nations (Ashcroft *et al.* 1995); how the national museums of such countries compare and contrast, and the implications of these distinctions for minority indigenous groups. Ethnicity as an increasingly self-confident term (Melucci 1989; Hall 1992) is redefining itself as a concept and in the process challenges notions of postcolonialism and the workings of postcolonial national museums. Within the museological literature, settler nations – Canada, Australia, New Zealand and the United States – dominate the debate as to the nature and complexities of the postcolonial museum (Simpson 2001; Boswell and Evans 1999). In these countries, discussion has centred on national museums and their efforts to incorporate minority indigenous peoples into the national narrative. It will here be argued that the experiences of other, less analysed postcolonial states, such as the non-settler nations of Ireland and India, disrupt the conventional debates around postcolonial theory and its application to national museums. These two countries constitute postcolonial nations in which the indigenous population remained in the majority and as such are rather different from the nations cited above.

In particular, the Republic of Ireland's complex form of postcolonialism could be argued to be one that identifies with a history of resistance to British imperialism, but yet in some ways was accomplice to the imperial project itself.

Postcolonialism and museums

Edward Said's *Orientalism* (1978) is often cited as the genesis of colonial and postcolonial studies (McLeod 2007; Carroll and King 2003). The discipline's initial critique of power relations and the construction of 'the Other' by Europeans continues to dominate the literature. However, postcolonial theory has become increasingly diffused into, and a mode of critical examination for, postmodernism more generally. As such, it now affects all areas of cultural studies. Its role in the analysis and understanding of museums is aided by John McLeod's attempt to relate the theory to the material world:

> Discontinuity, heterogeneity and diversity are perhaps the field's recurring markers. Yet postcolonial studies is not simply disparate and incoherent either; to be a field it must have some recurring elements or goals. One way of discerning the field's possible cohesion lies, I believe, in the configuration of that relationship between the material world and how we conceptualize it. Or, to put it differently, postcolonial studies requires us to recognize and explore the inseparable relationship between history and culture in the primary context of colonialism and its consequences.
>
> (McLeod 2007: 8)

The influence of postcolonial theory on the field of museum studies around issues of ethnic representation has been immense (Macdonald 2003; Hallam and Street 2000; Karp and Lavine 1991). In its earliest formulations, postcolonial theory highlighted the dominance of the colonized by the colonizer and the complex legacies of such contact following independence. One such legacy was the transformation of control of all aspects of political, economic and cultural life of the nation from a previous occupying force to what Whelan (2003: 94) describes as 'postcolonial paralysis' – the ossifying orthodoxy of the emergent nationalistic state which retains the institutional and ideological apparatus of the prior colonial state. Such 'ossifying orthodoxy' was all too clear in former colonies where the settler white population became the ethnic majority. Several national museums in such countries have been notable in their attempts to incorporate minority voices into the new national narrative, for example, the Canadian Museum of Civilization, Te Papa Tongarewa in New Zealand and the Australian National Museum (Simpson 2001; Casey 2001). However, the placement of settler nations in the context of postcolonial studies is in itself contentious and the ability of postcolonial theory to act as a guide for understanding ethnic relations in (all) postcolonial nations is increasingly being called into question. Many First Nation peoples within Canada, for example, are adamant that not only is the country not postcolonial but that it is in fact neo-colonial, as it never truly altered national hierarchies of power after the colonial period (Moss 2003). Brydon (2003) has adopted the counterview, arguing for Canada's place in postcolonial debates (and therefore by default other settler nations) by emphasizing the need of the country to disentangle itself from European history and influence, and by recognizing distinctions between both settler and indigenous postcolonial positions.

Ireland poses other serious challenges to current models of postcolonialism in that it was a majority 'white' ethnic indigenous colony:

> Ireland was the first of England's colonies, the training ground for the colonists to North America, and the context of the first English discourse on why and how to conquer and colonize. The Irish were also subject to representation as racially 'Other'. The resistance of the Irish to a violently imposed submission to English social, economic, and political structures made them inherently warlike and nomadic in the colonizers' characterizations of them as 'natural' slaves, which were in turn deployed to describe other colonized peoples.
>
> (Carroll 2003: 3)

Notions of the postcolonial condition are not uniform, even within former colonial states, where perspectives vary between different interest groups. How such variance has impacted on the development of national ethnic identities has been a key concern for many postcolonial theorists. Of particular interest is the concept of postcolonial hybridity and its theoretical application within the museum display practices of settler nations.

Homi Bhabha (1994, 1990) has had a significant influence on the representational practices of national museums such as the Canadian Museum of Civilization and Te Papa Tongarewa. Bhabha is credited with extending Said's binary notion of colonizer–colonized to the more problematic 'hybridity' of cultures. For Bhabha (1994: 1–2), the process of colonization resulted in the creation of new forms of identity in postcolonial states: 'It is in the emergence of the interstices – the overlap and displacement of domains of difference – that the intersubjective and collective experiences of nationness, community interest, or cultural value are negotiated.'

Bhabha's theory of national hybridity has been a seductive framework for post-colonial states trying to accommodate ethnic differences within their borders in attempts to create a more homogeneous, unified citizenship. It has been particularly useful for majority settler nations who are conscious of large minority indigenous peoples historically kept to the margins of the national story. The notion of hybridity and the sharing of a particular experience, such as colonialism, has been utilized in attempts to destabilize past animosities between settler and indigenous communities. A clear example of this is the case of Te Papa Tongarewa where the country's 'bicultural settlement' has been implemented at the museum to depict that 'Maori and Pakeha are joined together in the deep ahistorical space of land and nature' (Dyson 2005: 122). However, as Dyson goes on to highlight, in its attempts to project a national and international notion of unified identity, the museum continues to downplay past injustices. It would appear that Canada's national museums face similar problems in attempting to represent a unified idea of nation. Macdonald and

Alsford (1995: 286) state that 'ethnic and aboriginal museums remain, at this time, the main outlet for the interpretative perspectives of cultural minorities in Canada'. Increasingly, Bhabha's ideas concerning ethnic hybridity and shared legacies have come to be questioned, with many critics seeing in them a dilution of the voice of marginalized groups. San Juan has stated forcefully:

> The claim of postcolonial theory (as exemplified by assorted Australian, Indian, and Canadian ventriloquists of 'speechless' subalterns) to be more radical than avant-garde modernism is problematic especially for people of colour seeking to affirm their autochthonous traditions of resistance. What is at stake is their survival, their authentic and incommensurable dignities.
>
> (San Juan 1998: 30)

San Juan sees 'hybridism' and postcolonial theory as positions that fundamentally undermine the efforts of minorities to affirm rights and sentiments of difference. It is a view echoed in Bordewich's (1996) study of Native American struggles for cultural and sovereign independence. Postcolonial theory, which might have been thought to offer the best insights for post-independence constructions of identity, has found itself undermined by the ever-present and fundamental matter of ethnicity. Ethnicity as a power broker within national museums and constructions of nation would appear to be a force that demands a place at the negotiating table.

Ethnicity and national museums in non-settler nations

Ethnicity as a term and field of study is a recent development, appearing for the first time in the English language in the 1950s (Hutchinson and Smith 1996: 6) and defined as: 'a named human population with myths of common ancestry, shared historical memories, one or more elements of common culture, a link with a homeland and a sense of solidarity among at least some of its members'. It has been argued that human recourse to constructed notions of ethnic identity and nation is on the wane, with Benedict Anderson's *Imagined Communities* (2006) cited repeatedly as a bulwark against what many postmodernist thinkers would consider 'regressive' thinking. However, the continual ethnic strife and warfare that has erupted in various locations around the globe in the latter decades of the twentieth century and continuing into the twenty-first century would seem to contradict the assumed end of ethnic and ethno-national loyalties. Within the museological literature the continued importance of the subject is evidenced by numerous publications that deal with issues of identity, community and race (Watson 2007; Crooke 2007; Littler and Naidoo 2005). Postcolonial national museums have been, perhaps more than others, involved with the complexities of ethnic construction and representation since independence. First Nations have increasingly asserted their ethnic and cultural rights and national museums, so important to national recognition and legitimization, have been seen by many as key objectives.

Ethnicity as a concept divides opinion: in one sense identity-defining and in another discriminating, possibly even racist. Stuart Hall has been vocal in his support for the principle of difference and has sought to alter the parameters of the debate around ethnicity:

> The fact that this grounding of ethnicity in difference was deployed, in the discourse of racism, as a means of disavowing the realities of racism and repression does not mean that we can permit the term to be permanently colonized. That appropriation will have to be contested, the term disarticulated from its position in the discourse of 'multiculturalism' and transcoded, just as we previously had to recuperate the term 'black', from its place in a system of negative equivalence. The new politics of representation therefore also sets in motion an ideological contestation around the term 'ethnicity'.
>
> (Hall 1992: 162)

What, then, is the situation for national museums that confront increasing calls from self-identified ethnic groups for formal recognition within these national institutions, and can the experiences of postcolonial museums offer some ways forward? If attempts at 'hybridity' and 'biculturalism' have been notable projects within the national museums of settler nations, what have been the experiences of non-settler countries, and how do they differ?

It is significant that following independence, both the Irish and Indian governments assumed absolute control of their emerging national museums. Such events could not take place in settler nations. At the end of colonial rule in both countries (southern Ireland became the Irish Free State in 1922 and Republic of Ireland in 1948; India gained independence in 1947), the majority population was not one of 'invader-settlers' but the indigenous peoples. Unlike other postcolonial states, internal differences were predominantly constructed around religion, class and caste rather than purely racial distinctions. However, in spite of such distinctions, both countries still witnessed a coalescing of power in the hands of indigenous bourgeois elements that welded notions of national ethnic identity firmly to their specific concerns. In many ways the exclusionary politics of the new Irish and Indian middle classes was not that dissimilar from the departed colonial power. Mbembe describes similar developments in relation to many African postcolonial nations:

> Under colonial rule and beyond, a constellation of distinctively indigenous interests gradually came into being. Sometimes in competition with the evolues (the educated elite) and elements of the colonial bureaucracy, sometimes in symbiosis with them, this constellation of interests strongly influenced the shape of the independent states. This was also how – even though the postcolonial state's aim was to overcome 'old hierarchies' – relations of subjection were introduced and consolidated that broadly perpetuated those the colonial state had initiated.
>
> (Mbembe 2001: 40)

Ireland experienced a similar process:

> The successor state sponsored a nationalist project, constructed around the hegemonic bloc of the national bourgeoisie (agrarian and small business), and intertwining the state with the Church, education and media.
>
> (Whelan 2003: 94)

The new 'indigenous interests' that emerged in postcolonial nations with ethnic majorities such as Ireland, India and many African states were the construction of a small but significant elite within those countries. The new power brokers came to realize that recently liberated national museums could be utilized to project not only their ideas of the new nation but that as keepers of culture their hierarchical positions in the new society could be upheld to the exclusion of others. Overall, postcolonial theory has been surprisingly quiet in addressing the representation of ethnic minorities within indigenous-majority independent nations such as Ireland and India. As such, the representational practices of Irish and Indian national museums, particularly in the area of ethnic inclusion, have rarely been considered.

Ethnicity and the National Museum of Ireland

The modern-day National Museum of Ireland emerged under colonial rule as the Dublin Museum of Science and Arts in 1877. It was not officially designated a 'national' museum until 1908 (Wallace and O'Floinn 2002). This is not surprising as neither the British administrators nor the privileged class of Anglo-Irish interest groups involved in its formation intended that a truly nationalistic institution be inaugurated. Its purpose and function were first and foremost to further developments and appreciation in science and agriculture, which it was expected would translate into economic dividends from exploitation of the country's natural resources. A similar focus on natural history can be seen in the origins of what would become national museums in other non-settler nations such as South Africa (Goodnow 2006) and Kenya (Karanja Mirara 2002). Leading up to and following Irish independence in 1922, an important reorientation of the National Museum took place. The gold and bronze prehistoric collections that had previously been displayed in anonymous side galleries now took centre stage in the museum's main exhibition space. Thus began the National Museum's role in constructing the new Irish state as the direct descendant of an ancient Celtic culture (albeit with an interlude of several centuries). The transformation from colonial museum to National Museum is stark when prehistoric antiquities replaced 'field guns captured by Lord Gough from the Sikhs . . . and copies of classical statuary and ceremonial carriages' (Wallace and O'Floinn 2002: 11). The newly reoriented National Museum of Ireland and its re-evaluated prehistoric antiquities came to be instrumentally tied up with a refashioning of Irish ethnic identity following independence (Figure 28.1).

The satirical British publication *Punch* had infamously depicted Irish ethnicity as subhuman and simian in feature; a means to overcome the inconvenience of a shared

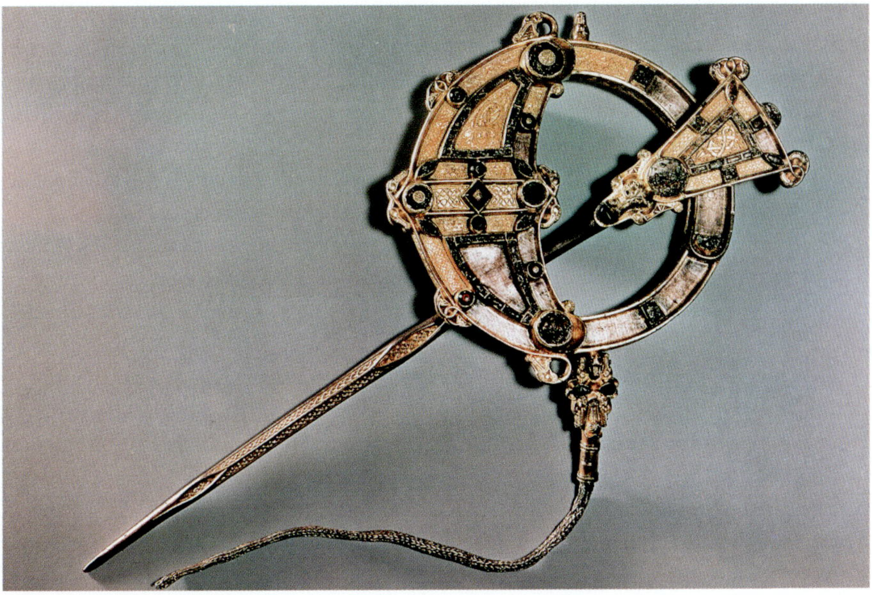

Figure 28.1 The Tara Brooch, dating from the early eighth century and unearthed in 1850, is considered one of Ireland's national treasures. Its decoration in the Celtic La Tène style saw the artefact become incorporated into concepts of Irish cultural and ethnic identity at the National Museum of Ireland in the late nineteenth century during the country's struggle for independence from British rule. Anglo-Saxon and Germanic design influences on the brooch were only later acknowledged following independence. Photo: Copyright National Museum of Ireland

'whiteness' amongst colonized and colonizer. It is not difficult to see why the post-colony sought to resurrect a romanticized and heroic Celtic ancestry that would not only support a desire for a distinct Irish identity but an identity based on an ancient and unassailable white European origin; 'to encounter the archaeological treasures of Ireland is to appreciate national pride in a unique identity rooted over millennia in the achievements of an ancient European culture' (Wallace and O'Floinn 2002: 2). This tacitly acknowledges the legacy of Scandinavian and Anglo-Norman influence on Irish craftsmanship but states, 'while acknowledging these various influences, the distinctiveness and originality of the national treasures define the greatness and, in European terms, the uniqueness of ancient Ireland'.

The common, and some may argue inevitable, move of postcolonial countries with ethno-majority populations like Ireland to rehabilitate the image of their ethnicity after colonialism can then set challenges for their future inclusiveness. In Ireland's case, the position of sizeable indigenous ethnic minority groups, for example, Anglo-Irish, Jewish, Traveller, has generally been overlooked in the bigger project to reconstitute a unified notion of national identity. None of the aforementioned Irish ethnic groups are represented in a coherent and inclusive manner within the current National Museum of Ireland. The inclusion and representation of newly emerging Irish ethnic communities, for example, Indian-Irish and Chinese-Irish, also challenge Ireland's own postcolonial mindset following independence. In a situation where a nation's national museums find it challenging to represent white ethnic diversity, how will they fare when it comes to the inclusion of non-white Irish ethnic communities? It is a problem for post-Soviet Europe and more generally prevalent across continental Europe where national boundaries have had considerable impermanence and where all nations have experienced conquest.

The National Museum of Ireland's ethnocentric display practices, which have sought to represent the majority, differ significantly from the more self-conscious construction of 'shared', bi-cultural visions of a national museum such as Te Papa Tongarewa in New Zealand. Both countries had vastly different experiences of colonialism and their postcolonial legacies reflect this. In turn, policies of representation at their respective national museums have evolved out of such experiences. If Te Papa's approach to ethnicity is taken to be representative of exhibitionary strategies in postcolonial settler nations, then the National Museum of Ireland finds its echoes in similar postcolonial non-settler nations, for instance, at the National Museum of India (Guha-Thakurta 2004). In fact, the similarity of experience of identity formation was felt to be so strong between Ireland and India that connections were made by nationalists both theoretically and physically prior to both countries' eventual independence. The most notable example is the friendship and mutual support of the Anglo-Irish writer William Butler Yeats and the Indian writer and artist Rabindranath Tagore (Lennon 2003). Both men were instrumental in harnessing the power of cultural nationalism, which took the form of the Celtic Revival in Ireland and Hindu spiritualism in India, to engage the Irish and Indian populations in an appreciation of a 'golden age' in their pre-colonial pasts. A significant element of this rallying of national support against the colonizer was the call to spiritual superiority. Irish and Indian independence movements made use of ideas around ancient mystical and spiritual ancestors that connected the now dispossessed indigenous peoples in a relationship to the land.

As Guha-Thakurta points out in relation to Indian art:

> Rejecting Western academic realism as a sign of mere material appearances, the non-naturalistic, stylized religious imagery of Indian sculpture and painting could be made the representative of a more elevated spiritual aesthetics.
> (Guha-Thakurta 2004: 155)

The exact same process took place in Irish art where Western canons of art history were usurped for a return to Celtic abstraction involving techniques of strapwork and interlacing designs. Although similar resistive developments occurred in relation to indigenous populations within settler nations, the national impact was minimal. This call to a spiritual unity has had an immense impact on Ireland and India, and as a consequence their national museums. The most obvious effect is the lasting legacy of tying a religious-spiritual element to an ethnic majority in a struggle for independence. The effects of such a historic event can be seen particularly in the homogenizing narratives of ethnicity and nation at the National Museum of Ireland and the National Museum of India. Safe in the comfort that the majority were once the 'colonized natives' and not the invader-settlers of places like New Zealand or Canada, the indigenous majority constructed its nation and did not consider ethnic-minority concerns within the overall framework of the new Irish and Indian states.

Representing ethnicity

Postcolonial museums have shown us that ethnicity is a deeply ingrained element in human lives and that its expression in a public space, particularly a national museum, produces diverse responses according to the colonial experience. In the case of non-settler nations, this presents a particular problem for inclusive interpretation as these nations seek a more essentialized understanding of the nation. This contrasts with the 'shared' or 'bi-cultural' strategies seen in other parts of the world, or the more inclusive tendencies of old colonial powers.

Much of the literature contends that the postcolonial museums of Canada and New Zealand have found effective strategies for representing ethnic difference, but the establishment of eco-museums and neighbourhood museums, by ethnic minority groups in settler nations, has begun to doubt such claims. Criticism directed at the National Museum of the American Indian, an institution often cited as a case study of community collaboration, highlights the intense resistance to what could be perceived as any homogenization of the national narrative (Jonaitis and Berlo 2008).

Bibliography

Anderson, B. (2006) *Imagined Communities: Reflection on the Origins and Spread of Nationalism*, London: Verso.

Ashcroft, B., Griffin, G. and Tiffin, H. (eds) (1995) *The Post-colonial Studies Reader,* London: Routledge.

Bhabha, H. (1990) *Nation and Narration*, New York: Routledge.

Bhabha, H. (1994) *The Location of Culture*, New York: Routledge.

Bordewich, F. M. (1996) *Killing the White Man's Indian: Reinventing Native Americans at the End of the Twentieth Century*, New York: Doubleday.

Boswell, E. D. and Evans, J. (eds) (1999) *Representing the Nation: A Reader: Histories, Heritage and Museums*, London: Routledge.

Brydon, D. (2003) 'Canada and postcolonialism: Questions, inventories, and futures', in L. Moss (ed.) *Is Canada Postcolonial? Unsettling Canadian Literature*, Waterloo, Canada: Wilfrid Laurier University Press, pp. 49–77.

Carroll, C. (2003) 'The nation and postcolonial theory', in C. Carroll and P. King (eds) *Ireland and Postcolonial Theory*, Cork: Cork University Press, pp. 1–15.

Carroll, C. and King, P. (eds) (2003) *Ireland and Postcolonial Theory*, Cork: Cork University Press.

Casey, D. (2001) 'Museums as agents of social and political change', in S. Watson (ed.) (2007) *Museums and Their Communities*, London: Routledge, pp. 292–99.

Crooke, E. (2007) *Museums and Community: Ideas, Issues and Challenges*, London: Routledge.

Dyson, L. (2005) 'Reinventing the nation: British heritage and the bicultural settlement in New Zealand', in J. Littler and R. Naidoo (eds) *The Politics of Heritage: The Legacies of 'Race'*, London: Routledge, pp. 115–30.

Goodnow, K. (2006) *Challenge and Transformation: Museums in Cape Town and Sydney,* Uckfield, UK: Beacon Press.

Guha-Thakurta, T. (2004) *Monuments, Objects, Histories: Institutions of Art in Colonial and Postcolonial India*, New York: Columbia University Press.

Hall, S. (1992) 'The new ethnicities', in J. Hutchinson and A. D. Smith (eds) (1996) *Ethnicity*, Oxford: Oxford University Press, pp. 161–63.

Hallam, E. and Street, B. V. (eds) (2000) *Cultural Encounters: Representing 'Otherness'*, London and New York: Routledge.

Hutchinson, J. and Smith, A. D. (eds) (1996) *Ethnicity*, Oxford: Oxford University Press.

Jonaitis, A. and Berlo, J. C. (2008) '"Indian Country" on the National Mall', in A. Lonetree and A.J. Cobb (eds) *The National Museum of the American Indian, Critical Conversations*, Lincoln: University of Nebraska Press, pp. 208–40.

Karanja Mirara, F. (2002) 'The national museums of Kenya and social responsibility: Working with street children', in R. Sandell (ed.) *Museums, Society, Inequality*, London: Routledge, pp. 175–81.

Karp, I. and Lavine, S. D. (eds) (1991) *Exhibiting Cultures: The Poetics and Politics of Museum Display*, Washington, DC: Smithsonian Institute.

Lennon, J. (2003) 'Irish orientalism: An overview', in C. Carroll and P. King (eds) *Ireland and Postcolonial Theory*, Cork: Cork University Press, pp. 129–57.

Littler, J. and Naidoo, R. (eds) (2005) *The Politics of Heritage: The Legacies of 'Race'*, London: Routledge.

Lloyd, D. (1993) *Anomalous States: Irish Writing and the Post-colonial Moment*, Dublin: Lilliput Press.

Macdonald, G. and Alsford, S. (1995) 'Canadian museums and the representation of culture in a multicultural nation', in S. Watson (ed.) (2007) *Museums and their Communities*, London: Routledge, pp. 276–91.

Macdonald, S. (2003) 'Museums, national, postnational and transcultural identities', *Museum and Society*, 1(1): 1–16.

Mbembe, A. (2001) *On the Postcolony*, Berkeley and Los Angeles: University of California Press.

McLeod, J. (ed.) (2007) *Postcolonial Studies: The Routledge Companion*, London: Routledge.

Melucci, A. (1989) 'The post-modern revival of ethnicity', in J. Hutchinson and A. D. Smith, (eds) (1996) *Ethnicity*, Oxford: Oxford University Press, pp. 367–70.

Moss, L. (ed.) (2003) *Is Canada Postcolonial? Unsettling Canadian Literature*, Waterloo, Canada: Wilfrid Laurier University Press.

Said, E. (1978) *Orientalism*, New York: Routledge.

San Juan, E. (1998) *Beyond Postcolonial Theory*, New York: St Martin's Press.

Simpson, M. G. (2001) *Making Representations: Museums in the Post-colonial Era*, London: Routledge.

Wallace, F. P. and O'Floinn, R. (eds) (2002) *Treasures of the National Museum of Ireland: Irish Antiquities*, Dublin: National Museum of Ireland/Gill & Macmillan.

Watson, S. (ed.) (2007) *Museums and their Communities*, London and New York: Routledge.

Whelan, K. (2003) 'Between filiation and affilation: The politics of postcolonial memory', in C. Carroll and P. King (eds) *Ireland and Postcolonial Theory*, Cork: Cork University Press, pp. 92–108.

Chapter 29

FACING UP TO DIVERSITY

Conversations at the National Museum of Colombia

Cristina Lleras

IN LATIN AMERICA, indigenous communities and movements had a profound impact on the adoption of multicultural policies and constitutions in the late 1980s and early 1990s (Gros 2000: 359). Colombia's 1991 Constitution, for example, states: 'The State recognizes and protects the ethnic and cultural diversity of the Colombian Nation.' This moment of change led to the decentralisation of certain political and administrative functions which permitted particular ethnic groups to preserve their cultures and identities through the use of their own languages, methods of education and religious practices. This was a radical change for Latin America. It meant that:

> modernity can coexist with the presence of different groups that share different cultures . . . that these groups have a future, and that the State does not need to work for the disappearance of the Indian, the Black and in favour of miscegenational culture, but on the contrary, has to recognize their particular rights, education, territory, their own forms of government.
> (Gros 2000: 359)

This chapter records some of the conversations that have taken place in the National Museum of Colombia (Figure 29.1) as it has considered the implications of multiculturalism and recognition of the specific rights of ethnic groups for the museum's future reimagining (Figure 29.2).

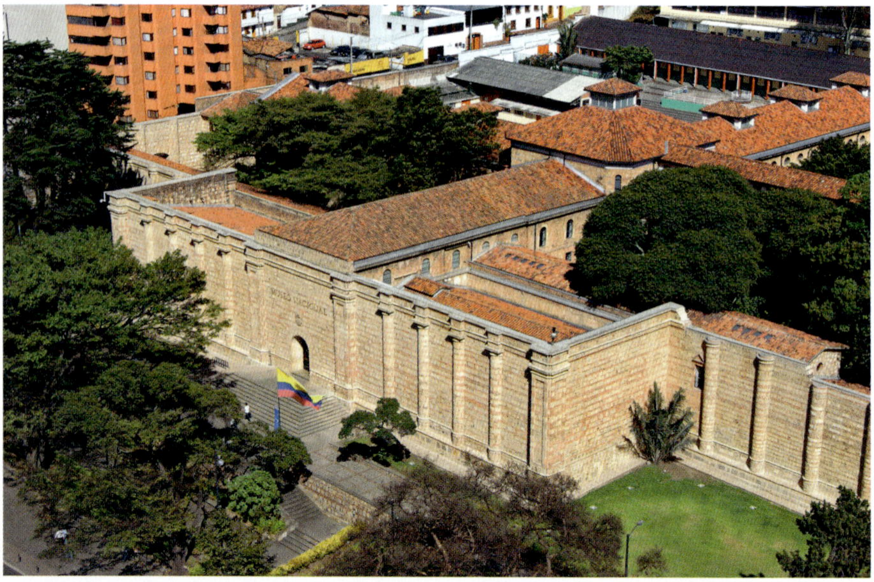

Figure 29.1 The National Museum of Colombia. Photo: Carlos Gustavo Suárez

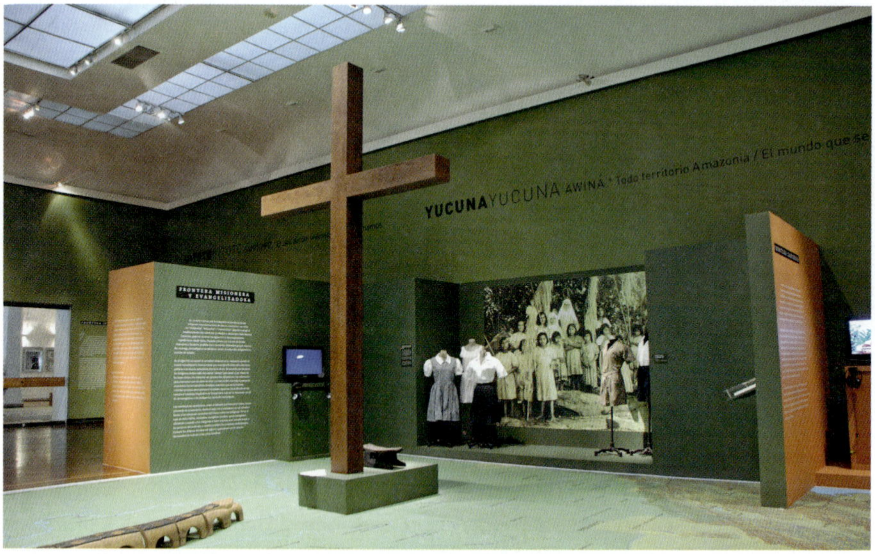

Figure 29.2 In 2009, the National Museum staged an exhibit on the Amazon. Here it highlighted the conflict between Christian religious beliefs and the spirituality and cosmology of indigenous communities. Photo: Carlos Gustavo Suárez

Multiculturalism in Colombia

Colombia is in many respects typical of settler nations established in Latin America, though it has its own distinctive aspects. It is a nation composed of descendants of native inhabitants, together with those of enslaved Africans and other immigrants (particularly those of Spanish, Jewish and Arab origin) who have both intermarried and yet preserved their own separate communities. Prior to the establishment of the 1991 Constitution, assimilation had been the prevailing social model. Here the State promoted the superiority of Whites and mestizos (those of mixed White–Amerindian birth). Miscegenation was a political and cultural project that denied, and hoped to obliterate, other identities (Black, mixed-Black and indigenous). The 1991 Constitution was, then, a remarkable step which has since been developed into a highly sophisticated constitutional and judicial framework; Colombia has, in this respect, become a social paradigm for other countries in the region (Bonilla 2006). As Hooker (2005: 286) has noted: 'only Brazil, Colombia, Ecuador, Guatemala, Honduras and Nicaragua extend (some) collective rights to Afro-Latinos'.

However, despite this quite fundamental political and legal shift there is still considerable work to be done to overcome discrimination – including work that can be done in museums. But the ease with which terms like 'discrimination' and 'multiculturalism' can be used rather underplays the ethnic and cultural complexity confronting organisations like the National Museum of Colombia. For example, while both indigenous groups and people of African descent suffer discrimination, each has been affected differently by the rise of multicultural citizenship. Indigenous peoples have been empowered through their cultural distinctiveness in ways that Afro-Colombians and the Black community have not. Hooker (2005: 307) has argued that this kind of intervention can shift the political debate 'privileg[ing] issues of cultural recognition over the struggle against racial discrimination as the basis of political mobilization'. It permits the construction of sanitised apolitical groups (Hooker 2005: 310), which can then find their way into the museum. It separates and diverts a shared politics of discrimination.

Recognition of the rights of different ethnic groups poses other problems for national museums – not least, the problem of reconciling the tension between unity and difference enshrined in the Constitution. It embraces two sets of principles: the first is constituted by the cultural rights granted to minority groups and the recognition of the multicultural character of the nation; the second is constituted by the values of unity, sovereignty and universal human dignity (Bonilla 2006). This second set of principles defends the neutrality of the state and the equality of all its citizens. In practice, however, the state promotes certain festivals, languages, curricula and national symbols which are not necessarily embraced by all. Museums possessing this multicultural mandate are confronted with the same concerns: how to display and narrate in a manner that both respects minorities, whose rights have been systematically ignored, and yet also project the unity of the nation. Colombia is not alone in confronting these issues – they affect all settler nations – and many of the debates that have recently taken place in the country echo those that have also taken place in New Zealand, Australia and the USA (see, for example, Dean and Ridder 2005; Message 2005). There is, however, a specificity to the issues each nation confronts and, indeed, to the manner in which a national museum must act which may contrast with the policy initiatives of government ministries.

Take, for example, the Ministry of Culture of Colombia's summary of its plans and programmes published in 2007, which defined its mission, vision and objectives. The document was 'to highlight the importance and worth of the different cultural manifestations that make up our culture' (Ministerio de Cultura 2007: 6). Although based on home-grown legislation, the Ministry articulated its actions in the rhetoric of UNESCO declarations and conventions, which gave a sense that culture could be a 'cure-all' in state policies (for which, see Message 2005: 477). In this document, Colombia's diversity was to be understood as a set of cultural assets: regions, ethnicities, languages and cultural manifestations. It was accompanied by a promotional video that exalted diversity: people from different regions and diverse landscapes, as well as national heroes including Nobel Prize-winning novelist, Gabriel García Márquez, great sporting figures and others. It emphasised the 'purity' of indigenous cultures in the Amazon, the authenticity of Afro-Colombian music and the strength of cowboys from the plains. It essentialised diversity, privileged great White men and represented other cultures only as communities. This is the vision; it is not simply political or marketing gloss. It has been suggested that 'governments have a vested interest in promoting a heritage that reassures and reconciles rather than disturbs and divides. Heritage is being used to soothe away our individual and collective stresses, leaving only contented well-balanced people in an all-inclusive harmonious society' (Ashworth *et al.* 2007: 53). This cannot – unless the national museum is merely to be an instrument of government – be a vision for a national museum. As Colombian anthropologist, Jaime Arocha, suggests, Colombia today possesses two competing models of multiculturalism: one promulgated by the state involves idealistic commodification while still permitting exclusion and discrimination; the other, existing at grassroots level, involves communities both defining and defending their territories while also, on a daily basis, engaging in intercultural exchange.[1]

Exhibiting Afro-Colombian culture

Founded in 1823, the National Museum is part of the Ministry of Culture but it works to its own strategic plan, which in 2003 stated that it should be 'the main educational resource on the history of Colombian culture in the country and a centre to disseminate research on the multiethnic and multicultural identity of the nation' (Ministerio de Cultura 2003: 11). The plan demanded that the museum demonstrate it was not an elitist institution; that it belonged to 'all Colombians' and can be a forum for the negotiation of difference. It was only in 2005 that the museum began, in close collaboration with scholars Jaime Arocha and Claudia Mosquera, three years of round-table discussions, visits, and research by an interdisciplinary group which resulted in the temporary exhibition, *Velorios y santos vivos. Comunidades negras, Afrocolombianas, raizales y palenqueras* (*Wakes and Living Saints amongst Black, Afro-Colombian, Maroon and Islander Communities*), examining the concept of ancestry as seen through funerary rituals and festivities devoted to deities (Figure 29.3).

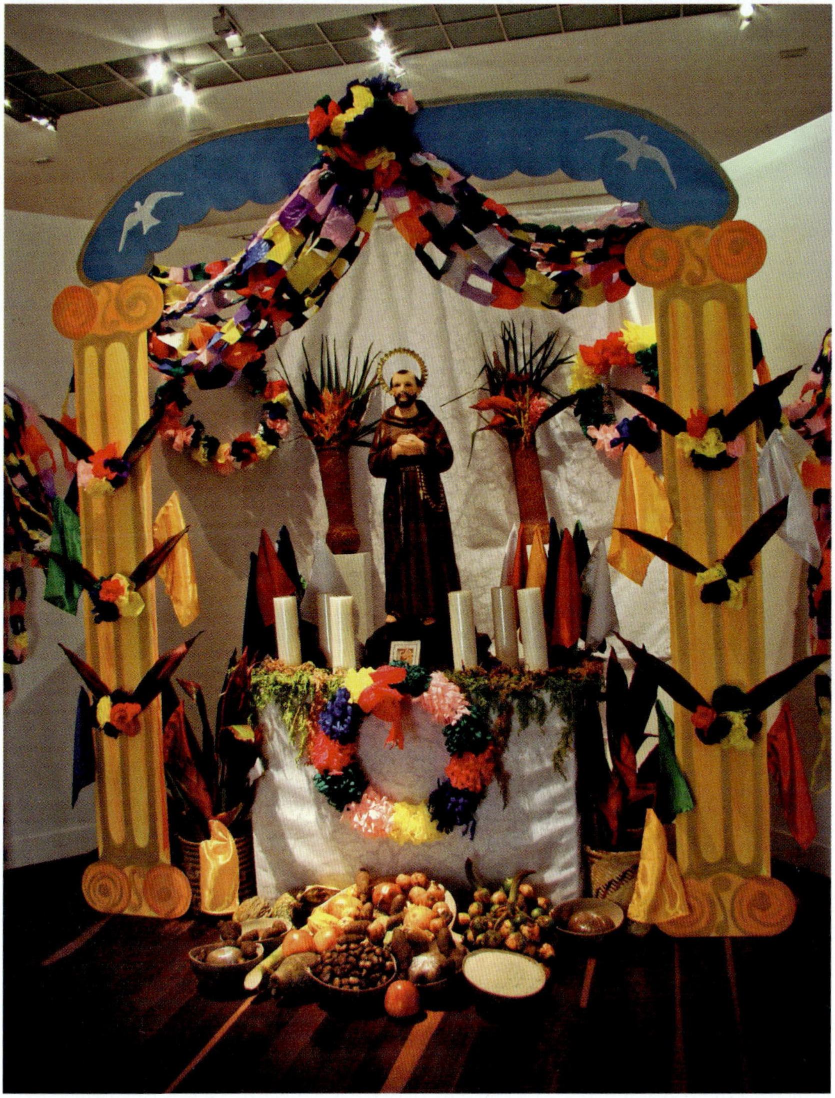

Figure 29.3 Altar to Saint Francis ('San Pacho') designed by Madolia de Diego in Quibdó (Chocó) and enhanced by Miguel Ángel Cuesta at the National Museum of Colombia. Photo: Carlos Gustavo Suárez

The exhibition opened at the National Museum of Colombia on 21 August 2008 and ran until 2 November.[2] Developed through close participation with seven communities, it sought to counter the invisibility to which these communities had long been subjected (Figures 29.4 and 29.5). During the exhibition, groups from the regions represented came to the museum and staged their own presentations, enriching the contents of the exhibition. Inevitably, there were a few objections, such as to the ethnicity of the curator and anthropologist, and the contents of the displays, but generally the exhibition was well received – though attendance by Afro-Colombians was very low. However, the aim of this chapter is not to review the exhibition so much as to reveal how it acted as a catalyst to discuss the future of the museum. It is based on interviews with visitors, focus groups with Afro-Colombians and a group interview with staff.[3] In my interviews with staff, I should note that I did not direct particular questions to staff according to the roles they occupied in the museum, but rather tried to get some sense of the 'culture of the museum' as a diverse and interactive social space. The following discussion records the conversations that took place in this museum as it confronted the legacy of the past still present in its permanent displays, and a future signalled by this temporary exhibition. These reveal, to a surprising degree, how similarly those within the museum and those without visualise this future.

Unifying narratives

When confronted with imagining the museum's future, it was the Assistant to the Curator of Ethnography and Archaeology who remarked:

> I think that a starting point would be common elements. Though we are a multicultural nation, there are things we do without it really mattering what group or community we belong to. So, for instance, if we are going to have an exhibit on family in Colombia, what is the family experience in the different communities in the country? . . . What are the themes that cover this nation or permeate us all but in very different ways?

Elsy, a member of one of the Afro-Colombian focus groups, thought similarly: 'Why not show these things where we converge. What I was saying was: in what moment are we different but then in what moment do we converge in social and cultural aspects that are part of a culture that composes everything that makes us Colombian.' The notion of creating a unifying account exploiting common themes has been called the 'salad bowl' model of heritage policy, where each group contributes to the whole while emphasising shared elements (Ashworth *et al.* 2007). The approach has been adopted by a number of museums around the world. It holds the potential for intercultural dialogue but runs the risk of promoting simplistic notions of culture.

The museum's Designer and Director both saw a rather different kind of unifying narrative, one that returned to the grand narrative but which was inclusive. The Director stated:

> We would have to take away much from the hegemonic history and bring into this other story people who are unknown. We are conscious that the nation is not composed of the gentlemen who are hung on the wall. . . . There is no correspondence between the country as it is told in the Museum and how I conceive the nation with all the inhabitants of a territory.

There are risks here that a museum might adopt an assimilation model or a 'core plus' model which leaves the strong central theme unchanged (Ashworth *et al.* 2007). Littler warned:

> the language of inclusion can also imply assimilation to a pre-existing set of national norms rather than a genuine diversity. . . . It can imply that there is only one already sealed and written national story, with room at the most for a little non-threatening difference, and to which 'newcomers' need to conform to gain acceptance.
>
> (Littler 2005: 12)

This seems not to have been in the minds of staff at the National Museum of Colombia as all staff acknowledged that the core of the museum was not representative of this truly diverse nation. There was a consensus that there were stories that have not been told and communities that remain unseen. Outside the museum's inner circle of staff, the academics and visitors shared these feelings. Manuel, a political science student, said: 'right now it's a museum of a few for the many'. Janine, one of the Afro-Colombian guides who worked on the exhibit, commented: 'here we only see one type of person, and the rest are excluded, as though they don't exist'. It should be renamed 'Museum of Colombia . . . so that it represents everything', Eduardo concluded. These external commentators recognised that the notion of diversity should not be confined to well-known and established ethnic groups. There was, however, a commitment to the grand view. One Afro-Colombian focus group suggested it should include 'elements of all the cultures in the nation'; 'everything that has to do with the cultural aspect of the country and the regions'. Visitors interviewed in the exhibition thought that the nation should be represented so that people 'see what is similar and different among Colombians', so that one could 'travel across the regions', 'giving relevance to what makes them Colombian', as two anthropology students commented. Half of the visitors surveyed were either born in another city or lived outside the capital, but focus groups also felt that place of birth and ancestry were key identifying attributes. A young Afro-Colombian communication student, Joven, said the museum should include:

> everything that is Colombia, and in that sense it should not exclude anyone. There are regions that one doesn't see in the Museum . . . we are being relegated and therefore we are not protagonists of history, and that doesn't mean that in reality we are not.

This feeling was universal amongst members of groups currently excluded from the museum's grand narrative: Janine commented that the museum 'will show all of the people who belong in the Colombian territory and it will develop points of view from the political, social, cultural and economic to include the most salient exponents of each ethnic group'.

While this sense of a more inclusive national narrative was strong, visitors also wished to know other cultures – to reach beyond local or national circumstances; to engage in what we might understand as interculturalism and yet build their own identities in the process. A Xerox machine technician wished to see an interchange of Afro-Colombian culture 'with our culture just as it happens in other countries, like in New Orleans'. A visitor from Medellín, Colombia's second most populous city, who described himself as 'half-indigenous', said that visitors should experience Colombia's rich cultural diversity: 'More of what Colombia is, because it has something very interesting. It can have 10 or 20 different dances; when you look at the Germans, they only have one.'

Representation

Staff were reminded that a strong component in Afro-Colombian communities claims reparation through narratives in the National Museum which come to grips with the legacies of enslavement and racism and give particular groups their own gallery spaces. Staff expected concerns about this apportionment of space, seeing it as discriminatory in an already fragmented society and, moreover, requiring spatial resources beyond those available to the museum. One member suggested this would establish the equivalent of ghettos in the museum. Staff appeared immediately to imagine the 'pillar model' of cultural heritage, where communities are set apart without communication (Ashworth *et al.* 2007).

Nevertheless, it was also recognised that having a space in the museum was part of a political process of recognition. The Deputy Director thought these reparation claims valid, but stopped short of accepting the need for an allocation of spaces:

> When you only hear their voice, as though they had been independent, it generates these discourses that are racist against others. So it's the greatness of the Black people versus the cruelty of the Whites. And it's again these tensions of Black and White, wonderful indigenous peoples versus Bad conquistadors. . . . I do accept that because of the exclusion to which they have been subjected, they should have more relevance but it should not be something exclusive.

The Director agreed: 'We want to make diversity visible but when we put it in boxes we are not being true to the reality, which is integrated.' The museum's designer saw other potential in the idea of separation; the possibility of liberating the museum narrative: 'We should leave the script up to them so that they choose what to do.'

Some non-Afro-Colombian visitors did directly refer to the issue of privilege when asked if any group should preferentially be given space. A woman from another city running errands with her husband in Bogotá said, 'Why privilege something? If we have so many things we should know them all, I can't think of anything specific. Why give something more importance if all our culture is important?' Some in the focus groups also supported this idea. A Black man who had lived in what he called 'the apartheid regime' created by the North American mining companies in the Colombian Pacific said that he thought that Black people should have access to politics or institutions on their merits; otherwise, he considered the quota system discriminatory: 'We are all equal, first we identify ourselves as people, but if we want to leap forward we need education.'

This sense of the fundamental importance of non-discriminatory narratives was also shared by focus group members. But young Black participants also saw merits in a separation of spaces: 'A space to see how different we are.' Eduardo, a sociologist and guide to the exhibition, drew attention to the need to have histories where different groups meet, but also histories that are particular to each community.

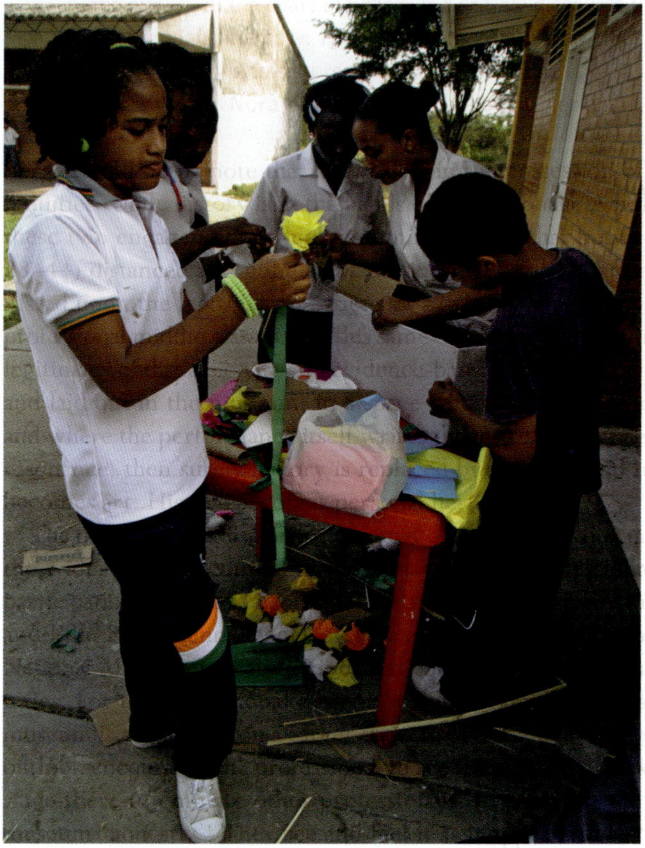

Figure 29.4
Velorios became a travelling exhibition and included conversations with people in the regions who participated in its construction. Photo: Sofia González

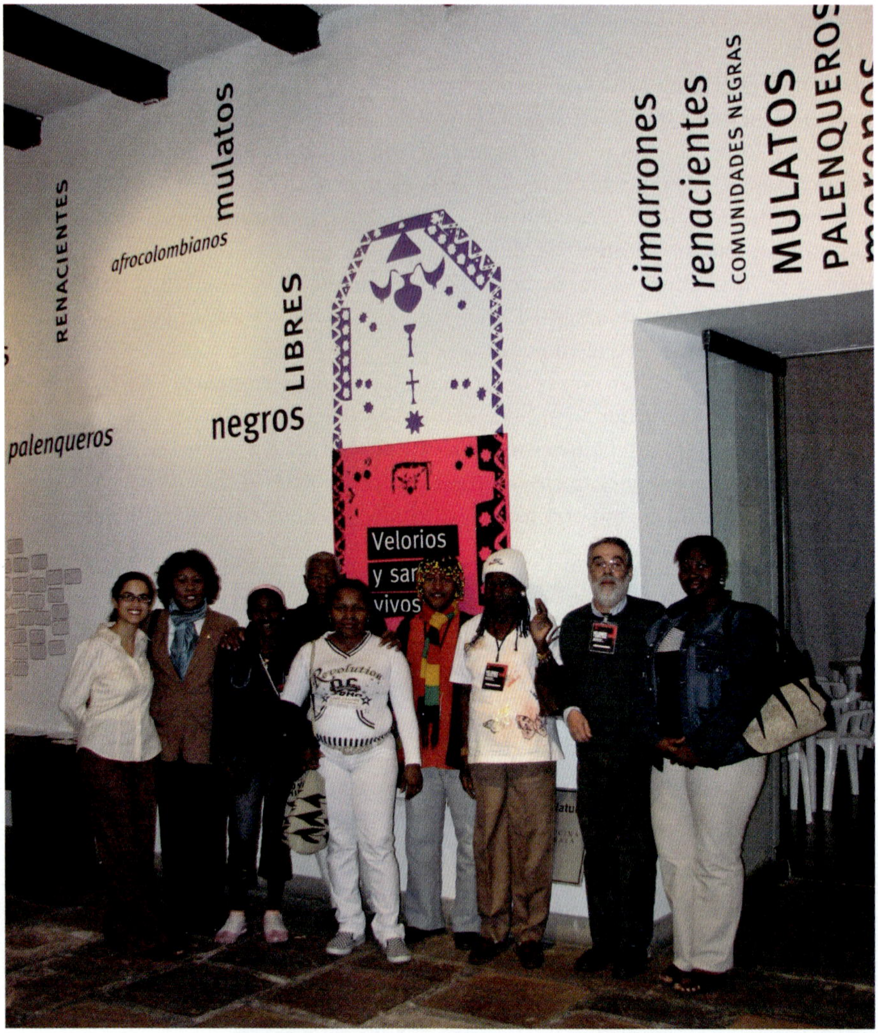

Figure 29.5 *Velorios* was made possible through intercultural dialogue. Here leaders from various communities meet with production team members. Photo: Carlos Gustavo Suárez

The conversations in the museum showed that most people were comfortable with differences only if communities were treated equally. Staff recognised that the possibility of being truly representative was impossible – diversity was beyond rational definition and categorisation. The Head of the Education Division questioned the authority of the museum in deciding issues of representation but also doubted that public consultation alone would resolve the matter. Nevertheless, members of the temporary exhibition's committee understood that the museum could not represent communities without community involvement. The Deputy

Director reflected on what had been achieved in the preparation of the temporary exhibition: 'Once that door is open, it's difficult to think that this is not the natural process that the museum is going to follow.' The museum administrator who manages the visitor book commented:

> It's not clear to me if it's the museum that has to put together a series of elements of representation and histories, to incorporate the communities that are not represented, or if it is the communities that come to the museum and present proposals to be seen as they want to be identified. . . . So I think that the museum should make a call to different communities to analyse how they see themselves.

For the visitors surveyed, the idea of participation was much more natural. They had no shortage of ideas of who could tell the story of the diverse nation. Eduardo listed both academics and knowledgeable people from communities and social movements. Catalina extended the list to include docents, children and the elderly; she saw the process as being fundamentally inclusive.

Conflict

The museum's strategic plan states that the museum should not hide memories of conflict but should be careful to analyse them and create a cultural dialogue that allows both historical and cultural perspectives. Histories were to be produced which were closer to the visitors' own experiences.

A video shown at the entrance to the *Velorios* exhibit put this into practice. It reflected on the economic, social and violent processes that are altering the rituals and communities represented in the exhibition. It opened the exhibition up to political debate and seemed to encourage the criticism of particular social agents and even the state. It provoked one visitor who works with a foundation that has projects with Afro-Colombian communities in the Pacific to comment:

> The video makes us think that Colombia is a multiethnic and multicultural nation but that in a few years this will only be true on paper because it's been disappearing and colonisers go into their territories with goods that supplant local ones and are putting an end to their nourishment security. And this makes one think, what is the State doing so that this doesn't happen?

A young woman of mixed ancestry visiting with her mother was provoked to doubt the multicultural objectives of the government:

> Though the Constitution says that Colombia is multiethnic, apparently it stays on paper . . . because up until now ethnic communities have not been given the recognition they deserve. . . . And I think that in the cultural realm it

> might be a credit they deserve, but it's always been about recognising indigenous and Afro-Colombian communities in the cultural sphere, but the political part, the social participation, in those spheres, it has not been granted.

Staff favoured the negotiation of contexts that included conflict between groups, or between groups and the state, but the museum's Director also acknowledged that the museum did not stand at a historical distance from these conflicts and was funded by an executive branch of government. The museum was not in any sense independent of the state.

Facing up to the future

These conversations revealed the degree to which all groups, regardless of age, gender, occupation and ethnicity, shared certain values regarding the way ahead for the museum. All recognised the deficiencies of the museum and saw a solution in a collective narrative produced through participation that tended towards and encouraged interculturalism. This seemed to reflect not the multiculturalism of state but rather that found at grassroots levels in Colombian society.

There was rather more of a separation between staff and the public on the matter of conflict. Clearly, visitors to the exhibition benefited and responded to the political issues it raised, particularly in the opening video. However, if they did not see the video – and the majority missed it – they were rather less inclined to see politics, conflict and contested rights as key elements in the reimagined museum. However, further probing might reveal a rather different picture. The museum's staff, however, clearly believed in the institutional mission and believed the negotiation of conflicts and difference to be quite central to the interpretive goals of the museum. Despite the constitutional ideal of a plural society, staff at the National Museum of Colombia felt they could not create a collective narrative without dissent. In this respect it seemed to differ from the Ministry of Culture's own vision. It was a grounded institution and it could not run from the realities it sought to interpret or the experiences of its visitors. The museum needs to be an invitation that requires visitors to think about their own standpoint and yet offer personal ways to find common ground; the staff now understood that only through intercultural dialogue could the museum challenge the old hegemonic discourse.

The criticism of the museum's use of temporary exhibitions as a kind of reparation for shortfalls and omissions is supported more generally by research and policy in other areas where social justice has been debated. In other areas of government, such practices would be considered inadequate. 'Special events' simply will not do: 'Multiculturalism has evolved and social justice issues (human rights issues, antiracist pedagogy, programs that address systemic discrimination, equity, inclusion, etc.) need to be firmly entrenched' (Chalmers 2002: 300). Museum staff differed in their beliefs regarding how rapidly the institution could change; some favoured radicalism, others were incrementalists. The more fundamental challenge was how to grapple with a more critical, rather than celebratory, multiculturalism in the museum.

Notes

1 Arocha, J. and Lleras, C. 'Armonía y conflicto. Representación e identidad en el patrimonio cultural', paper presented at the Cátedra de Historia Ernesto Restrepo Tirado in the Museo Nacional de Colombia, November 2008.

2 I was co-curator and coordinator of the visitor study. For a full account of the exhibition process, see Ministerio de Cultura 2008.

3 I conducted 45 semi-structured interviews between September and October 2008. In order to gather data on the ways Afro-Colombians had interpreted the exhibition, three focus groups (consisting of between six and ten individuals of mixed gender and age) were undertaken in November 2008. On 18 December 2008, I conducted a group interview with the museum's Committee for Temporary Exhibitions which met on a weekly basis. The Director was interviewed separately on 18 February 2009 using the same questionnaire. The names of those who participated in the study have been changed in order to protect their anonymity.

Bibliography

Ashworth, G. J., Graham, B. and Tunbridge, J. E. (2007) *Pluralising Pasts: Heritage, Identity and Place in Multicultural Societies*, London and Ann Arbor, MI: Pluto Press.

Bonilla, D. (2006) *La constitución multicultural*, Bogotá: Siglo del Hombre Editores, Universidad de los Andes-Facultad de Derecho, Pontificia Universidad Javeriana-Instituto Pensar.

Chalmers, F. G. (2002) '"Celebrating pluralism" six years later: Visual transculture/s, education, and critical multiculturalism', *Studies in Art Education,* (43) 4: 293–306.

Dean, D. and Rider, P. (2005) 'Museums, nation and political history in the Australian National Museum and the Canadian Museum of Civilization', *Museum and Society*, 3(1): 35–50.

Gros, C. (2000) 'De la nación mestiza a la nación plural: el nuevo discurso de las identidades en el contexto de la globalización', in G. Sánchez and M. E. Wills (eds) *Museo, memoria y nación: Misión de los museos nacionales para los ciudadanos del futuro*, Bogotá: Ministerio de Cultura.

Hooker, J. (2005) 'Indigenous inclusion/ black exclusion: race, ethnicity and multicultural citizenship in Latin America', *Journal of Latin American Studies*, 37: 285–310.

Littler, J. (2005) 'Introduction', in J. Littler and R. Naidoo (eds) *The Politics of Heritage, The Legacies of Race*, London and New York: Routledge.

Message, K. (2005) 'Representing cultural diversity in a global context: The Museum of New Zealand Te Papa Tongarewa and The National Museum of Australia', *International Journal of Cultural Studies*, 8: 465–85.

Ministerio de Cultura, Museo Nacional de Colombia (2003) *Plan estratégico 2001–2010. Bases para el Museo Nacional del futuro*, Bogotá: Ministerio de Cultura.

Ministerio de Cultura, Museo Nacional de Colombia (2008) *Velorios y Santos Vivos: Comunidades Negras, Afrocolombianas, raizales y palenqueras*, Bogotá: Ministerio de Cultura.

Ministerio de Cultura, República de Colombia (2007) *Un Ministerio de puertas abiertas*, Bogotá: Ministerio de Cultura.

INDEX